Sequential Methods and Their Applications

Sequential Methods and Their Applications

Nitis Mukhopadhyay
Basil M. de Silva

CRC Press is an imprint of the
Taylor & Francis Group, an **informa** business
A CHAPMAN & HALL BOOK

CRC Press
Taylor & Francis Group
6000 Broken Sound Parkway NW, Suite 300
Boca Raton, FL 33487-2742

First issued in paperback 2019

ISBN-13: 978-1-58488-102-5 (hbk)
ISBN-13: 978-0-367-38653-5 (pbk)

Library of Congress Cataloging-in-Publication Data

Mukhopadhyay, Nitis, 1950-
Sequential methods and their applications / Nitis Mukhopadhyay, Basil M. de Silva.
p. cm.
Includes bibliographical references and index.
ISBN 978-1-58488-102-5 (hardcover : acid-free paper)
1. Sequential analysis. I. Silva, Basil M. de. II. Title.

QA279.7.M85 2009
519.5'4--dc22 2008039015

Visit the Taylor & Francis Web site at
http://www.taylorandfrancis.com

and the CRC Press Web site at
http://www.crcpress.com

This Book is Dedicated to Our Families

Mahua, Shankha, and Ranjan

Lalitha and Bindusri

Preface

A number of specialized monographs covering various areas of sequential analysis have been available. But, we feel that a majority of these are highly mathematical for undergraduate, MS and beginning Ph.D. students or practicing statisticians. We believe that these monographs were essentially aimed at researchers. So, a handful of top-notch researchers could probably envision teaching a course on sequential analysis from one of these monographs.

Our textbook is different, completely different. It is a book on sequential methodologies with integrated software for all major topics that we have covered. And, we have indeed covered plenty. This way, everyone can interactively run simulations or may experiment with live data to make sequential analysis come alive. We have provided some live data in a number of chapters.

It is our hope that anyone interested in this field will be able to teach from this book at all levels: undergraduate, Masters or Ph.D. students. If someone is interested in teaching and learning at the same time, this book is suitable for that purpose too. Such a course can cater to students from practically any discipline and not simply from statistics or mathematical sciences. An appropriate course may be tailored, for example, for students specializing in agriculture, biosciences, computer science, economics, engineering, environmental science, finance, information management, marketing or public health. This textbook will be refreshing for a researcher or practitioner, both in or out of traditional classrooms.

Why did we write this book? What are some exclusive novelties? What is its coverage? What are the aims and scope? What audience is this book for? These and other pertinent questions have been addressed at length in Chapter 1, "Objectives, Coverage, and Hopes." We simply did not want to clutter the preface by making it too long.

We may, however, summarize a few pointers. We can safely add that our style of presentation is flexible and user-friendly. We have included plenty of mathematics in appropriate places, usually at the end of a chapter so that the readers will not get trampled by mathematics. We have emphasized the concepts surrounding the methodologies and their implementations. The interactive software will add new energy, new effort and new hopes in sequential analysis. There is no other book like it. No existing monograph comes close.

This book has been in the making for the better part of the past eight years. This project has challenged us in ways that nothing else has. Rob Calver, acquisitions editor (statistics) at Chapman & Hall/CRC, continued to encourage us and gave us hope at every turn throughout this long ordeal. Thank you, Rob. We also thank the production staff at Chapman & Hall/CRC for their professionalism.

We are thrilled to be able to finally present this textbook to the students, researchers and practitioners in all areas of statistical science. We earnestly hope that every statistician will find something interesting in the methodologies that are described in this book.

The reviewers have painstakingly gone through earlier versions of these chapters. We remain grateful for their numerous commentaries, which have been extremely helpful in guiding our thoughts in constructive ways. We sincerely thank these reviewers because their comments have indeed led to significant improvements.

We take this opportunity to thank the students and colleagues in the department of statistics at the University of Connecticut-Storrs, and in the School of Mathematical and Geospatial Sciences at The RMIT University-Melbourne for their kind understanding and support. Over the years, we had the honor and joy of collaborating with many researchers who kindly shared their knowledge and enthusiasm with us. We remain indebted to them. On numerous occasions, we have both taught courses and given lectures and seminars on sequential analysis at many places. During those encounters, we always learned something new. It is impossible to name everyone, but we earnestly thank all who continued to watch over us.

In fulfilling a project of this magnitude, immediate families endure unimaginable stress. We express sincerest gratitude and love to our families: Mahua, Shankha and Ranjan Mukhopadhyay as well as Lalitha and Bindusri de Silva. Without support and commitment from them, this book would have remained undone, period.

Nitis Mukhopadhyay
Department of Statistics
The University of Connecticut-Storrs, U.S.A.
and
Basil M. de Silva
School of Mathematical and Geospatial Sciences
The RMIT University-Melbourne, Australia

Contents

CHAPTER 1

Objectives, Coverage, and Hopes

1.1 Introduction

This introductory chapter is more like a preface, only longer. We address a key rhetorical question: Why have we embarked upon writing this book? Unfortunately, we have no one-line answer.

In this chapter, we raise a number of important issues, explain some crucial points, and give justifications for writing this book. We hope that teachers, students, practitioners, researchers and others who are generally interested in statistical science will find our rationale convincing and regard this book useful.

1.2 Back to the Origin

Mahalanobis (1940) showed the importance of sampling in steps and developed sampling designs for estimating the acreage of jute crop in the whole state of Bengal. This seminal development in large-scale survey sampling of national importance was regarded by many, including Abraham Wald, as the forerunner of sequential analysis. Wald and his collaborators systematically developed theory and methodology of sequential tests in the early 1940s to reduce the number of sampling inspections without compromising the reliability of the terminal decisions. This culminated into a true classic of our time (Wald, 1947).

Sequential analysis began its march with deep-rooted applied motivations in response to demands for efficient testing of anti-aircraft gunnery and other weapons during the World War II. Those methodologies were especially crucial then. Fewer sampling inspections with accurate outcome were essential to gain advantage in the front line. These developments were "classified" in the early to mid 1940s.

Methodological researchers caught on and began applying sequential analysis to solve a wide range of practical problems from inventory, queuing, reliability, life tests, quality control, designs of experiments and multiple comparisons, to name a few. In the 1960s through 1970s, researchers in clinical trials realized the relevance of emerging adaptive designs and optimal stopping rules. The area of clinical trials continues to be an important beneficiary of some of the basic research in sequential methodologies. The basic research in clinical trials has also enriched the area of sequential sampling designs. The following selection of books and monographs include important accounts: Armitage (1975), Ghosh and Sen (1991), Jennison and Turnbull (1999), Rosenberger and Lachin (2002), and Whitehead (1997).

A number of celebrated books and monographs already exist. We have mentioned Wald (1947) and Mukhopadhyay et al. (2004) before. Additionally, one will find other volumes including Bechhofer et al. (1968), Chernoff (1972), Chow et al. (1971), Ghosh (1970), Ghosh et al. (1997), Ghosh and Sen (1991), Gibbons et al. (1977), Govindarajulu (1981), Gupta and Panchapakesan (1979), Gut (1988), Mukhopadhyay and Solanky (1994), Schmitz (1972), Sen (1981,1985), Shiryaev (1978), Siegmund (1985), Wetherill (1975), and Woodroofe (1982). Some references, for example, Gibbons et al. (1977) and Gupta and Panchapakesan (1979) highlight sequential analysis only partly. Govindarajulu (2004) includes codes for some selected computer programs. Our book includes interactive and executable computer programs for nearly all the material that we have included.

1.3 Recent Upturn and Positive Feelings

We are noticing interesting and newer applications of sequential methodologies today. This is especially so in handling contemporary statistical challenges in agriculture, clinical trials, data mining, finance, gene mapping, micro-arrays, multiple comparisons, surveillance, tracking and others. The *Applied Sequential Methodologies*, edited by Mukhopadhyay et al. (2004), would support this sentiment and provide references to some recent initiatives.

We feel that the field of sequential analysis can become as applied and user-friendly as "applied statistics" gets. However, a superficial impression one may get from the large body of the literature is that this field looks too mathematical. But, that does not have to be the last word.

The fact is that in many problems, a sequential methodology is more efficient than a customary fixed-sample-size method. In other problems, sequential methodologies may be essential because no fixed-sample-size methodology would work. But, simple and positive messages get frequently lost in the face of heavy-duty mathematics. We believe that the theory and practice of sequential analysis should ideally move forward together as partners. While mathematics does have its place, the intrinsic value of sequential methodologies in practice deserves due emphasis beyond the glory of mathematical depth alone.

Our message is simple. Sequential methodologies are out there to be used by anyone regardless of one's level of mathematical sophistication. We make an honest attempt to make the area more accessible to the teachers, students, practitioners, researchers and others who may be generally interested in statistical science.

1.4 The Objectives

We were guided by a number of major considerations as we moved along.

1. This work should not be viewed as a handbook, an encyclopedia or a cookbook in sequential analysis.
2. Each chapter will tend to begin with a low key and the bar will rise slowly.
3. Any instructor, including someone who is willing to learn and teach the subject at the same time, should be able to offer a course on sequential analysis from this book. The concepts, methodologies, their implementations and properties should be explained in layman's terms so that a reader may learn the subject matter itself and not merely the number crunching.
4. The majority of chapters should provide essential computing tools so that a reader would be able to run simulations or perform experiments with live data interactively as each chapter progresses.
5. Some short derivations have to be integrated within the text so that it will not turn into a "cookbook". Other selected derivations ought to be placed in later sections within a chapter to aid advanced readers.

Put together these are surely tall orders, but we have tried our best to deliver nearly everything. If we have fallen short, we apologize and we hope that this will be brought to our attention. We will try to correct any mishaps in a later printing.

The item #4 is a true novelty in this book. Everyone is aware that some commercially marketed software are available in the area of sequential clinical trials. While that is clearly helpful, one should realize that sequential analysis has much more to offer than what one sees in the context of clinical trials alone! No general-purpose software is available that can be readily used for all kinds of problems under sequential analysis. Our interactive computer programs usable in a variety of statistical problems make a serious attempt to explore this field and appreciate its real potential through data analysis. We hope that the users would feel comfortable to try and apply some of the methodologies from this book in practice.

We will continue to upgrade our computing tools in order to make them even more user-friendly than they are now. We welcome constructive suggestions for improvement.

1.5 The Coverage

The book has sixteen chapters, some rather long and some short, followed by an appendix. We expect our readers to get a glimpse of some of the basic techniques of sampling and inference procedures. In some cases, we have deliberately stayed away from so-called "optimal procedure" for the ease of basic understanding. Easier derivations are often included within the text itself for a better grasp of the material. Most chapters include a section where advanced readers will find more involved derivations in one place. We emphasize that we do not include all available methodologies in addressing each problem.

When a new computer program is introduced, details including examples and interpretations of an associated output are furnished. The on-screen instructions would prompt users appropriately to take one action or another as relevant. An option is given to save the output of a computer-based exercise. In several statistical problems, we have addressed competing sampling methodologies. The readers would easily be able to compare the performances of competing methodologies for a variety of probability distributions using data analysis.

Chapter 2 gives an introduction to a number of interesting statistical problems. In some problems, a designed sequential methodology would beat the best fixed-sample-size methodology. In other cases, a designed sequential methodology would be essential because no fixed-sample-size methodology would solve the problems. We include typical problems from tests of hypotheses, confidence intervals, point estimation with a risk-bound, selection and ranking and multiple comparisons. **Chapter 3** develops the essential introductory tools for implementing Wald's *sequential probability ratio test* (SPRT). One will be able to readily compare the performances of an SPRT and the best fixed-sample-size test with similar errors using data analysis.

The methods of constructing sequential tests for composite hypotheses are varied and their properties are normally complicated. One may refer to Ghosh (1970) for an elaborate and more advanced systematic treatment. Instead of providing general machineries, in **Chapter 4**, we resort to a couple of specific examples of sequential tests for composite hypotheses. We include sequential tests for (i) the mean of a normal distribution with its variance unknown, (ii) the variance of a normal distribution with its mean known, (iii) the correlation coefficient in a bivariate normal distribution (Pradhan and Sathe, 1975) and (iv) the gamma shape parameter when the scale parameter is unknown (Phatarfod, 1971). These are not simple problems, but these do enjoy some simplicities. Even a first-time reader, we believe, will be able to relate to these specific problems.

Chapter 5 includes some nonparametric sequential tests of hypotheses. We first introduce nonparametric sequential tests (Darling and Robbins, 1968) with power one for the mean of a population having an unspecified distribution, discrete or continuous, when the population variance is known or unknown. Then, a sequential nonparametric test for a percentile of an unspecified continuous distribution is summarized from which, a test for the median follows. This chapter ends with a discussion of a sequential sign test (Hall, 1965).

Chapter 6 introduces basic estimation methodologies for the mean of a normal distribution whose variance is unknown. We begin with a fixed-width confidence interval problem and develop Stein's (1945,1949) two-stage methodology. Then, we successively walk through its modifications including multistage (for example, accelerated sequential and three-stage) and sequential methodologies (Ray, 1957, Hall 1981,1983, Mukhopadhyay, 1980,1990,1996, Mukhopadhyay and Duggan, 1997 and Mukhopadhyay and Solanky, 1991,1994). Next, we move to the bounded risk point estimation problem (Ghosh and Mukhopadhyay, 1976) for the mean for which, suitable two-stage and purely sequential methodologies are discussed. The chapter ends with an introduction to the minimum risk point estimation problem for the mean. In this context, suitable purely sequential (Robbins, 1959) and parallel piecewise sequential (Mukhopadhyay and Sen, 1993) methodologies are implemented. The concepts behind each methodology, along with their implementations, are emphasized. We have also explained the usefulness of a majority of the stated properties.

One will find an overwhelming presence of the methodologies from Chapter 6 in the literature. A curious mind may wonder whether there is another distribution where one may enjoy such a breadth of success stories. **Chapter 7** is important from this standpoint and considers a negative (or a two-parameter) exponential distribution with unknown location and scale parameters. We begin with a fixed-width confidence interval estimation problem for the location parameter and present both two-stage (Ghurye, 1958) and purely sequential (Mukhopadhyay, 1974a) sampling strategies. Then, we present a purely sequential sampling strategy for the minimum risk point estimation (Basu, 1971 and Mukhopadhyay, 1974a) of the location parameter involving its *maximum likelihood estimator* (MLE). Next, a minimum risk point estimation problem for the location parameter involving its *uniformly minimum variance unbiased estimator* (UMVUE) is briefly mentioned under an appropriate purely sequential sampling strategy (Ghosh and Mukhopadhyay, 1990). The readers will quickly realize that the majority of the methodologies and the associated results resemble those found in Chapter 6 even though the bottom-line estimators in these two chapters are very different from one another.

In **Chapter 8**, we pay attention to an exponential distribution with a single unknown scale parameter which is the mean. We begin with a purely sequential sampling strategy proposed by Starr and Woodroofe (1972) for the minimum risk point estimation problem for the mean. The situation here is different from the majority of sequential methodologies discussed in Chapters 6 and 7 since one cannot express the sequential risk involving moments of the sample size alone. Next, we summarize a recent two-stage bounded risk methodology developed by Mukhopadhyay and Pepe (2006) for the point estimation of the mean. This highlights a striking result that a preassigned *risk-bound* can be met exactly through a genuine Stein-type two-stage sampling strategy. An improved procedure of Zacks and Mukhopadhyay (2006a) is briefly mentioned afterward.

Prior to Chapter 8, we have introduced multistage sampling methodologies for fixed-width interval estimation of a parameter of interest in a variety of continuous probability distributions. But, we have hardly mentioned anything about fixed-width confidence intervals for a parameter in a discrete probability distribution. This shortcoming is addressed briefly in **Chapter 9**. For brevity, we focus on a probability distribution involving a single unknown parameter. We introduce a general purely sequential approach (Khan, 1969) that allows us to construct a fixed-width confidence interval for a parameter based on its MLE. We also include an accelerated version of the sequential fixed-width confidence interval. Operationally, an accelerated sequential fixed-width confidence interval procedure is more convenient to apply. We highlight examples from Bernoulli, Poisson, exponential and normal distributions.

In Chapters 6-9, we have handled a number of multistage estimation methods for fixed-width confidence interval as well as minimum risk and bounded risk problems. These problems revolved around estimating a parameter of a specific known distribution while the methodologies relied heavily upon certain special features of an underlying distribution. In **Chapter 10**, we approach distribution-free (or nonparametric) one-sample problems. We first introduce a fixed-width confidence interval problem for the mean of an unknown distribution function. We begin with the fundamental purely sequential procedure of Chow and Robbins (1965) that is followed by the modified two-stage procedure of Mukhopadhyay (1980). Next, we present minimum risk point estimation problems for the mean of an unknown distribution function and discuss asymptotically risk efficient purely sequential and two-stage methodologies (Mukhopadhyay, 1978,1980 and Mukhopadhyay and Duggan, 1997). This is followed by the fundamental purely sequential methodology due to Ghosh and Mukhopadhyay (1979). An important purely sequential bounded length confidence interval procedure proposed by Geertsema (1970) for the center of symmetry of an unknown distribution is also introduced.

In **Chapter 11**, we exclusively handle estimation problems for the mean vector in a p-dimensional normal distribution with an unknown but positive definite (p.d.) variance-covariance(or dispersion) matrix. We develop both fixed-size confidence regions as well as minimum risk point estimators for the mean vector. In some cases, we have assumed a certain special structure for the dispersion matrix (Mukhopadhyay and Al-Mousawi, 1986 and Wang, 1980). In other problems, we have assumed nothing about the dispersion matrix except that it is p.d. (Chatterjee, 1959a,b and Ghosh et al., 1976). We have introduced a number of multistage and sequential methodologies in a wide variety of multivariate problems.

Chapter 12 exclusively handles estimation problems for the regression parameters in a linear model. We do so largely under the normality assumption for the error distribution. We first develop multistage fixed-size confidence region methodologies in linear (Chatterjee, 1962a, Gleser (1965) and Srivastava, 1967,1971) and general linear (Finster 1983,1985) models. These are followed by an asymptotically risk efficient purely sequential methodology (Mukhopadhyay, 1974b and Mukhopadhyay and Abid, 1986b) and its accelerated version.

Next, we begin paying our attention to two-sample normal problems. We introduce fixed-width confidence intervals for comparing two population means in **Chapter 13**. This is followed by minimum risk point estimation problems for comparing the two distributions' means. We do so when (i) the variances are the same but unknown (Scheffé, 1943,1970 and Mukhopadhyay and Abid, 1986a) and (ii) the variances are unknown and unequal (Chapman, 1950, Ghosh and Mukhopadhyay, 1980, Robbins et al., 1967, Mukhopadhyay, 1975,1976a,1977 and Mukhopadhyay and Abid, 1986a). The case (ii) is commonly referred to as the Behrens-Fisher problem. Then, we touch upon analogous problems for (a) two independent negative exponential distributions, (b) two multivariate normal distributions, (c) k-sample comparisons with $k \geq 3$ and (d) two distribution-free populations.

The material on selection and ranking is developed briefly in **Chapter 14**. We discuss a selection problem to identify the "best" among $k(\geq 2)$ treatments distributed normally with a common unknown variance. Such problems are also handled in the area of multiple comparisons. However, basic philosophies and approaches differ significantly. We introduce Bechhofer's (1954) *indifference-zone* approach and develop Bechhofer et al.'s (1954) path-breaking two-stage selection methodology. This is followed by the pioneering sequential selection methodology of Robbins et al. (1968). See also Mukhopadhyay and Solanky (1994).

The area of sequential Bayesian estimation is complicated and unfortunately its presentation can quickly get out of hand requiring a much higher level of mathematical understanding than what we have assumed. A common sentiment expressed by researchers in this area points out that identifying the Bayes stopping rules explicitly is a daunting task in general. There is one exception where considerable simplifications occur which force the Bayes sequential estimation rules to coincide with the fixed-sample-size rules. In **Chapter 15**, we first provide an elementary exposition of sequential concepts under a Bayesian framework. This is followed by briefly mentioning the binomial and normal examples to highlight sequential Bayesian estimation in the light of Whittle and Lane (1967). An advanced reader may review this area from Ghosh et al. (1997, chapter 5), Baron (2004) and Tartakovsky and Veeravalli (2004).

Chapter 16 includes specific applications emphasizing both relevance and importance of sequential methodologies in contemporary areas of statistical science. They provide valuable approaches to run statistical experiments and facilitate real data analysis in today's statistical applications. In Section 16.2 on selected applications, we briefly include the important area of clinical trials. We also refer readers to the monographs of Armitage (1975), Jennison and Turnbull (1999), Whitehead (1997) and Rosenberger and Lachin (2002) and other sources. Next, we present a selection of other recent practical applications. We quickly go through integrated pest management problems and summarize the Mexican bean beetle data analysis borrowed from Mukhopadhyay and de Silva (2005) in Section 16.3. Then, in Section 16.4, an interesting problem from experimental psychology is introduced that dealt with a human's perception of 'distance'. Some data analyses are provided from the findings that were originally reported in Mukhopadhyay (2005b). This is followed by a problem that originated from horticulture. A detailed summary and data analyses, as well as some real data, are provided in Section 16.5. These have been borrowed from Mukhopadhyay et al. (2004). In the end, in Section 16.6, we also provide snapshots of practical applications from other contemporary areas of statistics where sequential and multistage sampling designs have made a difference (Mukhopadhyay et al., 1992, Mukhopadhyay, 2002, Chang and Martinsek, 2004, de Silva and Mukhopadhyay, 2004, Efromovich, 2004, Mukhopadhyay and Cicconetti, 2004b, Mulekar and Young, 2004, Young, 2004 and Tartakovsky et al., 2006). In addition, the edited volume of Mukhopadhyay et al. (2004) will serve as an invaluable resource.

In an **Appendix**, we have summarized some selected results from statistics and probability. However, we should caution that this is not an exhaustive list. Then, we include some of the standard statistical tables. These are followed by **References** where we have compiled a long list of publications. We end with an **Author Index** and a **Subject Index**.

1.5.1 The Novelties

True novelties can be found throughout this textbook. The first and foremost is the interactive nature of its software. A reader should be able to run simulations or experiments with live data easily as each chapter progresses. We have provided some live data. The second novelty is the book's user-friendly style and flexible presentation.

In Chapter 8 and in Chapters 10 through 14, we have not been able to incorporate some of the methodologies for related problems. In these chapters, instead of depriving our readers from more advanced topics completely, we have included separate sections on selected multistage procedures other than those that have been systematically introduced. This is a novel feature too. The readers will surely discover other specialties as they get ready to take a tour of this field.

1.5.2 Computing Tools

Except Chapter 16, all other chapters provide essential computing tools for implementing and analyzing the methodologies that we have introduced. No prior knowledge of statistical computing is assumed. Our customized software will run on PCs and hence statistical analyses will come alive in classrooms and outside alike. Each Chapter shows examples of input, output and their interpretations.

One will find an option to save some parts of an output. That way, one will have an opportunity to revisit some computer generated data for further examination with exploratory data analysis. No book in sequential analysis has any of these features.

An executable version of our customized software should be downloaded from the 'Downloads and Updates' section of the publishers's web-site. The software is made available exclusively for those who would use this specialty book for enhanced understanding of this field.

1.5.3 Style of Presentation

The basic style is user-friendly. It is our hope that one will not freeze by the fear of mathematics. The important properties of each methodology have been clearly explained in layman's terms. Many exercises given at the end of chapters come with hints. That way, readers will experience successes when solving exercises. The initial success will breed more success, we believe, and that will make readers even more confident to try and solve other challenging exercises.

An individual may be a top-notch researcher in sequential analysis or someone willing to teach and learn sequential analysis at the same time or a general user of statistics. This book can be adopted at any level by anyone. The real emphasis lies on methodologies and their implementations. We hope that everyone will find something interesting in this book. No book in sequential analysis has any of these features.

1.6 Aims and Scope

We have written this textbook with the hope that it will facilitate course offerings in sequential analysis at junior/senior level as well as the masters and Ph.D. levels. A colleague experienced in this field will be able to teach a one-semester or two-semesters course from this book easily. At the same time, any colleague having some cursory experience or no experience in this area will also be able to offer a one-semester or two-semesters course from this book easily. If a colleague looks to offer an innovative course on some out-of-the-ordinary topics, we believe that this textbook will fit the bill. Learning something new, something different, and involving the students can be refreshing. Everyone is welcomed to teach a course on sequential analysis from this textbook. The important point is that now everyone can.

1.6.1 Undergraduate Level

The **audience** would be undergraduate students majoring in any area (including statistics) with suggested minimum prerequisite.

The **minimum prerequisite** may include Calculus I and II plus a semester's worth of course-work on probability and statistics. Any exposure to linear algebra and applied statistics will be beneficial. No prior knowledge of statistical computing is assumed.

We suggest inclusion of the following **material**. Spend a substantial amount of time explaining the basic concepts and the methodologies from Chapters 2, 3, 6, 9, 13 and 16. Move slowly and cover those parts that look more interesting than others. Emphasize the software that comes with the book and statistical analyses of simulation runs. Interpret theoretical properties through simulations. Encourage simple class-projects where students may generate their own data and then feed them to a computer as instructed. Leave some parts as reading assignments. If time permits, cover selected parts from Chapters 12 and 14. If the material in Chapter 12 appears hard for students, specialize the material in the context of fitting a simple straight line.

1.6.2 Masters Level

The **audience** would be graduate students from any area (including statistics) with suggested minimum prerequisite.

The **minimum prerequisite** may include a semester's worth of course-work each on probability theory, statistical inference, linear models, and applied statistics at the masters level. Any exposure to multivariate statistical methods will be beneficial. No prior knowledge of statistical computing is assumed.

We suggest inclusion of the following **material**. Spend substantial amount of time explaining the basic concepts and the methodologies from Chapters 2 - 6, 9, 12, 14 and 16. Cover those parts that look more interesting than others, but the pace is faster. Emphasize the software that comes with the book and statistical analyses of simulation runs. Interpret theoretical properties through simulations. Encourage projects where students may generate their own data and feed them to a computer as instructed. Leave some parts as reading assignments, especially some of the easier derivations.

1.6.3 Ph.D. Level

The **audience** would be Ph.D. students from statistics or another discipline with comparable backgrounds.

The **minimum prerequisite** would include the guidelines set for the masters students and beyond. Any exposure to advanced inference, nonparametric statistics, multivariate analysis, and probability theory will be beneficial. No prior knowledge of statistical computing is assumed.

We suggest inclusion of the following **material**. Spend substantial amount of time explaining the basic concepts and the methodologies from Chapters 2 - 6, 9, 12, 14 - 16. Cover as much details as possible and go through the important derivations. Select material from the remaining Chapters 8, 10, 11, and 13 that look more interesting than others, but the pace should be rather fast. Emphasize the software that comes with the book and statistical analyses of simulation runs. Encourage projects where students may generate their own data. Leave many parts as reading assignments, especially some of the derivations. Assign problems based on the reading material.

1.6.4 Advanced Research Seminar

The **audience** would include senior Ph.D. students from statistics and other disciplines with comparable backgrounds.

We suggest inclusion of the following **material**. Devote one-half to two-thirds of a semester on more methodological aspects from this textbook and spend the remaining time on the advanced material from this book itself depending on the areas of research interests. Integrate statistical computing with the lecture material. Encourage projects where participants may generate their own data and make use of the software.

1.6.5 More on the Audience

We have written this textbook with the hope that it will facilitate course offerings in sequential analysis at all levels. The book can also be used by researchers and practitioners in all areas of statistics. For someone who is new in this area of statistics, the mathematics is not expected to dampen one's willingness to learn the basic methodologies for the sake of applications alone.

1.7 Final Thoughts

We may remind everyone that sequential or multistage sampling methodologies should cross one's mind in at least two types of circumstances. In some practical problems, an appropriately designed sequential or multistage sampling methodology will outperform the best conventional non-sequential methodology. In other situations, an appropriate sequential or multistage sampling methodology will be essential since no conventional non-sequential methodology will provide deliverables.

The included software will help all users of this textbook, we believe, to learn the subject matter and appreciate its real usefulness in practice. We also request and encourage all to implement the methodologies described in this book in field-trials of their own. We believe that there will be plenty of opportunities to do so.

Enjoy the grass-root practicality of sequential methodologies!

There is no reason not to.

CHAPTER 2

Why Sequential?

2.1 Introduction

In sequential analysis, an experimenter gathers information regarding an unknown parameter θ by observing random samples in successive steps. One may take one observation at a time or a few at a time, but a common characteristic among such sampling designs is that the total number of observations collected at termination is a positive integer valued random variable, usually denoted by N.

We may like to estimate θ by some statistic T_n, obtained from a random sample $X_1, \ldots, X_n$ of fixed size n which is known to the experimenter. This is referred to as a fixed-sample-size methodology. In contrast, a sequential experiment is associated with a genuine stopping variable N and random samples $X_1, \ldots, X_N$. Then, one may consider estimating θ by the corresponding randomly stopped estimator T_N. We motivate the necessity for adopting a sequential strategy such as the one having (N, T_N) with the help of a few important examples.

2.2 Tests of Hypotheses

Suppose that we have a sequence of independent and identically distributed (i.i.d.) random variables $X_1, X_2, \ldots$ from a universe having a $N(\mu, \sigma^2)$ distribution with $-\infty < \mu < \infty, 0 < \sigma < \infty$. For simplicity, let us assume that μ is unknown but σ is known. The problem is to test a simple null hypothesis $H_0 : \mu = \mu_0$ against a simple alternative hypothesis $H_1 : \mu = \mu_1 (> \mu_0)$ where μ_0, μ_1 are two specified real numbers. The most powerful (MP) test for H_0 versus H_1, based on a random sample $X_1, \ldots, X_n$ of size n, at the preassigned level α, $0 < \alpha < 1$, has the following form:

$$\text{Reject } H_0 \text{ if and only if } \frac{\sqrt{n}\,(\overline{X}_n - \mu_0)}{\sigma} > z_\alpha. \qquad (2.2.1)$$

Here, $\overline{X}_n = \frac{1}{n}\sum_{i=1}^{n} X_i$ is the sample mean and z_α is the upper $100\alpha\%$ point of the standard normal distribution. The fact that (2.2.1) is the MP test follows from the classical theory of Neyman and Pearson (1933).

The type II error probability, β is given by

$$\begin{aligned}
&\mathrm{P}(\text{Accepting } H_0 \text{ when } H_1 \text{ is true})\\
&= \mathrm{P}\left\{\frac{\sqrt{n}\,(\overline{X}_n - \mu_0)}{\sigma} \le z_\alpha \Big| \mu = \mu_1\right\}\\
&= \mathrm{P}\left\{\frac{\sqrt{n}\,(\overline{X}_n - \mu_1)}{\sigma} \le z_\alpha - \frac{\sqrt{n}(\mu_1 - \mu_0)}{\sigma} \Big| \mu = \mu_1\right\}\\
&= \mathrm{P}\left\{Z \le z_\alpha - \frac{\sqrt{n}\,(\mu_1 - \mu_0)}{\sigma}\right\}, \qquad (2.2.2)
\end{aligned}$$

where Z is the standard normal variable. Let us write $\phi(x)$ and $\Phi(x)$ respectively for the probability density function and the cumulative distribution function of a standard normal variable, that is

$$\phi(x) = (\sqrt{2\pi})^{-1}\exp(-x^2/2), \Phi(x) = \int_{-\infty}^{x} \phi(y)dy,$$

for $-\infty < x < \infty$. In other words, (2.2.2) is rewritten as:

$$\text{Type II error probability} = \Phi\left(z_\alpha - \frac{\sqrt{n}\,(\mu_1 - \mu_0)}{\sigma}\right). \qquad (2.2.3)$$

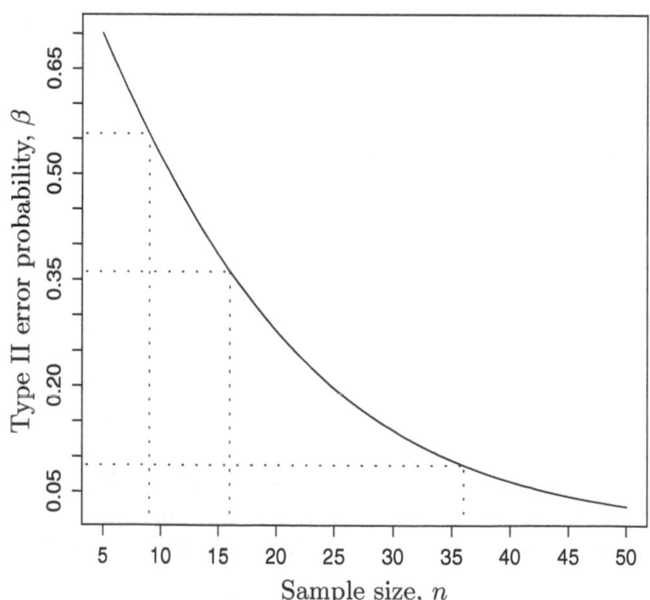

Figure 2.2.1. *Type II error probability* β:
$\alpha = 0.05, \mu_0 = 0, \mu_1 = 1, \sigma = 2$.

The type II error probability function (of n) is plotted in the Figure 2.2.1 when $\alpha = 0.05, \mu_0 = 0, \mu_1 = 1$ and $\sigma = 2$. It is clear that if we fix n in advance, the type II error probability of the best α-level test is also fixed, and moreover, it may not be very small. From (2.2.3), it follows easily that the type II error probability respectively equals $0.558, 0.361, 0.088$ when $n = 9, 16, 36$, and this substantiates our message. One may read off these type II error probabilities from the Figure 2.2.1 too.

An experimenter may, however, decide that the type II error probability should be no larger than β where $0 < \beta < 1$ is preassigned. Now, arbitrarily fixing the sample size n in advance is not necessarily going to help in achieving the goal. In view of (2.2.3), one requires

$$\Phi\left(z_\alpha - \frac{\sqrt{n}\,(\mu_1 - \mu_0)}{\sigma}\right) \leq \beta = \Phi(z_{1-\beta}) = \Phi\left(-z_\beta\right).$$

So, we must determine the appropriate sample size n satisfying the equation

$$z_\alpha - \frac{\sqrt{n}\,(\mu_1 - \mu_0)}{\sigma} \leq -z_\beta,$$

that is,

$$n \geq \left(\frac{z_\alpha + z_\beta}{\mu_1 - \mu_0}\right)^2 \sigma^2 = n^*, \text{ say.} \tag{2.2.4}$$

In other words, one should determine n as the smallest integer $\geq n^*$, and obtain random samples $X_1, \ldots, X_n$ in order to implement the best level α test given by (2.2.1). This test will also have its type II error probability $\leq \beta$.

One can compute the type II error probabilities or find the sample sizes required to achieve the preassigned β using the program **Power** that is built inside **Seq02.exe**. For more details refer to Section 2.5.

We will see in Chapter 3, that a properly executed sequential test would essentially satisfy both type I and type II error probability requirements at the preassigned levels α, β respectively. At the same time, the proposed sequential test achieves this feat with an average sample size that is nearly one-half of n^*. This is a situation where one may appropriately fix the sample size $n(\geq n^*)$ in advance and solve the problem that way, but a properly designed sequential solution would do better with significant savings in the sample size.

Sequential test can beat Neyman-Pearson's most powerful test with comparable type I and type II errors.

2.3 Estimation Problems

In estimation of parameters, there are many problems which cannot be solved by any fixed-sample-size methodology. But these can be solved by implementing appropriate sequential strategies, as can be seen, for example, in Chapters 6-13 and 16.

Before we give examples, we state a general result which helps in verifying the nonexistence of any fixed-sample-size procedure to solve many useful problems. This result was proved in Lehmann (1951). One may refer to Section 3.7 in Ghosh et al. (1997) for a proof.

Theorem 2.3.1 : *Let $X_1, \ldots, X_n$ be i.i.d. with a probability density function $\frac{1}{\sigma} f\left(\frac{x-\theta}{\sigma}\right)$ where $-\infty < \theta < \infty, 0 < \sigma < \infty$ are two unknown parameters. For estimating θ, let the loss function be given by $W\Big(\theta, \delta(\boldsymbol{x})\Big) = H\Big(|\delta(\boldsymbol{x}) - \theta|\Big)$ where $\boldsymbol{x} = (x_1, \ldots, x_n)$ is a realization of $\boldsymbol{X} = (X_1, \ldots, X_n)$. Assume that $H(|u|) \uparrow |u|$, and let $M = \sup_{-\infty<u<\infty} H(|u|)$, which may be infinite. Then, for any fixed $L < M$, there does not exist an estimator $\delta(\boldsymbol{X})$ such that $\sup_{\theta,\sigma} \mathrm{E}_{\theta,\sigma}\{W(\theta, \delta(\boldsymbol{X}))\} \le L$.*

This statement is similar to that of Theorem 3.7.1 in Ghosh et al. (1997). We omit its proof, but let us explore some applications.

Example 2.3.1 : Let $X_1, \ldots, X_n$ be i.i.d. $N(\mu, \sigma^2)$ where $-\infty < \mu < \infty, 0 < \sigma < \infty$ are both unknown parameters. Let $\overline{X}_n = \frac{1}{n}\sum_{i=1}^n X_i$, $S_n^2 = \frac{1}{n-1}\sum_{i=1}^n (X_i - \overline{X}_n)^2$, $n \ge 2$. Given $0 < \alpha < 1$, the usual $(1-\alpha)$ confidence interval for μ is:

$$J_n = \left[\overline{X}_n \pm a_{n-1}\frac{S_n}{\sqrt{n}}\right] \tag{2.3.1}$$

where a_{n-1} = the upper $50\alpha\%$ point of the Student's t-distribution with $(n-1)$ degrees of freedom. One notes that the length of J_n is given by

$$K_n = 2a_{n-1}\frac{S_n}{\sqrt{n}} \tag{2.3.2}$$

which is a random variable in itself. One should realize that K_n may be large, for fixed n, with positive probability, whatever be the unknown magnitude of σ.

We provide the program **ProbKn** that is built inside **Seq02.exe** to evaluate $\mathrm{P}_{\mu,\sigma}(K_n > c\sigma)$ for fixed σ, α and n. How large can K_n be? We have more to say in Section 2.5. The confidence interval J_n may indeed be too wide in some instances to be of any practical use. Thus, it would make sense to try and construct a $(1-\alpha)$ confidence interval for μ where the confidence interval has a preassigned width $2d(>0)$. Constructions of a fixed-width confidence interval corresponds to finding estimators under the $0-1$ loss function. The explanation follows.

Let $\boldsymbol{X} = (X_1, \ldots, X_n), \boldsymbol{x} = (x_1, \ldots, x_n)$ and $\delta(\boldsymbol{X})$ be an estimator of μ. Suppose that

$$W\Big(\mu, \delta(\boldsymbol{x})\Big) = \begin{cases} 0 & \text{if } |\delta(\boldsymbol{x}) - \mu| \le d \\ 1 & \text{if } |\delta(\boldsymbol{x}) - \mu| > d, \end{cases} \tag{2.3.3}$$

where $d(>0)$ is preassigned. Under the loss function (2.3.3) one obtains

$$\mathrm{E}_{\mu,\sigma}\Big\{W(\mu, \delta(\boldsymbol{X}))\Big\} = \mathrm{P}_{\mu,\sigma}\Big\{|\delta(\boldsymbol{X}) - \mu| > d\Big\}, \tag{2.3.4}$$

that is, the risk function (= expected value of the loss function) is actually 1 - confidence level.

In this example, note that $M = 1$ and hence for fixed $0 < \alpha < 1$, there does not exist any estimator $\delta(\boldsymbol{X})$ for which $\mathrm{P}_{\mu,\sigma}\{|\delta(\boldsymbol{X}) - \mu| > d\} \leq \alpha$ for all μ, σ. That is, there does not exist any estimator $\delta(\boldsymbol{X})$ for which

$$\mathrm{P}_{\mu,\sigma}\left\{|\delta(\boldsymbol{X}) - \mu| \leq d\right\} > 1 - \alpha$$

for all μ and σ. So, this fixed-width confidence interval problem cannot be solved by any fixed sample size procedure.

Observe that $\delta(\boldsymbol{X})$ is a very general estimator, not necessarily $\overline{X}_n$, and the result shows nonexistence of any appropriate $\delta(\boldsymbol{X})$. Dantzig (1940) had proved this result first when $\delta(\boldsymbol{X}) = \overline{X}_n$. In Chapter 6, we show how sampling sequentially or in stages can solve this fundamental problem.

Non-existence of fixed-sample-size methodologies that can deliver a preassigned estimation error.

Example 2.3.2 : Consider the $N(\mu, \sigma^2)$ universe as in Example 2.3.1 and suppose that we wish to construct a point estimator of μ based on $\boldsymbol{X} = (X_1, \ldots, X_n)$. Here, the sample size n is fixed in advance. For an estimator $\delta(\boldsymbol{X})$ of μ, suppose that the loss function is given by

$$W\Big(\mu, \delta(\boldsymbol{X})\Big) = \Big(\delta(\boldsymbol{X}) - \mu\Big)^2 \tag{2.3.5}$$

under which, the risk function becomes,

$$\mathrm{E}_{\mu,\sigma}\Big\{W\Big(\mu, \delta(\boldsymbol{X})\Big)\Big\} = \mathrm{E}_{\mu,\sigma}\Big\{\Big(\delta(\boldsymbol{X}) - \mu\Big)^2\Big\}. \tag{2.3.6}$$

Observe that $M = \infty$ in this case and hence fix any L, $0 < L < \infty$. But, there does not exist any estimator $\delta(\boldsymbol{X})$ for which one can conclude $\mathrm{E}_{\mu,\sigma}\{(\delta(\boldsymbol{X}) - \mu)^2\} \leq L$ for all μ and σ. In other words, if we wish to set a preassigned upper bound L for the mean square error of an estimator for all μ and σ, then it remains an unattainable goal. This refers to the so called bounded risk point estimation problem for μ. In Chapter 6, we show how sequential approaches can help in this framework.

In the two previous problems, Theorem 2.3.1 directly shows the nonexistence of any fixed-sample-size methodology. Often, minimum risk point and fixed-size confidence region estimation in multiparameter problems are also not solvable by any fixed-sample-size methodology. Chapters 11-12 introduce such problems and show the usefulness of sequential sampling strategies.

2.4 Selection and Ranking Problems

This area is related to one where we customarily handle *multiple comparison* problems. Some advanced books devoted exclusively to the area of multiple comparisons are available. For example, one may refer to Hochberg and Tamhane (1987) and Hsu (1996). The interface between sequential analyses and selection problems is available in the advanced book of Mukhopadhyay and Solanky (1994). Other important references are given in Chapter 14. In what follows, we mention only one specific selection problem for the purpose of introduction.

Suppose we have $\Pi_i, i = 1, \ldots, k(\geq 2)$, *independent, normally distributed* populations, having unknown mean μ_i and a common unknown variance σ^2. We assume that μ_i is a real number and σ is a positive real number, $i = 1, \ldots, k$. Let $\mu_{[1]} \leq \mu_{[2]} \leq \cdots \leq \mu_{[k]}$ be the ordered μ -values and the problem is to identify the population associated with the largest mean $\mu_{[k]}$, with some preassigned probability that the selection made via random samples from $\Pi_1, \ldots, \Pi_k$ is correct. We denote the population associated with the mean $\mu_{[i]}$ by $\Pi_{(i)}$ but we do not imply any known relationships between the Π_i's and $\Pi_{(i)}$'s, $i = 1, \ldots, k$.

Now, is that an important statistical problem? For appreciation let us suppose, for example, that we are comparing k varieties of wheat and Π_i corresponds to the ith wheat variety, $i = 1, \ldots, k$. Let X_i be the generic "yield" corresponding to the ith variety, $i = 1, \ldots, k$. One can easily verify that

$$\max_{1 \leq i \leq k} \mathrm{P}\Big\{X_i > c\Big\} = \mathrm{P}\Big\{X_j > c\Big\}$$

if and only if X_j corresponds to the variety with $\mu_j = \mu_{[k]}$ for every fixed real number c. That is, the variety associated with $\mu_{[k]}$ will *most likely* give rise to the largest yield. In this sense, we aim at selecting or identifying the "best" wheat variety. It should sound like a very reasonable goal to have.

Having recorded the i.i.d. observations $X_{i1}, \ldots, X_{in}$ from Π_i, let us write $\overline{X}_{in} = \frac{1}{n}\sum_{j=1}^{n} X_{ij}$, $i = 1, \ldots, k$. A very intuitive selection rule can now be stated as follows:

$$\begin{array}{l}\text{Select } \Pi_j \text{ as the population associated with } \mu_{[k]} \\ \text{if } \overline{X}_{jn} \text{ is the largest among } \overline{X}_{1n}, \overline{X}_{2n}, \ldots, \overline{X}_{kn}.\end{array} \tag{2.4.1}$$

In other words, we declare that the population associated with the largest sample mean has the mean $\mu_{[k]}$. But then the important question is this. What is the probability that our decision is correct? Let $\delta_i = \mu_{[k]} - \mu_{[i]}$, $\boldsymbol{\mu} = (\mu_1, \ldots, \mu_k)$, and $\overline{X}_{(in)}$ be the sample mean from $\Pi_{(i)}$. Now, we obtain

$$\begin{aligned}
&\mathrm{P}_n(\text{Correct Selection}) \\
&= \mathrm{P}_{\boldsymbol{\mu},\sigma}\left\{\overline{X}_{(kn)} > \overline{X}_{(in)}, i = 1, \ldots, k-1\right\} \\
&= \mathrm{P}_{\boldsymbol{\mu},\sigma}\Big\{\frac{\sqrt{n}}{\sigma}\left[\overline{X}_{(kn)} - \mu_{[k]}\right] + \frac{\sqrt{n}}{\sigma}\delta_i > \frac{\sqrt{n}}{\sigma}\left[\overline{X}_{(in)} - \mu_{[i]}\right], \\
&\qquad\qquad\qquad\qquad\qquad\qquad\qquad\qquad i = 1, \ldots, k-1\Big\} \\
&= \mathrm{P}_{\boldsymbol{\mu},\sigma}\left\{Y_{(k)} + \frac{\sqrt{n}}{\sigma}\delta_i > Y_{(i)}, i = 1, \ldots, k-1\right\},
\end{aligned} \tag{2.4.2}$$

where $Y_{(j)} = \dfrac{\sqrt{n}\left[\overline{X}_{(jn)} - \mu_{[j]}\right]}{\sigma}$, $j = 1, \ldots, k$, are i.i.d. standard normal variables. One can rewrite (2.4.2) as

$$\begin{aligned}
&\mathrm{P}_n(\text{Correct Selection}) \\
&= \int_{-\infty}^{\infty} \mathrm{P}\left\{Y_{(k)} + \frac{\sqrt{n}}{\sigma}\delta_i > Y_{(i)}, i = 1, \ldots, k-1 \middle| Y_{(k)} = y\right\} \times \\
&\qquad f_{Y_{(k)}}(y)dy \\
&= \int_{-\infty}^{\infty} \mathrm{P}\left\{y + \frac{\sqrt{n}}{\sigma}\delta_i > Y_{(i)}, i = 1, \ldots, k-1 \middle| Y_{(k)} = y\right\} \phi(y)dy \\
&= \int_{-\infty}^{\infty} \prod_{i=1}^{k-1} \mathrm{P}\left\{Y_{(i)} < y + \frac{\sqrt{n}}{\sigma}\delta_i\right\} \phi(y)dy \\
&= \int_{-\infty}^{\infty} \prod_{i=1}^{k-1} \Phi\left(y + \frac{\sqrt{n}}{\sigma}\delta_i\right) \phi(y)dy. \qquad (2.4.3)
\end{aligned}$$

In (2.4.3), we first used the conditional argument, and then used the fact that the conditional distribution of $(Y_{(1)}, \ldots, Y_{(k-1)})$ given $Y_{(k)}$ is the same as that of $(Y_{(1)}, \ldots, Y_{(k-1)})$ since $Y_{(j)}$'s are independent. Next, we used the fact that $Y_{(1)}, \ldots, Y_{(k)}$ are i.i.d. $N(0, 1)$ variables.

Now, consider the whole parameter space

$$\Omega = \{\boldsymbol{\mu} = (\mu_1, \ldots, \mu_k) : \mu_i \in I\!R, i = 1, \ldots, k\}.$$

In the space Ω, we have $\delta_i \geq 0$ for all $i = 1, \ldots, k$ and thus

$$\inf_{\boldsymbol{\mu} \in I\!R} \mathrm{P}_n(\text{Correct Selection}) = \int_{-\infty}^{\infty} \Phi^{k-1}(y)\phi(y)dy = \frac{1}{k}. \qquad (2.4.4)$$

That is, there will be no way to guarantee a high probability of correct selection over the whole parameter space Ω. In fact, the smallest probability of correct selection corresponds to picking any one of the k populations randomly with equal probability k^{-1}.

> Non-existence of fixed-sample-size methodologies in selecting the best treatment with preassigned probability of correct selection. But, two-stage and sequential methodologies can deliver.

Bechhofer's (1954) fundamental formulation asks one to select the population associated with $\mu_{[k]}$ as long as $\mu_{[k]}$ is sufficiently larger than the next largest, namely, $\mu_{[k-1]}$. Suppose $\delta^*(> 0)$ is preassigned and we are interested in the selection problem when $\boldsymbol{\mu}$ belongs to a subset of Ω defined as follows:

$$\Omega(\delta^*) = \left\{\boldsymbol{\mu} = (\mu_1, \ldots, \mu_k) : \mu_{[k]} - \mu_{[k-1]} \geq \delta^*\right\}. \qquad (2.4.5)$$

Thus, the whole parameter space Ω is viewed as $\Omega(\delta^*) \cup \Omega^c(\delta^*)$ where

$\Omega(\delta^*)$ is referred to as the *preference zone* and $\Omega^c(\delta^*)$ is called the *indifference zone.* The original goal is revised accordingly in order to achieve a high probability of correct selection. Our interest shifts to identify the population associated with $\mu_{[k]}$ if $\mu_{[k]}$ is at least δ^* units larger than $\mu_{[k-1]}$, that is, when $\boldsymbol{\mu}$ belongs to the preference zone. On the other hand, if $\boldsymbol{\mu} \in \Omega^c(\delta^*)$, then $\mu_{[k]}$ and $\mu_{[k-1]}$ are "practically" too close to each other, and hence in this situation we remain indifferent about selecting the population associated with $\mu_{[k]}$.

Indifference zone formulation: Bechhofer (1954)

This approach is referred to as the *indifference zone formulation* of Bechhofer (1954). For $\boldsymbol{\mu} \in \Omega(\delta^*)$, since $\delta_i \geq \delta^*$ for all $i = 1, \ldots, k-1$, from (2.4.3) one obtains

$$\inf_{\boldsymbol{\mu} \in \Omega(\delta^*)} \mathrm{P}_n(\text{Correct Selection}) = \int_{-\infty}^{\infty} \Phi^{k-1}\left(y + \frac{\sqrt{n}}{\sigma}\delta^*\right)\phi(y)dy, \quad (2.4.6)$$

as $\Phi(t)$ is a monotone increasing function of t. Now, the experimenter may wish to control the probability of correct selection at a preassigned level, say, $P^* \in (k^{-1}, 1)$ by equating the right hand side in (2.4.6) with P^*, and determining n accordingly. But, the solution of n would depend on the unknown σ. In fact, no fixed-sample-size methodology exists for this problem which will work for all σ. It was proved by Dudewicz (1971). Chapter 14 addresses this problem through appropriate sequential sampling.

2.5 Computer Programs

The kinds of computations mentioned in Sections 2.2 and 2.3 utilize two computer programs, namely, **Power** and **ProbKn**. These programs together with the programs **ProbGn** and **ProbHn** are combined into one program called **Seq02.exe**. To execute this program, one needs a computer equipped with MS-DOS* or Microsoft Windows* operating system. Copy the program into a suitable subdirectory. Then, select "Run" from the menu found in "Start" and use "Browse" to run **Seq02.exe**. Figure 2.5.1 shows the initial part of the screen output of this program. Then, select the program by typing the corresponding number given in the left hand side of the screen output. Note that the programs number 3-4 from the available menu, that is, programs **ProbGn** and **ProbHn** will be used in Exercises 2.3.1 and 2.3.2 respectively.

Now the program **Power** is executed by selecting 1 from the menu given in Figure 2.5.1. This program relates only to the problem discussed in Section 2.2. It computes

- the probability of type II error when the sample size is known, and

* MS-DOS and Microsoft Windows are registered trademarks of Microsoft Corporation.

- the required sample size to achieve a preassigned type II error probability β

for the given MP test (2.2.1) of hypotheses $H_0 : \mu = \mu_0$ versus $H_1 : \mu = \mu_1$ where $\mu_0 < \mu_1$.

```
          ************************************************
          *  Welcome to Computer Programs for the Text:  *
          *  SEQUENTIAL METHODOLOGIES AND APPLICATIONS    *
          *                    by                         *
          *  Nitis Mukhopadhayay and Basil M. de Silva    *
          ************************************************

   Do you want to save the results in a file (Y or N)? Y

   Note that the default results outfile name is:
   "SEQ02.OUT".

   If you have created this file earlier, data in the
   earlier file will be erased.

   Do you want to change the outfile name (Y or N)? N

          ************************************************
          *  Now you are required to select one of the   *
          *  following programs using the number given   *
          *  in the left hand side.  For more  details   *
          *  please refer to Section 2.5 of the text.    *
          ************************************************
          1   Power:  Compute beta  or n for a Normal test
          2   ProbKn: Probability of K(n) > c*sigma
          3   ProbGn: Probability of G(n) > c*sigma
          4   ProbHn: Probability of H(n) > c*sigma
          0   Exit

          Enter your choice :0

Press any key to continue
```

Figure 2.5.1. *The initial screen output of* ***Seq02.exe****.*

Figure 2.5.2 shows the screen output of this program for computation of β when n given in (2.2.3) is 16, for the MP test of hypotheses $H_0 : \mu = 0$ versus $H_1 : \mu = 1$ with $\sigma = 2$, $\alpha = 0.05$. The result of the computation follows:

$$\beta = 0.361 \text{ when } n = 16.$$

With the help of the program **Power**, one can examine how the type II error probability function changes with different values of α, n, μ_0, μ_1 and

σ. For fixed $\mu_0, \mu_1, \sigma, \alpha$ and n, the type II error probability is of course fixed but its magnitude may or may not be small. One can use the same program **Power** to determine the appropriate smallest sample size for which the MP level α test (2.2.1) will also have the type II error probability $\leq \beta$ for a preassigned $0 < \beta < 1$.

```
*******************************************************
*  This program considers the testing of hypotheses:  *
*    H0: mu = mu0 against H1: mu = mu1 at level alpha *
*    where mu is the mean of a normal population and  *
*    sigma is the known standard deviation.           *
*                                                     *
*  The program computes:                              *
*    (a) the probability of type II error for a       *
*        given sample size or                         *
*    (b) the sample size for a given probability      *
*        of type II error.                            *
*******************************************************

Now, input the parameter values for the above hypotheses
          mu0   = 0
          mu1   = 1
          sigma = 2
          alpha = .05

Do you want to change the input parameters (Y or N)? N

     To compute probability of the type II
     error (beta), input the sample size (n)
                        OR
     To compute the sample size (n), input the
     type II error probability (beta).

****************************************************
*  Choose one of the following options:            *
*    1 : Only one computation of beta or n         *
*    2 : Compute beta values for a set of n values *
*    3 : Compute n values for a set of beta values *
****************************************************

Enter your option : 1
  Enter a value for beta or n : 16

When n =    16 type II error probability, beta=0.361
```

Figure 2.5.2. *A screen output results from* ***Power***.

The program **ProbKn** evaluates

$$P(K_n > c\sigma) = P(\frac{K_n}{\sigma} > c) \tag{2.5.1}$$

for a fixed set of values of $\alpha, n, c(> 0)$ and σ where $K_n = 2a_{n-1}\frac{S_n}{\sqrt{n}}$ from (2.3.3). Note that this probability should not depend on a chosen value of σ. One should refer to (2.3.1) - (2.3.3). Since the length of the confidence interval in (2.3.1) is a random variable, it makes sense to investigate the behavior of the probability that this length K_n exceeds some given number, $c\sigma$ (equivalently, the probability that $\frac{K_n}{\sigma}$ exceeds c). Then, one may take different values of c and investigate the pattern of the values of $P(\frac{K_n}{\sigma} > c)$. Instead, one may fix some value of c and vary the sample size n to see the effect on the magnitude of $P(K_n/\sigma > c)$. Note that

$$\frac{n(n-1)K_n^2}{4a_{n-1}^2\sigma^2} = \frac{(n-1)S_n^2}{\sigma^2} \sim \chi^2_{n-1}.$$

```
*********************************************
*  This program computes the probability of  *
*  K(n) > c*sigma for given alpha and sample  *
*  size.                                      *
*********************************************

       Input following parameters:
           alpha =  .05
           sample size, n = 20

      Probability p = P(K(n)>c*sigma)
     alpha =.050 and Sample Size =  20
     ---------------------------------

   c     p           c    p              c     p
 .44 1.000         .80  .791           1.16  .063
 .48  .999         .84  .703           1.20  .038
 .52  .998         .88  .604           1.24  .022
 .56  .995         .92  .499           1.28  .012
 .60  .989         .96  .395           1.32  .006
 .64  .975        1.00  .300           1.36  .003
 .68  .952        1.04  .218           1.40  .002
 .72  .915        1.08  .151           1.44  .001
 .76  .862        1.12  .100           1.48  .000
```

Figure 2.5.3. *A screen output results from* ***ProbKn***.

```
     ************************************************
     *  Results from Chapter 2 Computer Programs    *
     *  SEQUENTIAL METHODOLOGIES AND APPLICATIONS   *
     *                     by                       *
     *  Nitis Mukhopadhyay and Basil M. de Silva    *
     ************************************************

OUTPUT FROM THE PROGRAM: Power
==============================
    For the Test H0: mu =   0
        against  H1: mu =   1      at alpha = .050
        where mu is the mean of normal with sigma =  2.0

        If n    =   16   then beta = .361

OUTPUT FROM THE PROGRAM: ProbKn
===============================
             Probability, p=P(K(n)>c*sigma)
            alpha =.050 and Sample Size =  20
            ---------------------------------


     c      p             c     p              c     p
   .44  1.000           .80  .791            1.16  .063
   .48   .999           .84  .703            1.20  .038
   .52   .998           .88  .604            1.24  .022
   .56   .995           .92  .499            1.28  .012
   .60   .989           .96  .395            1.32  .006
   .64   .975          1.00  .300            1.36  .003
   .68   .952          1.04  .218            1.40  .002
   .72   .915          1.08  .151            1.44  .001
   .76   .862          1.12  .100            1.48  .000
```

Figure 2.5.4. *Output file:* ***SEQ02.OUT***.

Therefore, the probability in (2.5.1) is computed using the integral of the chi-square density over the interval

$$\left(\frac{n(n-1)c^2}{4a_{n-1}^2}, \infty\right).$$

This program computes the probabilities $\mathrm{P}(\frac{K_n}{\sigma} > c)$ for fixed values of c. Figure 2.5.3 shows the output for $\alpha = 0.05$ and $n = 20$.

One can verify easily that

$$\mathrm{P}\left\{K_n > c\sigma\right\} = \mathrm{P}\left\{\chi_{n-1}^2 > \frac{n(n-1)c^2}{4a_{n-1}^2}\right\}$$

which depends only on n, c and α. This probability does not change with different values of σ. One may compare the output files from Figure 2.5.3 for different values of σ and come to the same conclusion.

We can summarize the results presented in Figure 2.5.3 as follows: *For a 95% confidence interval, its random length* K_n *will certainly exceed* 0.44σ, *but on the other hand* K_n *will rarely exceed* 1.48σ.

As shown in Figure 2.5.1, the program allows the user to write the results by default to a file named **SEQ02.OUT** or any other file with a name no more than nine characters long. This file can be printed on any printer using either **PRINT** or **COPY** DOS commands. Whenever the program is executed the earlier contents in the file **SEQ02.OUT** will be erased and the current output results will be written into the file. Figure 2.5.4 shows the **SEQ02.OUT** file listing after the programs **Power** and **ProbKn** executed in that order.

2.6 Exercises

Exercise 2.2.1: Suppose that $X_1, \ldots, X_n$ are i.i.d. random variables from a universe having the $N(\mu, \sigma^2)$ distribution with $-\infty < \mu < \infty$, $0 < \sigma < \infty$. We assume that μ is unknown but σ is known. Let us test a simple null hypothesis $H_0 : \mu = \mu_0$ versus a simple alternative hypothesis $H_1 : \mu = \mu_1 (< \mu_0)$ at the preassigned level α, $0 < \alpha < 1$. Denote $\overline{X}_n = \frac{1}{n}\sum_{i=1}^{n} X_i$, the sample mean, where the sample size n is fixed.

(i) Show that the MP level α test for H_0 versus H_1 would reject H_0 if and only if $\dfrac{\sqrt{n}(\overline{X}_n - \mu_0)}{\sigma} < -z_\alpha$.

(ii) Find the expression for the type II error probability associated with the MP test found in part (i).

(iii) Find the expression for the minimum fixed sample size n^* needed by the MP level α test from part (i) so that its type II error probability will not exceed β, with $0 < \beta < 1$ preassigned.

Exercise 2.2.2: Suppose that $X_1, \ldots, X_n$ are i.i.d. random variables from a universe having the $N(\mu_1, \sigma^2)$ distribution, $Y_1, \ldots, Y_n$ are i.i.d. random variables from a universe having the $N(\mu_2, \sigma^2)$ distribution with $-\infty < \mu_1, \mu_2 < \infty, 0 < \sigma < \infty$. Suppose also that the X's and Y's are independent. Let us assume that μ_1, μ_2 are both unknown but σ is known. We wish to test a null hypothesis $H_0 : \mu_1 = \mu_2$ versus an alternative hypothesis $H_1 : \mu_1 > \mu_2$ at the preassigned level α, $0 < \alpha < 1$. Consider the following level α test:

$$\text{Reject } H_0 \text{ if and only if } \frac{\sqrt{n}(\overline{X}_n - \overline{Y}_n)}{\sqrt{2}\sigma} > z_\alpha$$

where $\overline{X}_n = \frac{1}{n}\sum_{i=1}^{n} X_i, \overline{Y}_n = \frac{1}{n}\sum_{i=1}^{n} Y_i$.

(i) Find the expression for the type II error probability associated with the given level α test.

(ii) Find the expression for the minimum fixed sample size n^* needed by the given level α test so that its type II error probability will not exceed β, with $0 < \beta < 1$ preassigned.

Exercise 2.2.3: Suppose that $X_1, \ldots, X_n$ are i.i.d. random variables from a universe having the $N(\mu_1, \sigma^2)$ distribution, $Y_1, \ldots, Y_n$ are i.i.d. random variables from a universe having the $N(\mu_2, p\sigma^2)$ distribution, with $-\infty < \mu_1, \mu_2 < \infty, 0 < p, \sigma < \infty$. Suppose also that the X's and Y's are independent. Let us assume that μ_1, μ_2 are both unknown but p, σ are known. We wish to test a null hypothesis $H_0 : \mu_1 = \mu_2$ versus an alternative hypothesis $H_1 : \mu_1 < \mu_2$ at the preassigned level α, $0 < \alpha < 1$. Consider the following level α test:

$$\text{Reject } H_0 \text{ if and only if } \frac{\sqrt{n}\left(\overline{X}_n - \overline{Y}_n\right)}{\sqrt{1+p}\,\sigma} < -z_\alpha$$

where $\overline{X}_n = \frac{1}{n}\sum_{i=1}^n X_i, \overline{Y}_n = \frac{1}{n}\sum_{i=1}^n Y_i$.

(i) Find the expression for the type II error probability associated with the given level α test.

(ii) Find the expression for the minimum fixed sample size n^* needed by the given level α test so that its type II error probability will not exceed β, with $0 < \beta < 1$ preassigned.

Exercise 2.2.4: Suppose that $X_1, \ldots, X_n$ are i.i.d. random variables from a universe having the negative exponential distribution with the p.d.f.

$$f(x) = \frac{1}{\sigma}\exp\left\{-\frac{x-\mu}{\sigma}\right\} I(x > \mu)$$

where $-\infty < \mu < \infty, 0 < \sigma < \infty$. Here and elsewhere, $I(x > \mu) = 1$ or 0 according as $x > \mu$ or $x \le \mu$. We assume that μ is unknown but σ is known. Let us test a simple null hypothesis $H_0 : \mu = \mu_0$ versus a simple alternative hypothesis $H_1 : \mu = \mu_1 (> \mu_0)$ at the preassigned level α, $0 < \alpha < 1$. Denote $X_{n:1} = \min\{X_1, \ldots, X_n\}$, the minimum order statistic, where the sample size n is fixed.

(i) Show that $Y = \dfrac{n\{X_{n:1} - \mu\}}{\sigma}$ has the standard exponential distribution, that is, its p.d.f. is given by $g(y) = \exp(-y)I(y > 0)$.

(ii) Consider a test for H_0 versus H_1 which rejects H_0 if and only if $\dfrac{n\{X_{n:1} - \mu_0\}}{\sigma} > a$ with $a = -\ln(\alpha)$. Show that this test has level α.

(iii) Find the expression for the type II error probability associated with the level α test found in part (ii).

(iv) Find the expression for the minimum fixed sample size n^* needed by the level α test found in part (ii) so that its type II error probability will not exceed β, with $0 < \beta < 1$ preassigned.

Exercise 2.2.5: Suppose that $X_1, \ldots, X_n$ are i.i.d. random variables from a universe having the negative exponential distribution with the p.d.f.

$$f(x) = \frac{1}{\sigma} \exp\left\{-\frac{x-\mu}{\sigma}\right\} I(x > \mu)$$

where $-\infty < \mu < \infty, 0 < \sigma < \infty$. We assume that μ is unknown but σ is known. Let us test a simple null hypothesis $H_0 : \mu = \mu_0$ versus a simple alternative hypothesis $H_1 : \mu = \mu_1 (< \mu_0)$ at the preassigned level α, $0 < \alpha < 1$. Denote $X_{n:1} = \min\{X_1, \ldots, X_n\}$, the minimum order statistic, where the sample size n is fixed.

(i) Consider a test for H_0 versus H_1 which rejects H_0 if and only if $\dfrac{n\{X_{n:1} - \mu_0\}}{\sigma} < b$ with $b = -\ln(1-\alpha)$. Show that this test has level α.

(ii) Find the expression for the type II error probability associated with the level α test found in part (i).

(iii) Find the expression for the minimum fixed sample size n^* needed by the level α test found in part (i) so that its type II error probability will not exceed β, with $0 < \beta < 1$ preassigned.

Exercise 2.3.1: Suppose that $X_1, \ldots, X_n$ are i.i.d. random variables from a universe having the $N(\mu_1, \sigma^2)$ distribution, $Y_1, \ldots, Y_n$ are i.i.d. random variables from a universe having the $N(\mu_2, \sigma^2)$ distribution, with $-\infty < \mu_1, \mu_2 < \infty, 0 < \sigma < \infty$. Suppose also that the X's and Y's are independent. We assume that $\mu_1, \mu_2, \sigma_1, \sigma_2$ are all unknown parameters. Here, the sample size $n (\geq 2)$ is fixed. We denote $\overline{X}_n = \frac{1}{n}\sum_{i=1}^n X_i$, $\overline{Y}_n = \frac{1}{n}\sum_{i=1}^n Y_i$, $S_{1n}^2 = \frac{1}{n-1}\sum_{i=1}^n (X_i - \overline{X}_n)^2$, $S_{2n}^2 = \frac{1}{n-1}\sum_{i=1}^n (Y_i - \overline{Y}_n)^2$ and $S_{Pn}^2 = \frac{1}{2}\left(S_{1n}^2 + S_{2n}^2\right)$ which is the *pooled sample variance.* The problem is one of constructing a $1-\alpha$ confidence interval for $\mu_1 - \mu_2$ where $0 < \alpha < 1$ is preassigned. Let a_{2n-2} be the upper $50\alpha\%$ point of the Student's t distribution with $2(n-1)$ degrees of freedom.

(i) Show that $\dfrac{\sqrt{n}\{(\overline{X}_n - \overline{Y}_n) - (\mu_1 - \mu_2)\}}{\sqrt{2}S_{Pn}}$ has the Student's t distribution with $2(n-1)$ degrees of freedom.

(ii) Show that $J_n = \left[(\overline{X}_n - \overline{Y}_n) \pm a_{2n-2}\dfrac{\sqrt{2}S_{Pn}}{\sqrt{n}}\right]$ is a $(1-\alpha)$ confidence interval for $\mu_1 - \mu_2$.

(iii) Let G_n be the length of the confidence interval J_n given in part (ii). For $c > 0$, show that

$$\mathrm{P}\{G_n > c\sigma\} = \mathrm{P}\left\{\chi_{2n-2}^2 > \frac{c^2 n(n-1)}{4a_{2n-2}^2}\right\}.$$

Use the program **ProbGn** from **Seq02.exe** to evaluate this probability for a range of values of c.

Exercise 2.3.2: Suppose that $X_1, \ldots, X_n$ are i.i.d. random variables from a universe having the negative exponential distribution with the p.d.f.

$$f(x) = \frac{1}{\sigma} \exp\left\{-\frac{x-\mu}{\sigma}\right\} I(x > \mu)$$

where $-\infty < \mu < \infty, 0 < \sigma < \infty$. We assume that μ, σ are both unknown. The problem is one of constructing a $(1-\alpha)$ confidence interval for μ where $0 < \alpha < 1$ is preassigned. Let b_n be the upper $100\alpha\%$ point of the F distribution with $2, 2n-2$ degrees of freedom. Denote $X_{n:1} = \min\{X_1, \ldots, X_n\}, U_n = \frac{1}{n-1}\sum_{i=1}^{n}(X_i - X_{n:1})$ where the sample size $n(\geq 2)$ is fixed.

(i) Show that $\dfrac{n(X_{n:1}-\mu)}{U_n}$ has the F distribution with $2, 2n-2$ degrees of freedom.

(ii) Show that $J_n = \left[X_{n:1} - b_n \dfrac{U_n}{n}, X_{n:1}\right]$ is a $(1-\alpha)$ confidence interval for μ.

(iii) Let H_n be the length of the confidence interval J_n given in part (ii). For $c > 0$, show that

$$\mathrm{P}\{H_n > c\sigma\} = \mathrm{P}\left\{\chi^2_{2n-2} > \frac{2cn(n-1)}{b_n}\right\}.$$

Use the program **ProbHn** from **Seq02.exe** to evaluate this probability for a range of values of c.

Exercise 2.4.1: In the case of the normal example discussed in Section 2.4, show that

$$\max_{1 \leq i \leq k} \mathrm{P}\{X_i > c\} = \mathrm{P}\{X_j > c\}$$

if and only if X_j corresponds to the population with the mean $\mu_j = \mu_{[k]}$, the largest mean, whatever be the real number c.

Exercise 2.4.2: Suppose that $X_{i1}, \ldots, X_{in}$ are i.i.d. random variables from the universe Π_i having the negative exponential distribution with the p.d.f.

$$f_i(x) = \frac{1}{\sigma} \exp\left\{-\frac{x-\mu_i}{\sigma}\right\} I(x > \mu_i)$$

where $-\infty < \mu_i < \infty, 0 < \sigma < \infty, i = 1, \ldots, k(\geq 2)$. We assume that these populations are independent, $\mu_1, \ldots, \mu_k$ are all unknown but σ is known. Let $\mu_{[1]} \leq \mu_{[2]} \leq \ldots \leq \mu_{[k]}$ be the ordered values of μ. Let us denote $X_{n:1}^{(i)} = \min\{X_{i1}, \ldots, X_{in}\}, \delta_i = \mu_{[k]} - \mu_{[i]}, i = 1, \ldots, k, \boldsymbol{\mu} = (\mu_1, \ldots, \mu_k)$. Now, consider using the natural selection procedure:

> Select Π_j as the population associated with the largest location parameter $\mu_{[k]}$ if $X_{n:1}^{(j)} = \max\left\{X_{n:1}^{(1)}, X_{n:1}^{(2)}, \ldots, X_{n:1}^{(k)}\right\}$.

(i) Along the lines of Section 2.4, derive the expression for P_n\{Correct Selection\}.

(ii) Show that

$$\inf_{\boldsymbol{\mu}\in\mathbb{R}^k} \mathrm{P}_n\{\text{Correct Selection}\} = \frac{1}{k},$$

and the $\inf_{\boldsymbol{\mu}\in\mathbb{R}^k} \mathrm{P}_n$ is attained when $\delta_i = 0, i = 1, \ldots, k$. That is, P_n\{Correct Selection\} cannot be made arbitrarily large in the whole parameter space $\mathbb{R}^k$.

(iii) Consider the preference zone $\Omega(\delta^*)$ as in Section 2.4 with $\delta^* > 0$. Show that $\inf_{\boldsymbol{\mu}\in\Omega(\delta^*)} \mathrm{P}_n$\{Correct Selection\} can be made at least P^* where $P^* \in (k^{-1}, 1)$, as long as the sample size n is sufficiently large. The $\inf_{\boldsymbol{\mu}\in\Omega(\delta^*)} \mathrm{P}_n$ is attained when $\delta_i = \delta^*, i = 1, \ldots, k$.

Exercise 2.4.3 (Exercise 2.4.2 Continued): In the case of the negative exponential problem discussed in Exercise 2.4.2, for simplicity let us write X_i for a typical observation from the population $\Pi_i, i = 1, \ldots, k$. Then, show that

$$\max_{1\le i\le k} \mathrm{P}\{X_i > c\} = \mathrm{P}\{X_j > c\}$$

if and only if X_j corresponds to the population with location parameter $\mu_j = \mu_{[k]}$, the largest location parameter, whatever be the real number c.

Exercise 2.5.1 (Exercise 2.2.1 Continued): Suppose that $X_1, \ldots, X_n$ are i.i.d. random variables from a universe having the $N(\mu, \sigma^2)$ distribution with $-\infty < \mu < \infty, 0 < \sigma < \infty$. We assume that μ is unknown but σ is known. Let us test a simple null hypothesis $H_0 : \mu = \mu_0$ versus a simple alternative hypothesis $H_1 : \mu = \mu_1 (< \mu_0)$ at the preassigned level α, $0 < \alpha < 1$. Consider the MP level α test from Exercise 2.2.1.

(i) Use the program **Power** to calculate the type II error probability β, having fixed some values for n, α, σ, μ_0, and μ_1.

(ii) Use the program **Power** to calculate the minimum fixed sample size $n \equiv n^*$, having fixed some values for $\alpha, \beta, \sigma, \mu_0$, and μ_1.

\{*Hint*: The program **Power** computes the expression given by (2.2.3). All one has to do is to write down the present type II error probability exactly in that form and identify the appropriate parameters.\}

Exercise 2.5.2 (Exercise 2.2.2 Continued): Suppose that $X_1, \ldots, X_n$ are i.i.d. random variables from a universe having the $N(\mu_1, \sigma^2)$ distribution, $Y_1, \ldots, Y_n$ are i.i.d. random variables from a universe having the $N(\mu_2, \sigma^2)$ distribution, with $-\infty < \mu_1, \mu_2 < \infty, 0 < \sigma < \infty$. Suppose also that the X's and Y's are independent. We assume that μ_1, μ_2 are both unknown but σ is known. We wish to test a null hypothesis $H_0 : \mu_1 = \mu_2$ versus an alternative hypothesis $H_1 : \mu_1 > \mu_2$ at the preassigned level α, $0 < \alpha < 1$. Consider the level α test from Exercise 2.2.2.

(i) Use the program **Power** to calculate the type II error probability β, having fixed some values for n, α, σ, μ_1, and μ_2,

(ii) Use the program **Power** to calculate the minimum fixed sample size $n \equiv n^*$, having fixed some values for $\alpha, \beta, \sigma, \mu_1$, and μ_2.

{*Hint*: The program **Power** computes the expression given by (2.2.3). All one has to do is to write down the present type II error probability exactly in that form and identify the appropriate parameters.}

Exercise 2.5.3 (Exercise 2.2.3 Continued): Suppose that $X_1, \ldots, X_n$ are i.i.d. random variables from a universe having the $N(\mu_1, \sigma^2)$ distribution, $Y_1, \ldots, Y_n$ are i.i.d. random variables from a universe having the $N(\mu_2, p\sigma^2)$ distribution, with $-\infty < \mu_1, \mu_2 < \infty, 0 < p, \sigma < \infty$. Suppose also that the X's and Y's are independent. We assume that μ_1, μ_2 are both unknown but p, σ are known. We wish to test a null hypothesis $H_0 : \mu_1 = \mu_2$ versus an alternative hypothesis $H_1 : \mu_1 < \mu_2$ at the preassigned level α, $0 < \alpha < 1$. Consider the level α test from Exercise 2.2.3.

(i) Use the program **Power** to calculate the type II error probability β, having fixed some values for $n, \alpha, p, \sigma, \mu_1$, and μ_2.

(ii) Use the program **Power** to calculate the minimum fixed sample size $n \equiv n^*$, having fixed some values for $\alpha, \beta, p, \sigma, \mu_1$, and μ_2.

{*Hint*: The program **Power** computes the expression given by (2.2.3). All one has to do is to write down the present type II error probability exactly in that form and identify the appropriate parameters.}

Exercise 2.5.4 (Exercise 2.2.4 Continued): Suppose that $X_1, \ldots, X_n$ are i.i.d. random variables from a universe having the negative exponential distribution with the p.d.f.

$$f(x) = \frac{1}{\sigma} \exp\left\{-\frac{x-\mu}{\sigma}\right\} I(x > \mu)$$

where $-\infty < \mu < \infty, 0 < \sigma < \infty$. We assume that μ is unknown but σ is known. Let us test a simple null hypothesis $H_0 : \mu = \mu_0$ versus a simple alternative hypothesis $H_1 : \mu = \mu_1 (> \mu_0)$ at the preassigned level α, $0 < \alpha < 1$. Consider the level α test from Exercise 2.2.4.

(i) Derive the expression for the type II error probability β, having fixed some values for n, α, σ, μ_0, and μ_1.

(ii) Derive the expression for the minimum fixed sample size $n \equiv n^*$, having fixed some values for $\alpha, \beta, \sigma, \mu_0$, and μ_1.

{*Hint*: Note that under the alternative hypothesis, $\dfrac{n\left(X_{n:1}-\mu_1\right)}{\sigma}$ has the standard exponential distribution.}

Exercise 2.5.5 (Exercise 2.2.5 Continued): Suppose that $X_1, \ldots, X_n$ are i.i.d. random variables from a universe having the negative exponential distribution with the p.d.f.

$$f(x)=\frac{1}{\sigma}\exp\left\{-\frac{x-\mu}{\sigma}\right\}I(x>\mu)$$

where $-\infty<\mu<\infty, 0<\sigma<\infty$. We assume that μ is unknown but σ is known. Let us test a simple null hypothesis $H_0: \mu=\mu_0$ versus a simple alternative hypothesis $H_1: \mu=\mu_1(<\mu_0)$ at the preassigned level α, $0<\alpha<1$. Consider the level α test from Exercise 2.2.5.

(i) Derive the expression for the type II error probability β, having fixed some values for n, α, σ, μ_0, and μ_1.
(ii) Derive the expression for the minimum fixed-sample-size $n \equiv n^*$, having fixed some values for $\alpha, \beta, \sigma, \mu_0$, and μ_1.

{*Hint*: Note that under the alternative hypothesis, $\dfrac{n\left(X_{n:1}-\mu_1\right)}{\sigma}$ has the standard exponential distribution.}

CHAPTER 3

Sequential Probability Ratio Test

3.1 Introduction

Consider a sequence of i.i.d. random variables $X_1, X_2, \ldots$, having the common probability density function (p.d.f.) $f(x|\theta)$ for $x \in \mathfrak{X} \subseteq I\!R$ and $\theta \in \Theta \subseteq I\!R$. The methodology and the results would be laid out pretending that X's are continuous variables. However these are general enough to include discrete distributions as well. In the situations where X's are discrete, $f(x|\theta)$ would stand for the probability mass function (p.m.f.), and in the derivations, one should replace the *integrals* with the appropriate *sums* for a smooth transition.

The problem is one of testing a simple null hypothesis $H_0 : \theta = \theta_0$ versus a simple alternative hypothesis $H_1 : \theta = \theta_1$ where $\theta_0 \neq \theta_1$ are two specified numbers in the parameter space Θ. We also have two preassigned numbers $0 < \alpha, \beta < 1$ corresponding to the type I and type II error probabilities. *From a practical point of view, let us assume that* $\alpha + \beta < 1$.

Let us write $\boldsymbol{X}_j = (X_1, \ldots, X_j), j = 1, 2, \ldots$. Now, when θ is the true parameter value, the likelihood function of $\boldsymbol{X}_j$ is given by

$$\mathcal{L}(\theta; j, \boldsymbol{X}_j) = \prod_{l=1}^{j} f(X_l|\theta). \tag{3.1.1}$$

We denote $\mathcal{L}_0(j) = \mathcal{L}(\theta_0; j, \boldsymbol{X}_j)$ and $\mathcal{L}_1(j) = \mathcal{L}(\theta_1; j, \boldsymbol{X}_j)$ respectively under H_0 and H_1. Having observed $\boldsymbol{X}_j$, the *most powerful* (MP) level α test has the form

$$\text{Reject } H_0 \text{ if and only if } \frac{\mathcal{L}_1(j)}{\mathcal{L}_0(j)} \geq k \tag{3.1.2}$$

where $k(> 0)$ is to be determined appropriately.

One should keep in mind that in a discrete case, appropriate randomization would be mandatory to make sure that the size of this test is indeed exactly α.

The test given in (3.1.2) is the best among all fixed-sample-size $(= j)$ tests at level α. But its type II error probability can be quite a bit larger than β, the preassigned target. In order to meet both (α, β) requirements, in the early 1940's, Abraham Wald, among others, looked at this problem from a different angle. Having observed $\boldsymbol{X}_j$ let us consider the sequence of successive likelihood ratios:

$$\Lambda_j = \frac{\mathcal{L}_1(j)}{\mathcal{L}_0(j)}, \ j = 1, 2, \ldots.$$

In other words, we observe $X_1, X_2, X_3, \ldots$ one by one and accordingly obtain $\Lambda_1, \Lambda_2, \Lambda_3, \ldots$ in a sequential manner. As soon as we find one Λ_j in the sequence that appears *too small* or *too large* we quit sampling, and we decide in favor of H_0(or H_1) whenever Λ_j is too small (or too large), $j = 1, 2, \ldots$.

To be specific, Wald (1947) first fixed two real numbers A and B such that $0 < B < A < \infty$. Intuitively speaking, a likelihood ratio is interpreted too large (too small) if it falls at or above (at or below) the number $A(B)$. The *sequential probability ratio test* (**SPRT**) is then implemented as follows:

We continue observing $X_1, X_2, \ldots$ one by one in a sequence until the likelihood ratio Λ_j goes out of the interval (B, A) for the first time. In other words, let

$$N \text{ be the first integer } j (\geq 1) \text{ for which } \Lambda_j \leq B \ \text{ or } \ \Lambda_j \geq A. \qquad (3.1.3)$$

We decide in favor of H_0(or H_1) if $\Lambda_N \leq B$ (or $\Lambda_N \geq A$). Equivalently, we can express the same sampling strategy as follows. Define

$$\begin{aligned} N \equiv \ & \text{first integer } j (\geq 1) \text{ for which } \textstyle\sum_{i=1}^{j} Z_i \leq b \ \text{ or} \\ & \textstyle\sum_{i=1}^{j} Z_i \geq a \text{ where } a = \ln(A), b = \ln(B) \\ & \text{and } Z_i = \ln\{f(X_i|\theta_1)/f(X_i|\theta_0)\}. \end{aligned} \qquad (3.1.4)$$

We decide in favor of H_0(or H_1) if $\sum_{i=1}^{N} Z_i \leq b$ (or $\sum_{i=1}^{N} Z_i \geq a$). Note that $Z_1, Z_2, \ldots$ are i.i.d. random variables and hence from (3.1.4) it ought to be clear that N corresponds to the first-time crossing of certain boundaries by a random walk. One should realize, however, that the terminal decision rule would work provided N is finite with probability one. We will have more to say on this in the next section.

Sequential test (3.1.3) or equivalently (3.1.4) is the famous Wald's SPRT.

3.2 Termination and Determination of A and B

Wald's SPRT described by (3.1.3) or (3.1.4) is purely sequential in nature. So, it is crucial to show first that the sampling terminates with probability one under both H_0, H_1. Observe that

$$\begin{aligned} \mathrm{P}_\theta(N < \infty) &= \sum_{1 \leq i < \infty} \mathrm{P}_\theta(N = i) \\ &= \lim_{n \to \infty} \sum_{i=1}^{n} \mathrm{P}_\theta(N = i) = \lim_{n \to \infty} \mathrm{P}_\theta(N \leq n) \end{aligned}$$

and thus, we have

$$\mathrm{P}_\theta(N = \infty) = \lim_{n\to\infty} \mathrm{P}_\theta(N > n). \tag{3.2.1}$$

The stopping variable N defined in (3.1.4) has the exact same form as in Theorem 3.5.1. Now, $\mathrm{P}_\theta(Z_1 = 0) = \mathrm{P}_\theta\{f(X_1|\theta_1) = f(X_1|\theta_0)\}$ which should be zero in practical applications because $\theta_1 \neq \theta_0$. Hence, Theorem 3.5.1 would let us conclude that the SPRT terminates with probability one under H_0 and H_1 because

$$\mathrm{P}\{N > n|H_i\} \leq c_i r_i^n \quad \text{for all } n \geq 1 \tag{3.2.2}$$

with $c_i > 0$, $0 < r_i < 1$. This implies that

$$0 \leq \mathrm{P}\{N = \infty|H_i\} \leq c_i \lim_{n\to\infty}(r_i^n) = 0,$$

and hence $\mathrm{P}(N = \infty|H_i) = 0$, $i = 0, 1$.

In fact, (3.2.2) implies that the moment generating function (m.g.f.) of N is finite under H_0, H_1. This can be seen very easily as follows. We write, under H_i,

$$\begin{aligned} \mathrm{E}\left[e^{tN}|H_i\right] &= \sum_{n=1}^{\infty} e^{tn}\mathrm{P}(N = n|H_i) \\ &\leq \sum_{n=1}^{\infty} e^{tn}\mathrm{P}(N > n-1|H_i) \\ &\leq \sum_{n=1}^{\infty} e^{tn} c_i r_i^{n-1} = c_i e^t \sum_{n=1}^{\infty}\left(r_i e^t\right)^{n-1}, \end{aligned} \tag{3.2.3}$$

which converges if t is chosen so that $(0 <) r_i e^t < 1$, that is if $t < -\ln(r_i)$, $i = 0, 1$. This result in turn shows that all moments of N are finite under H_0 and H_1. From this point onward let us write P_i and E_i for the evaluation of probability and expectation respectively under the hypothesis H_i, $i = 0, 1$.

Theorem 3.2.1: $B \geq \dfrac{\beta}{(1-\alpha)}$ *and* $A \leq \dfrac{(1-\beta)}{\alpha}$.

Proof: Let C_i stand for the event of accepting the hypothesis H_i after termination, $i = 0, 1$. Let C_{ij} be the event that $N = j$ and H_i is accepted, $i = 0, 1$. Then $C_i = \bigcup_{j=1}^{\infty} C_{ij}$ where $C_{i1}, C_{i2}, \ldots$ are mutually exclusive events. Now we have

$$\begin{aligned} \beta &= \text{type II error probability} = \mathrm{P}_1(C_0) \\ &= \sum_{1\leq j<\infty} \int_{C_{0j}} \cdots \int \mathcal{L}_1(j)dx_1 \cdots dx_j. \end{aligned} \tag{3.2.4}$$

But, note that on the set C_{0j}, one has $\mathcal{L}_1(j)/\mathcal{L}_0(j) \leq B$ because we accept

H_0 upon termination. Thus, from (3.2.4), we obtain

$$\begin{aligned}\beta &\le \sum_{1\le j<\infty} \int_{C_{0j}}\cdots\int B\mathcal{L}_0(j)dx_1\cdots dx_j \\ &= B \sum_{1\le j<\infty} \int_{C_{0j}}\cdots\int \mathcal{L}_0(j)dx_1\cdots dx_j \\ &= B \sum_{1\le j<\infty} \mathrm{P}_0(C_{0j}) = B\ \mathrm{P}_0(C_0) = B(1-\alpha),\end{aligned}$$

since $\mathrm{P}_0(C_0)$ stands for the probability of accepting H_0 when H_0 is true, which means that $\mathrm{P}_0(C_0) = 1-\alpha$. Thus $B \ge \beta/(1-\alpha)$.

For the other part, the proof is similar. So, we show only the main steps without giving full explanations. We write

$$\begin{aligned}\alpha &= \mathrm{P}_0(C_1) \\ &= \sum_{1\le j<\infty} \int_{C_{1j}}\cdots\int \mathcal{L}_0(j)dx_1\cdots dx_j \\ &\le \frac{1}{A}\sum_{1\le j<\infty} \int_{C_{1j}}\cdots\int \mathcal{L}_1(j)dx_1\cdots dx_j = \frac{1}{A}\mathrm{P}_1(C_1),\end{aligned} \tag{3.2.5}$$

and the result follows since $\mathrm{P}_1(C_1) = 1-\beta$. ■

Wald (1947) recommended implementation of the SPRT given in (3.1.3) or (3.1.4) with $A = (1-\beta)/\alpha$ and $B = \beta/(1-\alpha)$. These are referred to as **Wald's approximations** for A and B. Our goal is to achieve the (α, β) requirement, but with these specific choices for A, B we may instead attain something different, (α', β'). One can, however, use Theorem 3.2.1 to prove that

$$\alpha' + \beta' < \alpha + \beta. \tag{3.2.6}$$

Refer to Exercise 3.2.5. The equation (3.2.6) shows that both error probabilities simultaneously cannot exceed the prescribed nominal levels.

The subprogram **SimAB** within the program **Seq03.exe** would estimate α' and β' by simulating the SPRT with the observations drawn from selected distributions listed in Figure 3.2.1. One would estimate α' by simulating the data from the distribution with $\theta = \theta_0$ and estimate β' by simulating the data with $\theta = \theta_1$. The average number of observations used in each simulation is shown in the program output.

This program can be executed by selecting **SimAB** from the program **Seq03.exe.** Figure 3.2.1 shows a set of input and output of the program. In this example, the program estimates α' and β' in the case of the SPRT for $H_0 : \theta = 1$ against $H_1 : \theta = 5$ with $\alpha = 0.05$ and $\beta = 0.1$ by simulating data from the $N(\theta, 25)$ distribution. Each simulation was replicated 1000 times using $\theta = 1$ and $\theta = 5$ respectively to obtained the estimates for α'

and β'. That is,

$$\text{estimate } \alpha' = \frac{\#\text{ rejections of } H_0 \text{ out of 1000 replications when } \theta = 1}{1000}$$

and

$$\text{estimate } \beta' = \frac{\#\text{ rejections of } H_1 \text{ out of 1000 replications when } \theta = 5}{1000}.$$

The Ave N is simply the average of 1000 observed values of the stopping variable which should estimate $\text{E}(N)$ under the appropriate hypothesis under consideration. Figure 3.2.1 gives three sets of simulation results. The results clearly validate that the estimated values of $\alpha' + \beta'$ are less than $\alpha + \beta$ which is 0.15. The example is presented only for the given normal population. With the help of the program **SimAB**, one should explore analogous results for the SPRT on an appropriate parameter in the case of other selected distributions.

Since the original SPRT is purely sequential in nature, it may continue sampling for quite some time, even though $\text{P}_i(N < \infty) = 1, i = 0, 1$. An experimenter, on the other hand, may decide to terminate sampling when the sample size reaches the limit, say, K. This will be the scenario, for example, when the sampling cost approaches the allocated budget, but the SPRT in (3.1.3) does not terminate sampling. Determination of K would often take into account factors such as resource management and time constraints. The SPRT given by (3.1.3) **would be truncated at K** as follows:

$$\begin{aligned} &\text{Implement (3.1.3) and let } T = \min\{N, K\}. \\ &\text{If } T = N, \text{ we decide in favor of } H_0(\text{ or } H_1) \\ &\quad \text{if } \Lambda_N \le B(\text{ or } \Lambda_N \ge A). \\ &\text{If } T = K, \text{ we decide in favor of } H_0(\text{ or } H_1) \\ &\quad \text{if } \Lambda_K < \tfrac{1}{2}(A+B) \left(\text{or } \Lambda_K > \tfrac{1}{2}(A+B)\right). \end{aligned} \tag{3.2.7}$$

In other words, once N reaches K but the original SPRT (3.1.3) still wants to continue sampling, we can stop sampling right there, and make the decision to accept H_0 or H_1 depending on the *closeness* of Λ_K, the likelihood ratio of $\boldsymbol{X}_K$, to B or A on an intuitive ground. Such truncation will obviously affect the magnitudes of the attained type I and type II error probabilities. The size of the impact on the type I and II error probabilities will largely depend upon the magnitude of K.

The subprogram **SPRT**, given within the program **Seq03.exe**, can also be used for testing $H_0 : \theta = \theta_0$ versus $H_1 : \theta = \theta_1$ with a set of real data which is supposedly being observed sequentially. One will see an option to run the truncated version (3.2.7) of the SPRT with an experimenter's choice of K. Figure 3.2.2 shows the layout of typical data input in the subprogram **SPRT** for a simulation study with the truncated SPRT. Figure 3.2.3 shows the output from the subprogram **SPRT** for the same input data given in the Figure 3.2.2.

```
Select one of the following programs:
        1   SimAB:  Simulation of alpha + beta for a test
        2   SPRT:   SPRT for given or simulated data
        3   OC:     Computation of OC and ASN functions
        0   Exit
        Enter your choice: 1

SPRT for HO: theta = theta0  versus H1: theta = theta1
*****************************************************
*     ID# DISTRIBUTION          Theta        Known    *
*                          (Parameter1) Parameter2    *
*      1  Normal                  mean     variance   *
*      2  Gamma                  scale        shape   *
*      3  Weibull                scale        shape   *
*      4  Erlang             exp. mean   k(integer)   *
*      5  Exponential             mean                *
*      6  Poisson                 mean                *
*      7  Bernoulli          P(Success)               *
*      8  Geometric          P(Success)               *
*      9  Negative Binomial       mean            k   *
*****************************************************
Input distribution  ID#: 1
Enter theta0 = 1
      theta1 = 5
      alpha  = .05
      beta   = .1
Do you want to change the input parameters (Y or N)? N

Number of replications for a simulation? 1000
Enter a positive integer(<10) to initialize
          the random number sequence: 1
Input variance of the Normal Distribution: 25
How many simulation runs required? 3
**************************************************

SPRT for  HO: theta =  1.0  versus  H1: theta = 5.00
   for normal distribution with variance = 25.0
   with alpha = .050 and beta = .100

                   Simulation Results
         N      N              N      N              alpha'
Run#    Ave   S.E.   alpha'  Ave   S.E.    beta'   +beta'
  1     8.2  0.190   0.032   9.5  0.199   0.063   0.095
  2     7.9  0.173   0.026   9.7  0.193   0.060   0.086
  3     8.0  0.183   0.032   9.2  0.194   0.054   0.086

              Input Values: alpha + beta = 0.150
```

Figure 3.2.1. *Screen input and output of* ***SimAB***.

```
    Select one of the following programs:
           1   SimAB:  Simulation of alpha + beta for a test
           2   SPRT:   SPRT for given or simulated data
           3   OC:     Computation of OC and ASN functions
           0   Exit
           Enter your choice: 2

    SPRT for H0: theta = theta0  versus H1: theta = theta1
    *****************************************************
    *    ID# DISTRIBUTION          Theta       Known    *
    *                          (Parameter1) Parameter2  *
    *     1  Normal                 mean     variance   *
    *     2  Gamma                 scale        shape   *
    *     3  Weibull               scale        shape   *
    *     4  Erlang            exp. mean  k(integer)    *
    *     5  Exponential            mean                *
    *     6  Poisson                mean                *
    *     7  Bernoulli         P(Success)               *
    *     8  Geometric         P(Success)               *
    *     9  Negative Binomial      mean            k   *
    *****************************************************
    Input distribution  ID#: 5
    Enter theta0 = 1
          theta1 = 2
          alpha  = 0.05
          beta   = 0.01
 Do you want to change the input parameters (Y or N)? N

    Would you like to truncate the SPRT when
    sample size reaches K (Y or N)? Y
    Enter a value for K: 50

    Do you want to simulate data or input real data?
    Type S (Simulation) or D (Real Data): S

    Do you want to store simulated sample sizes (Y or N)? Y
    Input  file name to store sample sizes
         (no more than 9 characters): Sprt1.dat

    Number of replications for a simulation? 1000
    Enter a positive integer(<10) to initialize
              the random number sequence: 5

    For the data simulation, input
         mean of the exponential distribution: 1.0
```

Figure 3.2.2. *Data input to subprogram **SPRT**.*

```
OUTPUT FROM THE PROGRAM: SPRT
=============================
    SPRT:  H0: theta =  1.00 versus  H1: theta =  2.00 at
           alpha =0.05 and beta =0.01  where theta is the
           mean of  exponential distribution

    Simulated sample sizes are given in file: Sprt1.dat

           Results for simulated data from
           exponential distribution with mean =  1.00
           ------------------------------------------

           No. of SPRTs truncated (with K =  50)        =   25
           No. of times H0 is accepted after truncation =   21

           Number of simulations            =  1000
           Number of times H0 is accepted =   966

           Average sample size (n bar)      =  23.11
           Std. dev. of n (s)               =  10.53
           Minimum sample size (min n)      =      3
           Maximum sample size (max n)      =     50
```

Figure 3.2.3. *Output of* ***SPRT***.

```
OUTPUT FROM THE PROGRAM: SPRT
=============================
    SPRT:  H0: theta =  1.00 versus  H1: theta =  2.00 at
           alpha =0.05 and beta =0.01  where theta is the
           mean of  exponential distribution

           Results for simulated data from
           exponential distribution with mean =  1.50
           ------------------------------------------

           No. of SPRTs truncated (with K =  50)        =  179
           No. of times H0 is accepted after truncation =   95

           Number of simulations            =  1000
           Number of times H0 is accepted =   262

           Average sample size (n bar)      =  26.35
           Std. dev. of n (s)               =  16.12
           Minimum sample size (min n)      =      1
           Maximum sample size (max n)      =     50
```

Figure 3.2.4. *Output of* ***SPRT***.

Figure 3.2.3 gives the results of 1000 simulated tests for testing H_0 : $\theta = 1$ versus $H_1 : \theta = 2$ with the probabilities of type I and II errors being respectively 0.05 and 0.01. The data for these tests are obtained by generating random numbers from the exponential population with mean 1.0 using the seed number 5. The results indicate that 25 out of 1000 simulated tests have been truncated and among the truncated tests, 21 decided in favor of H_0. From the total analysis, 966 tests out of 1000 favor H_0. The program also gives information about the sample sizes.

Next, we used the subprogram **SPRT** to test the above hypotheses using random data generated from an exponential population with mean 1.5. The results are given in Figure 3.2.4. Note that the number of truncated tests and the average sample size are both higher in this case than in the previous case. In the previous case, the statistical model was generated assuming that H_0 was true, that is with $\theta = 1$. In the second situation, the statistical model was generated with $\theta = 1.5$, that is neither H_0 nor H_1 was really true. In the second case, the SPRT needed more observations to test $H_0 : \theta = 1$ versus $H_1 : \theta = 2$. This should be quite natural.

Let $n_1, \ldots, n_J$ be the observed values of the stopping variable N where J is the number of replications. In Figures 3.2.3-3.2.4, we fixed $J = 1000$. Now, let us denote

$$\overline{n} = \frac{1}{J}\sum_{i=1}^{J} n_i \text{ and } s^2 = \frac{1}{J-1}\sum_{i=1}^{J}(n_i - \overline{n})^2.$$

The expressions $\overline{n}$, otherwise called Ave(N), and s^2 are the sample mean and sample variance of the observed n values. These respectively estimate $\mathrm{E}(N)$ and $\mathrm{V}(N)$ relevant for an appropriate hypothesis under consideration. The estimated standard error of $\overline{n}$ can be readily computed as $s/\sqrt{J}$.

Wald's fundamental identity in sequential analysis: Theorem 3.5.2

3.3 ASN Function and OC Function

Consider the SPRT from (3.1.4) and assume that we can find a real number $t_0 \neq 0$ such that $M_\theta(t_0) = 1$ where $M_\theta(t_0) = \mathrm{E}_\theta\left[e^{t_0 Z_1}\right]$ is the moment generating function (m.g.f.) of Z_1 when θ is the true parameter value. This t_0 will depend on θ. Note that even though we are testing simple hypotheses H_0 versus H_1, we consider here all θ values in the parameter space Θ. Let

$$\begin{aligned} L(\theta) &= \text{Probability of accepting } H_0 \\ &\text{when } \theta \text{ is the true parameter value}, \theta \in \Theta. \end{aligned} \tag{3.3.1}$$

It is clear that $L(\theta_0) = 1 - \alpha$ and $L(\theta_1) = \beta$. If one plots the points $(\theta, L(\theta))$ for all $\theta \in \Theta$, the corresponding plot is called the *operational characteristic* (OC) function or curve. Here and elsewhere, let $I(\cdot)$ stand for the indicator of $(\cdot)$.

Using the fundamental identity of sequential analysis (Theorem 3.5.2), we can write

$$\begin{aligned} 1 &= \mathrm{E}_\theta\left[e^{t_0 S_N}\right] \\ &= \mathrm{E}_\theta\left[e^{t_0 S_N} I\left(S_N \le b\right)\right] + \mathrm{E}_\theta\left[e^{t_0 S_N} I\left(S_N \ge a\right)\right] \\ &\qquad\qquad + \mathrm{E}_\theta\left[e^{t_0 S_N} I\left(b \le S_N \le a\right)\right] \\ &= \mathrm{E}_\theta\left[e^{t_0 S_N} I\left(S_N \le b\right)\right] + \mathrm{E}_\theta\left[e^{t_0 S_N} I\left(S_N \ge a\right)\right]. \end{aligned} \tag{3.3.2}$$

We can claim this by arguing as follows. First, indeed $\mathrm{P}_\theta(Z_1 = 0) < 1$. So, the set where $S_N \le b$ or $S_N \ge a$ at the stopped stage should have P_θ-probability zero. Hence, the term $\mathrm{E}_\theta\left[e^{t_0 S_N} I\left(b \le S_N \le a\right)\right]$ which appears in the third line in (3.3.2) is zero.

Wald's first equation: Theorem 3.5.4
Wald's second equation: Theorem 3.5.5

If we assume that the excess of S_N over the respective boundaries a or b is negligible, then from (3.3.2) we can claim that

$$\begin{aligned} 1 &\approx e^{t_0 b}\mathrm{P}_\theta(S_N \le b) + e^{t_0 a}\mathrm{P}_\theta(S_N \ge a) \\ &= e^{t_0 b}L(\theta) + e^{t_0 a}\left\{1 - L(\theta)\right\}. \end{aligned}$$

This provides the following approximation for the OC function:

$$L(\theta) \approx \begin{cases} \dfrac{e^{t_0 a} - 1}{e^{t_0 a} - e^{t_0 b}} & \text{if } t_0 \neq 0 \\ \dfrac{a}{a-b} & \text{if } t_0 = 0. \end{cases} \tag{3.3.3}$$

In the case of $t_0 = 0$, one can easily check that

$$\lim_{t_0 \to 0} \frac{e^{t_0 a} - 1}{e^{t_0 a} - e^{t_0 b}} = \frac{a}{a-b},$$

by applying the L'Hôpital's rule of differentiation. This is the expression that is displayed in (3.3.3) when $t_0 = 0$.

The *average sample number* (**ASN**) is defined as $\mathrm{E}_\theta(N)$ when θ is the true parameter value. We consider the ASN function for all $\theta \in \Theta$.

Using *Wald's first equation* (Theorem 3.5.4), when $\mathrm{E}_\theta(Z_1) \neq 0$, we can write

$$\begin{aligned} \mathrm{E}_\theta(N) &= \frac{\mathrm{E}_\theta(S_N)}{\mathrm{E}_\theta(Z_1)} = \frac{\mathrm{E}_\theta\left\{S_N I(S_N \le b)\right\} + \mathrm{E}_\theta\left\{S_N I(S_N \ge a)\right\}}{\mathrm{E}_\theta(Z_1)} \\ &\approx \frac{bL(\theta) + a\left\{1 - L(\theta)\right\}}{\mathrm{E}_\theta(Z_1)}, \end{aligned}$$

neglecting the excess over the boundaries at the stopped stage.

In the case where $\mathrm{E}_\theta(Z_1) = 0$, we use *Wald's second equation* (Theorem 3.5.5) instead to write

$$\mathrm{E}_\theta(N) = \frac{\mathrm{E}_\theta(S_N^2)}{\mathrm{E}_\theta(Z_1^2)} \approx \frac{b^2 L(\theta) + a^2\{1 - L(\theta)\}}{\mathrm{E}_\theta(Z_1^2)} \approx \frac{-ab}{\mathrm{E}_\theta(Z_1^2)},$$

again neglecting the excess over the boundaries.

In passing, let us merely mention that $t_0(\theta)$ and $\mathrm{E}_\theta(Z_1)$ are of opposite signs whenever $t_0 \neq 0$. This follows from Theorem 3.5.3.

The approximation to the ASN function can be summarized as follows:

$$\mathrm{E}_\theta(N) \approx \begin{cases} \dfrac{bL(\theta) + a(1 - L(\theta))}{\mathrm{E}_\theta(Z_1)} & \text{if } \mathrm{E}_\theta(Z_1) \neq 0 \\ \dfrac{-ab}{\mathrm{E}_\theta(Z_1^2)} & \text{if } \mathrm{E}_\theta(Z_1) = 0 \end{cases} \tag{3.3.4}$$

where the appropriate expression for $L(\theta)$ is plugged in from (3.3.3). This approximation further reduces to

$$\begin{aligned} \mathrm{E}_{\theta_0}(N) &\approx \frac{(1-\alpha)\ln(\beta/(1-\alpha)) + \alpha\ln((1-\beta)/\alpha)}{\mathrm{E}_{\theta_0}(Z_1)}, \\ \mathrm{E}_{\theta_1}(N) &\approx \frac{\beta\ln(\beta/(1-\alpha)) + (1-\beta)\ln((1-\beta)/\alpha)}{\mathrm{E}_{\theta_1}(Z_1)}, \end{aligned} \tag{3.3.5}$$

respectively under H_0 and H_1.

From Hoeffding (1953), one can claim that the approximate expressions given on the right hand side of (3.3.5) are in fact lower bounds for $\mathrm{E}_\theta(N)$ under H_0 and H_1 respectively.

One of the most celebrated results regarding the SPRT in (3.1.4) is referred to as its **optimality property** among all comparable tests, including the fixed-sample-size most powerful test. It is a very deep result, but its practical implications are far reaching and easy to understand. The original result was proved in Wald and Wolfowitz (1948).

Theorem 3.3.1 [Optimal Property of the SPRT]: *Consider testing $H_0 : \theta = \theta_0$ versus $H_1 : \theta = \theta_1$. Then, among all tests, fixed-sample-size or sequential, for which the type I error probability $\leq \alpha$, the type II error probability $\leq \beta$ and for which $\mathrm{E}_\theta(N) < \infty$ when $\theta = \theta_0, \theta_1$, the SPRT from (3.1.4) with error probabilities α and β minimizes $\mathrm{E}_\theta(N)$ when $\theta = \theta_0, \theta_1$ and $\alpha + \beta < 1$.*

The main import of this result can be explained as follows. Among all reasonable tests, including the fixed-sample-size most powerful test, having type I error probability $\leq \alpha$ and type II error probability $\leq \beta$, from a practical point of view it turns out that the SPRT with error probabilities α, β would on an average require the minimum number of observations at termination. One would have to assume that $\alpha + \beta < 1$ for the validity of the optimality result in general. But, from a practical point of view, α and β would both be small numbers between zero and unity, and hence the required condition that $\alpha + \beta$ be smaller than one would be automatically satisfied.

Wald's SPRT beats Neyman-Pearson's most powerful test with comparable type I and type II errors.

3.4 Examples and Implementation

First, we describe explicitly the SPRT for the unknown mean of a normal distribution whose variance is known. Next, we describe the nature of the SPRT in the case of a one-parameter exponential family. This latter approach opens up the possibility of similar techniques in tests for the parameter in a binomial, Poisson or exponential distribution, for example.

Example 3.4.1 : Let $X_1, X_2, \ldots$ be i.i.d. $N(\theta, \sigma^2)$ where $\theta \in \mathbb{R}$, $\sigma \in \mathbb{R}^+$ with θ unknown but σ^2 known. We wish to test $H_0 : \theta = \theta_0$ versus $H_1 : \theta = \theta_1 (> \theta_0)$ having preassigned $0 < \alpha, \beta < 1$. Let $A = (1-\beta)/\alpha, B = \beta/(1-\alpha)$. Now, $Z_i = \ln\{f(X_i|\theta_1)/f(X_i|\theta_0)\} = \sigma^{-2}(\theta_1 - \theta_0)X_i - \frac{1}{2}\sigma^{-2}(\theta_1^2 - \theta_0^2), i = 1, 2, \ldots$. The SPRT summarized by (3.1.4) would simplify to the following: Let $Y(n) = \sum_{i=1}^n X_i$ and

$$N \equiv \begin{array}{l} \text{first integer } n(\geq 1) \text{ for which the inequality} \\ Y(n) \leq c_1 + dn \text{ or } Y(n) \geq c_2 + dn \text{ holds with} \\ c_1 = \sigma^2 \ln(B)/(\theta_1 - \theta_0), c_2 = \sigma^2 \ln(A)/(\theta_1 - \theta_0) \\ \text{and } d = \frac{1}{2}(\theta_1 - \theta_0). \text{ At the stopped stage,} \\ \text{if } Y(N) \leq c_1 + dN, \text{ we accept } H_0, \text{ but} \\ \text{if } Y(N) \geq c_2 + dN, \text{ we accept } H_1. \end{array} \tag{3.4.1}$$

Figure 3.4.1 shows the acceptance and rejection regions for H_0 for the special case where $\theta_0 = 0, \theta_1 = 1, \sigma^2 = 4, \alpha = 0.05$ and $\beta = 0.01$. Now, from (3.4.1) we have $c_1 = -18.215, c_2 = 11.943$ and $d = 0.5$. That is, if $Y(N) \leq 0.5N - 18.215$, we accept H_0 and if $Y(N) \geq 0.5N + 11.943$, we accept H_1. At the intermediate stages, we continue sampling.

Observe that Z_1 satisfies all the conditions stated in Theorem 3.5.3 and hence, we can find a unique $t_0 \equiv t_0(\theta)$, which is non-zero, such that $\mathrm{E}_\theta\left[e^{t_0 Z_1}\right] = 1$. Note that $\mathrm{E}_\theta(Z_1) = \sigma^{-2}(\theta_1 - \theta_0)\left[\theta - \frac{1}{2}(\theta_1 + \theta_0)\right]$ and $\mathrm{V}_\theta(Z_1) = \sigma^{-2}(\theta_1 - \theta_0)^2$. In fact, we have

$$\begin{aligned} \mathrm{E}_\theta\left[e^{tZ_1}\right] &= \exp\left\{-\frac{1}{2}t\frac{(\theta_1^2 - \theta_0^2)}{\sigma^2}\right\}\mathrm{E}_\theta\left\{\exp\left[t\frac{(\theta_1 - \theta_0)}{\sigma^2}X_1\right]\right\} \\ &= \exp\left[-\frac{1}{2}t\frac{(\theta_1^2 - \theta_0^2)}{\sigma^2} + t\frac{(\theta_1 - \theta_0)\theta}{\sigma^2} + \frac{1}{2}t^2\frac{(\theta_1 - \theta_0)^2}{\sigma^2}\right] \end{aligned} \tag{3.4.2}$$

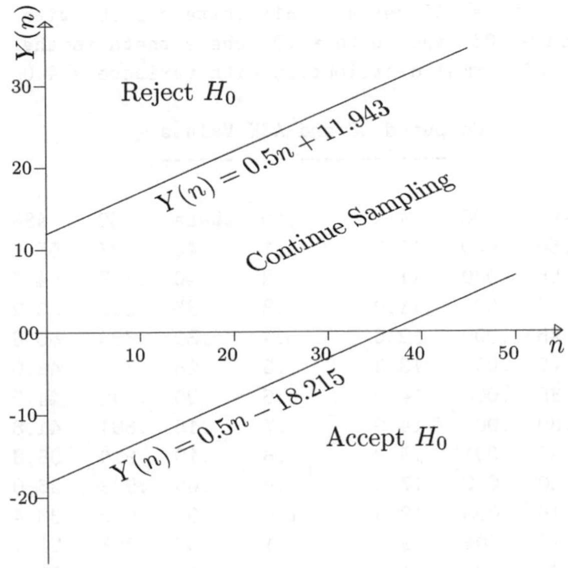

Figure 3.4.1. *Acceptance and rejection regions for* H_0 *when* $\theta_0 = 0, \theta_1 = 1, \sigma^2 = 4$, $\alpha = 0.05$ *and* $\beta = 0.01$.

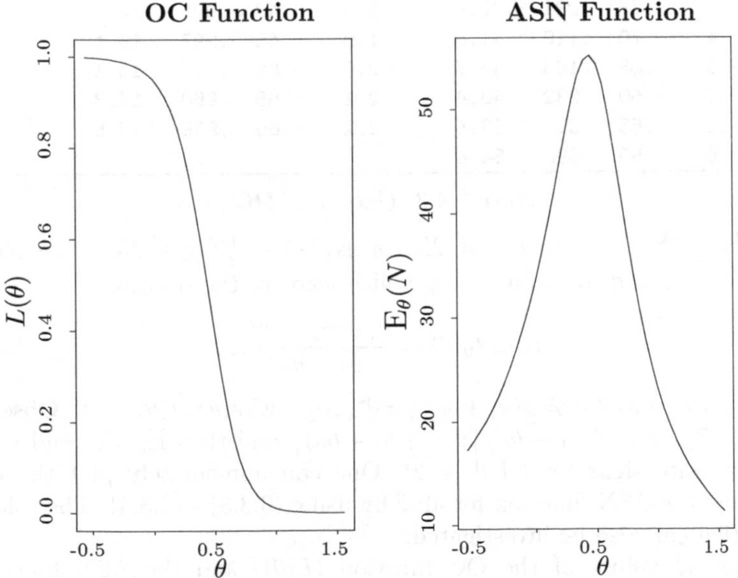

Figure 3.4.2. *OC function and ASN function for testing normal mean,* $\theta = 0$ *against* $\theta = 1$, *with* $\alpha = 0.05$, $\beta = 0.01$, $\sigma^2 = 4$.

```
SPRT:   H0: theta = .00 versus   H1: theta = 1.00  with
        alpha = .05  and  beta = .01 where theta is the
        mean of normal distribution with variance = 4.0

                 Computed OC and ASN Values
                 --------------------------
```

t0	theta	OC	ASN	t0	theta	OC	ASN
-2.2	1.60	.000	10.9	.1	.45	.487	55.2
-2.1	1.55	.000	11.4	.2	.40	.577	54.7
-2.0	1.50	.000	11.9	.3	.35	.660	53.2
-1.9	1.45	.000	12.6	.4	.30	.733	50.8
-1.8	1.40	.000	13.3	.5	.25	.794	48.0
-1.7	1.35	.000	14.0	.6	.20	.842	44.9
-1.6	1.30	.001	14.9	.7	.15	.881	41.8
-1.5	1.25	.001	15.9	.8	.10	.910	38.8
-1.4	1.20	.002	17.0	.9	.05	.933	36.0
-1.3	1.15	.003	18.3	1.0	.00	.950	33.4
-1.2	1.10	.004	19.7	1.1	-.05	.963	31.1
-1.1	1.05	.006	21.4	1.2	-.10	.972	29.0
-1.0	1.00	.010	23.3	1.3	-.15	.979	27.1
-.9	.95	.015	25.5	1.4	-.20	.985	25.4
-.8	.90	.024	28.1	1.5	-.25	.989	23.8
-.7	.85	.036	31.0	1.6	-.30	.992	22.5
-.6	.80	.055	34.3	1.7	-.35	.994	21.2
-.5	.75	.081	37.9	1.8	-.40	.995	20.1
-.4	.70	.119	41.8	1.9	-.45	.997	19.1
-.3	.65	.168	45.7	2.0	-.50	.997	18.1
-.2	.60	.232	49.4	2.1	-.55	.998	17.3
-.1	.55	.309	52.4	2.2	-.60	.999	16.5
.0	.50	.396	54.4				

Figure 3.4.3. *Output of* ***OC***.

since $\mathrm{E}_\theta\left[e^{sX_1}\right]$, the m.g.f. of X_1, is $\exp(s\theta+\frac{1}{2}s^2\sigma^2)$. Now, we write $\mathrm{E}_\theta\left[e^{tZ_1}\right]=1$ and solve for $t=t_0$, which leads to the solution.

$$t_0 \equiv t_0(\theta) = \frac{\theta_1+\theta_0-2\theta}{\theta_1-\theta_0}, \tag{3.4.3}$$

which is non-zero if $\theta \neq \frac{1}{2}(\theta_1+\theta_0)$ $(=\theta^*$, say), whereas $t(\theta^*)=0$. Observe that $\mathrm{E}_\theta(Z_1)=\sigma^{-2}(\theta_1-\theta_0)\left\{\theta-\frac{1}{2}(\theta_1+\theta_0)\right\}$ and thus $\mathrm{E}_\theta(Z_1)$ and $t_0(\theta)$ have opposite signs for all $\theta\neq\theta^*$. One can immediately plot the OC curve and the ASN function for all θ by using (3.3.3) - (3.3.4). The role of truncation can also be investigated.

Numerical values of the OC function ($L(\theta)$) and the ASN function ($\mathrm{E}_\theta(N)$) for a given SPRT among the nine specified population distributions, can be easily obtained using the subprogram **OC** found within the program **Seq03.exe**. An output from this program for the special case of the normal mean example when one tests $\theta=0$ versus $\theta=1$ with $\alpha=0.05$,

$\beta = 0.01$ and $\sigma^2 = 4$ is given in Figure 3.4.3. The t_0 values used in the computation of $L(\theta)$ and $\mathrm{E}_\theta(N)$ are also shown so that the results can be easily verified. Here, the OC function and the ASN function, are computed using the approximations given by (3.3.3) and (3.3.4) respectively. Figure 3.4.2 displays these numerical values graphically.

Example 3.4.2: Let $X_1, X_2, \ldots$ be i.i.d. real valued random variables having the common p.d.f. or p.m.f. $f(x|\theta)$ belonging to a **one parameter exponential family**. To be specific, suppose that

$$f(x|\theta) = \begin{cases} \exp\{p(\theta)K(x) + S(x) + q(\theta)\}, & x \in \mathfrak{X}, \theta \in \Theta \\ 0, & \text{otherwise} \end{cases} \tag{3.4.4}$$

where $\mathfrak{X}$ is an appropriate known subset of $I\!R$ and Θ is a known subinterval of $I\!R$. Whether $\mathfrak{X}$ is a discrete space or a continuum, in what follows we pretend that $f(x|\theta)$ is a p.d.f. according to our convention. Now, while testing $H_0 : \theta = \theta_0$ versus $H_1 : \theta = \theta_1$, we have

$$Z_i = \{p(\theta_1) - p(\theta_0)\} K(X_i) + \{q(\theta_1) - q(\theta_0)\}, \tag{3.4.5}$$

$i = 1, 2, \ldots$. Hence, the SPRT in (3.1.4) would again correspond to crossing two straight line boundaries similar to the situation portrayed in Figure 3.4.1.

It is easy to see that Example 3.4.1 is a special case of the present example where $p(\theta) = \theta/\sigma^2$, $q(\theta) = -\frac{1}{2}\theta^2/\sigma^2, K(x) = x$ and $S(x) = -\frac{1}{2}x^2/\sigma^2$ with $\sigma(>0)$ known, $\mathfrak{X} = \Theta = I\!R$, the real line.

Example 3.4.3: Let $X_1, X_2, \ldots$ be i.i.d. Bernoulli $(\theta), 0 < \theta < 1$, that is $\Theta = (0, 1)$ and $\mathfrak{X} = \{0, 1\}$. It is easy to check that $f(x|\theta)$ coincides with (3.4.4) with $p(\theta) = \ln\{\theta/(1-\theta)\}, K(x) = x, S(x) = 0$, $q(\theta) = \ln(1-\theta)$. Thus (3.4.5) reduces to

$$Z_i = X_i \ln\left\{\frac{\theta_1(1-\theta_0)}{\theta_0(1-\theta_1)}\right\} + \ln\left\{\frac{1-\theta_1}{1-\theta_0}\right\}, \tag{3.4.6}$$

and now the SPRT (3.1.4) for $H_0 : \theta = \theta_0$ versus $H_1 : \theta = \theta_1(>\theta_0)$ can be easily implemented. The role of truncation can also investigated. We denote

$N \equiv$ first integer $n(\geq 1)$ for which the inequality $Y(n) \leq c_1 + dn$ or $Y(n) \geq c_2 + dn$ holds with $Y(n) = \sum_{i=1}^{n} X_i$,

with

$$c_1 = \frac{\ln(B)}{\ln\left(\dfrac{\theta_1(1-\theta_0)}{\theta_0(1-\theta_1)}\right)}, c_2 = \frac{\ln(A)}{\ln\left(\dfrac{\theta_1(1-\theta_0)}{\theta_0(1-\theta_1)}\right)} \text{ and } d = \frac{\ln\left(\dfrac{1-\theta_0}{1-\theta_1}\right)}{\ln\left(\dfrac{\theta_1(1-\theta_0)}{\theta_0(1-\theta_1)}\right)}.$$

At the stopped stage, if $Y(N) \leq c_1 + dN$, we accept H_0, but if $Y(N) \geq c_2 + dN$, we accept H_1.

Note that since we have $\theta_1 > \theta_0$, $\ln\left(\dfrac{\theta_1(1-\theta_0)}{\theta_0(1-\theta_1)}\right)$ is positive. Next, in order to obtain the OC curve and the ASN function, first we find the m.g.f. of Z_1 and write

$$\begin{aligned}\mathrm{E}_\theta\left[e^{tZ_1}\right] &= \left(\frac{1-\theta_1}{1-\theta_0}\right)^t \mathrm{E}_\theta\left[\left\{\frac{\theta_1(1-\theta_0)}{\theta_0(1-\theta_1)}\right\}^{tX_1}\right] \\ &= \left(\frac{1-\theta_1}{1-\theta_0}\right)^t \left[(1-\theta)+\theta\left\{\frac{\theta_1(1-\theta_0)}{\theta_0(1-\theta_1)}\right\}^t\right]. \end{aligned} \tag{3.4.7}$$

Then equating the expression from (3.4.7) with unity at $t = t_0$, we obtain

$$\theta = \frac{\left(\dfrac{1-\theta_0}{1-\theta_1}\right)^{t_0} - 1}{\left(\dfrac{\theta_1(1-\theta_0)}{\theta_0(1-\theta_1)}\right)^{t_0} - 1}. \tag{3.4.8}$$

Also from (3.3.3), we obtain

$$L(\theta) \approx \begin{cases} \dfrac{\left(\dfrac{1-\beta}{\alpha}\right)^{t_0} - 1}{\left(\dfrac{1-\beta}{\alpha}\right)^{t_0} - \left(\dfrac{\beta}{1-\alpha}\right)^{t_0}} & \text{if } \theta \neq \theta^* \\ \dfrac{\ln\left(\dfrac{1-\beta}{\alpha}\right)}{\ln\left(\dfrac{(1-\beta)(1-\alpha)}{\alpha\beta}\right)} & \text{if } \theta = \theta^* \end{cases} \tag{3.4.9}$$

with $\theta^* = \dfrac{\ln\Big\{(1-\theta_0)/(1-\theta_1)\Big\}}{\ln\Big\{\theta_1(1-\theta_0)/\left(\theta_0(1-\theta_1)\right)\Big\}}$. Observe that $\mathrm{E}_\theta(Z_1) = 0$ if and only if $\theta = \theta^*$ which is equivalent to saying that $t_0(\theta^*) = 0$. By taking the limit of the expression in the right hand side of (3.4.7) as $t_0 \to 0$, one can see that θ would then reduce to θ^*. Note that

$$\begin{aligned}\mathrm{E}_\theta\left(Z_1\right) &= \theta \ln\left\{\frac{\theta_1(1-\theta_0)}{\theta_0(1-\theta_1)}\right\} + \ln\left\{\frac{1-\theta_1}{1-\theta_0}\right\}, \\ \mathrm{V}_\theta\left(Z_1\right) &= \theta(1-\theta)\left[\ln\left\{\frac{\theta_1(1-\theta_0)}{\theta_0(1-\theta_1)}\right\}\right]^2, \end{aligned}$$

and hence the ASN function given by (3.3.4) can be evaluated easily.

Now let us examine a special case of this when $\theta_0 = 0.5, \theta_1 = 0.7$ and $\alpha = \beta = 0.05$. Then from the above results, one obtains

$$c_2 = -c_1 = 3.475 \text{ and } d = 0.603.$$

That is, if $Y(N) \leq 0.603N - 3.475$ we accept H_0 and if $Y(N) \geq 0.603N + 3.475$ we reject H_0. Furthermore, as long as $0.603N - 3.475 < Y(N) < 0.603N + 3.475$, we continue to sample from the population.

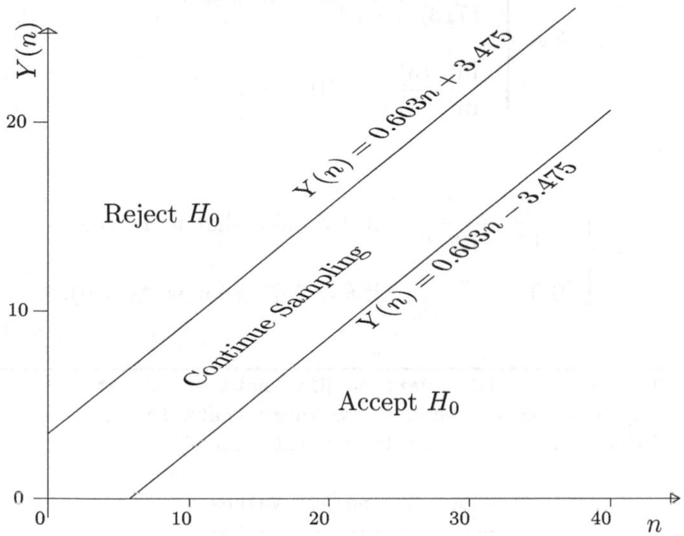

Figure 3.4.4. *Acceptance and rejection regions for* H_0 *when* $\theta_0 = 0.5, \theta_1 = 0.7$ *and* $\alpha = \beta = 0.05$.

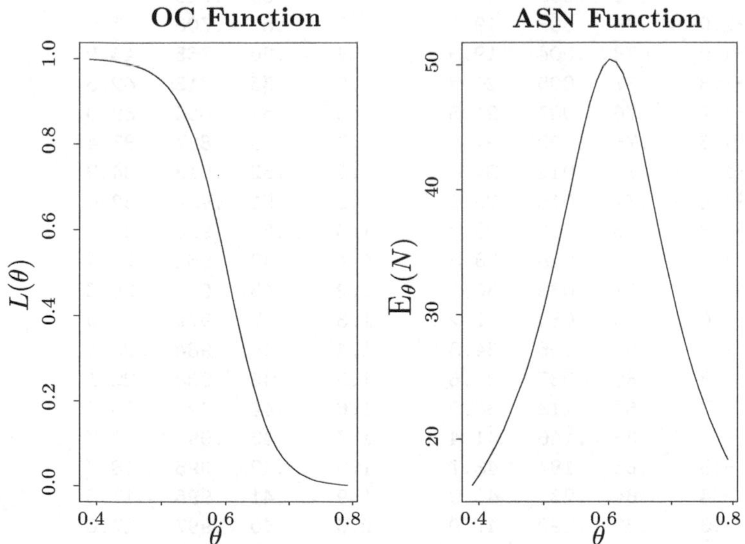

Figure 3.4.5. *OC function and ASN function for testing Bernoulli probability,* $\theta = 0.5$ *against* $\theta = 0.7$ *with* $\alpha = \beta = 0.05$.

A graphical representation of these acceptance and rejection regions are given in Figure 3.4.4. In this special situation, the expressions for θ and $L(\theta)$ given by (3.4.8) and (3.4.9) are reduced to

$$\theta = \begin{cases} \dfrac{(5/3)^{t_0} - 1}{(7/3)^{t_0} - 1} & \text{if } t_0 \neq 0 \\ \\ \dfrac{\ln(5/3)}{\ln(7/3)} = 0.60 & \text{if } t_0 = 0 \end{cases}$$

and

$$L(\theta) = \begin{cases} \dfrac{19^{t_0} - 1}{19^{t_0} - 19^{-t_0}} & \text{if } \theta \neq 0.60 \text{ (that is, } t_0 \neq 0) \\ \\ 0.5 & \text{if } \theta = 0.60 \text{ (that is, } t_0 = 0). \end{cases}$$

```
SPRT:  H0: theta = .50    versus  H1: theta = .70  at
       alpha = .05 and beta = .05 where theta is the
       P(Success) in Bernoulli distribution

                Computed OC and ASN Values
                --------------------------
```

t0	theta	OC	ASN	t0	theta	OC	ASN
-2.2	.80	.002	17.7	.1	.59	.573	49.9
-2.1	.79	.002	18.3	.2	.58	.643	48.8
-2.0	.78	.003	19.1	.3	.57	.708	47.0
-1.9	.78	.004	19.9	.4	.56	.765	44.9
-1.8	.77	.005	20.8	.5	.55	.813	42.5
-1.7	.76	.007	21.8	.6	.54	.854	39.9
-1.6	.75	.009	22.8	.7	.53	.887	37.4
-1.5	.74	.012	24.0	.8	.52	.913	34.9
-1.4	.74	.016	25.4	.9	.51	.934	32.6
-1.3	.73	.021	26.8	1.0	.50	.950	30.4
-1.2	.72	.028	28.5	1.1	.49	.962	28.4
-1.1	.71	.038	30.3	1.2	.48	.972	26.6
-1.0	.70	.050	32.2	1.3	.47	.979	24.9
-.9	.69	.066	34.3	1.4	.46	.984	23.4
-.8	.68	.087	36.6	1.5	.45	.988	22.1
-.7	.67	.113	38.9	1.6	.44	.991	20.8
-.6	.66	.146	41.4	1.7	.43	.993	19.7
-.5	.65	.187	43.7	1.8	.42	.995	18.7
-.4	.64	.235	46.0	1.9	.41	.996	17.8
-.3	.63	.292	47.9	2.0	.40	.997	17.0
-.2	.62	.357	49.3	2.1	.39	.998	16.3
-.1	.61	.427	50.2	2.2	.38	.998	15.6
.0	.60	.500	50.4				

Figure 3.4.6. *Output of **OC**.*

The subprogram **OC** computes θ, $L(\theta)$ and $\mathrm{E}_\theta(N)$ for $-2.1 \le t_0 \le 2.1$ with 0.1 increment. The results are listed in Figure 3.4.6 and they are graphically displayed in Figure 3.4.5.

Remark 3.4.1: Distributions such as Poisson (θ) or exponential with a single scale parameter $\theta, \theta > 0$, belong to the exponential family defined by (3.4.4). The associated SPRTs for $H_0 : \theta = \theta_0$ versus $H_0 : \theta = \theta_1 (> \theta_0)$ would follow readily from Example 3.4.2. Details are left out as exercises.

Remark 3.4.2: In the examples 3.4.1-3.4.3, the SPRTs for $H_0 : \theta = \theta_0$ versus $H_1 : \theta = \theta_1 (< \theta_0)$ can be immediately obtained from the proposed tests by essentially reversing the roles of θ_0 and θ_1. Similar comments hold, in spirit, in the context of Remark 3.4.1 by reversing the roles of θ_0 and θ_1 while testing $H_0 : \theta = \theta_0$ versus $H_1 : \theta = \theta_1 (< \theta_0)$.

3.5 Auxiliary Results

In this section we state some general results without giving proofs. The proofs are often complicated as well as lengthy and these may be found from the original cited sources. These are also readily available in Chapter 2 of Ghosh et al. (1997). The emphasis here is to apply these results.

Theorem 3.5.1 (Stein, 1956): *Let $U_1, U_2, \ldots$ be a sequence of i.i.d. real valued random variables such that $\mathrm{P}(U_1 = 0) < 1$. Let a and b be real numbers, $a < b$, and N be the smallest integer $j (\ge 1)$ for which the inequality $a < \sum_{l=1}^{j} U_l < b$ is violated. Then there exists constants $c (> 0)$ and $0 < r < 1$ such that*

$$\mathrm{P}(N > n) \le cr^n \text{ for } n \ge 1.$$

The following result goes by the name **Fundamental Identity of Sequential Analysis** or **Wald's Fundamental Identity**. An elegant proof was provided by Bahadur (1958). A proof of this fundamental result is outlined in Exercise 3.5.4.

Theorem 3.5.2: *Let N be defined exactly as in Theorem 3.5.1 and assume that $\mathrm{P}(U_1 = 0) < 1$. Define $M(t) = \mathrm{E}\left[e^{tU_1}\right]$, the m.g.f. of U_1, for some $t \in \mathbb{R}$. Then,*

$$\mathrm{E}\left[e^{tW_N} \{M(t)\}^{-N}\right] = 1$$

for all t such that $M(t)$ is finite, where $W_N = \sum_{i=1}^{N} U_i$.

In many circles, this result is known as a likelihood ratio identity. This identity has been used extensively in order to handle many types of statistical problems. The recent discussion paper of Lai (2004a) is filled with an incredible breadth of information.

The fundamental identity is often used when $M(t) = 1$ for some $t \ne 0$. The idea is to use a simpler version such as $\mathrm{E}\left[e^{tW_N}\right] = 1$ with $t \ne 0$ such that $M(t) = 1$. The following results give a set of sufficient conditions under which one can claim that $M(t) = 1$ for a unique $t \ne 0$.

Theorem 3.5.3 (Wald, 1947): *Let U be a random variable such that (i)* $\mathrm{P}(U > 0) > 0$, *(ii)* $\mathrm{P}(U < 0) > 0$, *(iii)* $\mathrm{E}(U) \neq 0$ *and (iv)* $M(t)$, *the m.g.f. of U, is finite for all t. Then, there exists a unique non-zero real number t_0 such that $M(t_0) = 1$.*

The following two results are known as **Wald's First Equation** and **Wald's Second Equation** respectively. These give the forms of the expectation and the second moment of a randomly stopped sum of i.i.d. random variables. These theorems are stated in a general fashion along the lines of Chow et al. (1965). In the case of the fixed-sample analyses, one can fill in $N = n$ with probability one and then the two results obviously hold. Most beginning statistics texts include such versions with N replaced by n. Proofs for these two fundamental results are outlined in Exercise 3.5.5 assuming a finite m.g.f. for U's.

Theorem 3.5.4 [Wald's First Equation]: *Let $U_1, U_2, \ldots$ be i.i.d. real valued random variables and N be a stopping variable such that (i)* $\mathrm{E}(|U_1|) < \infty$, *(ii) the event $N \geq j$ depends only on $U_1, U_2, \ldots, U_{j-1}$ for all $j = 1, 2, \ldots$ and (iii)* $\mathrm{E}(N) < \infty$. *Then,*

$$\mathrm{E}(W_N) = \mu \mathrm{E}(N)$$

where $\mu = \mathrm{E}(U_1)$ *and* $W_N = \sum_{i=1}^{N} U_i$.

Theorem 3.5.5 [Wald's Second Equation]: *Let $U_1, U_2, \ldots,$ the stopping variable N and W_N be as in Theorem 3.5.4. Suppose that* (i) $\mathrm{E}(U_1^2) < \infty$ *and the conditions* $(ii) - (iii)$ *from Theorem 3.5.4 hold. Then,*

$$\mathrm{E}\left[(W_N - N\mu)^2\right] = \sigma^2 \mathrm{E}(N)$$

where $\mu = \mathrm{E}(U_1)$ *and* $\sigma^2 = \mathrm{V}(U_1)$.

3.6 Exercises

Exercise 3.1.1: Suppose that $X_1, \ldots, X_n, \ldots$ are i.i.d. random variables from a $N(\theta, \sigma^2)$ distribution with $-\infty < \theta < \infty, 0 < \sigma < \infty$. We assume that θ is unknown but σ is known. Let us test a simple null hypothesis $H_0 : \theta = \theta_0$ versus a simple alternative hypothesis $H_1 : \theta = \theta_1 (< \theta_0)$ with preassigned type I and type II error probabilities $\alpha, \beta, 0 < \alpha, \beta < 1$. Write down the SPRT from (3.1.3) explicitly and show the different regions geometrically.

Exercise 3.1.2: Suppose that $X_1, \ldots, X_n, \ldots$ are i.i.d. random variables from the Bernoulli (θ) distribution with $0 < \theta < 1$. We assume that θ is unknown. Let us test a simple null hypothesis $H_0 : \theta = \theta_0$ versus a simple alternative hypothesis $H_1 : \theta = \theta_1 (< \theta_0)$ with preassigned type I and type II error probabilities $\alpha, \beta, 0 < \alpha, \beta < 1$. Write down the SPRT from (3.1.3) explicitly and show the different regions geometrically.

Exercise 3.1.3: Suppose that $X_1, \ldots, X_n, \ldots$ are i.i.d. random variables from a Poisson(θ) distribution with $0 < \theta < \infty$. We assume that θ is unknown. Let us test a simple null hypothesis $H_0 : \theta = \theta_0$ versus a simple alternative hypothesis $H_1 : \theta = \theta_1 (> \theta_0)$ with preassigned type I and type II error probabilities $\alpha, \beta, 0 < \alpha, \beta < 1$. Write down the SPRT from (3.1.3) explicitly and show the different regions geometrically. How would this SPRT change if the alternative hypothesis happened to be $H_1 : \theta = \theta_1 (< \theta_0)$ instead?

Exercise 3.1.4: Suppose that $X_1, \ldots, X_n, \ldots$ are i.i.d. random variables from a universe having an exponential distribution with the unknown mean $\theta, 0 < \theta < \infty$. Let us test a simple null hypothesis $H_0 : \theta = \theta_0$ versus a simple alternative hypothesis $H_1 : \theta = \theta_1 (> \theta_0)$ with preassigned type I and type II error probabilities $\alpha, \beta, 0 < \alpha, \beta < 1$. Write down the SPRT from (3.1.3) explicitly and show the different regions geometrically. How would this SPRT change if the alternative hypothesis happened to be $H_1 : \theta = \theta_1 (< \theta_0)$ instead?

Exercise 3.2.1 (Exercise 3.1.1 Continued): Suppose that $X_1, \ldots, X_n, \ldots$ are i.i.d. random variables from a $N(\theta, \sigma^2)$ distribution with $-\infty < \theta < \infty, 0 < \sigma < \infty$. We assume that θ is unknown but σ is known. Let us test a simple null hypothesis $H_0 : \theta = \theta_0$ versus a simple alternative hypothesis $H_1 : \theta = \theta_1 (< \theta_0)$ with preassigned type I and type II error probabilities $\alpha, \beta, 0 < \alpha, \beta < 1$. Use the program **SimAB** to simulate the SPRT for a range of values of $\alpha, \beta, n, \theta_0, \theta_1$ and σ. Also, use the program **SimAB** for simulating the truncated version of this SPRT along the lines of (3.2.7) with $K = n^*$. Here, the expression of n^* is borrowed from Exercise 2.2.1, part (iii).

Exercise 3.2.2 (Exercise 3.1.2 Continued): Suppose that $X_1, \ldots, X_n, \ldots$ are i.i.d. random variables from the Bernoulli(θ) distribution with $0 < \theta < 1$. We assume that θ is unknown. Let us test a simple null hypothesis $H_0 : \theta = \theta_0$ versus a simple alternative hypothesis $H_1 : \theta = \theta_1 (< \theta_0)$ with preassigned type I and type II error probabilities $\alpha, \beta, 0 < \alpha, \beta < 1$. Use the program **SimAB** to simulate the SPRT for a range of values of $\alpha, \beta, n, \theta_0$ and θ_1. Also, use the program **SimAB** for simulating the truncated version of this SPRT along the lines of (3.2.7) with a range of values of K.

Exercise 3.2.3 (Exercise 3.1.3 Continued): Suppose that $X_1, \ldots, X_n, \ldots$ are i.i.d. random variables from a Poisson(θ) distribution with $0 < \theta < \infty$. We assume that θ is unknown. Let us test a simple null hypothesis $H_0 : \theta = \theta_0$ versus a simple alternative hypothesis $H_1 : \theta = \theta_1 (> \theta_0)$ with preassigned type I and type II error probabilities $\alpha, \beta, 0 < \alpha, \beta < 1$. Use the program **SimAB** to simulate the SPRT for a range of values of $\alpha, \beta, n, \theta_0$ and θ_1. Also, use the program **SimAB** for simulating the truncated version of this SPRT along the lines of (3.2.7) with a range of values of K. Repeat those simulations when the alternative hypothesis happens to be $H_1 : \theta = \theta_1(< \theta_0)$ instead.

Exercise 3.2.4: Suppose that $X_1, \ldots, X_n, \ldots$ are i.i.d. random variables from an exponential distribution with the unknown mean $\theta, 0 < \theta < \infty$. Let us test a simple null hypothesis $H_0 : \theta = \theta_0$ versus a simple alternative hypothesis $H_1 : \theta = \theta_1(> \theta_0)$ with preassigned type I and type II error probabilities $\alpha, \beta, 0 < \alpha, \beta < 1$. Use the program **SimAB** to simulate the SPRT for a range of values of $\alpha, \beta, n, \theta_0$ and θ_1. Also, use the program **SimAB** for simulating the truncated version of this SPRT along the lines of (3.2.7) with a range of values of K. Repeat those simulations when the alternative hypothesis happens to be $H_1 : \theta = \theta_1(< \theta_0)$ instead.

Exercise 3.2.5: Using Theorem 3.2.1, verify the inequality given in the equation (3.2.6).

{*Hint*: Assume that $B = \beta/(1-\alpha, A = (1-\beta)/\alpha$. Then, using Theorem 3.2.1, one would claim that $B \geq \beta'/(1-\alpha')$ and $A \leq (1-\beta')/\alpha'$. By cross-multiplying, one will obtain two inequalities: $\beta - \beta\alpha' \geq \beta' - \beta'\alpha$ and $\alpha' - \beta\alpha' \leq \alpha - \beta'\alpha$. Now, one would simply subtract the second inequality from the first inequality.}

Exercise 3.3.1: In (3.3.3), apply L'Hôpital's rule to show that

$$\lim_{t_0 \to 0} \frac{e^{t_0 a} - 1}{e^{t_0 a} - e^{t_0 b}} = \frac{a}{a-b}.$$

Exercise 3.3.2: Consider Wald's approximations for $\mathrm{E}_\theta(N)$ from (3.3.5) under both the hypotheses H_0 and H_1. Show that these approximate expressions are positive.

{*Hint*: The approximate expression for $\mathrm{E}_{\theta_0}(N)$ happens to be $\{b(1-\alpha) + a\alpha\}/\mathrm{E}_{\theta_0}(Z_1)$. Since the function $\ln(y), y > 0$, is strictly concave, one must have

$$\begin{aligned}
&(1-\alpha)\ln\left(\frac{\beta}{1-\alpha}\right) + \alpha\ln\left(\frac{1-\beta}{\alpha}\right) \\
&< \ln\left((1-\alpha)\left(\frac{\beta}{1-\alpha}\right) + \alpha\left(\frac{1-\beta}{\alpha}\right)\right) \\
&= \ln(1) = 0.
\end{aligned}$$

Also, by Jensen's inequality, one can see that

$$\begin{aligned} \mathrm{E}_{\theta_0}(Z_1) &= \int_{\mathcal{X}} \ln\left(\frac{f(x_1|\theta_1)}{f(x_1|\theta_0)}\right) f(x_1|\theta_0)dx_1 \\ &< \ln\left(\int_{\mathcal{X}} \left[\frac{f(x_1|\theta_1)}{f(x_1|\theta_0)}\right] f(x_1|\theta_0)dx_1\right) \\ &= \ln\left[\int_{\mathcal{X}} f(x_1|\theta_1)dx_1\right] = \ln(1) = 0. \end{aligned}$$

In other words, $\{b(1-\alpha)+a\alpha\}/\mathrm{E}_{\theta_0}(Z_1)$ happens to be a ratio of two negative numbers. Modify these arguments suitably under H_1.}

Exercise 3.3.3 (Exercise 3.2.3 Continued): Suppose that $X_1, \ldots, X_n, \ldots$ are i.i.d. random variables from a Poisson(θ) distribution with $0 < \theta < \infty$. We assume that θ is unknown. Consider Wald's SPRT for a simple null hypothesis $H_0 : \theta = \theta_0$ versus a simple alternative hypothesis $H_1 : \theta = \theta_1 (> \theta_0)$ with preassigned type I and type II error probabilities $\alpha, \beta, 0 < \alpha, \beta < 1$. Obtain the expression for $t_0 \equiv t_0(\theta)$ such that $\mathrm{E}_\theta\left[e^{t_0 Z_1}\right] = 1, \theta > 0$. Next, write down the expression of Wald's approximation of the OC function, $L(\theta)$ from (3.3.3) and the ASN function, $\mathrm{E}_\theta(N)$ from (3.3.5).

Exercise 3.4.1 (Exercise 3.2.1 Continued): Suppose that $X_1, \ldots, X_n, \ldots$ are i.i.d. random variables from a $N(\theta, \sigma^2)$ distribution with $-\infty < \theta < \infty, 0 < \sigma < \infty$. We assume that θ is unknown but σ is known. Let us test a simple null hypothesis $H_0 : \theta = \theta_0$ versus a simple alternative hypothesis $H_1 : \theta = \theta_1 (< \theta_0)$ with preassigned type I and type II error probabilities $\alpha, \beta, 0 < \alpha, \beta < 1$. Graphically represent the SPRT as in Figure 3.4.4. Use the program **OC** to compute Wald's approximations of the OC and ASN functions in this special case along the lines of Figure 3.4.3.

Exercise 3.4.2 (Exercise 3.2.2 Continued): Suppose that $X_1, \ldots, X_n, \ldots$ are i.i.d. random variables from the Bernoulli(θ) distribution with $0 < \theta < 1$. We assume that θ is unknown. Let us test a simple null hypothesis $H_0 : \theta = \theta_0$ versus a simple alternative hypothesis $H_1 : \theta = \theta_1 (< \theta_0)$ with preassigned type I and type II error probabilities $\alpha, \beta, 0 < \alpha, \beta < 1$. Graphically represent the SPRT as in Figure 3.4.4. Use the program **OC** to compute Wald's approximations of the OC and ASN functions in this special case along the lines of Figure 3.4.6.

Exercise 3.4.3 (Exercise 3.2.3 Continued): Suppose that $X_1, \ldots, X_n, \ldots$ are i.i.d. random variables from a Poisson(θ) distribution with $0 < \theta < \infty$. We assume that θ is unknown. Let us test a simple null hypothesis $H_0 : \theta = \theta_0$ versus a simple alternative hypothesis $H_1 : \theta = \theta_1 (> \theta_0)$ with preassigned type I and type II error probabilities $\alpha, \beta, 0 < \alpha, \beta < 1$. Graphically represent the SPRT. Use the program **OC** to compute Wald's approximations of the OC and ASN functions in this special case.

Exercise 3.4.4 (Exercise 3.2.4 Continued): Suppose that $X_1, \ldots, X_n, \ldots$ are i.i.d. random variables from an exponential distribution with an unknown mean $\theta, 0 < \theta < \infty$. Let us test a simple null hypothesis $H_0 : \theta = \theta_0$ versus a simple alternative hypothesis $H_1 : \theta = \theta_1 (> \theta_0)$ with preassigned type I and type II error probabilities $\alpha, \beta, 0 < \alpha, \beta < 1$. Graphically represent the SPRT. Use the program **OC** to compute Wald's approximations of the OC and ASN functions in this special case.

Exercise 3.4.5: In a hospital, the staff is recording the sex of the newborn babies. The staff records the number 1 for a girl and the number 0 for a boy. The hospital administrator is contemplating a repetition of i.i.d. Bernoulli trials where θ stands for the unknown probability that a newborn is a girl. We wish to test $H_0 : \theta = 0.5$ versus $H_1 : \theta = 0.7$ with type I and type II error probabilities $0.05, 0.01$ respectively. The following data have been recorded on the sex of the newborn during the next twenty deliveries:

$$
\begin{array}{lllll}
x_1 = 1 & x_5 = 0 & x_9 = 1 & x_{13} = 1 & x_{17} = 1 \\
x_2 = 1 & x_6 = 0 & x_{10} = 1 & x_{14} = 1 & x_{18} = 1 \\
x_3 = 1 & x_7 = 0 & x_{11} = 1 & x_{15} = 0 & x_{19} = 1 \\
x_4 = 0 & x_8 = 0 & x_{12} = 0 & x_{16} = 0 & x_{20} = 0
\end{array}
$$

Use the program **SPRT** to feed this real data one by one in a sequence, namely, first enter x_1, then enter x_2 if needed, followed by x_3 as needed, and so on. Recommend whether the data gathering should have terminated (along with the decision in favor or against H_0) before collecting all twenty observations or whether the collected data are not yet enough to reach any decision.

Exercise 3.4.6: The length of time a bus takes to reach the city center from the airport is *normally* distributed with an unknown mean θ (in minutes) and standard deviation 10 minutes. We wish to test $H_0 : \theta = 45$ versus $H_1 : \theta = 60$ with type I and type II error probabilities $0.01, 0.05$ respectively. The following data have been recorded on the length of time a bus took to reach the city center from the airport during the next sixteen comparable runs:

$$
\begin{array}{llll}
x_1 = 63.1 & x_5 = 25.5 & x_9 = 61.7 & x_{13} = 44.5 \\
x_2 = 52.7 & x_6 = 50.6 & x_{10} = 53.4 & x_{14} = 43.1 \\
x_3 = 40.6 & x_7 = 48.4 & x_{11} = 39.1 & x_{15} = 35.7 \\
x_4 = 53.0 & x_8 = 40.0 & x_{12} = 58.7 & x_{16} = 50.1
\end{array}
$$

Use the program **SPRT** to feed this real data one by one in a sequence, namely, first enter x_1, then enter x_2 if needed, followed by x_3 as needed, and so on. Recommend whether the data gathering should have terminated (along with the decision in favor or against H_0) before collecting all sixteen observations or whether the collected data are not yet enough to reach any decision.

Exercise 3.4.7: The number of telephone calls arriving at the switch board of a large department store within the first five minutes of its opening in the morning has a Poisson distribution with an unknown average number of calls, θ. We wish to test $H_0 : \theta = 45$ versus $H_1 : \theta = 30$ with both type I and type II error probabilities 0.05. The following data have been recorded on the number of calls over a period of twenty comparable days:

$$\begin{array}{ccccc} x_1 = 30 & x_5 = 29 & x_9 = 23 & x_{13} = 18 & x_{17} = 27 \\ x_2 = 26 & x_6 = 35 & x_{10} = 30 & x_{14} = 29 & x_{18} = 33 \\ x_3 = 33 & x_7 = 21 & x_{11} = 31 & x_{15} = 28 & x_{19} = 37 \\ x_4 = 28 & x_8 = 28 & x_{12} = 26 & x_{16} = 33 & x_{20} = 28 \end{array}$$

Use the program **SPRT** to feed this real data one by one in a sequence, namely, first enter x_1, then enter x_2 if needed, followed by x_3 as needed, and so on. Recommend whether the data gathering should have terminated (along with the decision in favor or against H_0) before collecting all twenty observations or whether the collected data are not yet enough to reach any decision.

Exercise 3.5.1: Under the conditions of Theorem 3.5.1, show that the m.g.f. $M_N(t)$ of the stopping variable N is finite for an appropriate range of values of t.

{*Hint*:

$$\begin{aligned} M_N(t) = \mathrm{E}[e^{tN}] &= \sum_{n=1}^{\infty} e^{tn}\mathrm{P}\{N = n\} \\ &\le \sum_{n=1}^{\infty} e^{tn}\mathrm{P}\{N > n-1\} \le \sum_{n=1}^{\infty} e^{tn}cr^{n-1}, \end{aligned}$$

by Theorem 3.5.1. Thus, $M_N(t) \le ce^t \sum_{n=1}^{\infty}(e^t r)^{n-1}$. But, this upper bound is a geometric series which is finite if $t < -\ln(r)$.}

Exercise 3.5.2 (Exercise 3.5.1 Continued): Under the conditions of Theorem 3.5.1, show that all the moments, in particular the mean and the variance, of the stopping variable N are finite.

{*Hint*: Note that $0 < \mathrm{E}[N^k] < 1$ for all $k < 0$. With $0 < t < -\ln(r)$, for any positive integral k, one has $N^k \le t^{-k}k!e^{tN}$ with probability one. Thus, $\mathrm{E}[N^k] \le t^{-k}k!\mathrm{E}[e^{tN}] = t^{-k}k!M_N(t)$, which is finite using the Exercise 3.5.1. How should one handle a positive non-integral value of k?}

Exercise 3.5.3: Suppose that $U = Z_1 (= \ln[f(X_1|\theta_1)/f(X_1|\theta_0)])$ in the context of the SPRT from (3.1.4). Show that the sufficient conditions of Theorem 3.5.3 hold when

(i) $X_1, X_2, \ldots$ are i.i.d. $N(\theta, \sigma^2)$ with $-\infty < \theta < \infty$ unknown but $0 < \sigma < \infty$ known.

(ii) $X_1, X_2, \ldots$ are i.i.d. Bernoulli(θ) with $0 < \theta < 1$ unknown.

(iii) $X_1, X_2, \ldots$ are i.i.d. Poisson(θ) with $0 < \theta < \infty$ unknown.

(iv) $X_1, X_2, \ldots$ are i.i.d. exponential with unknown mean $0 < \theta < \infty$.

(v) $X_1, X_2, \ldots$ are i.i.d. $N(\mu, \theta^2)$ with $-\infty < \mu < \infty$ known but $0 < \theta < \infty$ unknown.

Exercise 3.5.4: Prove Wald's Fundamental Identity (Theorem 3.5.2).

{*Hint*: Suppose that the U's are i.i.d. with a common p.d.f. $p(u)$. Now, one may express $\mathrm{E}\left[e^{tW_N}\{M(t)\}^{-N}\right]$ as

$$\begin{aligned}
&\textstyle\sum_{n=1}^{\infty} \mathrm{E}\left[e^{tW_N}\{M(t)\}^{-N} I(N=n)\right] \\
&= \textstyle\sum_{n=1}^{\infty} \mathrm{E}\left[e^{tW_n}\{M(t)\}^{-n} I(N=n)\right] \\
&= \textstyle\sum_{n=1}^{\infty} \int \cdots \int_{N=n} e^{t\Sigma_{i=1}^{n} u_i}\{M(t)\}^{-n} \Pi_{i=1}^{n} p(u_i) \Pi_{j=1}^{n} du_j \\
&= \textstyle\sum_{n=1}^{\infty} \int \cdots \int_{N=n} \Pi_{i=1}^{n}\{M(t)\}^{-1} e^{tu_i} p(u_i) \Pi_{j=1}^{n} du_j \\
&= \textstyle\sum_{n=1}^{\infty} \int \cdots \int_{N=n} \Pi_{i=1}^{n} p_t(u_i) \Pi_{j=1}^{n} du_j,
\end{aligned}$$

where we let $p_t(u) = \{M(t)\}^{-1} e^{tu} p(u)$ to be a new common p.d.f. of the U's when t is held fixed. Now, denoting the probability (measure) corresponding to the p.d.f. p_t by P_t, we can claim:

$$\mathrm{E}\left[e^{tW_N}\{M(t)\}^{-N}\right] = \mathrm{P}_t(N < \infty).$$

The proof is then completed by appealing to Stein's result, Theorem 3.5.1. This elegant proof is due to Bahadur (1968).}

Exercise 3.5.5: Prove the two Theorems 3.5.4 and 3.5.5 assuming that Wald's Fundamental Identity (Theorem 3.5.2) holds.

{*Hint*: Assume Wald's Fundamental Identity (Theorem 3.5.2). For $t \in \mathbb{R}$, we have $\mathrm{E}\left[e^{tW_N}\{M(t)\}^{-N}\right] = 1$ where $W_n = \sum_{i=1}^{n} U_i$, $n \geq 1$. By differentiating with respect to t and taking the derivative inside the expectation, one gets:

$$0 = \mathrm{E}\left[e^{tW_N} W_N \{M(t)\}^{-N} + e^{tW_N}(-N)\{M(t)\}^{-N-1} M'(t)\right]$$

Then, one plugs in $t = 0$ and $M(0) = 1, M'(0) = \mathrm{E}[U_1]$. Hence, we have $0 = \mathrm{E}\left[W_N - N\mathrm{E}(U_1)\right]$ which implies that $\mathrm{E}\left[W_N\right] = \mathrm{E}\left[N\right]\mathrm{E}(U_1)$. This proves Wald's first equation. In order to prove Wald's second equation, one needs the second derivative of the Fundamental Identity and proceed analogously.}

CHAPTER 4

Sequential Tests for Composite Hypotheses

4.1 Introduction

Sequential tests for composite hypotheses in general form a topic of fairly complex nature. In order to review some of these advanced topics, one may look at the books of Wald (1947), Ghosh (1970), Govindarajulu (1981) and Siegmund (1985), among others. Our objective here is very modest. We simply want to expose the readers to sequential tests of composite hypotheses, only in a few special but interesting cases with the help of examples. We deliberately do not discuss any optimality issues. We include sequential tests for (i) σ^2 in a $N(\mu, \sigma^2)$ distribution, (ii) μ in a $N(\mu, \sigma^2)$ distribution when σ^2 is unknown, (iii) the correlation coefficient ρ in a bivariate normal distribution with all parameters unknown and (iv) the shape parameter p in a gamma distribution when the scale parameter μ is unknown.

In the literature, sequential analogs of the t-test, χ^2-text and F-test are available. Ideas of sequential t (and t^2) tests were developed by Rushton (1950), Arnold (1951), Hoel (1954), Ghosh (1960a,b) and Sacks (1965). The sequential χ^2, t and F tests can be viewed as special cases from a larger class of tests known as the invariant SPRTs. Hall et al. (1965) and Wijsman (1971) gave excellent accounts of this area. Details can be obtained from Ghosh (1970), Govindarajulu (1981) and Siegmund (1985), among others. We merely focus on building some appreciation of the breadth of useful problems which can be resolved with sequential sampling.

4.2 Test for the Variance

Let $Y_1, Y_2, \ldots$ be i.i.d. random variables having a $N(\mu, \sigma^2)$ distribution with $-\infty < \mu < \infty, 0 < \sigma < \infty$. **If μ were known**, the customary estimator of σ^2 would be $n^{-1}\sum_{i=1}^{n}(Y_i - \mu)^2$, based on a random sample $Y_1, \ldots, Y_n$ for all $n \geq 1$. In this case, in order to test a simple null hypothesis $H_0 : \sigma^2 = \sigma_0^2$ versus a simple alternative hypothesis $H_1 : \sigma^2 = \sigma_1^2 (\sigma_0 \neq \sigma_1)$, we can define $X_i = (Y_i - \mu)^2, i = 1, 2, \ldots$. Now, given α, β, we can apply the SPRT defined in (3.1.4) by identifying $\theta = \sigma^2$ and noting that the X_i's are i.i.d. with the p.d.f. $f(x|\theta) = (2\pi\theta)^{1/2}\exp(-\frac{1}{2}x/\theta)x^{-1/2}$ for $x > 0$.

This constitutes an immediate application of the earlier methodology and one can easily obtain the OC curve and the ASN function associated with the corresponding SPRT for all values of $\sigma^2(> 0)$.

The program **Seq04.exe** can be used for this purpose and other composite SPRTs given in this chapter. This program has two subprograms, namely **CompSPRT** and **CompOC**. These subprograms are respectively similar to the subprograms **SPRT** and **OC** described in Chapter 3. The subprogram **CompSPRT** can be used for testing $H_0 : \theta = \theta_0$ versus $H_1 : \theta = \theta_1$ with the help of a set of real data observed sequentially or via simulation study where sequential data are generated from a specified probability distribution. **CompOC** gives the numerical values of the associated OC and ASN functions.

Example 4.2.1 : Let θ be the unknown variance of a normal population with mean $\mu = 2$. Consider the SPRT for testing $H_0 : \theta = 1$ versus $H_1 : \theta = 2$ with preassigned values $\alpha = 0.1$ and $\beta = 0.05$ using the subprogram **CompSPRT**. Figure 4.2.1 shows an output of **CompSPRT** with initial sample size $m = 10$ and random number sequence 2.

```
OUTPUT FROM THE PROGRAM: CompSPRT
=================================

 SPRT: H0: theta =  1.00  versus  H1: theta =  2.00 with
       alpha = .10 and beta = .05  where theta is the
       variance of a normal distribution with mean = 2.0

       Initial sample size = 10

       Results for simulated data from
       Normal distribution with mean =  2.0
       and variance =  1.0
       -------------------------------------

       No. of SPRTs truncated (with K =  50)         =  80
       No. of times H0 is accepted after truncation =  61

       Number of simulations          =  1000
       Number of times H0 is accepted =   939

       Average sample size (n bar)    =  26.54
       Std. dev. of n (s)             =  11.45
       Minimum sample size (min n)    =     10
       Maximum sample size (max n)    =     50
```

Figure 4.2.1. *Simulation results from* ***CompSPRT****.*

The initial sample size option was not used in the SPRTs given in Chapter 3. In those SPRTs, we drew one observation in the beginning of the procedure. Therefore, in Chapter 3, the process assumed the initial sample size for those SPRTs to be one. However, in some composite SPRTs, the initial sample size m needs to be at least two. Also, in practice, one may like to start the sampling process with five or ten observations, for example. Therefore, throughout this chapter we begin all composite SPRTs with an initial sample size specified by a user. Apart from this additional initial sample size option, the computations carried out in **CompSPRT** are similar to those carried out by the subprogram **SPRT** described in Chapter 3. Further, the option of data truncation, during the real data input as well as a simulation study, is still available in **CompSPRT**.

In this example, we have used the initial sample size $m = 10$ and each simulation run was truncated at $K = 50$. This implies that the maximum number of observations is limited to fifty. The program can be executed with an input value of K. Theoretically, the sample size N required for a particular simulation run to stop can be large even though we know that $P(N < \infty) = 1$. Hence, the truncation option is an important feature.

Figure 4.2.1 shows that 80 out of 1000 simulation of SPRTs have been truncated at $K = 50$ and among them we decided in favor of the null hypothesis $H_0 : \theta = 1$ on 61 occasions. But overall, we concluded in favor of H_0 on 939 occasions out of 1000 simulation runs. Since the data were simulated from a $N(2, 1)$ distribution (that is using $\theta = 1$) the estimated probability of type I error is $\alpha' = 1 - 0.939 = 0.061$ which is less than the preset value $\alpha = 0.1$. Also, on an average this SPRT required 26.54 observations with a standard deviation 11.45. In other words, the interpretations are similar to those discussed in Figure 3.2.4.

Next we compute the OC and the ASN functions of the above SPRT for different values of θ using the subprogram **CompOC**. The associated computation and the output are similar to those of the subprogram **OC** introduced in Chapter 3 (Figure 3.4.3). Figure 4.2.2 gives the computed values of the OC and the ASN functions for the SPRT in this example.

We face a more interesting situation when we need to test a composite null hypothesis $H_0 : \sigma^2 = \sigma_0^2$ versus a composite alternative hypothesis $H_1 : \sigma^2 = \sigma_1^2$ where $\sigma_0 \neq \sigma_1$ are two specified positive numbers. Here, **the mean** μ of the distribution **is assumed unknown** and it is often called a **nuisance parameter**. We need to get rid of any impact of this nuisance parameter on the proposed sequential test. We recall that σ^2 is customarily estimated by the sample variance $S_n^2 = (n-1)^{-1}\sum_{i=1}^{n}(Y_i - \overline{Y}_n)^2$ with $\overline{Y}_n = n^{-1}\sum_{i=1}^{n} Y_i$, based on the random samples $Y_1, \ldots, Y_n$ for all $n \geq 2$. Let us define

$$X_i = \{Y_1 + \cdots + Y_i - iY_{i+1}\} \,/\, \{i(i+1)\}^{1/2}, \qquad (4.2.1)$$

$i = 1, 2, \ldots, n-1, \ldots$. From Theorem 4.7.1, we know that the observable random variables $X_1, X_2, \ldots, X_{n-1}, \ldots$ are distributed as i.i.d. $N(0, \sigma^2)$ and $S_n^2 = (n-1)^{-1}\sum_{i=1}^{n-1} X_i^2$ for all $n \geq 2$.

```
OUTPUT FROM THE PROGRAM: CompOC
===============================

 SPRT:  H0: theta =  1.00 versus H1: theta =  2.00 with
        alpha = .10 and beta = .05  where theta is the
        variance of a normal distribution with mean = 2.0

                 Computed OC and ASN Values
                 --------------------------

      t0  theta    OC    ASN        t0  theta    OC    ASN
    -2.4   3.57  .001    4.1        .1   1.34  .501   27.9
    -2.3   3.41  .001    4.4        .2   1.29  .564   28.3
    -2.2   3.27  .002    4.8        .3   1.25  .625   28.5
    -2.1   3.13  .002    5.1        .4   1.21  .681   28.4
    -2.0   3.00  .003    5.5        .5   1.17  .731   28.1
    -1.9   2.88  .004    6.0        .6   1.13  .776   27.6
    -1.8   2.76  .005    6.5        .7   1.10  .815   27.0
    -1.7   2.65  .007    7.0        .8   1.06  .849   26.2
    -1.6   2.54  .010    7.6        .9   1.03  .877   25.4
    -1.5   2.44  .013    8.3       1.0   1.00  .900   24.6
    -1.4   2.34  .017    9.1       1.1    .97  .919   23.8
    -1.3   2.25  .022    9.9       1.2    .94  .935   23.0
    -1.2   2.16  .029   10.8       1.3    .91  .948   22.2
    -1.1   2.08  .038   11.9       1.4    .89  .958   21.4
    -1.0   2.00  .050   13.0       1.5    .86  .966   20.7
     -.9   1.92  .065   14.2       1.6    .84  .973   20.1
     -.8   1.85  .084   15.6       1.7    .81  .978   19.4
     -.7   1.78  .108   17.1       1.8    .79  .983   18.9
     -.6   1.72  .137   18.6       1.9    .77  .986   18.3
     -.5   1.66  .172   20.2       2.0    .75  .989   17.8
     -.4   1.60  .214   21.8       2.1    .73  .991   17.3
     -.3   1.54  .262   23.3       2.2    .71  .993   16.9
     -.2   1.49  .317   24.8       2.3    .69  .994   16.5
     -.1   1.44  .376   26.0       2.4    .68  .996   16.1
      .0   1.39  .438   27.1
```

Figure 4.2.2. *OC and ASN values computed from* ***CompOC***.

Note that H_0, H_1 are composite hypothesis to begin with, but in terms of the transformed sequence of the X's, these may be loosely viewed as simple hypotheses. We exploited the idea of location invariance, thereby getting rid of any influence of the unknown μ when statistical inferences are based on the X's. But, observe that from $\{Y_1, \ldots, Y_n\}$ we come up with $\{X_1, \ldots, X_{n-1}\}$ for all $n \geq 2$. In other words, we have lost one degree of freedom in order to get rid of any influence of the nuisance parameter on inferences regarding the parameter σ. This phenomenon is closely tied with the degree of freedom $(n-1)$ associated with the sample variance S_n^2.

Now, having observed $Y_1, \ldots, Y_n, \ldots$ sequentially, we first construct the X-sequence. Then, given α and β, we apply the SPRT from (3.1.4). Let us use Z_i defined by (3.4.5) where $\theta = \sigma^2$, $\theta_0 = \sigma_0^2$, $\theta_1 = \sigma_1^2$, $p(\theta) = -1/(2\theta)$, $q(\theta) = -\frac{1}{2}\ln(2\pi\theta)$, $K(x) = x^2$ and $S(x) = 0$, $\mathfrak{X} = I\!R$, $\Theta = I\!R^+$. See also (3.4.4). The stopping variable N would correspond to the number of observed X's. In other words, the sample size from the original population corresponding to the observed Y's would be $N+1$. The approximations to the OC and ASN functions will follow from (3.3.3) - (3.3.4).

One should explore this SPRT using both real and simulated data with the help of the subprogram **CompSPRT** by selecting the **id# 2** which tests for the **variance of a normal population (mean unknown)**. To enable the transformation of $Y_1, Y_2, Y_3, \ldots$ to $X_1, X_2, \ldots$, the initial sample size m is assumed to be at least two. Next, by selecting the options similar to those in the subprogram **CompOC**, one can evaluate the OC and the ASN functions of this SPRT.

Sequential test for σ^2 is really Wald's SPRT based on i.i.d. Helmert random variables found in (4.2.1).

4.3 Test for the Mean

Let $X_1, X_2, \ldots$ be i.i.d. $N(\mu, \sigma^2)$ random variables where $\mu \in I\!R$, $\sigma \in I\!R^+$ are unknown parameters. Robbins and Starr (1965) proposed a purely sequential procedure for testing a composite null hypothesis $H_0 : \mu = \mu_0$ versus a composite alternative hypothesis $H_1 : \mu = \mu_1 (\neq \mu_0)$, where μ_0, μ_1 are two specified numbers. The test has type I and type II error probabilities respectively close to two preassigned numbers α and β, $0 < \alpha, \beta < 1$. As before, let z_p stand for the upper $100p\%$ point for the standard normal distribution, $0 < p < 1$. That is, $\mathrm{P}(Z > z_p) = \mathrm{P}(Z \le -z_p) = p$ for any fixed p $(0 < p < 1)$ where Z is a standard normal variable. Let us write $\theta = (\mu, \sigma^2)$.

For the time being, let us *pretend* that σ^2 is known. Having recorded $X_1, \ldots, X_n$ let us denote

$$\overline{X}_n = n^{-1}\sum_{i=1}^{n} X_i,\ S_n^2 = (n-1)^{-1}\sum_{i=1}^{n}(X_i - \overline{X}_n)^2,\ n \ge 2.$$

Case 1 $\mu_1 > \mu_0$: Now, we would reject H_0 if $\overline{X}_n > k$ for some appropriate real number k such that

$$\begin{aligned} \mathrm{P}_\theta\left\{\overline{X}_n > k \text{ when } \mu = \mu_0\right\} &= \alpha, \\ \mathrm{P}_\theta\left\{\overline{X}_n \le k \text{ when } \mu = \mu_1\right\} &= \beta. \end{aligned} \tag{4.3.1}$$

Thus, we have

$$n^{\frac{1}{2}}(k - \mu_0)/\sigma = z_\alpha,\ \ n^{\frac{1}{2}}(k - \mu_1)/\sigma = -z_\beta. \tag{4.3.2}$$

In other words, we obtain

$$k = \frac{z_\alpha \mu_1 + z_\beta \mu_0}{z_\alpha + z_\beta} \quad \text{and} \quad n = \frac{(z_\alpha + z_\beta)^2 \sigma^2}{(\mu_1 - \mu_0)^2} \; (= n^*, \text{ say}). \tag{4.3.3}$$

That is, if σ were known, we could take a sample of size n, the smallest integer $\geq n^*$, and would reject H_0 if $\overline{X}_n > k$ with k given by (4.3.3). But since σ is unknown, Robbins and Starr (1965) aimed at estimating n^* and proposed the following sequential procedure:

Let $m(\geq 2)$ be the initial sample size and then proceed by taking one additional observation X at a time according to the stopping time

$$N = \text{ first integer } n(\geq m) \text{ for which } n \geq \frac{(z_\alpha + z_\beta)^2 S_n^2}{(\mu_1 - \mu_0)^2}. \tag{4.3.4}$$

Finally, we reject H_0 in favor H_1 if and only if $\overline{X}_N > k$ where k is defined in (4.3.3). Whatever be μ and σ, this sequential sampling procedure terminates with probability one. Hence, (4.3.4) provides a genuine sequential test of the hypotheses H_0 versus H_1.

Case 2 $\mu_1 < \mu_0$**:** In this situation, the fixed-sample-size test would reject H_0 if and only if $\overline{X}_n < k$. But, one can easily show that both k and n must satisfy (4.3.3) and thus the stopping rule given in (4.3.4) remains valid for all $\mu_0 \neq \mu_1$. However, we reject H_0 in favor H_1 if and only if $\overline{X}_N < k$ when $\mu_0 < \mu_1$.

The third selection, **Normal mean with unknown variance**, from the subprogram **CompSPRT** can be used to implement this test procedure using either real or simulated data.

Observe that $S_n^2 \xrightarrow{\text{P}_\theta} \sigma^2$ as $n \to \infty$ and both S_N^2, S_{N-1}^2 converge to σ^2 in P_θ-probability as $\delta(= \mu_1 - \mu_0) \to 0$, that is when H_0 and H_1 are closer to each other. In such a situation, the random sample size N required to select one of the hypotheses with preassigned errors α and β diverges to ∞ with probability one. Along with these facts, we state the following results from Robbins and Starr (1965): As $\delta \to 0$,

$$\frac{N}{n^*} \xrightarrow{\text{P}_\theta} 1 \quad \text{and} \quad \text{E}_\theta\left(\frac{N}{n^*}\right) \to 1. \tag{4.3.5}$$

That is, the random sample size N gets very close to n^*, both in P_θ-probability and mean, where n^* is defined in (4.3.3). It should be obvious, however, that the distribution of N depends only on σ^2.

It is clear that the event $[N = n]$ depends only on $(S_m^2, \ldots, S_n^2)$ for all fixed $n(\geq m)$. But, in view of Theorem 4.7.1, $\overline{X}_n$ and $(S_m^2, \ldots, S_n^2)$ are independently distributed all fixed $n(\geq m)$ since the parent population is assumed normal. Thus, we claim that $\overline{X}_n$ and $I[N = n]$ are independent for all fixed $n(\geq m)$. Now, observe that the type I error probability is expressed as

$$\begin{aligned}
&\mathrm{P}(\text{Rejecting } H_0 \text{ when } H_0 \text{ is true}) \\
&\quad = \mathrm{P}\left\{\overline{X}_N > k \text{ when } \mu = \mu_0\right\} \\
&\quad = \sum_{n=m}^{\infty} \mathrm{P}\left\{\overline{X}_N > k \text{ when } \mu = \mu_0 | N = n\right\} \mathrm{P}(N = n) \\
&\quad = \sum_{n=m}^{\infty} \mathrm{P}\left\{\overline{X}_n > k \text{ when } \mu = \mu_0 | N = n\right\} \mathrm{P}(N = n) \\
&\quad = \sum_{n=m}^{\infty} \mathrm{P}\left\{\overline{X}_n > k \text{ when } \mu = \mu_0\right\} \mathrm{P}(N = n) \\
&\quad = \sum_{n=m}^{\infty} \left\{1 - \Phi\left(n^{1/2}[k - \mu_0]/\sigma\right)\right\} \mathrm{P}(N = n) \\
&\quad = \mathrm{E}_{\sigma^2}\left[1 - \Phi\left(\frac{N^{1/2} z_\alpha (\mu_1 - \mu_0)}{(z_\alpha + z_\beta)\sigma}\right)\right]. \qquad (4.3.6)
\end{aligned}$$

In a similar fashion, we can also express the type II error probability as

$$\begin{aligned}
&\mathrm{P}\left\{\text{Accepting } H_0 \text{ when } H_1 \text{ true}\right\} \\
&\quad = \mathrm{P}\left\{\overline{X}_N \le k \text{ when } \mu = \mu_1\right\} \\
&\quad = \mathrm{E}_{\sigma^2}\left[\Phi\left(-\frac{N^{1/2} z_\beta (\mu_1 - \mu_0)}{(z_\alpha + z_\beta)\sigma}\right)\right]. \qquad (4.3.7)
\end{aligned}$$

Theorem 4.3.1 (Robbins and Starr, 1965): *As $\delta(= \mu_1 - \mu_0) \to 0$, the type I and type II error probabilities for the sequential test (4.3.4) respectively converge to α and β.*

Proof: From (4.3.6) - (4.3.7), we can rewrite the type I and II error probabilities as $\mathrm{E}_{\sigma^2}\left[1 - \Phi\left((N/n^*)^{1/2} z_\alpha\right)\right]$ and $\mathrm{E}_{\sigma^2}\left[\Phi\left(-(N/n^*)^{1/2} z_\beta\right)\right]$. Recall from (4.3.5) that $N/n^* \xrightarrow{\mathrm{P}_\theta} 1$. Also, $\Phi(\cdot)$ lies between zero and unity, that is $\Phi(\cdot)$ is bounded. Now, using the dominated convergence theorem (Theorem A.3.12 from Appendix A), we can claim that

$$\begin{aligned}
&\lim_{\delta \to 0} \mathrm{E}_{\sigma^2}\left[1 - \Phi\left((N/n^*)^{1/2} z_\alpha\right)\right] \\
&\quad = \mathrm{E}_{\sigma^2}\left[1 - \Phi\left(z_\alpha \lim_{\delta \to 0} (N/n^*)^{1/2}\right)\right] \\
&\quad = \mathrm{E}\left[1 - \Phi(z_\alpha)\right] \\
&\quad = 1 - (1 - \alpha) = \alpha, \qquad (4.3.8)
\end{aligned}$$

and similarly

$$\lim_{\delta \to 0} \mathrm{E}_{\sigma^2}\left[\Phi\left(-(N/n^*)^{1/2} z_\beta\right)\right] = \mathrm{E}\left[\Phi\left(-z_\beta\right)\right] = \beta. \qquad (4.3.9)$$

The proof is complete. ■

Sequential test (4.3.4) came from an unpublished technical report of Robbins and Starr (1965).

4.4 Test for the Correlation Coefficient

We consider $\{(U_j, V_j); j \geq 1\}$, a sequence of i.i.d. random variables distributed as bivariate normal having means μ_1 and μ_2, variances σ_1^2 and σ_2^2, and the correlation coefficient ρ. Such a distribution is denoted by $N_2(\mu_1, \mu_2, \sigma_1^2, \sigma_2^2, \rho)$ where we assume $-\infty < \mu_1, \mu_2 < \infty$, $0 < \sigma_1, \sigma_2 < \infty$, $-1 < \rho < 1$ and σ_1^2, σ_2^2 and ρ are unknown. Our goal is to test a composite null hypothesis $H_0 : \rho = \rho_0$ versus a composite alternative hypothesis $H_1 : \rho = \rho_1$. Here, $-1 < \rho_0 \neq \rho_1 < 1$ are specified and type I, type II error probabilities are respectively α, β.

First, let us focus on the situation where the **means** μ_1, μ_2 **are both known**. In this case, one can subtract μ_1, μ_2 from U, V respectively and consider a new sequence of i.i.d. $N_2(0, 0, \sigma_1^2, \sigma_2^2, \rho)$ random pairs of variables. So, without any loss of generality, let us suppose that $\mu_1 = \mu_2 = 0$. In order to propose a suitable sequential test, the basic ideas of Kowalski (1971) and Pradhan and Sathe (1975) revolved around finding a sequence of independent random pairs of variables whose common distribution depended only on ρ. In other words, they found statistics whose distributions were not affected by the *unknown nuisance parameters* σ_1^2 and σ_2^2. Then, they utilized a SPRT-like test for deciding between the hypotheses H_0 and H_1 with preassigned error probabilities α, β. Let $\boldsymbol{\theta} = (\sigma_1, \sigma_2, \rho)$.

The following techniques are borrowed from Pradhan and Sathe (1975). Define

$$X_i = \begin{cases} 1 & \text{if } U_{2i-1}V_{2i-1} + U_{2i}V_{2i} > 0 \\ 0 & \text{otherwise,} \end{cases} \tag{4.4.1}$$

$i = 1, 2, \ldots$. At every step we observe two pairs of random variables (U_{2i-1}, V_{2i-1}) and (U_{2i}, V_{2i}), both $N_2(0, 0, \sigma_1^2, \sigma_2^2, \rho)$ and form the X_i sequence according to (4.4.1), $i = 1, 2, \ldots$. Now, $X_1, X_2, \ldots$ certainly form a sequence of i.i.d. random variables, but what is the common distribution? It is very simple to show that the X_i's are i.i.d. having a Bernoulli(p) distribution where

$$p = \mathrm{P}_{\boldsymbol{\theta}}\{X_1 = 1\} = \mathrm{P}_{\boldsymbol{\theta}}\{U_1V_1 + U_2V_2 > 0\} = \mathrm{P}_{\boldsymbol{\theta}}\left\{\frac{U_1}{\sigma_1}\frac{V_1}{\sigma_2} + \frac{U_2}{\sigma_1}\frac{V_2}{\sigma_2} > 0\right\}. \tag{4.4.2}$$

Hence, it is apparent that the parameter p cannot depend on the unknown values of σ_1^2, σ_2^2. That is, p would functionally depend only on the parameter ρ.

Let us rewrite p as $\mathrm{P}_\rho\{U_1^*V_1^* + U_2^*V_2^* > 0\}$ where (U_1^*, V_1^*) and (U_2^*, V_2^*)

are i.i.d. $N_2(0,0,1,1,\rho)$. Now, observe that

$$\begin{aligned}
&U_1^*V_1^* + U_2^*V_2^* \\
&= \frac{1}{4}\left[(U_1^* + V_1^*)^2 + (U_2^* + V_2^*)^2\right] - \frac{1}{4}\left[(U_1^* - V_1^*)^2 + (U_2^* - V_2^*)^2\right] \\
&= U_0 - V_0, \text{ say.} \qquad (4.4.3)
\end{aligned}$$

One easily verifies that $U_i^*+V_i^* \sim N(0, 2(1+\rho))$, $U_i^*-V_i^* \sim N(0, 2(1-\rho))$ and $U_i^* + V_i^*$ is independent of $U_i^* - V_i^*$, $1 \le i \ne j \le 2$. That is, $U_0^* = 2U_0/(1+\rho)$ and $V_0^* = 2V_0/(1-\rho)$ are i.i.d. χ_2^2 random variables. Thus, we combine (4.4.2) - (4.4.3) to rewrite

$$p = \mathrm{P}_\rho\{U_0 - V_0 > 0\} = \mathrm{P}_\rho\left\{\frac{U_0}{V_0} > 1\right\} = \mathrm{P}_\rho\left\{\frac{U_0^*}{V_0^*} > \frac{1-\rho}{1+\rho}\right\}. \qquad (4.4.4)$$

But U_0^*/V_0^* has F-distribution with degrees of freedom (2,2). Hence from (4.4.4), with $c = (1-\rho)/(1+\rho)$ and using the p.d.f. of $F_{2,2}$, we can express

$$p = \int_c^\infty \frac{dy}{(1+y)^2} = \left[-\frac{1}{1+y}\right]_c^\infty = \frac{1}{1+c} = \frac{1}{2}(1+\rho). \qquad (4.4.5)$$

Thus, in terms of the X-sequence, one should test $H_0 : p = p_0$ versus $H_1 : p = p_1$ where $p_i = \frac{1}{2}(1+\rho_i)$, $i = 0, 1$. Hence, one will implement a SPRT along the lines of the Example 3.4.3. Further details are omitted for brevity.

Note, however, that the random sample size N obtained at the termination of such a SPRT would correspond to N observations of the X's defined in (4.4.1). In turn, that would mean $2N$ observations from the original (U, V) sequence having the $N_2(0,0,\sigma_1^2,\sigma_2^2,\rho)$ distribution with $(\sigma_1, \sigma_2, \rho)$ unknown.

Sequential test based on X's in (4.4.1) and relying upon Wald's SPRT came from Pradhan and Sathe (1975).

Kowalski's (1971) SPRT is slightly different and it can be described as follows. We observe one pair (U_i, V_i) at a time and define

$$X_i = \begin{cases} 1 & \text{if } U_iV_i > 0 \\ 0 & \text{otherwise} \end{cases} \qquad (4.4.6)$$

$i = 1, 2, \ldots$. It can be easily verified that these X's are also i.i.d. having the Bernoulli(p) distribution where

$$p = \frac{1}{2} + \frac{1}{\pi}\sin^{-1}(\rho). \qquad (4.4.7)$$

See Exercise 4.4.1. As in the previous formulation, here again we can test $H_0 : \rho = \rho_0$ versus $H_1 : \rho = \rho_1$ by means of testing $H_0 : p = p_0$ versus $H_1 : p = p_1$ with $p_i = \frac{1}{2} + \frac{1}{\pi}\sin^{-1}(\rho_i)$, $i = 0, 1$, via SPRT along the lines of Example 3.4.3 and using the X-sequence from (4.4.6). The random sample size N obtained at the termination of such a SPRT would correspond to the number of observations on the X's defined in (4.4.6) and it will coincide with the number of observed pairs (U, V) from the original sequence of $N_2(0, 0, \sigma_1^2, \sigma_2^2, \rho)$ variables with $(\sigma_1, \sigma_2, \rho)$ unknown.

Pradhan and Sathe(1975) have shown that the SPRT defined via (4.4.1) is more efficient than Kowalski's (1971) SPRT defined via (4.4.6), for a wide range of values of $-1 < \rho_0 \neq \rho_1 < 1$.

Now, suppose that μ_1, μ_2, σ_1^2, σ_2^2 and ρ **are all unknown**. Having recorded $(U_1, V_1), \ldots, (U_n, V_n)$ from the $N_2(\mu_1, \mu_2, \sigma_1^2, \sigma_2^2, \rho)$ population, let us first define Helmert's transformed variables along the lines of (4.7.1) as follows:

$$\begin{aligned} U_i^* &= \{U_1 + \cdots + U_i - iU_{i+1}\}/(i^2+i)^{1/2}, \\ V_i^* &= \{V_1 + \cdots + V_i - iV_{i+1}\}/(i^2+i)^{1/2}, \end{aligned} \tag{4.4.8}$$

$i = 1, 2, \ldots, n-1$ and $n \geq 2$. We leave it as an exercise to verify that the new sequence $\{(U_j^*, V_j^*), j \geq 1\}$consists of i.i.d. random variables from a $N_2(0, 0, \sigma_1^2, \ \sigma_2^2, \rho)$ distribution with $(\sigma_1, \sigma_2, \rho)$ unknown. In other words, as we keep observing from the (U, V) sequence, we keep obtaining new pairs from (U^*, V^*) sequence. Hence, we can define a SPRT through the X's, as in (4.4.1) or (4.4.6). We would simply replace the (U, V)'s with the corresponding (U^*, V^*)'s. Further details would be redundant and hence these are omitted.

The subprograms **CompSPRT** and **CompOC** can be used to aid in further exploration of the Pradhan-Sathe procedure and the Kowalski procedure.

Sequential test based on X's in (4.4.6) and relying upon Wald's SPRT came from Kowalski (1971).

Remark 4.4.1: Choi (1971) gave a sequential test for $H_0 : \rho = \rho_0$ versus $H_1 : \rho = \rho_1$ when $\sigma_1^2 = \sigma_2^2 = \sigma^2$, but the common variance is unknown with $\sigma \in I\!R^+$. That particular test is omitted from our discussion for brevity.

Remark 4.4.2: In general, it is hard to construct unbiased estimators for the correlation coefficient ρ. The sample correlation coefficient, also known as Pearson's correlation coefficient, is not an unbiased estimator of ρ. But, Pradhan and Sathe's (1975) construction introduced in (4.4.1) easily provides a consistent and unbiased estimator of ρ. Suppose that we observe $(U_1, V_1), \ldots, (U_n, V_n)$ and define

$$X_{ij} = \begin{cases} 1 & \text{if } U_iV_i + U_jV_j > 0 \\ 0 & \text{otherwise,} \end{cases} \tag{4.4.9}$$

$1 \leq i < j \leq n$. Then we write $\overline{X}_n = \sum_{1 \leq i < j \leq n} X_{ij} / \binom{n}{2}$, the sample mean obtained from all X's. In view of (4.4.5), Pradhan and Sathe (1975) had proposed the estimator

$$\widehat{\rho}_n = 2\overline{X}_n - 1 \tag{4.4.10}$$

for ρ. This estimator is obviously unbiased for ρ and one can show that it is consistent for ρ too.

4.5 Test for the Gamma Shape Parameter

The gamma distribution is widely used in reliability and survival analyses. Testing and estimation procedures for the scale parameter in a gamma distribution are quite plentiful when the shape parameter is known. See the recent volume, exclusively devoted to the exponential and gamma distributions, which has been edited by Balakrishnan and Basu (1995). Let us consider $Y_1, Y_2, \ldots$, a sequence of i.i.d. random variables with the common p.d.f.

$$f(y|\mu,p) = \begin{cases} \{\Gamma(p)\}^{-1}\mu^p e^{-\mu y} y^{p-1} & \text{if } y > 0 \\ 0 & \text{otherwise,} \end{cases}$$

where the scale parameter μ and the shape parameter p are both assumed positive. We assume that **both parameters are unknown.** The problem is one of testing a composite null hypothesis $H_0 : p = p_0$ versus a composite alternative hypothesis $H_1 : p = p_1$ with specified error probabilities α and β. Here $p_0 \neq p_1$ are two specified positive numbers. Phatarfod (1971) gave a sequential test procedure for this problem, based on the idea of removing the effect of the nuisance scale parameter μ on the procedure.

Having recorded $Y_1, Y_2, \ldots, Y_n, \ldots$ from a gamma distribution, one defines

$$X_i = Y_{i+1}/(Y_1 + \cdots + Y_{i+1}), \tag{4.5.1}$$

$i = 1, 2, \ldots, n-1, \ldots$. From random samples $Y_1, \ldots, Y_n$, one obtains $X_1, \ldots, X_{n-1}$ for $n \geq 2$. It is well known that $X_1, X_2, \ldots$ is a sequence of independent random variables and the p.d.f. of X_i is given by

$$f_i(x|p) = \begin{cases} \{\mathfrak{B}(p, ip)\}^{-1} x^{p-1}(1-x)^{ip-1} & \text{if } 0 < x < 1 \\ 0 & \text{otherwise} \end{cases} \tag{4.5.2}$$

$i = 1, 2, \ldots$, with the beta function $\mathfrak{B}(r,s) = \Gamma(r)\Gamma(s)/\Gamma(r+s)$ for $r > 0, s > 0$. The distribution given by (4.5.2) is customarily known as a **beta distribution** with parameters p and ip.

Now, the likelihood function of $X_1, \ldots, X_{n-1}$ given p, is

$$\mathcal{L}(p, \boldsymbol{X}_{n-1}) = \Gamma(np)\{\Gamma(p)\}^{-n} \prod_{i=1}^{n-1} X_i^{p-1}(1 - X_i)^{ip-1}. \tag{4.5.3}$$

Here, $\boldsymbol{X}_{n-1} = (X_1, \ldots, X_{n-1})$ belongs to the positive orthant $\mathbb{R}^{+n-1}$ of the $(n-1)$-dimensional space, that is,

$$\mathbb{R}^{+n-1} = \{(x_1, \ldots, x_{n-1}) : x_i > 0 \text{ for all } i = 1, ..., n-1\}, n \geq 2.$$

Hence, the likelihood ratio is

$$\begin{aligned} \Lambda_{n-1} &= \frac{\mathcal{L}(p_1, \boldsymbol{X}_{n-1})}{\mathcal{L}(p_0, \boldsymbol{X}_{n-1})} \\ &= \frac{\Gamma(np_1)}{\Gamma(np_0)} \left\{\frac{\Gamma(p_0)}{\Gamma(p_1)}\right\}^n \prod_{i=1}^{n-1} X_i^{(p_1-p_0)} (1 - X_i)^{i(p_1-p_0)}. \end{aligned} \tag{4.5.4}$$

Now, along the lines of Wald's SPRT in (3.1.3), Phatarfod (1971) first selected two numbers B and A, $0 < B < A < \infty$. Then, one starts with Y_1, Y_2 (which gives rise to X_1) and then continues with one more Y according to the following stopping rule: Let

$$N = \text{ the first integer } n(\geq 2) \text{ for which } \Lambda_{n-1} \leq B \text{ or } \Lambda_{n-1} \geq A. \quad (4.5.5)$$

We decide in favor of H_0 (or H_1) if $\Lambda_{N-1} \leq B$ (or $\Lambda_{N-1} \geq A$) at the stage of *termination*. Note that here N corresponds to the random number of Y's which are observed sequentially.

Given preassigned α and β, one can justify the same choices of A and B as given by Theorem 3.2.1. Even though the X's here are only independent, but not identically distributed, the proof of Theorem 3.2.1 can be modified. Phatarfod (1971) had recommended choosing $A = (1-\beta)/\alpha$ and $B = \beta/(1-\alpha)$. The sequential procedure (4.5.5) does terminate with probability one under H_0 and H_1 . The corresponding OC and ASN function's approximations were also derived in Phatarfod (1971). The subprogram **CompSPRT** given within the program **Seq04.exe** can be used to perform the test and conduct a simulation study.

Sequential test (4.5.5) for the shape parameter of a gamma distribution with an unknown scale parameter came from Phatarfod (1971).

4.6 Two-Sample Problem: Comparing the Means

Consider two independent populations, $N(\mu_1, \sigma_1^2)$ and $N(\mu_2, \sigma_2^2)$. Suppose that the means are unknown and we wish to test $H_0 : \mu_1 - \mu_2 = D_0$ versus $H_1 : \mu_1 - \mu_2 = D_1 (> D_0)$ where D_0, D_1 are specified numbers. Also, we have two kinds of preassigned errors, namely α and β. A simple sequential approach can be given for testing H_0 versus H_1 when we allow a pair of observations, one from $N(\mu_1, \sigma_1^2)$ and another from $N(\mu_2, \sigma_2^2)$ at a time. Let us observe $(U_1, V_1), (U_2, V_2), \ldots$, a sequence of independent random variables where U's are i.i.d. $N(\mu_1, \sigma_1^2)$ and V's are i.i.d. $N(\mu_2, \sigma_2^2)$. Let us define $X_i = U_i - V_i$ and thus $X_1, X_2, \ldots$ is a sequence of i.i.d. random variables with the distribution $N(\nu, \sigma^2)$ where $\nu = \mu_1 - \mu_2$ and $\sigma^2 = \sigma_1^2 + \sigma_2^2$. Let us restate our original testing problem as a choice to be made between $H_0 : \nu = D_0$ versus $H_1 : \nu = D_1$. In principle, this resembles a paired experiment. One may refer to the books of Wald (1947) and Ghosh (1970).

Now, **if σ^2 is known**, we would be back to a SPRT considered in the Example 3.4.1. **If σ^2 is unknown**, then we can fall back on the material discussed in Section 4.3. These methodologies would make sense even if (U_i, V_i)'s were i.i.d. $N_2(\mu_1, \mu_2, \sigma_1^2, \sigma_2^2, \rho)$ instead.

On the other hand, if σ_1 and σ_2 are unequal, one may contemplate using sequential sampling with **unequal sample sizes** by suitably extending the techniques from Section 4.3. In order to keep our presentation simple, we exclude such discussions. These techniques would resemble the procedures we introduce later in Chapter 13. Several exact and asymptotically optimal tests were developed by Aoshima et al. (1996) and Dudewicz and Ahmed (1998,1999)

4.7 Auxiliary Results

Theorem 4.7.1 : *Let $X_1, \ldots, X_n$ be i.i.d. $N(\mu, \sigma^2)$, $\overline{X}_k = k^{-1}\sum_{i=1}^{k} X_i$, $S_k^2 = (k-1)^{-1}\sum_{i=1}^{k}(X_i - \overline{X}_k)^2$, $2 \le k \le n$. Then, $\overline{X}_n$ and $(S_2^2, \ldots, S_n^2)$ are distributed independently.*

Proof: Consider what is called the Helmert's transformation of the random variables $X_1, \ldots, X_n$. Let us define

$$\begin{aligned}
Y_0 &= (X_1 + \cdots + X_n)/n^{1/2} \\
Y_1 &= (X_1 - X_2)/2^{1/2} \\
Y_2 &= (X_1 + X_2 - 2X_3)/6^{1/2} \\
&\vdots \\
Y_{n-1} &= \{X_1 + \cdots + X_{n-1} - (n-1)X_n\}/(n^2 - n)^{1/2} \qquad (4.7.1)
\end{aligned}$$

and denote

$$A = \begin{pmatrix}
\frac{1}{\sqrt{n}} & \frac{1}{\sqrt{n}} & \frac{1}{\sqrt{n}} & \cdots & \frac{1}{\sqrt{n}} \\
\frac{1}{\sqrt{2}} & -\frac{1}{\sqrt{2}} & 0 & \cdots & 0 \\
\frac{1}{\sqrt{6}} & \frac{1}{\sqrt{6}} & -\frac{2}{\sqrt{6}} & \cdots & 0 \\
 & & & \ddots & \\
\frac{1}{\sqrt{n^2-n}} & \frac{1}{\sqrt{n^2-n}} & \frac{1}{\sqrt{n^2-n}} & \cdots & -\frac{n-1}{\sqrt{n^2-n}}
\end{pmatrix},$$

$$\boldsymbol{X} = \begin{pmatrix} X_1 \\ X_2 \\ \vdots \\ X_n \end{pmatrix}, \quad \boldsymbol{Y} = \begin{pmatrix} Y_0 \\ Y_1 \\ \vdots \\ Y_{n-1} \end{pmatrix}.$$

Here, A is a $n \times n$ orthogonal matrix, that is, $AA' = A'A = I_{n \times n}$ and $\boldsymbol{Y} = A\boldsymbol{X}$. Since A' is the inverse of A, we have $\boldsymbol{X} = A'\boldsymbol{Y}$. Thus, we have $\boldsymbol{X}'\boldsymbol{X} = \boldsymbol{Y}'\boldsymbol{Y}$, that is $\sum_{i=1}^{n} X_i^2 = \sum_{i=0}^{n-1} Y_i^2$. Observe that

$$\left| \frac{\partial(x_1, \cdots, x_n)}{\partial(y_0, \cdots, y_{n-1})} \right| = |A'| = \pm 1,$$

and write the joint p.d.f. of $X_1, \ldots, X_n$:

$$f_{X_1,\ldots,X_n}(x_1, \ldots, x_n) = \frac{1}{(\sigma\sqrt{2\pi})^n} e^{-\frac{1}{2\sigma^2}\{\sum_{i=1}^n x_i^2 - 2\mu\sum_{i=1}^n x_i + n\mu^2\}} \quad (4.7.2)$$

for $-\infty < x_1, \ldots, x_n < \infty$ which provides the joint p.d.f. of $Y_0, \ldots, Y_{n-1}$ by using the transformation technique:

$$f_{Y_0,\ldots,Y_{n-1}}(y_0, \ldots, y_{n-1}) = \frac{1}{(\sigma\sqrt{2\pi})^n} e^{-\frac{1}{2\sigma^2}\{\sum_{i=0}^{n-1} y_i^2 - 2\mu\sqrt{n}y_0 + n\mu^2\}} \quad (4.7.3)$$

for $-\infty < y_0, y_1, \ldots, y_{n-1} < \infty$, since the absolute value of the Jacobian of this transformation is unity. Next, we rewrite

$$f_{Y_0,\ldots,Y_{n-1}}(y_0, \ldots, y_{n-1}) = \frac{1}{\sigma\sqrt{2\pi}} e^{-\frac{1}{2\sigma^2}(y_0 - \mu\sqrt{n})^2} \prod_{i=1}^{n-1} \left\{ \frac{1}{\sigma\sqrt{2\pi}} e^{-\frac{1}{2\sigma^2}y_i^2} \right\}, \quad (4.7.4)$$

for $-\infty < y_0, y_1, \ldots, y_{n-1} < \infty$.

From (4.7.4) it follows that $Y_0, Y_1, \ldots, Y_{n-1}$ are independent random variables; $\overline{X}_n = Y_0/n^{1/2} \sim N(\mu, \sigma^2/n)$; $Y_1, \ldots, Y_{n-1}$ are i.i.d. $N(0, \sigma^2)$.

The beauty of the Helmert's transformation is realized when we observe that

$$(n-1)S_n^2 = \sum_{i=1}^n X_i^2 - n\overline{X}_n^2 = \sum_{i=0}^{n-1} Y_i^2 - Y_0^2 = \sum_{i=1}^{n-1} Y_i^2. \quad (4.7.5)$$

Clearly, $\overline{X}_n$ depends only on Y_0 and S_n^2 depends only on the random vector $(Y_1, \ldots, Y_{n-1})$. But Y_0 is independent of $(Y_1, \ldots, Y_{n-1})$. Thus, $\overline{X}_n$ is independent of S_n^2. From (4.7.1) and (4.7.5), it is also clear that

$$(k-1)S_k^2 = \sum_{i=1}^{k-1} Y_i^2 \text{ for all fixed } k, 2 \leq k \leq n, \quad (4.7.6)$$

and thus the whole random vector $(S_2^2, \ldots, S_n^2)$ depends only on $(Y_1, \ldots, Y_{n-1})$ and hence the result follows. ■

As a by-product, from (4.7.5) it follows that

$$(n-1)S_n^2/\sigma^2 = \sum_{i=1}^{n-1} (Y_i/\sigma)^2 \sim \chi_{n-1}^2 \quad (4.7.7)$$

since $Y_1/\sigma, \ldots, Y_{n-1}/\sigma$ are i.i.d. $N(0,1)$ so that $(Y_i/\sigma)^2, i = 1, \ldots, n-1$, are i.i.d. χ_1^2.

4.8 Exercises

Exercise 4.2.1: Let $X_1, \ldots, X_n, \ldots$ be i.i.d. random variables from a $N(0, \sigma^2)$ distribution with $0 < \sigma < \infty$. We assume that σ is unknown. Consider testing a simple null hypothesis $H_0 : \sigma = \sigma_0$ versus a simple alternative hypothesis $H_1 : \sigma = \sigma_1 (0 < \sigma_1 < \sigma_0)$ with the preassigned type

I and type II error probabilities α, β, $0 < \alpha, \beta < 1$. Write down the SPRT from (3.1.3) explicitly in this special case.

Exercise 4.2.2: Suppose that $U_1, \ldots, U_n, \ldots$ are i.i.d. random variables from a $N(\mu, \sigma^2)$ distribution, $V_1, \ldots, V_n, \ldots$ are i.i.d. random variables from a $N(\theta, \sigma^2)$ distribution with $-\infty < \mu, \theta < \infty$, $0 < \sigma < \infty$. We assume that the U's and V's are independent and the parameters μ, θ and σ are all unknown. Consider testing a null hypothesis $H_0 : \sigma = \sigma_0$ versus an alternative hypothesis $H_1 : \sigma = \sigma_1 (> \sigma_0 > 0)$ with the preassigned type I and type II error probabilities α, β, $0 < \alpha, \beta < 1$. Suppose that we observe a pair (U, V) at a time. Use Helmert transformation to generate $2(n-1)$ i.i.d. $N(0, \sigma^2)$ observations from $U_1, \ldots, U_n, V_1, \ldots, V_n$. Then, apply the methodology from Section 4.2 to propose an appropriate sequential test. Use the program **CompSPRT** to simulate the scenario for a range of values of $\alpha, \beta, n, \sigma_0, \sigma_1, \mu$ and θ.

Exercise 4.2.3 (Exercise 4.2.2 Continued): Consider testing a null hypothesis $H_0 : \sigma = \sigma_0$ versus an alternative hypothesis $H_1 : \sigma = \sigma_1$ $(0 < \sigma_1 < \sigma_0)$ with the preassigned type I and type II error probabilities $\alpha, \beta, 0 < \alpha, \beta < 1$. Modify the methodology from the Exercise 4.2.2 to propose an appropriate sequential test in the present situation. Use the program **CompSPRT** to simulate the scenario for a range of values of $\alpha, \beta, n, \sigma_0, \sigma_1, \mu$ and θ.

Exercise 4.2.4: Suppose that $(X_1^*, Y_1^*), \ldots, (X_n^*, Y_n^*), \ldots$ are i.i.d. random samples from a $N_2(\mu, \theta, \sigma^2, \sigma^2, \rho)$ distribution with $-\infty < \mu, \theta < \infty$, $0 < \sigma < \infty$, $-1 < \rho < 1$. We assume that the parameters μ, θ, σ and ρ are all unknown. Consider testing a null hypothesis $H_0 : \sigma = \sigma_0$ versus an alternative hypothesis $H_1 : \sigma = \sigma_1 (> \sigma_0 > 0)$ with the preassigned type I and type II error probabilities $\alpha, \beta, 0 < \alpha, \beta < 1$. Define $U_i = X_i^* + Y_i^*$, $V_i = X_i^* - Y_i^*, i = 1, \ldots, n, \ldots$ so that these U's and V's are similar to those given in Exercise 4.2.2. Next, apply the methodology from Section 4.2 to propose an appropriate sequential test. Use the program **CompSPRT** to simulate the scenario for a range of values of $\alpha, \beta, n, \sigma_0, \sigma_1, \mu, \theta$ and ρ.

Exercise 4.2.5 (Exercise 4.2.4 Continued): Consider testing a null hypothesis $H_0 : \sigma = \sigma_0$ versus an alternative hypothesis $H_1 : \sigma = \sigma_1 (0 < \sigma_1 < \sigma_0)$ with the preassigned type I and type II error probabilities α, β, $0 < \alpha, \beta < 1$. Modify the methodology from Exercise 4.2.4 to propose an appropriate sequential test in the present situation. Use the program **CompSPRT** to simulate the scenario for a range of values of $\alpha, \beta, n, \sigma_0, \sigma_1, \mu, \theta$ and ρ.

Exercise 4.2.6: The length of time the airport bus takes to reach the city center from the airport is *normally* distributed with an average of

45 minutes but unknown standard deviation of σ minutes. We wish to test $H_0 : \sigma = 10$ versus $H_1 : \sigma = 20$ with the type I and type II error probabilities $0.01, 0.05$ respectively. The following data has been recorded on the length of time the airport bus took to reach the city center from the airport during the sixteen comparable runs:

$$\begin{array}{llll} x_1 = 63.1 & x_5 = 25.5 & x_9 = 61.7 & x_{13} = 44.5 \\ x_2 = 52.7 & x_6 = 50.6 & x_{10} = 53.4 & x_{14} = 43.1 \\ x_3 = 40.6 & x_7 = 48.4 & x_{11} = 39.1 & x_{15} = 35.7 \\ x_4 = 53.0 & x_8 = 40.0 & x_{12} = 58.7 & x_{16} = 50.1 \end{array}$$

Use the program **CompSPRT** to feed this real data one by one in a sequence, namely, first enter x_1, then enter x_2 if needed, followed by x_3 as needed, and so on. Recommend whether the data gathering should have terminated (along with the decision in favor or against H_0) before collecting all sixteen observations or whether the collected data is not yet enough to reach any decision. How about truncating the sampling process at $K = 16$?

Exercise 4.2.7 (Exercise 4.2.6 Continued): The length of time the airport bus takes to reach the city center from the airport is *normally* distributed with an unknown average of θ minutes and unknown standard deviation of σ minutes. We wish to test $H_0 : \sigma = 10$ versus $H_1 : \sigma = 20$ with the type I and type II error probabilities $0.01, 0.05$ respectively. Use the program **CompSPRT** to feed the data from Exercise 4.2.6 one by one in a sequence, namely, first enter x_1, then enter x_2 if needed, followed by x_3 as needed, and so on. Recommend whether the data gathering should have terminated (along with the decision in favor or against H_0) before collecting all sixteen observations or whether the collected data is not yet enough to reach any decision. How about truncating the sampling process at $K = 16$?

Exercise 4.2.8: In a poultry firm, the chicken's weight (Y, in pounds) is *normally* distributed with unknown average weight (μ, in pounds) and unknown standard deviation (σ, in pounds). We wish to test $H_0 : \sigma = 0.1$ versus $H_1 : \sigma = 0.2$ with the type I and type II error probabilities $0.05, 0.01$ respectively. The following data was recorded on the weights of thirty randomly selected chickens:

$$\begin{array}{llllll} x_1 = 3.10 & x_6 = 3.05 & x_{11} = 2.87 & x_{16} = 3.04 & x_{21} = 3.04 & x_{26} = 3.01 \\ x_2 = 2.96 & x_7 = 2.80 & x_{12} = 2.99 & x_{17} = 3.05 & x_{22} = 3.00 & x_{27} = 2.98 \\ x_3 = 2.96 & x_8 = 3.12 & x_{13} = 2.96 & x_{18} = 2.93 & x_{23} = 2.93 & x_{28} = 3.10 \\ x_4 = 3.11 & x_9 = 2.83 & x_{14} = 3.27 & x_{19} = 3.20 & x_{24} = 3.00 & x_{29} = 3.12 \\ x_5 = 2.97 & x_{10} = 3.11 & x_{15} = 2.93 & x_{20} = 2.96 & x_{25} = 2.93 & x_{30} = 2.99 \end{array}$$

Use the program **CompSPRT** to feed this real data one by one in a sequence, namely, first enter x_1, then enter x_2 if needed, followed by x_3

as needed, and so on. Recommend whether the data gathering should have terminated (along with the decision in favor or against H_0) before collecting all thirty observations or whether the collected data was not yet enough to reach any decision. How about truncating the sampling process at $K = 30$?

Exercise 4.3.1: Suppose that $U_1, \ldots, U_n, \ldots$ are i.i.d. random variables from a $N(\mu, \sigma^2)$ distribution, $V_1, \ldots, V_n, \ldots$ are i.i.d. random variables from a $N(\theta, \sigma^2)$ distribution with $-\infty < \mu, \theta < \infty$, $0 < \sigma < \infty$. We assume that the U's and V's are independent and the parameters μ, θ and σ are all unknown. Consider testing a null hypothesis $H_0 : \mu = \theta$ versus an alternative hypothesis $H_1 : \mu = \theta + \delta$ with the preassigned type I and type II error probabilities α, β, $0 < \alpha, \beta < 1$ where δ is a known positive number. Suppose that we observe a pair (U, V) at a time as needed. Consider $X_i = U_i - V_i, i = 1, 2, \ldots, n, \ldots$ and then apply the methodology from Section 4.3 to propose an appropriate sequential test. Use the program **CompSPRT** to simulate the scenario for a range of values of $\alpha, \beta, n, \mu, \theta$ and δ.

Exercise 4.3.2: Suppose that $X_1, \ldots, X_n, \ldots$ are i.i.d. random variables from a negative exponential p.d.f. given by

$$f(x; \theta, \sigma) = \begin{cases} \sigma^{-1} \exp\{-(x - \theta)/\sigma\} & \text{if } x > \theta \\ 0 & \text{otherwise} \end{cases}$$

with $-\infty < \theta < \infty$, $0 < \sigma < \infty$. We assume that the parameters θ and σ are both unknown. Consider testing a null hypothesis $H_0 : \theta = \theta_0$ versus an alternative hypothesis $H_1 : \theta = \theta_1 (> \theta_0)$ with the preassigned type I and type II error probabilities $\alpha, \beta, 0 < \alpha, \beta < 1$. Denote $X_{n:1} = \min\{X_1, \ldots, X_n\}$, $n \geq 1$ and suppose that

$$\text{we reject } H_0 \text{ if and only if } X_{n:1} > k.$$

(i) First obtain the expressions of k and $n = n^*$ such that the corresponding test via $X_{n^*:1}$ meets the two error requirements.

(ii) Next, propose a sequential test via $X_{n:1}$ along the lines of (4.3.4). Then, verify results similar to those found in Theorem 4.3.1.

Exercise 4.3.3 (Exercise 4.2.8 Continued): In a poultry firm, the chicken's weight (Y, in pounds) is *normally* distributed with unknown average weight (μ, in pounds) and unknown standard deviation (σ, in pounds). We wanted to make an inference about the average weight of the chickens in the firm by testing $H_0 : \mu = 3.0$ versus $H_1 : \mu = 3.2$ with the type I and type II error probabilities $0.01, 0.05$ respectively. Use the program **CompSPRT** to feed the real data from Exercise 4.2.8 one by one in a sequence, namely, first enter x_1, then enter x_2 if needed, followed by x_3 as needed, and so on according to the methodology described in Section 4.3. Recommend whether the data gathering should have terminated (along with the decision in favor or against H_0) before collecting all thirty observations or whether the collected data was not enough to reach any decision. How about truncating the sampling process at $K = 30$?

Exercise 4.4.1: Verify (4.4.7).

Exercise 4.4.2: Show that the estimator $\hat{\rho}_n$ from (4.4.10) is unbiased and consistent for ρ.

Exercise 4.5.1: Use transformation techniques to prove the result given by (4.5.2).

Exercise 4.6.1 (Exercise 4.3.1 Continued): Consider testing a null hypothesis $H_0 : \mu = \theta$ versus an alternative hypothesis $H_1 : \mu = \theta + \delta$with the preassigned type I and type II error probabilities α, β, $0 < \alpha, \beta < 1$ where δ is a known negative number. Suppose that we observe a pair (U, V) at a time as needed. Modify the methodology from Exercise 4.3.1 to propose an appropriate sequential test in the present situation. Use the program **CompSPRT** to simulate the scenario for a range of values of $\alpha, \beta, n, \mu, \theta$ and δ.

Exercise 4.6.2 (Exercise 4.3.1 Continued): Suppose that (U_1, V_1), ..., (U_n, V_n), ... are i.i.d. random samples from a $N_2(\mu, \theta, \sigma^2, \sigma^2, \rho)$ distribution with $-\infty < \mu, \theta < \infty$, $0 < \sigma < \infty$, $-1 < \rho < 1$. We assume that the parameters μ, θ, σ and ρ are all unknown. Consider testing a null hypothesis $H_0 : \mu = \theta$ versus an alternative hypothesis $H_1 : \mu = \theta + \delta$ with the preassigned type I and type II error probabilities α, β, $0 < \alpha, \beta < 1$ where δ is a known positive number. Suppose that we observe a pair (U, V) at a time as needed. Consider the sequence of i.i.d. observations $X_i = U_i - V_i, i = 1, 2, \ldots, n, \ldots$ and then apply the methodology from Exercise 4.3.1 to propose an appropriate sequential test in the present situation. Use the program **CompSPRT** to simulate the scenario for a range of values of $\alpha, \beta, n, \mu, \theta, \sigma, \rho$ and δ.

Exercise 4.6.3: Thirty students took the midterm and final examinations in a junior level course on statistical methods. The midterm exam score (U) and the final exam score (V) of each student are given below:

$$
\begin{array}{lll}
u_1 = 72, v_1 = 77 & u_6 = 76, v_6 = 74 & u_{11} = 80, v_{11} = 82 \\
u_2 = 81, v_2 = 84 & u_7 = 71, v_7 = 76 & u_{12} = 74, v_{12} = 82 \\
u_3 = 80, v_3 = 82 & u_8 = 82, v_8 = 82 & u_{13} = 77, v_{13} = 80 \\
u_4 = 77, v_4 = 82 & u_9 = 84, v_9 = 84 & u_{14} = 75, v_{14} = 80 \\
u_5 = 79, v_5 = 85 & u_{10} = 77, v_{10} = 81 & u_{15} = 80, v_{15} = 80
\end{array}
$$

$$\begin{array}{lll} u_{16}=79, v_{16}=82 & u_{21}=76, v_{21}=84 & u_{26}=80, v_{26}=82 \\ u_{17}=78, v_{17}=82 & u_{22}=74, v_{22}=80 & u_{27}=75, v_{27}=72 \\ u_{18}=79, v_{18}=85 & u_{23}=81, v_{23}=90 & u_{28}=87, v_{28}=86 \\ u_{19}=77, v_{19}=74 & u_{24}=74, v_{24}=75 & u_{29}=73, v_{29}=71 \\ u_{20}=77, v_{20}=74 & u_{25}=78, v_{25}=77 & u_{30}=80, v_{30}=77 \end{array}$$

Suppose that $(U_1, V_1), \ldots, (U_n, V_n), \ldots$ are i.i.d. random samples from a $N_2(\mu, \theta, \sigma^2, \sigma^2, \rho)$ distribution with $-\infty < \mu, \theta < \infty$, $0 < \sigma < \infty$, $-1 < \rho < 1$. We assume that the parameters μ, θ, σ and ρ are all unknown. Consider testing a null hypothesis $H_0 : \theta - \mu = 0$ versus an alternative hypothesis $H_1 : \theta - \mu = 2$, that is on the average a student scored 2 extra points in the final, with both the type I and type II error probabilities 0.05. Suppose that we observe a pair (U, V) at a time as needed. Recommend whether the data gathering should have terminated (along with the decision in favor or against H_0) before collecting all thirty observations or whether the collected data was not enough to reach any decision. How about truncating the sampling process at $K = 30$?

Consider the sequence of i.i.d. observations $X_i = U_i - V_i, i = 1, 2, \ldots, n, \ldots$ and then use the program **CompSPRT** to feed the real data one by one in a sequence, namely, first enter x_1, then enter x_2 if needed, followed by x_3 as needed, and so on. Recommend whether the data gathering should have terminated (along with the decision in favor or against H_0) before collecting all thirty observations or whether the collected data was not enough to reach any decision. Apply the methodology from Exercise 4.3.1 to propose an appropriate sequential test in the present situation.

Exercise 4.6.4 (Exercise 4.6.3 Continued): Thirty students took the midterm and final examinations in a junior level course on statistical methods. The midterm exam score (U) and the final exam score (V) of each student were supplied in Exercise 4.6.3. Suppose that $(U_1, V_1), \ldots, (U_n, V_n), \ldots$ are i.i.d. random samples from a $N_2(\mu, \theta, \sigma^2, \sigma^2, \rho)$ distribution with $-\infty < \mu, \theta < \infty$, $0 < \sigma < \infty$, $-1 < \rho < 1$. We assume that the parameters μ, θ, σ and ρ are all unknown. Use the construction from Remark 4.4.1 to find an unbiased estimate of the population correlation coefficient ρ.

Exercise 4.6.5 (Exercise 4.6.3 Continued): Thirty students took the midterm and final examinations in a junior level course on statistical methods. The midterm exam score (U) and the final exam score (V) of each student were supplied in Exercise 4.6.3. Suppose that $(U_1, V_1), \ldots, (U_n, V_n), \ldots$ are i.i.d. random samples from a $N_2(\mu, \theta, \sigma^2, \sigma^2, \rho)$ distribution with $-\infty < \mu, \theta < \infty$, $0 < \sigma < \infty$, $-1 < \rho < 1$. We assume that the parameters μ, θ, σ and ρ are all unknown. Consider testing a null hypothesis $H_0 : \rho = \frac{1}{2}$ versus an alternative hypothesis $H_1 : \rho = \frac{1}{4}$ with both the type I and type II error probabilities 0.01. Suppose that we observe a pair (U, V) at a time as needed. Consider the coded random samples X_{ij}'s defined by (4.4.9) and then use the program **CompSPRT** to feed the real data one by one in a sequence. Recommend whether the data gathering should have terminated (along with the decision in favor or against H_0) before collecting all thirty observations or whether the collected data was not yet enough to reach any decision. Apply the methodology from Remark 4.4.2 to propose an appropriate sequential test in the present situation.

CHAPTER 5

Sequential Nonparametric Tests

5.1 Introduction

We introduced sequential tests in a variety of parametric problems earlier in Chapters 2, 3 and 4. These were heavily dependent on specific distributional features. Nonparametric methods, however, may not require as many specifications regarding a probability distribution as their parametric counterparts. Hence, we present some nonparametric methods of sequential tests in this chapter.

But, sequential nonparametrics cover a vast area. Also, the mathematical mastery needed to appreciate many intricate details is beyond the scope of this book. Hence, we touch upon a small number of selected nonparametric sequential tests. We present this material mainly to instill some appreciation. One may obtain a broader perspective of this field from Sen (1981,1985), Govindarajulu (1981,2004) and other sources.

In many statistical problems, continuous monitoring of ongoing production processes takes the center stage. This involves some appropriate sampling inspection scheme. Under normal circumstances, an ongoing process produces items whose overall quality conforms to strict specifications and protocols. A null hypothesis H_0 postulates that a production process is *in control*, that is, the overall quality of products meets specifications. On the other hand, an alternative hypothesis H_1 postulates that a production process is *out of control.*

Under routine monitoring, a quality control engineer may not tinker with a process that continues to produce within specifications. If one does, then that will lead to significant loss in production of good items. So, we may think of observations $X_1, X_2, \ldots$ arriving in a sequence and sampling inspections being performed sequentially. One necessarily stops at some stage (that is, $N < \infty$) if and only if a production line seems to be out of control. Inspection sampling or monitoring continues (that is, $N = \infty$) if and only if a production line seems to be within specifications. That is,

$$\begin{array}{rcl} N < \infty & \Leftrightarrow & \text{Accept } H_1 \\ N = \infty & \Leftrightarrow & \text{Accept } H_0 \end{array}$$

Our goal is two-fold: We would like

(i) type I error probability to be small, namely $\mathrm{P}(N < \infty \mid H_0 \text{ is true}) \leq \alpha$ or $\approx \alpha$ where $0 < \alpha < 1$ is a small preassigned number;

(ii) to employ a power one test, namely $\mathrm{P}(N < \infty \mid H_1 \text{ is true}) = 1$.

Darling and Robbins (1968) developed ingenious methods to design sequential tests of power one with a small preassigned level or size α by exploiting the *law of the iterated logarithm.*

Robbins (1970) included follow-up thoughts. Some details may be found in Sen (1981, Chapter 9, pp. 233-243) and Govindarajulu (1981, Sections 3.13-3.14; 2004, Section 3.9).

We first discuss some one-sample problems. Suppose that we have available a sequence of independent observations $X_1, \ldots, X_n, \ldots$ from a population with a distribution function F. That is, $F(x) \equiv \mathrm{P}(X \leq x), x \in \mathbb{R}$, the real line. We denote the mean and the variance as follows:

$$\begin{aligned} &\text{Mean:} && \mu \equiv \mu(F) = \mathrm{E}_F[X] = \textstyle\int_{-\infty}^{\infty} x dF(x); \\ &\text{Variance:} && \sigma^2 \equiv \sigma^2(F) = \mathrm{E}_F[(X-\mu)^2] = \textstyle\int_{-\infty}^{\infty} (x-\mu)^2 dF(x). \end{aligned} \tag{5.1.1}$$

We include more specific assumptions regarding F in respective sections that follow.

Now, having recorded n observations $X_1, \ldots, X_n$, we denote

$$\begin{aligned} &\text{Sample Mean:} && \overline{X}_n = n^{-1} \textstyle\sum_{i=1}^{n} X_i, \\ &\text{Sample Variance:} && S_n^2 = (n-1)^{-1} \textstyle\sum_{i=1}^{n} (X_i - \overline{X}_n)^2, n \geq 2. \end{aligned}$$

We first introduce nonparametric power one sequential tests for the mean of a population with an unspecified F, discrete or continuous. Sections 5.2 and 5.3 introduce the Darling-Robbins methodologies when the population variance is known or unknown respectively. A sequential nonparametric test for the $100p^{\text{th}}$ percentile of a continuous distribution function F is included in Section 5.4. A sequential nonparametric test for the median follows as a special case when we fix $p = 0.5$. Next, a sequential sign test due to Hall (1965) is introduced in Section 5.5. Section 5.6 is devoted to some data analyses and concluding thoughts.

5.2 A Test for the Mean: Known Variance

Let us assume that F is such that $\mathrm{E}_F\left[|X|^r\right] < \infty$ for some $r > 2$. The unspecified distribution function F may correspond to either a discrete or a continuous population. Let us also assume that σ^2 is finite, positive and known. Obviously, the population mean μ is finite and we are interested in the following testing problem:

$$H_0 : \mu = \mu_0 \text{ vs. } H_1 : \mu > \mu_0, \text{ with } \mu_0 \text{ specified.} \tag{5.2.1}$$

Having recorded n independent observations $X_1, \ldots, X_n$, we denote the test statistic

$$T_n \equiv \frac{\overline{X}_n - \mu_0}{\sigma}, n \geq 1. \tag{5.2.2}$$

Now, let $X_1, \ldots, X_m$ be the pilot observations where m is an initial sample size. From this point onward, one will sample one additional observation X at-a-time according to the following stopping rule:

$$\begin{aligned} N &= \text{the smallest integer } n(\geq m) \text{ for which } \sqrt{n}T_n \geq \sqrt{2\ln\ln(n)} \\ &= \infty, \text{ if no such } n \text{ exists.} \end{aligned} \tag{5.2.3}$$

The test procedure associated with (5.2.3) is:

$$\text{Reject } H_0 \text{ if and only if (5.2.3) terminates, that is, } N < \infty. \tag{5.2.4}$$

The sequential test (5.2.3) - (5.2.4) is a classic Darling-Robbins power one test for the following reason:

$$\limsup_{n\to\infty} \frac{\sqrt{n}\left|\overline{X}_n - \mu\right|}{\sigma\sqrt{2\ln\ln(n)}} = 1 \quad \text{with probability 1 when } F, \mu \text{ and } \sigma \text{ are true.} \tag{5.2.5}$$

The limiting result in (5.2.5) is a restatement of the law of the iterated logarithm.

However, an expression for type I error probability associated with the power one test (5.2.3) -(5.2.4) is not readily accessible.

Sequential test (5.2.3)-(5.2.4) with power one came from Darling and Robbins (1968).

5.2.1 Another Test Procedure

Let us again assume that $\mathrm{E}_F\left[|X|^r\right] < \infty$ for some $r > 2$. Another nonparametric test may be proposed as follows. With some fixed $c > 0$, the stopping time is:

$$\begin{aligned} N(c) &= \text{the smallest integer } n(\geq m) \text{ for which } \sqrt{n}T_n \geq \sqrt{c\ln(n)} \\ &= \infty, \text{ if no such } n \text{ exists.} \end{aligned} \tag{5.2.6}$$

The test procedure associated with (5.2.6) is:

$$\text{Reject } H_0 \text{ if and only if (5.2.6) terminates, that is, } N < \infty. \tag{5.2.7}$$

The type I error probability associated with the test (5.2.6) - (5.2.7) is approximately given by:

$$\alpha \approx \left[1 - \Phi\left(\sqrt{c\ln(m)}\right)\right]\left[\frac{c+1}{c} + \ln(m)\right]. \tag{5.2.8}$$

See Sen (1981, p. 238) for more details.

Remark 5.2.1: A nonparametric sequential test analogous to (5.2.3) - (5.2.4) or (5.2.6) - (5.2.7) for the two-sided alternative hypothesis, $H_1 : \mu \neq \mu_0$, with μ_0 specified, can be easily constructed. For brevity, we omit further details.

5.3 A Test for the Mean: Unknown Variance

Let us assume that $\mathrm{E}_F\left[|X|^r\right] < \infty$ for some $r > 4$. The unspecified F may correspond to either a discrete or a continuous population. Let us also assume that σ^2 is finite, positive and unknown. Obviously, the population mean μ is finite and we are interested in the following testing problem:

$$H_0 : \mu = \mu_0 \text{ vs. } H_1 : \mu > \mu_0, \text{ with } \mu_0 \text{ specified.} \tag{5.3.1}$$

Having recorded n independent observations $X_1, \ldots, X_n$, we denote the test statistic

$$T_n \equiv \frac{\overline{X}_n - \mu_0}{S_n}, n \geq 1. \tag{5.3.2}$$

Now, let $X_1, \ldots, X_m$ be the pilot observations where m is an initial sample size. From this point onward, one will sample one additional observation X at-a-time according to the following stopping rule:

$$\begin{aligned} N &= \text{the smallest integer } n(\geq m) \text{ for which } \sqrt{n}T_n \geq \sqrt{2\ln\ln(n)} \\ &= \infty, \text{ if no such } n \text{ exists.} \end{aligned} \tag{5.3.3}$$

The test procedure associated with (5.3.3) is:

$$\text{Reject } H_0 \text{ if and only if (5.3.3) terminates, that is, } N < \infty. \tag{5.3.4}$$

The sequential test (5.3.3) - (5.3.4) is again a classic Darling-Robbins power one test for the following reason:

$$\limsup_{n\to\infty} \frac{\sqrt{n}\left|\overline{X}_n - \mu\right|}{S_n\sqrt{2\ln\ln(n)}} = 1 \quad \begin{array}{c}\text{with probability 1 when} \\ F, \mu \text{ and } \sigma \text{ are true.}\end{array} \tag{5.3.5}$$

The limiting result in (5.3.5) again follows from a combination of the *strong law of large numbers* and the *law of the iterated logarithm.*

An expression for the type I error probability associated with the power one test (5.3.3) - (5.3.4) is not readily accessible.

Sequential test (5.3.3)-(5.3.4) with power one came from Darling and Robbins (1968).

5.3.1 Another Test Procedure

Let us again assume that $\mathrm{E}_F\left[|X|^r\right] < \infty$ for some $r > 4$. Another nonparametric test may be proposed as follows: With some fixed $c > 0$, the stopping time is:

$$\begin{aligned} N(c) &= \text{the smallest integer } n(\geq m) \text{ for which } \sqrt{n}T_n \geq \sqrt{c\ln(n)} \\ &= \infty, \text{ if no such } n \text{ exists.} \end{aligned} \tag{5.3.6}$$

The test procedure associated with (5.3.6) is:

$$\text{Reject } H_0 \text{ if and only if (5.3.6) terminates, that is, } N < \infty. \tag{5.3.7}$$

The type I error probability associated with the test (5.3.6) - (5.3.7) is approximately given by:

$$\alpha \approx \left[1 - \Phi\left(\sqrt{c\ln(m)}\right)\right]\left[\frac{c+1}{c} + \ln(m)\right]. \qquad (5.3.8)$$

See Sen (1981, p. 238) for more details.

Remark 5.3.1: A nonparametric sequential test analogous to (5.3.3) - (5.3.4) or (5.3.6) - (5.3.7) for the two-sided alternative hypothesis, $H_1 : \mu \neq \mu_0$, with μ_0 specified, can be easily constructed. For brevity, we omit further details.

5.4 A Test for the Percentile

Let us continue to assume that F is unspecified. Additionally, let F be continuous so that its percentiles would be uniquely defined.

Let us denote the $100p\%$ point or the $100p^{\text{th}}$ percentile of F by ξ_p. That is, $F(\xi_p) = p, 0 < p < 1$. When we fix $p = 0.5$, we end up with the 50th percentile, namely $\xi_{0.50}$. It happens to be the population median. Now, we are interested in the following testing problem:

$$H_0 : \xi_p = \xi_{p0} \text{ vs. } H_1 : \xi_p > \xi_{p0}, \text{ with } \xi_{p0} \text{ specified.} \qquad (5.4.1)$$

From the original sequence of independent observations $X_1, \ldots, X_n, \ldots$ from F, we first define the independent but coded observations $Y_1, \ldots, Y_n, \ldots$ as:

$$Y_i = \begin{cases} 1 & \text{if } X_i \leq \xi_{p0} \\ 0 & \text{if } X_i > \xi_{p0} \end{cases}, i = 1, \ldots, n, \ldots \qquad (5.4.2)$$

Clearly, $Y_1, \ldots, Y_n, \ldots$ are i.i.d. Bernoulli(p) observations under H_0.

Now, having recorded $Y_1, \ldots, Y_n$, we denote $\overline{Y}_n = n^{-1}\sum_{i=1}^{n} Y_i$ and $\sigma^2 = p(1-p)$. Then, we define a test statistic:

$$T_n \equiv \frac{\overline{Y}_n - p}{\sqrt{p(1-p)}}, n \geq 1. \qquad (5.4.3)$$

We start with the pilot observations $X_1, \ldots, X_m$ and hence $Y_1, \ldots, Y_m$ where m is an initial sample size. From this point onward, we sample one additional observation X and hence one additional Y at-a-time according to the following stopping rule:

$$\begin{aligned} N &= \text{the smallest integer } n(\geq m) \text{ for which } \sqrt{n}T_n \geq \sqrt{2\ln\ln(n)} \\ &= \infty, \text{ if no such } n \text{ exists.} \end{aligned} \qquad (5.4.4)$$

The test procedure associated with (5.4.4) is:

Reject H_0 if and only if (5.4.4) terminates, that is, $N < \infty$. (5.4.5)

The sequential test (5.4.4) - (5.4.5) is again a classic Darling-Robbins power one test. The choice of m is a crucial input in investigating the behavior of this test's type I error probability.

Sequential percentile test (5.4.4)-(5.4.5) is a binomial test with power one that was due to Darling and Robbins (1968).

5.5 A Sign Test

We suppose that X and Y are independent continuous random variables with distribution functions $F(x) = \mathrm{P}(X \le x)$ and $G(y) = \mathrm{P}(Y \le y)$ respectively, $-\infty < x, y < \infty$. We suppose that X, Y have same support, A, which may be $\mathbb{R}$ or a subinterval of $\mathbb{R}$. We also assume that both F and G are unspecified. We denote:

$$p = \mathrm{P}_{F,G}(X < Y).$$

Now, suppose that we have available a sequence of i.i.d. pairs of observations

$$(X_1, Y_1), \ldots, (X_n, Y_n), \ldots$$

and we are interested in the following testing problem:

$$\begin{array}{l} H_0 : F(x) = G(x) \text{ for all } x \text{ in } A \text{ vs.} \\ H_1 : F(x) > G(x) \text{ for all } x \text{ in } A. \end{array} \tag{5.5.1}$$

Note that we may express p as follows:

$$\begin{aligned} P(X < Y) &= \textstyle\int_{\mathbb{R}} P(X < Y \mid Y = y) dG(y) = \int_{\mathbb{R}} P(X < y \mid Y = y) dG(y) \\ &= \textstyle\int_{\mathbb{R}} P(X < y) dG(y) = \int_{\mathbb{R}} P(X < y \mid Y = y) dG(y) \\ &= \textstyle\int_{\mathbb{R}} F(y) dG(y). \end{aligned}$$

Thus, under H_0, we have:

$$p = \int_{-\infty}^{\infty} F(y) dG(y) = \int_{-\infty}^{\infty} G(y) dG(y) = \int_0^1 u du = \frac{1}{2}.$$

Under H_1, we have:

$$p = \int_{-\infty}^{\infty} F(y) dG(y) > \int_{-\infty}^{\infty} G(y) dG(y) = \int_0^1 u du = \frac{1}{2}.$$

Now the same problem may be equivalently expressed as a choice to be made between the following hypotheses:

$$H_0 : p = \frac{1}{2} \text{ vs. } H_1 : p > \frac{1}{2}. \tag{5.5.2}$$

Hall (1965) proposed a sequential sign test that we are about to summarize in this section. From the original sequence of pairs of observations $(X_1, Y_1), \ldots, (X_n, Y_n), \ldots$, we define the coded observations $U_1, \ldots, U_n, \ldots$ as:

$$U_i = \begin{cases} 1 & \text{if } X_i < Y_i \\ 0 & \text{if } X_i \ge Y_i \end{cases}, i = 1, \ldots, n, \ldots \tag{5.5.3}$$

From the point of view of invariance and in the light of Hall et al. (1965), this data reduction may look very reasonable. Note that $U_1, \ldots, U_n, \ldots$ are i.i.d. Bernoulli(p) observations under (F, G). Hence, Hall (1965) proposed to use Wald's SPRT for a success probability p in a Bernoulli experiment.

We need to test

$$H_0 : p = \frac{1}{2} \text{ vs. } H_1 : p = p_a(> \frac{1}{2}) \tag{5.5.4}$$

with type I and II error probabilities α, β respectively, $0 < \alpha + \beta < 1$. Obviously, the number $p_a(> \frac{1}{2})$ is specified. Now, one can implement Wald's SPRT from Example 3.4.3.

Now, in the light of Example 3.4.3, Hall's (1965) sequential sign test will run as follows: We keep observing $U_1, \ldots, U_n, \ldots$ from (5.5.3), one observation at-a-time, according to the stopping rule

$$N = \begin{array}{l} \text{the smallest integer } n(\geq 1) \text{ for which the inequality} \\ \sum_{i=1}^{n} U_i \leq c_1 + dn \text{ or } \sum_{i=1}^{n} U_i \geq c_2 + dn \text{ holds} \end{array} \tag{5.5.5}$$

where $c_1 = \dfrac{\ln(B)}{\ln\left(\dfrac{p_a}{1-p_a}\right)}, c_2 = \dfrac{\ln(A)}{\ln\left(\dfrac{p_a}{1-p_a}\right)}, d = \dfrac{-\ln(2-2p_a)}{\ln\left(\dfrac{p_a}{1-p_a}\right)}$

and $B = \dfrac{\beta}{1-\alpha}, A = \dfrac{1-\beta}{\alpha}$.

At the stopped stage, the following decision rule is employed:

$$\text{Accept } H_0 \text{ if } \sum_{i=1}^{N} U_i \leq c_1 + dN, \text{Accept } H_1 \text{ if } \sum_{i=1}^{N} U_i \geq c_2 + dN. \tag{5.5.6}$$

The sequential sign test (5.5.5) - (5.5.6) is obviously distribution-free in the following senses: (i) Its implementation does not require any knowledge about F or G, and (ii) the type I and II error probabilities are rather tightly held around preassigned levels α, β respectively without any knowledge about F or G.

Sequential sign test (5.5.5)-(5.5.6) came from Hall (1965).

5.6 Data Analyses and Conclusions

Performances of the tests introduced in this chapter can be evaluated by using the program **Seq05.exe**. An example of the screen output of this program is shown in Figure 5.6.1. Program number 1 in **Seq05.exe** corresponds to tests for the mean with known variance given in Sections 5.2 and 5.2.1. The test described in Section 5.2.1 requires a positive input for c in (5.2.6). If a default "$c = 0$" is entered, as explained in Figure 5.6.1, the program implements the test from Section 5.2.

```
     Select a sequential nonparametric test
     using the numbers given in the left hand
     side. Refer Chapter 5 for details

********************************************
* Number          Program                   *
*  1   Test for Mean: Known Variance        *
*  2   Test for Mean: Unknown Variance      *
*  3   Test for the Percentile              *
*  4   Sign Test                            *
*  0   Exit                                 *
********************************************
     Enter the program number: 1

Do you want to save the results in a file(Y or N)? y
Note that your default outfile name is "SEQ05.OUT".
Do you want to change the outfile name (Y or N)? N

Nonparametric Sequential Test for H0: mu = mu0  vs.
H1: mu > mu0 where mu is the mean of the population
with known variance.
Enter mu0 = 0
 variance = 4

There are two procedures for this test.
See (5.2.3)-(5.2.4) and Section 5.2.1 for details.
To use the test procedure given in Section 5.2.1,
enter a positive value for c, otherwise enter 0: 0

Note: The test will be truncated at N= 1062
      and accept H0.

Do you want to change the truncation point (Y or N)? N
Enter initial sample size, m(>= 5) = 5

Do you want to simulate data or input real data?
Type S (Simulation) or D (Real Data): S

Do you want to store simulated sample sizes (Y or N)? N
Number of replications for a simulation? 10000

Enter a positive integer(<10) to initialize
        the random number sequence: 8

Select one of the following distributions:
****************************************************
*    ID#  DISTRIBUTION    Parameter 1  Parameter 2 *
*     1   Normal                mean    variance   *
*     2   Gamma                scale       shape   *
*     3   Weibull              scale       shape   *
*     4   Erlang           exp. mean  k(integer)   *
*     5   Exponential           mean               *
*     6   Poisson               mean               *
*     7   Bernoulli          P(Success)            *
*     8   Geometric          P(Success)            *
*     9   Negative Binomial     mean           k   *
****************************************************
Input distribution  ID#: 1

For the selected distribution enter
     the mean: 0
     variance: 4
```

Figure 5.6.1. *A screen output of* ***Seq05.exe****.*

Alternatively, the two tests for the mean in the unknown variance case, discussed in Sections 5.3 and 5.3.1, can be carried out by selecting the program number 2. This program is capable to run either test. If a default "$c = 0$" is entered, it implements the test from Section 5.3 and a positive input for c would identify the test from Section 5.3.1.

Program numbers 3 and 4 would implement the test for a percentile and the sign test described in (5.4.4) - (5.4.5) and (5.5.5) - (5.5.6) respectively.

5.6.1 Performances of Nonparametric Mean Tests

We recall that the nonparametric tests with power one given in Sections 5.2 - 5.3 are not supposed to terminate if the null hypothesis H_0 is true. However, in practice we interpret this as follows: If one of these tests happens to continue for a "very long" time, we will be inclined to accept H_0. In practice, we cannot keep running a test indefinitely! Hence, for practical implementation of these tests, it is essential to terminate the tests at some finite but large N if H_0 is true. The program numbers 1 and 2 compute a suitable truncation point for a given test internally, but allows an option to change this truncation point if desired.

The program **Seq05.exe** can be used to perform simulation studies as well as tests with a set of live data. Figure 5.6.2 shows the output obtained (given the input information from Figure 5.6.1) by running a simulated test of $H_0 : \mu = 0$ against $H_1 : \mu > 0$ when $\sigma^2 = 4$. Data were generated from $N(0, 4)$ and simulations were replicated 10,000 times. The results show that H_0 is accepted 5707 out of 10,000 tests with the truncation point "$N = 1062$".

```
Nonparametric Sequential Test for HO: mu =    0.00
vs. H1: mu >  0.00 where mu is the mean of the
population with variance   4.00
---------------------------------------------------

 The test is truncated at N = 1062 and accepted HO.
 Initial Sample size =      5

 Results for data simulated from Normal distribution
 with mean =  0.00  and variance =  4.00

        Number of simulations           =  10000
        Average sample size (n bar)     = 639.08
        Std. dev. of N (s)              = 501.03
        Minimum sample size (min N)     =      5
        Maximum sample size (max N)     =   1062
        Number of times HO is accepted  =   5707
        Number of times H1 is accepted  =   4293
```

Figure 5.6.2. *Result output for the test $H_0 : \mu = 0$ vs. $H_1 : \mu > 0$ with known variance, for data simulated from $N(0, 4)$.*

Since the simulations were carried out under H_0, we can estimate $\alpha = \mathrm{P}\{\text{type I error}\}$ as follows:

$$\widehat{\alpha} = \frac{\text{Number of times } H_1 \text{ is accepted}}{\text{Number of simulations}} = \frac{4293}{10000} = 0.4293. \tag{5.6.1}$$

On the other hand, when data were simulated from $N(1,4)$ instead of $N(0,4)$, Figure 5.6.3 shows that all 10,000 replications ended with the acceptance of H_1. That is, the estimated power of the test is 1, as expected.

```
Nonparametric Sequential Test for H0: mu =    0.00
vs. H1: mu >  0.00 where mu is the mean of the
population with variance   4.00
------------------------------------------------

 The test is truncated at N = 1500 and accepted H0.
 Initial Sample size =    5

 Results for data simulated from Normal distribution
 with mean =  1.00  and variance =  4.00

        Number of simulations            = 10000
        Average sample size (n bar)      =  9.85
        Std. dev. of N (s)               =  9.46
        Minimum sample size (min N)      =     5
        Maximum sample size (max N)      =   112
        Number of times H0 is accepted   =     0
        Number of times H1 is accepted   = 10000
```

Figure 5.6.3. *Result output for the test $H_0 : \mu = 0$ vs. $H_1 : \mu > 0$ with known variance for data simulated from $N(1,4)$.*

```
Nonparametric Sequential Test for H0: mu =    0.00
vs. H1: mu >  0.00 where mu is the mean of the
population with variance   4.00
------------------------------------------------

 The test is truncated at N = 2014 and accepted H0.
 Initial Sample size =    5

 Constant c (input):  2.00
 Approximate alpha =  0.113

 Results for data simulated from Normal distribution
 with mean =  0.00  and variance =  4.00

        Number of simulations            = 10000
        Average sample size (n bar)      =    1849.45
        Std. dev. of N (s)               =     548.80
        Minimum sample size (min N)      =          5
        Maximum sample size (max N)      =   2014
        Number of times H0 is accepted   =   9173
        Number of times H1 is accepted   =    827
```

Figure 5.6.4. *Result output for the test $H_0 : \mu = 0$ vs. $H_1 : \mu > 0$ with known variance for data simulated from $N(0,4)$.*

Figure 5.6.4 gives the results for a simulation study of the test $H_0 : \mu = 0$ vs. $H_1 : \mu > 0$ using the methodology from Section 5.2.1 with $c = 2$. The results show that $\alpha \approx 0.113$ when computed from the approximate formula given in (5.2.8). Since this simulated test was conduced under H_0, that is, the data were generated from $N(0,4)$, we have an estimate of α given by

$$\widehat{\alpha} = \frac{\text{Number of times } H_1 \text{ is accepted}}{\text{Number of simulations}} = \frac{827}{10000} = 0.0827. \qquad (5.6.2)$$

Clearly, $\widehat{\alpha}$ in (5.6.2) is considerably smaller than the value of $\widehat{\alpha}$ in (5.6.1).

Table 5.6.1: Simulated $\overline{n}$, S_N and α of the test
$H_0 : \mu = 0$ versus $H_1 : \mu > 0$ with variance 4

c	$\overline{n}$	S_N	$\widehat{\alpha}$
0.25	389.69	528.36	0.718
0.50	426.91	406.10	0.533
0.75	872.13	648.71	0.387
1.00	810.55	471.01	0.273
1.25	1183.94	588.11	0.208
1.50	1016.66	409.80	0.147
1.75	1422.25	492.87	0.111
2.00	1849.45	548.80	0.083
2.25	2288.95	585.59	0.063
2.50	2740.73	597.97	0.046
2.75	2032.18	383.05	0.035
3.00	2258.42	389.57	0.029
3.25	2495.64	367.61	0.021
3.50	2728.35	357.63	0.017
3.75	2964.50	339.17	0.013
4.00	3200.60	319.32	0.010
4.25	3434.28	319.10	0.009
4.50	3673.95	287.38	0.006
4.75	3913.94	265.61	0.005
5.00	4151.35	262.70	0.004

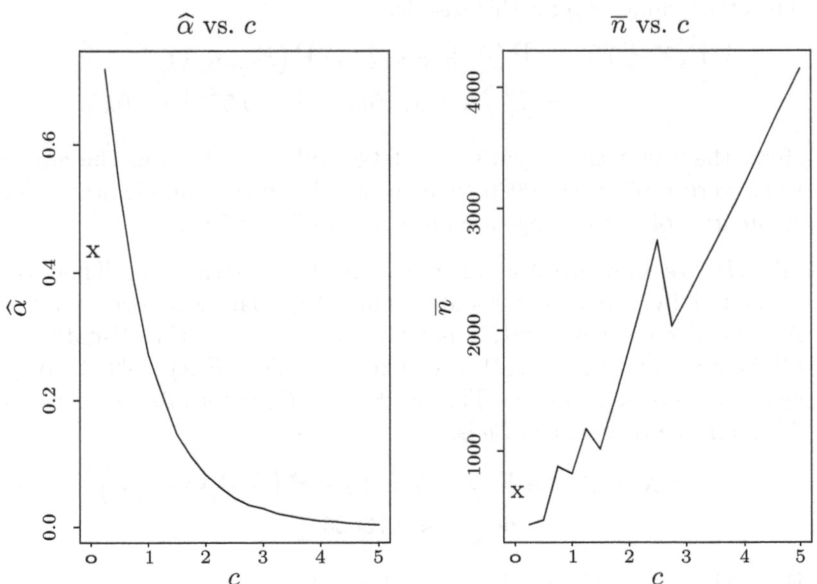

Figure 5.6.5. *Simulated* $\widehat{\alpha}$ *and* $\overline{n}$ *for*
$H_0 : \mu = 0$ *versus* $H_1 : \mu > 0$ *with variance* 4.

To examine the behavior of $\widehat{\alpha}$ further, the simulation study was repeated with $c = 0.25, 0.50, \ldots, 5.0$. The results for the test in (5.2.6), are shown briefly in Table 5.6.1 and Figure 5.6.5.

Next we simulate this test using the procedure in (5.2.3) are as follows:

$$\overline{n} = 639.08, \;\; S_N = 501.03 \text{ and } \widehat{\alpha} = 0.429$$

These values are denoted by symbol 'x' in Figure 5.6.5. In conclusion we can claim that the test procedure in (5.2.6) has lower level of significance than the procedure in (5.2.3) for higher values of c. However this lower level of significance is achieved by increased in sample size.

5.6.2 Analysis of the Sign Test

The sign test is evaluated with an arbitrary, but specified, value $p_a \in (0.5, 1)$. Theoretically, p_a can be any value in the specified range. Program number 3 in **Seq05.exe** can explore the behavior of the test for different p_a values. Let us consider the following simulation studies which are significantly different from one another:

Study I: We simulate X-data from an Expo(mean=3) distribution. Then, α is estimated by generating Y-data from $G(y)$ that also corresponds to Expo(mean=3) distribution. Next, β is estimated by generating Y-data from $G(y)$ corresponding to an Expo(mean=4) distribution. Since $X \sim$ Expo(mean =3) and $Y \sim$ Expo(mean =4), $F(x) = 1 - \exp(-x/3)$, and $G(y) = 1 - \exp(-y/4), x > 0, y > 0$. That is, $F(x) > G(x)$ for all $x > 0$. Thus true value of p for this case is:

$$\begin{aligned} \mathrm{P}(X < Y) &= \mathrm{P}\left(\tfrac{X}{2} \div \tfrac{Y}{4} < \tfrac{4}{3}\right) = \mathrm{P}\left(F_{2,2} < \tfrac{4}{3}\right) \\ &= \int_0^{4/3} (1+u)^{-2} du = \tfrac{4}{7} \approx 0.57143 (> 0.5). \end{aligned}$$

Here, the test was set with $\alpha = 0.05$ and $\beta = 0.1$ and the simulations were replicated 10000 times using the initial sample size 5. The summaries of our findings are presented in Table 5.6.2.

Study II: We simulate X-data from a $N(0,1)$ distribution. Then, α is estimated by generating Y-data from $G(y)$ that also corresponds to $N(0,1)$ distribution. Next, β is estimated by generating Y-data from $G(y)$ corresponding to a $N(1,1)$ distribution. Thus $F(x) = \Phi(x), G(y) = \Phi(y-1), -\infty < x, y < \infty$. That is, $F(x) > G(x)$ for all $-\infty < x < \infty$. Therefor the true value of p is:

$$\begin{aligned} \mathrm{P}(X < Y) &= \mathrm{P}(X - Y < 0) = \mathrm{P}\left(\tfrac{X-Y+1}{\sqrt{2}} < \tfrac{1}{\sqrt{2}}\right) \\ &= \Phi(\tfrac{1}{\sqrt{2}}) \approx 0.76025. \end{aligned}$$

Here, the test was set with $\alpha = \beta = 0.05$, $m = 5$ and the simulations were replicated 10000 times. The summaries of our findings are presented in Table 5.6.3.

Table 5.6.2: Simulation results of the test
$H_0 : F(x) = G(x)$ versus $H_1 : F(x) < G(x)$
with $\alpha = 0.05$, $\beta = 0.1$, $m = 5$

	$F(x) = G(x)$			$F(x) < G(x)$		
p_a	$\overline{n}$	S_N	$\widehat{\alpha}$	$\overline{n}$	S_N	$\widehat{\beta}$
0.55	410.45	333.78	0.049	305.78	176.23	0.014
0.60	101.40	78.67	0.047	161.21	125.03	0.293
0.65	45.50	35.96	0.046	75.14	61.36	0.605
0.70	25.64	19.75	0.043	39.71	32.93	0.752
0.75	16.38	12.47	0.043	23.13	19.03	0.820
0.80	11.63	8.37	0.039	15.32	11.67	0.863
0.85	8.37	5.19	0.041	10.48	7.50	0.880
0.90	6.76	3.25	0.046	7.92	4.65	0.888
0.95	5.65	1.86	0.039	6.13	2.63	0.915

Table 5.6.3: Simulation results of the test
$H_0 : F(x) = G(x)$ versus $H_1 : F(x) < G(x)$
with $\alpha = 0.05$, $\beta = 0.05$, $m = 5$

	$F(x) = G(x)$			$F(x) < G(x)$		
p_a	$\overline{n}$	S_N	$\widehat{\alpha}$	$\overline{n}$	S_N	$\widehat{\beta}$
0.55	537.83	388.78	0.049	63.21	14.44	0.000
0.60	135.23	96.67	0.044	35.88	12.27	0.000
0.65	59.49	42.31	0.044	26.81	11.86	0.000
0.70	33.26	23.37	0.043	23.04	12.83	0.005
0.75	20.85	14.85	0.041	20.60	13.07	0.027
0.80	13.89	9.68	0.038	19.11	12.83	0.095
0.85	10.23	6.49	0.043	16.22	11.86	0.201
0.90	7.57	4.34	0.034	12.98	9.09	0.360
0.95	6.41	2.39	0.038	9.77	6.15	0.506

The entries in Tables 5.6.2-5.6.3 indicate that α may not be greatly effected by the choice of p_a since the estimated $\widehat{\alpha}$ values appear close to the preset level α . However, the results show that estimated final sample size decreases with increasing p_a, as a result, β increases with p_a. Further when p_a is close to its upper limit, 1, β is unacceptably high and when it is close to its lower bound, 0.5, β is smaller than its preset value. From these two and other similar studies, it appears that when p_a is close to 0.6, the test performs according to the its preset parameter values.

5.7 Exercises

Exercise 5.2.1: Suppose that $X_1, \ldots, X_n, \ldots$ are i.i.d. having a common Geometric(p) distribution with an unknown parameter p, $0 < p < 1$. Evaluate

$$\limsup_{n\to\infty} \frac{\sqrt{n}\left|\overline{X}_n - p^{-1}\right|}{\sqrt{2\ln\ln(n)}}.$$

Use this result to propose a power one sequential test along the lines of (5.2.3) - (5.2.4).

Exercise 5.2.2: Suppose that $X_1, \ldots, X_n, \ldots$ are i.i.d. having a common $N(\mu, \sigma^2)$ distribution with an unknown mean μ, but known variance σ^2, $0 < \sigma < \infty$. Evaluate

$$\limsup_{n\to\infty} \frac{\sqrt{n}\left|\overline{X}_n - \mu\right|}{\sqrt{\ln\ln(n)}}.$$

Use this result to propose a power one sequential test along the lines of (5.2.3) - (5.2.4). How does this test compare with regard to α and β when compared with those for the SPRT?

Exercise 5.2.3: Suppose that $X_1, \ldots, X_n, \ldots$ are i.i.d. having a common $N(\mu, \sigma^2)$ distribution with the unknown mean μ and unknown variance σ^2, $0 < \sigma < \infty$. Evaluate

$$\limsup_{n\to\infty} \frac{\sqrt{n}\left|S_n^2 - \sigma^2\right|}{\sqrt{\ln\ln(n)}}$$

where S_n^2 is the customary sample variance.

Exercise 5.2.4: Fix some α and an initial sample size m. Now, find numerically $c(> 0)$ so the expression in (5.2.8) matches with α up to four decimal places. Now, repeat this for $\alpha = 0.01, 0.05, 0.10$ and $m = 5, 10, 15, 20, 25$. What can you say about the behavior of c?

Exercise 5.2.5: In (5.2.6), if c increases, what happens to the stopping variable N under H_0? Under H_1? What changes take place if c decreases to zero?

Exercise 5.3.1: Suppose that $X_1, \ldots, X_n, \ldots$ are i.i.d. having a common $N(\mu, \sigma^2)$ distribution with the unknown mean μ and unknown variance

σ^2, $0 < \sigma < \infty$. Evaluate

$$\limsup_{n\to\infty} \frac{\sqrt{n}\,|\overline{X}_n - \mu|}{S_n\sqrt{\ln\ln(n)}}$$

where S_n^2 is the customary sample variance.

Exercise 5.3.2: Suppose that $X_1, \ldots, X_n, \ldots$ are i.i.d. having a common $N(\mu, \sigma^2)$ distribution with the unknown mean μ and unknown variance σ^2, $0 < \sigma < \infty$. Evaluate

$$\limsup_{n\to\infty} \frac{\sqrt{n}\,|S_n^2 - \sigma^2|}{S_n\sqrt{\ln\ln(n)}}$$

where S_n^2 is the customary sample variance.

Exercise 5.4.1: Suppose that $X_1, \ldots, X_n, \ldots$ are i.i.d. having a common $N(\mu, \sigma^2)$ distribution with the unknown mean μ and unknown variance σ^2, $0 < \sigma < \infty$. Suppose that we want to test $H_0 : \mu = 0$ vs. $H_1 : \mu = \delta(> 0)$. Express the test from (5.4.4) specifically for this problem of testing for the median μ. Use computer simulations to compare this test with the SPRT?

Exercise 5.4.2: Suppose that $X_1, \ldots, X_n, \ldots$ are i.i.d. having a common $N(\mu, \sigma^2)$ distribution with the unknown mean μ, but known variance σ^2, $0 < \sigma < \infty$. Let $\xi_{0.95}$ be this distribution's 95th percentile. Suppose that we want to test $H_0 : \xi_p = 3$ vs. $H_1 : \xi_p > 3$. Express the test from (5.4.4) specifically for this problem. Could SPRT be also used in this problem? If so, use computer simulations to compare this test with the SPRT?

Exercise 5.5.1: We observe a sequence of pairs of observations (X_1, Y_1), $\ldots$, $(X_n, Y_n), \ldots$ which are i.i.d. having a common bivariate normal distribution with the means $\mu, \nu, \mu > \nu$, the same variance σ^2 and a zero correlation. Let $F(x), G(x)$ be the distribution functions of X, Y respectively. Prove or disprove that $F(x) < G(x)$ for all $-\infty < x < \infty$. Now, we pretend that we do not know whether we should go for $H_0 : F(x) = G(x)$ for all x or $H_1 : F(x) < G(x)$ for all x. Will the sign test defined in (5.5.5) - (5.5.6) apply?

Exercise 5.5.2: We observe a sequence of pairs of observations (X_1, Y_1), $\ldots$, $(X_n, Y_n), \ldots$ which are i.i.d. having a common bivariate normal distribution with the zero means, the variances $\sigma^2, \tau^2, \sigma > \tau$ and a zero correlation. Let $F(x), G(x)$ be the distribution functions of X, Y respectively. Prove or disprove that $F(x) < G(x)$ for all $-\infty < x < \infty$. Now, we pretend that we do not know whether we should go for $H_0 : F(x) = G(x)$ for all x or $H_1 : F(x) < G(x)$ for all x. Will the sign test defined in (5.5.5) - (5.5.6) apply?

Exercise 5.5.3: We observe a sequence of pairs of observations (X_1, Y_1), $\ldots$, $(X_n, Y_n), \ldots$ which are i.i.d. having a common bivariate normal distribution with the zero means, the variances $\sigma^2, \tau^2, \sigma < \tau$ and a zero correlation. Let $F(x), G(x)$ be the distribution functions of X, Y respectively. Prove or disprove that $F(x) < G(x)$ for all $-\infty < x < \infty$. Now, we pretend that we do not know whether we should go for $H_0 : F(x) = G(x)$ for all x or $H_1 : F(x) < G(x)$ for all x. Now, we pretend that we do not know whether we should go for $H_0 : F(x) = G(x)$ for all x or $H_1 : F(x) < G(x)$ for all x. Will the sign test from (5.5.5)-(5.5.6) apply? If not, suggest an easy alternative method of testing.

Exercise 5.5.4: We observe a sequence of pairs of observations (X_1, Y_1), $\ldots, (X_n, Y_n), \ldots$ which are i.i.d., the X, Y being independent. Suppose that the random variables X, Y are respectively distributed as exponentials with means $\mu > 0, \nu > 0, \mu > \nu$. Let $F(x), G(x)$ be the distribution functions of X, Y respectively. Prove or disprove that $F(x) < G(x)$ for all $-\infty < x < \infty$. Now, we pretend that we do not know whether we should go for $H_0 : F(x) = G(x)$ for all x or $H_1 : F(x) < G(x)$ for all x. Will the sign test from (5.5.5) - (5.5.6) apply?

Exercise 5.5.5: We observe a sequence of pairs of observations (X_1, Y_1), $\ldots, (X_n, Y_n), \ldots$ which are i.i.d., the X, Y being independent. Suppose that the random variables X, Y are respectively distributed as exponentials with means $\mu > 0, \nu > 0, \mu < \nu$. Let $F(x), G(x)$ be the distribution functions of X, Y respectively. Prove or disprove that $F(x) < G(x)$ for all $-\infty < x < \infty$. Now, we pretend that we do not know whether we should go for $H_0 : F(x) = G(x)$ for all x or $H_1 : F(x) < G(x)$ for all x. Will the sign test from (5.5.5) - (5.5.6) apply? If not, suggest an easy alternative method of testing.

Exercise 5.6.1: The data given below is a random sample obtained from a population with unknown mean μ and variance $\sigma^2 = 4$.

$$5.2,\ 5.2,\ 12.2,\ 5.0,\ 5.9,\ 5.7, 19.1,\ 5.4,\ 5.5,\ 5.4$$

Test the hypotheses $H_0 : \mu = 6$ vs. $H_1 : \mu > 6$ using the program **Seq05.exe** with an initial sample size 5 by implementing

(i) procedure given in Section 5.2;
(ii) procedure given in Section 5.2.1 with $c = 1$.

Exercise 5.6.2: Consider the test $H_0 : \mu = 5$ vs. $H_1 : \mu > 5$ with unknown variance. Generate data from an exponential population with mean 5. Use the program number 2 in **Seq05.exe** to perform simulation studies of the two test procedures given in Sections 5.3 and 5.3.1 with $c = 1.5, 2.0$ and 3. Repeat each simulation 100 times, estimate α, and comment on the results.

Exercise 5.6.3: Let α be type I error probability of the test $H_0 : \xi_{0.5} = 5$ vs. $H_1 : \xi_{0.5} > 5$. Use the program number 3 in **Seq05.exe** simulate the test by generating data from $N(5, 4)$. Replicate the simulation 1000 times with the initial sample size $m = 5, 10, 15, 20$ and show that α decreases with increasing m.

CHAPTER 6

Estimation of the Mean of a Normal Population

6.1 Introduction

This chapter introduces a number of basic estimation methodologies for the mean of a normal distribution whose variance is unknown. It is safe to say that the literature on sequential estimation of the mean of a normal population is vast. Also, the variety of problems that have been handled is plentiful. Hence, it will be practically impossible to address all conceivable problems, but we make an attempt to introduce a large majority of the important developments.

It is fair to say before anything else that the two papers, Stein (1945) and Robbins (1959), are regarded as truly seminal contributions in this field. These two contributions have invigorated this area like no other.

Section 6.2 begins with a classical fixed-width confidence interval problem from Section 2.3 and develops Stein's (1945) two-stage methodology. Then, we successively walk through its modifications including multistage and sequential methodologies of Ray(1957), Chow and Robbins (1965), Hall (1981,1983), Mukhopadhyay (1980,1990,1996) and Mukhopadhyay and Solanky (1991,1994). We have included both accelerated sequential and three-stage methodologies in this discussion. In doing so, we have carefully highlighted the distinction between the asymptotic concepts of first-order and second-order properties. We have emphasized the notion of asymptotic second-order efficiency that was introduced by Ghosh and Mukhopadhyay (1981).

Next, in Section 6.3, we move to the bounded risk point estimation problem from Example 2.3.2 (Ghosh and Mukhopadhyay, 1976) for which suitable two-stage and purely sequential methodologies are discussed.

Section 6.4 includes an introduction to the minimum risk point estimation problem. In this context, suitable purely sequential (Robbins, 1959) and parallel piecewise sequential (Mukhopadhyay and Sen, 1993) methodologies are discussed.

Again, we have tried our best to emphasize overall as many kinds of estimation methodologies as possible. In every section, the underlying concepts behind each methodology and the implementation of each methodology are emphasized. The computer software has been used very extensively and interactively for a large majority of the problems and the associated methodologies under consideration.

6.2 Fixed-Width Confidence Intervals

We go back to Section 2.3 where we first discussed the fixed-width confidence interval estimation problem for an unknown mean. We begin with a $N(\mu, \sigma^2)$ distribution where the mean μ and the variance σ^2 are both unknown with $-\infty < \mu < \infty$ and $0 < \sigma < \infty$. Given two preassigned numbers $d(> 0)$ and $0 < \alpha < 1$, we wish to construct a confidence interval J for μ such that:

$$\text{the length of } J \text{ is } 2d \text{ and } \mathrm{P}_{\mu,\sigma}\{\mu \in J\} \geq 1 - \alpha \text{ for all } \mu, \sigma.$$

We discussed this problem in Example 2.3.1 and mentioned that there did *not* exist (Dantzig, 1940) any fixed-sample-size procedure to solve this problem.

Mahalanobis (1940) introduced the path-breaking idea of using pilot samples in large-scale surveys to control margin of error. He was driven by practical aspects of statistical methods. Stein (1945,1949) created the foundation of two-stage sampling, also referred to as *double* sampling, that led to an *exact* solution for a fundamental problem in statistical inference.

The literature includes sampling strategies generally described as *multi-stage* procedures. First, we discuss Stein's ingenious method, followed by variations involving the three-stage and purely sequential sampling. These techniques and the associated fixed-width interval J for μ tend to have some different theoretical properties. We focus on explaining the methodologies along with their important properties and various ways to compare them.

Suppose that $X_1, \ldots, X_n$ are i.i.d. random variables from a $N(\mu, \sigma^2)$ distribution, $n \geq 2$. Let

$$\overline{X}_n = \frac{1}{n}\sum_{i=1}^{n} X_i \text{ and } S_n^2 = \frac{1}{n-1}\sum_{i=1}^{n}\left(X_i - \overline{X}_n\right)^2$$

respectively stand for the sample mean and the sample variance. Recall that $a_{n-1} \equiv a_{n-1,\alpha/2}$ is the upper $50\alpha\%$ point of the Student's t distribution with $n-1$ degrees of freedom.

Now, we consider the fixed-width confidence interval

$$J_n = \left[\overline{X}_n - d, \overline{X}_n + d\right] \tag{6.2.1}$$

for the unknown mean μ. The associated confidence coefficient is expressed as follows:

$$\begin{aligned} \mathrm{P}_{\mu,\sigma}\{\mu \in J_n\} &= \mathrm{P}_{\mu,\sigma}\{|\overline{X}_n - \mu| \le d\} \\ &= \mathrm{P}_{\mu,\sigma}\left\{\frac{\sqrt{n}\,|\overline{X}_n - \mu|}{\sigma} \le \frac{\sqrt{n}d}{\sigma}\right\} \\ &= 2\Phi\left(\sqrt{n}d/\sigma\right) - 1 \end{aligned} \tag{6.2.2}$$

since $\sqrt{n}(\overline{X}_n - \mu)/\sigma$ has a standard normal distribution. Again, we write $\Phi(z) = \int_{-\infty}^{z} \phi(y)dy$ where $\phi(z) = \frac{1}{\sqrt{2\pi}}\exp(-\frac{1}{2}z^2)$ with $-\infty < z < \infty$. That is, $\phi(z)$ and $\Phi(z)$ are respectively the probability density function and the distribution function of a standard normal variable.

Observe that J_n already has a fixed-width $2d$. Now, we also require that the associated confidence coefficient must be at least $1-\alpha$. Thus, utilizing (6.2.2), we can write

$$2\Phi\left(\sqrt{n}d/\sigma\right) - 1 \ge 1 - \alpha = 2\Phi(a) - 1 \tag{6.2.3}$$

where a is the upper $50\alpha\%$ point of a standard normal distribution. But, since $\Phi(z)$ is monotonically increasing in z, from (6.2.3) one can claim that

$$\frac{\sqrt{n}d}{\sigma} \ge a \Rightarrow n \ge \frac{a^2\sigma^2}{d^2} = C, \text{ say.} \tag{6.2.4}$$

In other words, the confidence coefficient associated with the interval J_n will be at least $1-\alpha$,

if and only if n is the smallest integer $\ge C$.

> In this sense, C is interpreted as the *optimal fixed sample size required* to construct the corresponding fixed-width confidence interval for μ, had σ been known.

Of course the numbers a and d are fixed in advance. On top of this, if σ were known too, then one will simply evaluate C. Now, for example, if $C = 18$ then one will take $n = 18$ observations whereas if $C = 32.1$ then one will take $n = 33$ observations, and construct the corresponding confidence interval J_n based on $n = 18$ or 33 observations respectively.

However, the fact is that the expression for C is known but its magnitude is unknown since σ^2 is assumed unknown.

> The important question is this: How many observations one must take, that is what should be the sample size n in order to construct the confidence interval J_n with its confidence coefficient $\ge 1-\alpha$?

In what follows, we propose a number of multistage sampling designs in the order of their original development.

6.2.1 Two-Stage Procedure

Stein (1945,1949) proposed taking observations in two stages. The observations at the first stage of size $m(\geq 2)$ are the pilot samples and m is called the pilot size. Now, based on $X_1, \ldots, X_m$, one determines the sample variance S_m^2 which estimates the unknown σ^2, and hence C is estimated. The smallest integer $N \geq \widehat{C}$ will be the designated final sample size and $J_N = [\overline{X}_N \pm d]$ would be the proposed interval for μ. This idea is the core of Stein's ingenious double sampling methodology. Let us denote

$$\langle u \rangle = \text{ the largest integer } < u, \text{ for } u > 0. \tag{6.2.5}$$

Recall that $a_{m-1} \equiv a_{m-1,\alpha/2}$ is the upper $50\alpha\%$ point of the Student's t distribution with $m-1$ degrees of freedom. We define the final sample size as

$$N \equiv N(d) = \max\left\{m, \left\langle \frac{a_{m-1}^2 S_m^2}{d^2} \right\rangle + 1\right\}. \tag{6.2.6}$$

We may use a notation N or $N(d)$ interchangeably because we may vary d and examine how $N(d)$ behaves as a function of d. It is easy to see that N is finite with probability one.

From (6.2.4), recall that $C = a^2\sigma^2/d^2$. In (6.2.6), observe that C is estimated by $a_{m-1}^2 S_m^2/d^2$ based on pilot observations by replacing σ^2, a^2 in C with S_m^2, a_{m-1}^2 respectively. In fact, we view N itself as an estimator of C.

> The positive integer valued random variable N is referred to as a *stopping time* or *stopping variable.*

The two-stage procedure (6.2.6) is implemented as follows:

Case 1: If $N = m$, that is when m is larger than the estimator of C, it indicates that we already have too many observations at the pilot stage. Hence, we do not need any more observations at the second stage.

Case 2: If $N > m$, that is when m is smaller than the estimator of C, it indicates that we have started with too few observations at the pilot stage. Hence, we sample the difference at the second stage. That is, we take new observations $X_{m+1}, \ldots, X_N$ at the second stage.

$$\begin{aligned} &\text{Case 1: If } N = m, \text{ the final dataset is } X_1, \ldots, X_m \\ &\text{Case 2: If } N > m, \text{ the final dataset is } X_1, \ldots, X_m, X_{m+1}, \ldots, X_N \end{aligned} \tag{6.2.7}$$

Combining the two possibilities from (6.2.7), one can say that the final dataset is composed of N and $X_1, \ldots, X_N$. This gives rise to the sample mean $\overline{X}_N$ and the associated fixed-width interval $J_N = [\overline{X}_N \pm d]$ along the lines of (6.2.1).

Observe that the confidence interval $J_N = [\overline{X}_N \pm d]$ is defined from a random sample $X_1, \ldots, X_N$ of random size N.

It is clear that (i) the event $\{N = n\}$ depends only on the random variable S^2_m, and (ii) $\overline{X}_n, S^2_m$ are independent random variables, for all fixed $n(\geq m)$. So, any event defined only through $\overline{X}_n$ must be independent of the event $\{N = n\}$. See Theorem 4.7.1 or Exercises 6.2.1-6.2.2.

Now, we express the confidence coefficient associated with J_N as:

$$\begin{aligned}
\mathrm{P}_{\mu,\sigma}\{\mu \in J_N\} &= \sum_{n=m}^{\infty} \mathrm{P}_{\mu,\sigma}\{|\overline{X}_N - \mu| \leq d \cap N = n\} \\
&= \sum_{n=m}^{\infty} \mathrm{P}_{\mu,\sigma}\{|\overline{X}_n - \mu| \leq d \cap N = n\} \\
&= \sum_{n=m}^{\infty} \mathrm{P}_{\mu,\sigma}\left\{|\overline{X}_n - \mu| \leq d\right\} \mathrm{P}_{\mu,\sigma}(N = n) \\
&\qquad \text{using the independence of } \overline{X}_n \text{ and } I(N = n) \\
&= \sum_{n=m}^{\infty} \left[2\Phi\left(\sqrt{n}d/\sigma\right) - 1\right] \mathrm{P}_{\mu,\sigma}(N = n), \quad \text{using (6.2.2)} \\
&= \mathrm{E}_{\mu,\sigma}\left[2\Phi\left(\sqrt{N}d/\sigma\right) - 1\right]. \qquad (6.2.8)
\end{aligned}$$

In the last step and elsewhere, we continue to write $\mathrm{E}_{\mu,\sigma}(\cdot)$ even though the distribution N involves σ^2 only.

Now, let us summarize some elementary properties of the procedure (6.2.6).

Theorem 6.2.1 : *For the stopping variable N defined by (6.2.6), for all fixed μ, σ, d and α, one has:*

(i) $a^2_{m-1}\sigma^2 d^{-2} \leq \mathrm{E}_{\mu,\sigma}(N) \leq m + a^2_{m-1}\sigma^2 d^{-2}$;

(ii) $\overline{X}_N$ *is an unbiased estimator of* μ *and its variance is* $\sigma^2\mathrm{E}_{\mu,\sigma}(N^{-1})$;

(iii) $Q \equiv \sqrt{N}(\overline{X}_N - \mu)/\sigma$ *has a standard normal distribution.*

Proof: From (6.2.6), observe the *basic inequality*

$$\frac{a^2_{m-1}S^2_m}{d^2} \leq N \leq m + \frac{a^2_{m-1}S^2_m}{d^2} \text{ w.p. } 1, \qquad (6.2.9)$$

which implies that

$$\frac{a^2_{m-1}\mathrm{E}_{\mu,\sigma}(S^2_m)}{d^2} \leq \mathrm{E}_{\mu,\sigma}(N) \leq m + \frac{a^2_{m-1}\mathrm{E}_{\mu,\sigma}(S^2_m)}{d^2}. \qquad (6.2.10)$$

Part (i) follows from (6.2.10) since $\mathrm{E}_{\mu,\sigma}(S^2_m) = \sigma^2$.

In order to prove part (ii), let us first write:

$$
\begin{aligned}
\mathrm{E}_{\mu,\sigma}(\overline{X}_N) &= \textstyle\sum_{n=m}^{\infty} \mathrm{E}_{\mu,\sigma}[\overline{X}_N | N=n] \mathrm{P}_{\mu,\sigma}(N=n) \\
&= \textstyle\sum_{n=m}^{\infty} \mathrm{E}_{\mu,\sigma}[\overline{X}_n | N=n] \mathrm{P}_{\mu,\sigma}(N=n) \\
&= \textstyle\sum_{n=m}^{\infty} \mathrm{E}_{\mu,\sigma}[\overline{X}_n] \mathrm{P}_{\mu,\sigma}(N=n) \\
&\qquad\qquad \text{since } \overline{X}_n \text{ and } I(N=n) \text{are independent} \\
&= \mu \mathrm{P}_{\mu,\sigma}(N<\infty) \ = \mu, \ \text{ since } \mathrm{P}_{\mu,\sigma}(N<\infty)=1.
\end{aligned}
$$

The expression for the variance of $\overline{X}_N$ can be verified analogously.

In order to prove part (iii), we start with an arbitrary real number z and proceed along the lines of (6.2.8) to write:

$$
\begin{aligned}
\mathrm{P}_{\mu,\sigma}\{Q \le z\} &= \sum_{n=m}^{\infty} \mathrm{P}_{\mu,\sigma}\left\{ \frac{\sqrt{N}(\overline{X}_N-\mu)}{\sigma} \le z \cap N=n \right\} \\
&= \sum_{n=m}^{\infty} \mathrm{P}_{\mu,\sigma}\left\{ \frac{\sqrt{n}(\overline{X}_n-\mu)}{\sigma} \le z \cap N=n \right\} \\
&= \sum_{n=m}^{\infty} \mathrm{P}_{\mu,\sigma}\left\{ \frac{\sqrt{n}(\overline{X}_n-\mu)}{\sigma} \le z \right\} \mathrm{P}_{\mu,\sigma}(N=n) \\
&\qquad\qquad \text{using the independence of } \overline{X}_n \text{ and } I(N=n) \\
&= \sum_{n=m}^{\infty} \Phi(z)\mathrm{P}_{\mu,\sigma}(N=n) \ = \Phi(z).
\end{aligned}
$$

That is, the distribution function of Q is the same as that for a standard normal variable. Now the proof is complete. ■

The main result is given in the following theorem:

Theorem 6.2.2: *For the stopping variable N defined by (6.2.6), for all fixed μ, σ, d and α, one has:*

$$
\mathrm{P}_{\mu,\sigma}\{\mu \in J_N\} \ge 1-\alpha. \quad [\textit{Consistency or Exact Consistency}]
$$

Proof: First note that fixing S_m^2 is equivalent to fixing a value of N. That is, conditionally given S_m^2, we can claim from Theorem 6.2.1, part (iii) that the distribution of Q continues to be $N(0,1)$. Since this conditional distribution of Q does not involve S_m^2, we can claim that Q and S_m^2 are independent. In other words,

$$
\frac{\sqrt{N}(\overline{X}_N-\mu)}{S_m} = \frac{Q}{\sqrt{Y/(m-1)}}
$$

where Q is $N(0,1)$, Y is χ^2_{m-1}, whereas Q and Y are independent. Hence,

$$
\frac{\sqrt{N}(\overline{X}_N-\mu)}{S_m} \text{ has the Student's } t_{m-1} \text{ distribution.} \tag{6.2.11}
$$

Next, using similar arguments, observe that

$$\begin{aligned} \mathrm{P}_{\mu,\sigma}\{\mu \in J_N\} &= \mathrm{P}_{\mu,\sigma}\{|\overline{X}_N - \mu| \le d\} \\ &= \mathrm{P}_{\mu,\sigma}\left\{|Q| \le \sqrt{N}d/\sigma\right\}. \end{aligned} \tag{6.2.12}$$

But, from the lower bound in (6.2.9), we can claim that

$$\sqrt{N}d/\sigma \ge a_{m-1}(S_m/\sigma) \text{ w.p. } 1.$$

Now, $|Q| \le a_{m-1}(S_m/\sigma)$ implies that $|Q| \le \sqrt{N}d/\sigma$. Hence, from (6.2.11)-(6.2.12), we obtain:

$$\begin{aligned} \mathrm{P}_{\mu,\sigma}\{\mu \in J_N\} &\ge \mathrm{P}_{\mu,\sigma}\{|Q| \le a_{m-1}(S_m/\sigma)\} \\ &= \mathrm{P}_{\mu,\sigma}\left\{\left|\tfrac{\sqrt{N}(\overline{X}_N - \mu)}{S_m}\right| \le a_{m-1}\right\} \end{aligned}$$

which is exactly $1-\alpha$ since a_{m-1} is chosen so that

$$\mathrm{P}\{|t_{m-1}| \le a_{m-1}\} = 1 - \alpha. \qquad \blacksquare$$

Theorem 6.2.2 is regarded as a **fundamental breakthrough** in this area. This result showed that Stein's two-stage estimation procedure could give an exact solution to a statistical problem for which there was no fixed-sample-size solution.

Now, some other comments are in order. From Theorem 6.2.1, part (i), we note that

$$\mathrm{E}_{\mu,\sigma}(N) \ge \frac{a_{m-1}^2\sigma^2}{d^2} > C \text{ since } a_{m-1} > a. \tag{6.2.13}$$

In other words, on an average, the Stein procedure requires more observations than the optimal fixed sample size C. This should not be surprising because C is "optimal" when we pretend that σ is known! But, in fact σ is unknown and we end up taking more than C observations.

So far we have discussed only the exact properties. But, one may ask: what are the associated large-sample properties? Now, we should think carefully about how we can make the random sample size N approach to infinity. We can make N approach to infinity (w.p. 1) by making d tend to zero or by letting σ tend to infinity. As d becomes smaller, we obtain shorter confidence intervals for μ with at least $1-\alpha$ coverage, and hence N will tend to increase. On the other hand, as σ becomes larger, since we continue demand the fixed-width confidence interval to have at least $1-\alpha$ coverage, N will tend to increase. In order to state large-sample properties precisely, one customarily lets $d \to 0$ or $\sigma \to \infty$.

Throughout this chapter, we consider limiting expressions as $d \to 0$.

From Theorem 6.2.1, part (i), it follows easily that

$$\lim_{d\to 0} \mathrm{E}_{\mu,\sigma}\left(\frac{N}{C}\right) = \frac{a_{m-1}^2}{a^2} > 1 \text{ [\textit{Asymptotic Inefficiency or Asymptotic First-Order Inefficiency}]} \quad (6.2.14)$$

This property is not as easily disposed of as we did in the case of (6.2.13). When $d \to 0$, we know that $N \to \infty$ w.p. 1 and $C \to \infty$, but on the other hand (6.2.14) points out that even asymptotically $\mathrm{E}_{\mu,\sigma}(N)$ exceeds the optimal fixed-sample size C! That is, $\mathrm{E}_{\mu,\sigma}(N)$ and C do not agree in the large-sample case, and hence the latter property is more disturbing than what we saw in (6.2.13). We may summarize by saying that the Stein procedure is *asymptotically (first-order) inefficient.*

This procedure remains inefficient, even asymptotically, perhaps because the pilot size m is held fixed but C is allowed to grow! Since the experimenter does not know C, the tendency may be to choose m small. Another issue one may raise is that σ^2 is estimated once and this estimator may be unstable if m is chosen too small. That is, the performance of the procedure will depend heavily on m.

Another interesting large-sample property may be summarized as follows:

$$\lim_{d\to 0} \mathrm{P}_{\mu,\sigma}\{\mu \in J_N\} = 1 - \alpha. \quad \text{[\textit{Asymptotic Consistency}]} \quad (6.2.15)$$

This property assures us that for a large sample size, the confidence coefficient will stay in the vicinity of the target, $1 - \alpha$.

> A word of caution may be in order: The *asymptotic consistency property* should not be confused with the *exact consistency property* claimed in Theorem 6.2.2.

Seelbinder (1953) and Moshman (1958) developed some criteria to suggest a reasonable choice of m. Starr (1966a) investigated some of the related issues fully. These approaches, however, relied upon nearly precise background information about σ^2. But, if one could rely upon such precise prior knowledge about σ^2, then one could practically implement a single-stage procedure instead of double sampling! Mukhopadhyay (2005a) has developed an information-based approach to suggest a reasonable pilot sample size for two-stage sampling without assuming a prior knowledge about σ^2.

Computations and Simulations

Program **Seq06.exe** obtains fixed-width confidence intervals for a normal mean using a number of multistage procedures described in this section. First, it provides the values of C for given σ^2, α and d. It can handle exercises with live data entry or it can help in running simulations. When the program **Seq06.exe** is executed, it will prompt us to select one of the nine programs developed exclusively for use in this chapter.

```
*****************************************************
* Number         Program                            *
*  1   Exact value of C for given parameters        *
*  2   Fixed-Width Confidence Interval Estimation   *
*  3   E(N) and Confidence Coefficient for Two-Stage *
*  4   Role Gamma in Modified Two-Stage Procedure   *
*  5   Second Order Terms of Purely Sequential      *
*  6   Second Order Terms of Accelerated Sequential *
*  7   Second Order Terms of Three-Stage Procedure  *
*  8   Bounded Risk Point Estimation                *
*  9   Minimum Risk Point Estimation                *
*  0   Exit                                         *
*****************************************************

     Enter the program number: 1

Do you want to save the results in a file(Y or N)?Y
Note that your default outfile name is "SEQ06.OUT".

Do you want to change the outfile name?
[Y(yes) or N(no)] N

This program computes the values of C for the
range of values of d, alpha and sigma.

For each of these parameter, enter
     number of values (k) and
     value of the parameter if k=1;
     minimum and maximum values if k > 1

This program computes C for given parameter values

      For d (half length):
               number of values = 3
               minimum  value   = 0.5
               maximum  value   = 1.0
Do you want to change the above values  (type Y or N): N

      For alpha (1- confidence coefficient):
               number of values = 2
               minimum  value   = 0.05
               maximum  value   = 0.10
Do you want to change the above values  (type Y or N): N

      For standard deviation (sigma):
               number of values = 1
               enter the value  = 1.0
Do you want to change the above values  (type Y or N): N
```

Figure 6.2.1. *Screen input for* ***Seq06.exe***.

The program number 1 computes the optimal sample size C from (6.2.4) for given σ^2, α and d. The program may use either a single value or a range of values for these parameters to compute C. One may input a set of values of a parameter by specifying a minimum, a maximum and the number of required values. A sample input session for the computation of C is given in Figures 6.2.1 and Figure 6.2.2 gives the corresponding output of the program.

```
Computation of C values for sigma, alpha and d
----------------------------------------------
Input values:
                 minimum  maximum  number
       Sigma       1.00     1.00       1
       Alpha        .05      .10       2
       d            .00     1.00       3
       ------------------------------

       Sigma      Alpha          d  Computed C
        1.00        .05        .50     15.37
        1.00        .05        .75      6.83
        1.00        .05       1.00      3.84

        1.00        .10        .50     10.82
        1.00        .10        .75      4.81
        1.00        .10       1.00      2.71
```

Figure 6.2.2. *Value of C for given σ, α and d.*

Program number 2 in **Seq06.exe** provides fixed-width intervals using one of the five multistage procedures described in this section. This program provides fixed-width intervals for the normal mean μ either using live data input or by running simulations with given values of d, α and m. In a simulation study, the program gives an option of saving simulated values of the sample sizes (N) into an ASCII data file.

Figure 6.2.3 shows the summary results from simulating the Stein procedure for a 90% confidence interval with fixed-width $2d = 1$. Data for the simulation were generated from a $N(5, 4)$ distribution and it was replicated 2000 times. Per Theorem 6.2.2, the output shows that the procedure is consistent, that is the coverage probability is at least 0.90. Further, the average simulated sample size $\overline{n} = 54.29$ which is larger than the corresponding optimal fixed-sample-size, $C = 43.29$. One will note that $\alpha = 0.10, a = 1.645, m = 10$ and $a_{m-1} = 1.8331$ so that $a_{m-1}^2/a^2 \approx 1.2418$. From Figure 6.2.3, we have $\overline{n}/C \approx 1.2541$, which is rather close to a_{m-1}^2/a^2. We may put this another way: An estimated or simulated oversampling percentage is 25.4% whereas theoretically one would expect nearly 24% oversampling! The simulation study seems to confirm our general feelings expressed by (6.2.13) in reality. Also, an estimated coverage probability $\overline{p}$ is 0.900 which validates the (exact) consistency result (Theorem 6.2.2) in practice.

```
     Fixed-Width Confidence Interval
        Using Two-Stage Procedure
        =========================

Simulation study for  90% confidence interval for
the normal mean. Data were generated from normal
distribution with mean =  5.00 and variance = 4.00

Number of simulation replications      =  2000
Width of the confidence interval (2d) =  1.00

Initial sample size = 10

    Optimal sample size (C)         =   43.29

    Average sample size (n_bar)     =   54.29
    Std. dev. of n (s)              =   24.61
    Minimum sample size (min n)     =   10
    Maximum sample size (max n)     =  183

    Coverage probability (p_bar)    = 0.900
    Standard error of p_bar         = 0.007
```

Figure 6.2.3. *Fixed-width confidence interval using Stein's two-stage procedure.*

Probability Distribution of the Final Sample Size N

Since the final sample size N given by (6.2.6) is a random variable, it makes sense to determine its exact distribution.

Observe that the distribution of N does not depend upon μ.

Let us denote $Y = (m-1)S_m^2/\sigma^2$ which has a χ^2_{m-1} distribution. One may easily verify that

$$
\begin{aligned}
\mathrm{P}_\sigma(N=m) &= \mathrm{P}_\sigma\left\{\left\langle \frac{a_{m-1}^2 S_m^2}{d^2}\right\rangle + 1 \le m\right\} \\
&= \mathrm{P}_\sigma\left\{0 < \frac{a_{m-1}^2 S_m^2}{d^2} \le m\right\} = \mathrm{P}_\sigma\left\{0 < Y \le \frac{m(m-1)d^2}{\sigma^2 a_{m-1}^2}\right\},
\end{aligned}
$$

and similarly for any positive integer k,

$$
\begin{aligned}
\mathrm{P}_\sigma(N=m+k) &= \mathrm{P}_\sigma\left\{\left\langle \frac{a_{m-1}^2 S_m^2}{d^2}\right\rangle = m+k-1\right\} \\
&= \mathrm{P}_\sigma\left\{m+k-1 < \frac{a_{m-1}^2 S_m^2}{d^2} \le m+k\right\} \\
&= \mathrm{P}_\sigma\left\{\frac{(m+k-1)(m-1)d^2}{\sigma^2 a_{m-1}^2} < Y \le \frac{(m+k)(m-1)d^2}{\sigma^2 a_{m-1}^2}\right\}.
\end{aligned}
$$

Let us write

$$g_m(y) = \left\{2^{(m-1)/2}\Gamma\left(\frac{m-1}{2}\right)\right\}^{-1} \exp[-y/2]y^{(m-3)/2}, y > 0;$$

$$A_k = \frac{(m+k)(m-1)d^2}{\sigma^2 a^2_{m-1}}, \ k = 0, 1, 2, \ldots.$$

Then,

$$\begin{aligned}
&\text{P}_\sigma(N = m) = \text{P}_\sigma\left\{0 < \chi^2_{m-1} \leq A_0\right\} = \textstyle\int_0^{A_0} g_m(y)dy; \\
&\text{P}_\sigma(N = m + k) = \text{P}_\sigma\left\{A_{k-1} < \chi^2_{m-1} \leq A_k\right\} = \textstyle\int_{A_{k-1}}^{A_k} g_m(y)dy,
\end{aligned} \tag{6.2.16}$$

for $k = 1, 2, 3, \ldots$.

Once the exact distribution of N is found, pretending that σ were known, one surely evaluate the expected value and the variance of N as well as the confidence coefficient associated with the fixed-width interval J_N given by (6.2.8).

```
E(N), Var(N) and the Associated Confidence
   Coefficient for Two-Stage Procedure
   ====================================

  Input Parameters:
       alpha                     =  0.05
       d (half length)           =  0.50
       Initial sample size, m =  5
       Population Variance       =   4.0

  Computed Results:
       Optimal sample size, C =   61.5
       Expected value of N       =  123.8
       Variance of N             = 7598.8

   Associated Confidence Coefficient = 0.951
```

Figure 6.2.4. E(N), V(N) *and associated confidence coefficient.*

Program number 3 in **Seq06.exe** helps in evaluating E(N), V(N) and the exact confidence coefficient associated with J_N. Figure 6.2.4 shows these results obtained for a 95% confidence interval with $d = 0.5$ and $m = 5$. Here we assumed that $\sigma^2 = 4$. These numbers confirm the results found earlier:

- (6.2.13), that is, E$(N) = 123.8 > C = 61.5$ and
- (6.2.8), that is, the exact confidence coefficient,

$$\text{P}(\mu \in J_N) = 0.951 \approx 0.95 = 1 - \alpha.$$

Since E(N) is not very small, the result from (6.2.13) may suggest that

$\mathrm{E}(N)/C$ ought to be expected to lie in a vicinity of a_{m-1}^2/a^2. With $m = 5, \alpha = 0.05$, we find $a_{m-1} = 2.7764$ and $a = 1.96$. Thus, $a_{m-1}^2/a^2 = 2.0066$ whereas $\mathrm{E}(N)/C = 2.013$ and these values are pretty close to each other! In other words, the result from (6.2.13) may not be as asymptotic as it portrays itself to be.

6.2.2 Purely Sequential Procedure

In order to reduce oversampling by the two-stage procedure, one may estimate σ^2 successively in a sequential manner. We begin with $m(\geq 2)$ observations and then continue to take one additional observation at a time, but terminate sampling when we have gathered "enough observations". How would we know whether we have reached that stage? Now, successively estimating σ^2 will translate into sequentially estimating C, and thus we will keep checking whether the sample size exceeds the estimate of C at every step. As soon as the sample size exceeds the estimate of C, the sampling is terminated and the confidence interval is constructed. This idea, which has been pursued by Anscombe (1952), Ray (1957), Chow and Robbins (1965) and others, forms the core of a purely sequential methodology.

Recall that $C = a^2\sigma^2/d^2$ where σ^2 is unknown. We start with $X_1, \ldots, X_m$, a pilot sample of size $m(\geq 2)$. Then, we take one additional X at a time as needed. The stopping time is defined as follows:

$$N \equiv N(d) \quad \text{is the smallest integer } n(\geq m) \text{ for which we observe } n \geq a^2 S_n^2/d^2. \tag{6.2.17}$$

This procedure is implemented as follows. In the initial stage, we obtain S_m^2 based on $X_1, \ldots, X_m$ and check whether $m \geq a^2 S_m^2/d^2$, an estimator of C at this stage. If $m \geq a^2 S_m^2/d^2$, then sampling terminates right here and our final sample size is m. But, if $m < a^2 S_m^2/d^2$, we take one additional observation X_{m+1} and update the sample variance by S_{m+1}^2 based on $X_1, \ldots, X_m, X_{m+1}$. Next, we check whether $m+1 \geq a^2 S_{m+1}^2/d^2$, an estimator of C at this stage. If $m+1 \geq a^2 S_{m+1}^2/d^2$, sampling terminates right here and our final sample size is $m+1$. But, if $m+1 < a^2 S_{m+1}^2/d^2$, then we take one additional observation X_{m+2} and update the sample variance by S_{m+2}^2 based on $X_1, \ldots, X_{m+1}, X_{m+2}$. This continues until for the first time we arrive at a sample size n which is at least as large as the associated estimator of C, namely $a^2 S_n^2/d^2$, constructed from $X_1, \ldots, X_n$.

We should note one crucial point here. Suppose, for example, $m = 10$ and the sequential sampling stopped with $N = 15$. What this means is that we *did not stop* with 10, 11, 12, 13 or 14 observations, but we *stopped* with 15 observations. In other words, one should realize that

> the event $\{N = 15\}$ *is not* the same as the event $\{15 \geq a^2 S_{15}^2/d^2\}$. Instead, the event $\{N = 15\}$ *is* the same as the event $\{n < a^2 S_n^2/d^2$ for $n = 10, 11, 12, 13, 14$, but $15 \geq a^2 S_{15}^2/d^2\}$.

This is the reason why the exact probability distribution of N is hard to find even if σ^2 were known! Starr (1966a) gave an algorithm, originally stated in Robbins (1959), which provided an exact distribution of N when C was small.

We may summarize the following results. For all fixed μ, σ, d and α, one can conclude:

The sampling terminates with probability one, that is $\mathrm{P}_{\mu,\sigma}(N < \infty) = 1$; $\mathrm{E}_{\mu,\sigma}(N)$ and $\mathrm{V}_{\mu,\sigma}(N)$ are finite; *$\overline{X}_N$ is an unbiased estimator of μ and its variance is $\sigma^2\mathrm{E}_{\mu,\sigma}(N^{-1})$.* (6.2.18)

Once the purely sequential procedure (6.2.17) stops sampling, we propose the fixed-width confidence interval $J_N = [\overline{X}_N \pm d]$ for μ where $\overline{X}_N$ is the sample mean based on N and $X_1, \ldots, X_N$.

Observe that for all fixed $n(\geq m)$, the event $\{N = n\}$ depends only on the random vector variable $(S_m^2, S_{m+1}^2, \ldots, S_n^2)$, but the sample mean $\overline{X}_n$ and $(S_m^2, S_{m+1}^2, \ldots, S_n^2)$ are independent. So, any event defined only through $\overline{X}_n$ must be independent of the event $\{N = n\}$. See Exercises 6.2.1-6.2.2 in this context. Hence, along the lines of (6.2.8), we express the confidence coefficient associated with J_N as

$$\mathrm{P}_{\mu,\sigma}\{\mu \in J_N\} = \mathrm{E}_{\mu,\sigma}\left[2\Phi\left(\sqrt{N}d/\sigma\right) - 1\right], \tag{6.2.19}$$

where N is the purely sequential stopping time from (6.2.17). From Chow and Robbins (1965), we summarize the following asymptotic properties:

$$\begin{aligned} &\lim_{d\to 0}\mathrm{E}_{\mu,\sigma}\left(\tfrac{N}{C}\right) = 1 \quad [\textit{Asymptotic Efficiency or}\\ &\qquad\qquad\qquad\qquad \textit{Asymptotic First-Order Efficiency}];\\ &\lim_{d\to 0}\mathrm{P}_{\mu,\sigma}\{\mu \in J_N\} = 1 - \alpha \quad [\textit{Asymptotic Consistency}]. \end{aligned} \tag{6.2.20}$$

Now, we contrast the properties of the two-stage and purely sequential procedures. From (6.2.14) it was clear that asymptotically the two-stage procedure (6.2.6) oversampled on an average compared with C. But, from the first part of (6.2.20) we conclude that the purely sequential procedure (6.2.17) samples on an average close to C observations. This property of a sequential methodology is described as *asymptotic efficiency*:

In practice, *asymptotic efficiency* or *asymptotic first-order efficiency* means: $\mathrm{E}_{\mu,\sigma}\left(\frac{N}{C}\right) \approx 1$ in the large-sample case. (6.2.21)

This is a good property to have, but notice that we have now claimed *only* asymptotic consistency property in (6.2.20) and *not* the exact consistency property of the Stein procedure. So, in a purely sequential procedure, we can only claim that for a large sample size, the confidence coefficient will be in the vicinity of the target, $1 - \alpha$.

The major success of Stein's (1945,1949) two-stage procedure is intertwined with its exact consistency property (Theorem 6.2.2).

Table 6.2.1: Asymptotic properties of two-stage and purely sequential procedures

Property	Two-Stage Procedure (6.2.6)	Purely Sequential Procedure (6.2.17)
(Exact) Consistency	Yes	No
Asymptotic Consistency	Yes	Yes
Asymptotic Efficiency	No	Yes

On the other hand, the purely sequential strategy does not oversample as much but it fails to deliver the most crucial exact consistency property. This situation has bothered many researchers. Simons (1968) proved the following remarkable properties associated with the purely sequential methodology (6.2.17). For all fixed μ, σ, d and α, one can claim:

(i) $\mathrm{E}_{\mu,\sigma}(N) \le C + m$;

(ii) There *exists* a *non-negative* integer k such that if one obtains k additional observations beyond the termination of the rule (6.2.17), and then constructs the confidence interval $J_{N+k} = [\overline{X}_{N+k} \pm d]$ based on N and $X_1, \ldots, X_N, \ldots, X_{N+k}$, then $\mathrm{P}_{\mu,\sigma}\{\mu \in J_{N+k}\} \ge 1 - \alpha$. Here, k does *not* depend upon μ, σ and d (6.2.22)

Let us understand these important results. In the beginning, we were ignorant about σ and so we started with a pilot of size m. The property (i) from (6.2.22) is remarkable. It shows that once we recover from this initial stage of ignorance (about σ), by updating the information about σ successively, we end up taking on an average not more than C observations. That is, apart from the pilot samples, the purely sequential strategy does at least as well (on an average) as the optimal fixed-sample-size strategy. Hence, Simons (1968) referred to m as the "cost of ignorance of σ".

The Exercise 6.2.4 shows how to prove a somewhat weaker result: $\mathrm{E}_{\mu,\sigma}(N) \le C + m + 2$, for all fixed μ, σ, and α

Property (ii) from (6.2.22) is also remarkable and let us explain why it is so. Suppose that one finds a non-negative integer k as in Simons (1968), and based on N and $X_1, \ldots, X_N, \ldots, X_{N+k}$ one proposes the confidence interval $J_{N+k} = [\overline{X}_{N+k} \pm d]$ for μ. This is a revised version of the original sequential procedure (6.2.17) due to Simons. Then, we may summarize by saying:

> Simons' (1968) procedure is *consistent*, *asymptotically consistent* and *asymptotically efficient.* It may appear that this revised procedure has captured the best properties of both two-stage and sequential procedures. *Not quite,* because the magnitude of the non-negative integer k remains *unknown*!

Simons (1968) had shown the *existence* of such a non-negative integer k which gave the property (ii) in (6.2.22) with the help of very intricate arguments. We leave these derivations out for brevity.

Computations and Simulations

After selecting procedure number 2 from program number 2 within **Seq06.exe**, one can obtain a fixed-width confidence interval for μ using a purely sequential procedure. The same selection can also be used to perform a simulation study of the purely sequential procedure. Figures 6.2.5 and 6.2.6 respectively show output of a simulation study and for a set of live input data. An added feature in this part of the program is that the user can input additional observations at the end of the sequential procedure and recompute the confidence interval including the additional observations. This would overcome the undersampling problem that is often inherent in a sequential procedure.

Now, we draw one's attention to (6.2.33) and (6.2.34) where we have defined an expression $\eta(p), p = 1, 2, \ldots$. Clearly, we have $\eta(1) < 0$. Hence, for the sequential procedure (6.2.17), we obtain from (6.2.34):

$$\mathrm{E}_{\mu,\sigma}(N - C) < 0 \text{ and } \mathrm{P}_{\mu,\sigma}\{\mu \in J_N\} < 1 - \alpha.$$

This explains why one may want to add extra observations at the conclusion of a purely sequential procedure. The simulations handles this situation by adding extra observations after termination so that the coverage probability may get closer to $(1 - \alpha)$. Currently, the program has a set upper limit of 50 additional observations beyond termination. However, in most cases, only a few additional observations may be required. For the confidence interval exercise seen in Figure 6.2.5, on an average seven additional observations were needed to achieve the preset confidence level.

When a fixed-with confidence interval is constructed using live data, the program will accept observations from a nominated ASCII data file. This ASCII data file name ought to have nine or fewer characters. Alternatively, one may input observations one at time using the keyboard. If the data file does not include a sufficiently long series of observations, the program will give an option to enter more data using the keyboard. At the end of computation of the confidence interval, the user may nominate the number of extra observations needed to overcome the undersampling problem. Figure 6.2.6 first shows the computed 90% confidence interval with fixed-width 2. Then, it highlights recomputation of the confidence interval after adding 5 new observations upon termination. The 5 additional input observations are also shown in Figure 6.2.6.

```
           Fixed-Width Confidence Interval
          Using Purely Sequential Procedure
          =================================

Simulation study for  90% confidence interval for
the normal mean. Data were generated from normal
distribution with mean = 25.00 and variance =  9.00

Number of simulation replications      =  1000
Width of the confidence interval (2d) =  2.00

Initial sample size =  5

    Optimal sample size (C)          =   24.35

    Average sample size (n_bar)      =   21.43
    Std. dev. of n (s)               =    8.80
    Minimum sample size (min n)      =    5
    Maximum sample size (max n)      =   42

    Coverage probability (p_bar)     = 0.842
    Standard error of p_bar          = 0.012

    k  = No. observations added
    cp = Coverage probability
    se = Standard error of the coverage probability

             k    cp      se
             1  0.861  0.011
             2  0.862  0.011
             3  0.886  0.010
             4  0.891  0.010
             5  0.887  0.010
             6  0.898  0.010
             7  0.900  0.009
```

Figure 6.2.5. *Simulated sample sizes and additional k observations.*

```
          Fixed-Width Confidence Interval
         Using Purely Sequential Procedure
         =================================

Initial sample size = 10

Initial Sample:
  1.60   2.20   2.00   4.20  11.30   5.10   2.80  -3.50
  3.70   5.90

Sequential Data:
  9.10   1.60   7.60   6.40  11.10   8.80  -5.60   2.00
  2.70   8.60   6.60  14.20   0.60   0.50   3.80  10.80
  5.50   3.90  -2.20   4.30   6.60  -4.40   5.70   7.90
  9.30   9.70   9.90   4.70  11.10  -1.30  10.40  12.20
 11.40  -0.60   2.20  -0.30   8.10   4.60   9.70  -0.40
  5.90   1.80  -0.20   6.60   6.40   5.70


  Sample size =    56
  Sample mean =  4.97
  Sample variance = 20.49

  90% confidence interval for the mean is:
                 (3.97,  5.97)

    Do you want add extra observations and
    re-compute the confidence interval
    Enter Y(yes) or N(no): Y

    Enter number of observations, k = 5

    Additional data entered:
      1.60  6.60  6.50  7.80  8.70

    Sample size =    61
    Sample mean =  5.07
    Sample variance = 19.41

     90% confidence interval for the mean is:
                       ( 4.07,  6.07)
```

Figure 6.2.6. *Purely sequential fixed-with confidence interval using given data.*

6.2.3 Modified Two-Stage Procedure

Is there a two-stage procedure which will capture the best properties of both two-stage and sequential procedures? Is there a two-stage procedure which is consistent, asymptotically consistent, and asymptotically efficient at the same time? The answer is in the affirmative. Mukhopadhyay (1980) developed a two-stage procedure that is summarized as follows.

Recall that $C = a^2\sigma^2/d^2$ where σ^2 is unknown. Let us choose and fix $0 < \gamma < \infty$ and define a pilot sample size

$$m \equiv m(d) = \max\left\{2, \left\langle \left(\frac{a}{d}\right)^{2/(1+\gamma)} \right\rangle + 1\right\}. \tag{6.2.23}$$

Once m is determined, we start with the pilot observations $X_1, \ldots, X_m$. Next, we define the final sample size

$$N \equiv N(d) = \max\left\{m, \left\langle \frac{a_{m-1}^2 S_m^2}{d^2} \right\rangle + 1\right\}, \tag{6.2.24}$$

and propose the fixed-width interval $J_N = \left[\overline{X}_N \pm d\right]$ for μ.

This modified two-stage procedure is implemented in exactly the same fashion as the original Stein procedure (6.2.6). But, there is one important difference. We have now given a specific choice for a pilot size m.

For fixed μ, σ, d and α, the modified two-stage procedure (6.2.23) - (6.2.24) would behave the same way as the Stein procedure (6.2.6) because then m from (6.2.23) remains fixed.

Hence, the modified two-stage procedure is *consistent.*

Construction of Confidence Intervals and the Role of γ

From (6.2.23) it is evident that the pilot size m is inversely affected by γ. Program number 4 in **Seq06.exe** evaluates m for given α, d and γ. Theoretically, γ may be assigned any finite positive value. But, for γ large enough, (6.2.23) will give $m = 2$ since eventually $(a/d)^{2/(1+\gamma)}$ will go under 2. Therefore, to obtain a suitable m, one should fix a reasonably small but positive value for γ. Figure 6.2.7 shows the computed m values using this program when $\alpha = 0.01$ and $d = 0.5$. The purpose of the program is to help a user to choose a suitable value of γ.

```
Role of gamma in the Modified Two-Stage Procedure
-------------------------------------------------
      Input values:
                 minimum  maximum  Number
          alpha    0.01     0.01      1
          d        0.50     0.50      1
          gamma    0.10     1.00     10
          -------------------------------

          alpha       d       gamma     m
          0.010     0.50      0.10     20
          0.010     0.50      0.20     16
          0.010     0.50      0.30     13
          0.010     0.50      0.40     11
          0.010     0.50      0.50      9
          0.010     0.50      0.60      8
          0.010     0.50      0.70      7
          0.010     0.50      0.80      7
          0.010     0.50      0.90      6
          0.010     0.50      1.00      6
```

Figure 6.2.7. *Initial sample size m versus γ.*

After deciding on a suitable value for γ, the user can obtain a confidence interval using the modified two-stage procedure by selecting procedure number 3 within program number 2. This selection will allow the user to obtain a fixed-width interval either with live data or simulations. Figure 6.2.8 shows results from a simulation study for a 99% confidence interval with $2d = 1$ and $\gamma = 0.2$. Data were generated from a $N(25, 16)$ distribution.

```
        Fixed-Width Confidence Interval
      Using Modified Two-Stage Procedure
      ==================================

Simulation study for  99% confidence interval for
the normal mean. Data were generated from normal
distribution with mean = 25.00 and variance = 16.00

Number of simulation replications      =  2000
Width of the confidence interval (2d) =  1.00

Initial sample size =  16
is obtained using gamma =  0.20

    Optimal sample size (C)          =  424.63

    Average sample size (n_bar)      =  559.76
    Std. dev. of n (s)               =  199.26
    Minimum sample size (min n)      =   86
    Maximum sample size (max n)      = 1375

    Coverage probability (p_bar)     = 0.994
    Standard error of p_bar          = 0.002
```

Figure 6.2.8. *Simulated sample size for modified two-stage procedure.*

How About the Large-Sample Properties?

First, let us explain the role of the particular choice of m. It is easy to see that

$$\lim_{d \to 0} m = \infty, \text{ but}$$

$$\lim_{d \to 0} \left(\frac{m}{C}\right) = \sigma^{-2} \lim_{d \to 0} \left(\frac{d}{a}\right)^{2\gamma/(1+\gamma)} = 0. \tag{6.2.25}$$

In other words, when d is small, both C and N would grow, and hence we should allow m to go up. If N is expected to be large, what is the point in starting with 10 or 15 observations? Instead, if N is expected to be large, it may be more sensible to work with a large value of m. But, then the question is how large should we allow m to be? Certainly, we cannot allow m to exceed C which is unknown!

The second part in (6.2.25) guarantees that m stays low compared with

C even though both m, C increase as d becomes small. Also, for large degrees of freedom, the Student's t distribution is approximated by a standard normal distribution. This leads to:

$$\text{As } d \to 0, \text{ we can claim that } a_{m-1} \to a. \tag{6.2.26}$$

From Theorem 6.2.1, part (i), we recall that

$$\frac{a_{m-1}^2 \sigma^2}{d^2} \le \mathrm{E}_{\mu,\sigma}(N) \le m + \frac{a_{m-1}^2 \sigma^2}{d^2}. \tag{6.2.27}$$

It can be rewritten as:

$$\frac{a_{m-1}^2}{a^2} \le \mathrm{E}_{\mu,\sigma}\left(\frac{N}{C}\right) \le \frac{m}{C} + \frac{a_{m-1}^2}{a^2}. \tag{6.2.28}$$

Next, taking limits throughout (6.2.28) as $d \to 0$ and combining with (6.2.26) - (6.2.27), we obtain the following result:

$$\lim_{d\to 0} \mathrm{E}_{\mu,\sigma}\left(\frac{N}{C}\right) = 1 \quad [\textit{Asymptotic Efficiency or Asymptotic First-Order Efficiency}]; \tag{6.2.29}$$

We leave it as an exercise to show that

$$\lim_{d\to 0} \mathrm{P}_{\mu,\sigma}\{\mu \in J_N\} = 1 - \alpha \quad [\textit{Asymptotic Consistency}]. \tag{6.2.30}$$

One should note that the modified two-stage procedure has combined the best properties of the Stein and the sequential procedures. On the other hand, operationally a modified two-stage procedure is certainly more convenient than one-by-one purely sequential strategy.

So, a natural question is this: What is the gain if we implement a sequential strategy instead of a modified two-stage scheme? What we have seen so far seems to indicate that a modified two-stage procedure should be preferred on the basis of its properties and operational simplicity. So, what is that property which gives a sequential procedure a definite edge?

Ghosh and Mukhopadhyay (1981) investigated this issue and originated the concept of *asymptotic second-order efficiency* property to draw a line between the modified two-stage and sequential strategies. Let us firm up some elementary ideas before anything else.

Definition 6.2.1: *Let $g(d)$ be a real valued function of d. We say that $g(d) = O(d^\kappa)$ as $d \to 0$ if and only if $\lim_{d\to 0}(g(d)/d^\kappa)$ is finite. We say that $g(d) = o(d^\kappa)$ as $d \to 0$ if and only if $\lim_{d\to 0}(g(d)/d^\kappa)$ is zero.*

Example 6.2.1: Suppose that $g(d) = \sigma d^2 + 2\sqrt{\alpha} d$. When $d \to 0$, the term involving d^2 becomes negligible faster than the term involving d. So, for small d, we would say that $g(d) \approx 2\sqrt{\alpha} d$. We may also say that $g(d) = O(d)$. One may also claim, for example, that $g(d) = o(\sqrt{d})$.

Now, let us use Cornish-Fisher expansion (Johnson and Kotz, 1970, p.

102) of a_{m-1}, a Student's t percentage point, in terms of a, the standard normal percentage point. For small d, or equivalently, large m:

$$a_{m-1} = a + \frac{a(a^2+1)}{4(m-1)} + O(m^{-2}). \tag{6.2.31}$$

Hence, from the lower bound in (6.2.28) we can write

$$\begin{aligned} \lim_{d\to 0}\inf \mathrm{E}_{\mu,\sigma}(N-C) &\geq \lim_{d\to 0}\left\{\left(a_{m-1}^2 - a^2\right)\frac{\sigma^2}{d^2}\right\} \\ &= O\left(d^{-2\gamma/(1+\gamma)}\right) \\ &= +\infty \text{ as } d \to 0. \end{aligned} \tag{6.2.32}$$

Now, contrast (6.2.29) with (6.2.32). For the modified two-stage procedure, we know that $\mathrm{E}_{\mu,\sigma}\left(\frac{N}{C}\right) \approx 1$ for small d. But, the difference between N and C on an average, namely $\mathrm{E}_{\mu,\sigma}(N-C)$, explodes for small d!

What happens in the case of a purely sequential procedure? It turns out that $\lim_{d\to 0}\mathrm{E}_{\mu,\sigma}(N-C)$ is a small finite number. That is, the difference between N and C on an average is tightly held for small values of d.

Definition 6.2.2 : *A procedure is called asymptotically first-order efficient if and only if* $\lim_{d\to 0}\mathrm{E}_{\mu,\sigma}\left(\frac{N}{C}\right) = 1$. *A procedure is called asymptotically second-order efficient if and only if* $\lim_{d\to 0}\mathrm{E}_{\mu,\sigma}(N-C)$ *is finite.*

The asymptotic second-order efficiency property is stronger than the asymptotic first-order efficiency property.

An asymptotically second-order efficient procedure certainly enjoys asymptotically first-order efficiency property, but the converse is not necessarily true. Note that a modified two-stage procedure is asymptotically first-order efficient, but it is *not* asymptotically second-order efficient.

Woodroofe (1977,1982), Lai and Siegmund (1977,1979) and Siegmund (1985) developed the machinery of *non-linear renewal theory.* Such powerful tools are essential to check desirable second-order properties. One is also referred to Mukhopadhyay and Solanky (1994, Chapter 2) and Ghosh et al. (1997) for many details.

Let us denote:

$$\begin{aligned} h(p) &= \textstyle\sum_{n=1}^{\infty}\frac{1}{n}\mathrm{E}\left[\max\left\{0, \chi_{np}^2 - 2np\right\}\right], \\ \eta(p) &= \tfrac{1}{2} - \tfrac{1}{p} - \tfrac{1}{p}h(p), \\ f(x;p) &= \left[2^{p/2}\Gamma(p/2)\right]^{-1} e^{-x/2}x^{(p-2)/2}, \\ &\quad \text{for } x>0 \text{ and } p = 1,2,\ldots. \end{aligned} \tag{6.2.33}$$

Notice that $f(x;p)$ is the p.d.f. of a χ_p^2 random variable. The following results are now summarized from Woodroofe (1977) for the purely sequential procedure (6.2.17). As $d \to 0$:

$$
\begin{aligned}
\mathrm{E}_{\mu,\sigma}(N-C) &= \eta(1) + o(1) = -1.1825 + o(1) \text{ if } m \geq 4; \\
\frac{N-C}{\sqrt{C}} &\text{ converges to } N(0,2) \text{ in distribution;} \\
\mathrm{V}_{\mu,\sigma}(N) &= 2C + o(C) \text{ if } m \geq 4; \\
\mathrm{P}_{\mu,\sigma}\{\mu \in J_N\} &= 1 - \alpha + \frac{a^2}{2C}\left\{2\eta(1) - 1 - a^2\right\} f(a^2;1) \\
&\quad + o(C^{-1}) \text{ if } m \geq 7.
\end{aligned}
\tag{6.2.34}
$$

From (6.2.34) we can make the following conclusion.

> The purely sequential procedure (6.2.17) is both *asymptotically consistent* and *asymptotically second-order efficient.*

For a small number d, that is for a large sample size, let us interpret these expansions and and explain how they may guide one in practice. From the first part, we can say that $\mathrm{E}_{\mu,\sigma}(N-C)$ will lie in a close proximity of $\eta(1)$ which is a computable number. The third part says that $\mathrm{V}_{\mu,\sigma}(N)$ should be nearly $2C$ while we already know $\mathrm{E}_{\mu,\sigma}(N)$ is expected to be around $C + \eta(1)$. The normalized stopping variable, namely $\frac{N-C}{\sqrt{C}}$, would have an approximate normal distribution. From the last expression in (6.2.34), we get an idea about the departure of the confidence coefficient from its target $1 - \alpha$.

Program number 5 within **Seq06.exe** would help one to explore these remarks further. This program evaluates theoretical expressions such as $\eta(1)$ and $f(a^2;1)$. It computes n_bar $-C$, the difference between the average simulated sample size $\overline{n}$ and C given α, d and σ^2. The n_bar $-C$ values ought to be compared with $\eta(1)$. These should be reasonably close to each other.

Further, this program evaluates the second-order term,

$$
\text{T\_so} = \frac{a^2}{2C}\left\{2\eta(1) - 1 - a^2\right\} f(a^2;1),
$$

found in (6.2.34) and also computes P_diff $= \overline{p} - (1-\alpha)$ where $\overline{p}$ is the average simulated coverage probability. T_so values ought to be compared with P_diff values. These should be reasonably close to each other.

Results shown in Figure 6.2.9 are largely consistent with general sentiments expressed earlier. As sample size increased, admittedly the findings became more consistent with general sentiments as expected. For larger sample sizes, al the values n_bar $-C$, P_diff, and T_so came out negative validating the theory. The P_diff and T_so values appear rather close to each other as one should expect. Also, the n_bar $-C$ values and $\eta(1)$ are not too far once one factors in the estimated standard deviations of $\overline{n}$ validating the theory. It is also clear that the estimated values of $\mathrm{V}_{\mu,\sigma}(N)/C$ stays close to 2 even for a moderate value of $C(= 109.27)$.

```
            Theoretical and Simulated Second-Order Terms
                 of the Purely Sequential Procedure
                 ----------------------------------

            Input values:
                        minimum   maximum   Number
                 sigma     1.00      4.00        2
                 alpha    0.050     0.050        1
                 d         0.25      1.00        4
                 -------------------------------

           Number of Simulation Replications =  2000
           Initial Sample size (m) = 10
           eta(1) =    -1.183

           P_diff   = p_bar-(1-alpha)
           T_so     = second-order term

  sigma = 1.00;      alpha =0.050;    f(a^2;1) = 0.0298

     d       C    n_bar     var_n  n_bar-C  var_n/C   P_diff      T_so
  0.25   61.46    60.00    144.16    -1.46   2.3454  -0.0140   -0.0067
  0.50   15.37    15.07     22.01    -0.30   1.4327  -0.0050   -0.0269
  0.75    6.83    10.35      0.96     3.52   0.1401   0.0300   -0.0604
  1.00    3.84    10.01      0.01     6.16   0.0016   0.0475   -0.1075

  sigma = 4.00;      alpha =0.050;    f(a^2;1) = 0.0298

     d       C    n_bar     var_n  n_bar-C  var_n/C   P_diff      T_so
  0.25  983.41   981.97   2050.07    -1.44   2.0846  -0.0035   -0.0004
  0.50  245.85   245.23    506.27    -0.62   2.0592  -0.0005   -0.0017
  0.75  109.27   108.03    239.14    -1.24   2.1885  -0.0090   -0.0038
  1.00   61.46    59.89    154.03    -1.58   2.5060  -0.0125   -0.0067
```

Figure 6.2.9. *Second-order terms of the purely sequential procedure.*

The reader should explore more of these and other features by running program number 5 within **Seq06.exe**. That will truly help in understanding many intricate issues.

6.2.4 Accelerated Sequential Procedure

We have seen that the purely sequential procedure is asymptotically second-order efficient. But, it can be inconvenient to record observations one-by-one until the process terminates. Thus, it will be useful in practice if we can offer a methodology terminating *quickly* with its sample size comparing favorably with that for a purely sequential strategy. That is, we would like to preserve second-order properties, but have some added operational convenience over a sequential strategy.

An accelerated sequential procedure will really fit the bill. This methodology first proceeds purely sequentially but continues only part of the way, followed by a batch of the remaining observations gathered in one step. Let us remember this: The final sample size N has to be close to C on an average! It should be understood that we are not about to reduce sample size. We merely wish to reach a sample size close to C quicker with fewer sampling operations compared with a sequential strategy.

Hall (1983) first developed an accelerated sequential estimation technique for this problem. Mukhopadhyay and Solanky (1991) built a unified theory of accelerated sequential sampling. Later, Mukhopadhyay (1996) proposed a simpler and improved accelerated sequential sampling technique. Now, we present this improved sampling strategy from Mukhopadhyay (1996).

Again, recall that $C = a^2\sigma^2/d^2$ where σ^2 is unknown. Let us choose and fix a number $\rho, 0 < \rho < 1$. We start with $X_1, \ldots, X_m$, a pilot sample of size $m(\geq 2)$ and followed by one additional X at a time as needed. The stopping time is defined as follows:

$$\begin{aligned} &t \equiv t(d) \text{ is the smallest integer } n(\geq m) \\ &\text{for which we observe } n \geq \rho \tfrac{a^2 S_n^2}{d^2}. \end{aligned} \tag{6.2.35}$$

We continue sampling purely sequentially until we arrive at the observations $X_1, \ldots, X_t$. But, observe the boundary condition in (6.2.35) and compare it with that in the original purely sequential procedure (6.2.17). In (6.2.17), N estimated C, and hence the present stopping variable t must be estimating ρC, a certain fraction of C.

At this point, we should estimate C with $\rho^{-1}t$. This would clearly indicate how many additional observations we may need beyond $X_1, \ldots, X_t$ so that the final sample size N may stay in a vicinity of C. The "additional observations" are gathered at the same time in one single batch, thus reducing the operational time by approximately $100(1-\rho)\%$. One can visualize that if the original purely sequential procedure (6.2.17) takes h minutes to quit sampling, then the present sampling strategy will require nearly ρh minutes to quit sampling! The explicit formulation and the associated results follow.

Having obtained t from (6.2.35), we define the final sample size as:

$$N \equiv N(d) = \left\langle \frac{t}{\rho} + q \right\rangle + 1 \text{ with } q = \frac{1}{2\rho}(5 + a^2). \tag{6.2.36}$$

and we gather the additional $(N-t)$ observations $X_{t+1}, \ldots, X_N$, all in *one batch*. The final data would consist of $X_1, \ldots, X_t, X_{t+1}, \ldots, X_N$. Then, we propose the fixed-width confidence interval $J_N = [\overline{X}_N \pm d]$ for μ. One can show that N is finite with probability one, and thus $\overline{X}_N$ and J_N are genuine estimators.

If ρ is chosen near zero, the accelerated procedure would clearly behave more like Stein's two-stage procedure. But, if ρ is chosen near one, the accelerated procedure would behave more like a purely sequential procedure.

Thus, an accelerated procedure is often implemented with $\rho = 0.4, 0.5$ or 0.6. In numerous problems, one tends to use $\rho = 0.5$.

Hall (1983) and Mukhopadhyay and Solanky (1991) gave asymptotic second-order results for an analogous but slightly more complicated acceleration technique. For accelerated sequential sampling in (6.2.35)-(6.2.36), we begin with some *asymptotic first-order* results, summarized from Mukhopadhyay (1996):

$$\begin{aligned}
&\lim_{d\to 0} \mathrm{E}_{\mu,\sigma}\left(\frac{N}{C}\right) = 1 \quad [\textit{Asymptotic Efficiency or} \\
&\qquad\qquad\qquad\qquad \textit{Asymptotic First-Order Efficiency}]; \\
&\lim_{d\to 0} \mathrm{P}_{\mu,\sigma}\{\mu \in J_N\} = 1 - \alpha \quad [\textit{Asymptotic Consistency}].
\end{aligned} \tag{6.2.37}$$

The *asymptotic second-order* results are little more cumbersome, and these are also summarized from Mukhopadhyay (1996) in the following.
As $d \to 0$:

$$\begin{aligned}
q - \frac{2}{\rho} + o(1) \le \mathrm{E}_{\mu,\sigma}(N - C) \le q - \frac{2}{\rho} + 1 + o(1) \text{ if } m \ge 6; \\
\frac{N - C}{\sqrt{C}} \text{ converges to } N(0, \frac{2}{\rho}) \text{ in distribution}; \\
\mathrm{V}_{\mu,\sigma}(N) = \frac{2}{\rho}C + o(C) \text{ if } m \ge 4; \\
\mathrm{P}_{\mu,\sigma}\{\mu \in J_N\} \ge 1 - \alpha + o(C^{-1}) \text{ if } m \ge 7.
\end{aligned} \tag{6.2.38}$$

From (6.2.38), we may summarize the basic findings as follows:

> The accelerated sequential procedure (6.2.35)-(6.2.36) is *asymptotically consistent* and *asymptotically second-order efficient.*

If d is small, that is when the sample size is large, we can explain how these expansions may guide one in practice. In the case of the first part, we can say that $\mathrm{E}_{\mu,\sigma}(N - C)$ may lie in a close proximity of $q - \frac{2}{\rho}(> 0)$ which is a computable number. The third part suggests that $\mathrm{V}_{\mu,\sigma}(N)$ may be nearly $\frac{2}{\rho}C$. The normalized stopping variable, namely $\frac{N-C}{\sqrt{C}}$, would have an approximate normal distribution. From the last expression in (6.2.38), we get an idea about the departure of the confidence coefficient from its target $1 - \alpha$.

Right away, we want everyone to realize an important point. The variance of N under accelerated sampling is approximately $\frac{2}{\rho}C$ which is strictly larger than the variance ($\approx 2C$) of N under sequential sampling. This may sound like bad news, but the news in itself should surprise no one. The increased variance under accelerated sampling is a direct result of reducing sampling operations. In practice, one should balance the gain from operational convenience and the added uncertainty from increased variability in N. Unfortunately, there is no one single or "optimal" way to achieve any desired balance.

Program number 6 within **Seq06.exe** would help one to explore these remarks further.

```
Simulated Second-Order Terms of the
  Accelerated Sequential Procedure
 ---------------------------------

Input values:
                minimum   maximum   number
       sigma     4.000     4.000       1
       alpha     0.050     0.050       1
       d          0.50      1.00       2
       rho        0.20      0.80       4
       ---------------------------------

Number of simulation replications = 2000
Initial sample size (m) = 10

   P_diff    = p_bar-(1-alpha)

sigma = 4.00; alpha = 0.05; d = 0.50; C = 245.85

 rho    n_bar    var_n  n_bar-C  var_n/C   P_diff
0.20  261.30  3266.72    15.45  13.2873  -0.0105
0.40  252.23  1397.41     6.37   5.6839  -0.0015
0.60  251.34   884.22     5.49   3.5965  -0.0095
0.80  251.04   592.56     5.18   2.4102   0.0040

sigma = 4.00; alpha = 0.05; d = 1.00; C =  61.46

 rho    n_bar    var_n  n_bar-C  var_n/C   P_diff
0.20   87.20   272.13    25.73   4.4275   0.0310
0.40   69.21   356.64     7.75   5.8026   0.0070
0.60   65.90   293.03     4.43   4.7676   0.0040
0.80   64.64   230.17     3.18   3.7449  -0.0010
```

Figure 6.2.10. *Second-order terms of the accelerated sequential procedure.*

Figure 6.2.10 shows an output from the said program for $\alpha = 0.05$, $d = 0.5, 1.0, \sigma = 4$ and $\rho = 0.2, 0.4, 0.6, 0.8$. The program obtains the values of n_bar -C, P_diff, and var_n/C in the same spirit as it did so in the purely sequential case. But, now n_bar - C ought to be compared with q^* and q^*+1 where $q^* \equiv q - 2\rho^{-1} = \frac{1}{2}\rho^{-1}(1+a^2)$. We find $q^* = 12.104, 6.052, 4.035$ and 3.026 when $\rho = 0.2, 0.4, 0.6$ and 0.8 respectively. The second-order result cited in (6.2.38) suggests that the values of n_bar -C may be expected to lie in a close proximity of the interval $(q^*, q^* + 1)$ when the sample size is large. Overall, this appears to be the case, especially when the estimated standard deviation of $\overline{n}$ is factored in. Clearly, an accelerated procedure would tend to oversample on an average compared with C, but the second-order term q^* seems to provide a good check on the magnitude of oversampling.

Also, ρ var_n values may be checked against $2C$. From Figure 6.2.10, however, we find that ρvar_n values do not appear to be close to $2C$. From the column of values for var_n/C shown in Figure 6.2.10, one will have the same feeling. We guess that the rate of convergence that may lead one to suggest something like ρvar_n $\approx 2C$ is rather slow!

We note that all P_diff ($= \overline{p} - 0.95$) values are near zero, but P_diff value has turned out negative a number of times. That is, the estimated coverage probability $\overline{p}$ fell under the target (=0.95) a number of times. We saw a similar feature for a sequential procedure earlier.

The reader should explore more of these and other features by running program number 6 within **Seq06.exe**. That will truly help in understanding many intricate issues.

6.2.5 Three-Stage Procedure

In the previous section, we learned that an accelerated procedure was asymptotically second-order efficient while it saved approximately $100(1-\rho)\%$ sampling operations compared with a sequential strategy. But, an accelerated strategy may appear inconvenient sometimes too because it starts with one-by-one sequential sampling. It may be useful in practice to design a methodology that will terminate "very quickly" with a sample size comparing favorably with that for accelerated sampling. That is, we will prefer (i) preserving second-order properties and (ii) having additional operational convenience beyond simple acceleration.

We will show that a three-stage sampling procedure fits the bill. This methodology starts with a pilot sample of size m. With the help of pilot data, one would estimate ρC, a fraction of C. Up to this step, this resembles two-stage sampling. Next, using the data from both first and second stages, one will estimate C and sample the difference as needed in the third stage. Using the combined dataset from all three stages, one will now propose a fixed-width interval for μ. This way, we wish to reach a sample size N close to C more quickly compared with an accelerated sequential strategy.

The basic idea of triple sampling was first introduced in Mukhopadhyay (1976b) to obtain fixed-width confidence intervals for the mean of a normal distribution when the variance was unknown. Triple sampling estimation strategies were taken to a new height by Hall (1981). These were designed by combining the operational savings made possible via batch sampling in three steps and the efficiency of sequential procedures. A unified framework was developed by Mukhopadhyay (1990) for general triple sampling procedures. He also laid out the theory for their associated asymptotic second-order properties. See also Mukhopadhyay and Solanky (1994, Chapter 2) and Ghosh et al. (1997, Chapter 6) for many details.

Again, we recall that $C = a^2\sigma^2/d^2$ where σ^2 is unknown. Let us choose and fix a number $\rho, 0 < \rho < 1$. We begin with $X_1, \ldots, X_m$, a pilot of size $m(\geq 2)$, and then we define:

$$T \equiv T(d) = \max\left\{m, \left\langle \frac{\rho a^2 S_m^2}{d^2} \right\rangle + 1\right\}. \tag{6.2.39}$$

Observe that T estimates ρC, a fraction of C. If $T = m$, we do not take any more observations at the second stage. But, we sample the difference $T - m$ at the second stage if $T > m$. Denote the combined data from both stages, $X_1, \ldots, X_T$.

At this point, one would estimate σ^2 and C by S_T^2 and $a^2 S_T^2 d^{-2}$ respectively. More precisely, we define the final sample size:

$$N \equiv N(d) = \max\left\{T, \left\langle \frac{a^2 S_T^2}{d^2} + \varepsilon \right\rangle + 1\right\} \text{ with } \varepsilon = \frac{5 + a^2 - \rho}{2\rho}. \quad (6.2.40)$$

Observe that N estimates C. The third leg of the three-stage strategy (6.2.39)-(6.2.40) is implemented as follows.

If $N = T$, that is when T is larger than the second-stage estimator of C, then it indicates that we have already taken too many observations in the first two stages. Hence, if $N = T$, we do not need any more observations at the third stage.

But, if $N > T$, that is when T is smaller than the second-stage estimator of C, then it indicates that we have recorded too few observations in the first two stages combined. Hence, if $N > T$, we sample the difference $N - T$ at the third stage to compensate for the deficiency. In other words, if $N > T$, we take additional observations $X_{T+1}, \ldots, X_N$ at the third stage.

Combining observations from all three steps in this strategy, we write down the final dataset, N and $X_1, \ldots, X_N$. These provide the sample mean $\overline{X}_N$ and the associated fixed-width interval $J_N = \left[\overline{X}_N \pm d\right]$. It is easy to see that N is finite with probability one, and thus $\overline{X}_N$ and J_N are genuine estimators.

If ρ is chosen near zero or one, a three-stage procedure would clearly behave more like Stein's two-stage procedure. Thus, a three-stage procedure is often implemented with $\rho = 0.4, 0.5$ or 0.6. In numerous problems, one tends to use $\rho = 0.5$.

We start with the following *asymptotic first-order results*, summarized from Mukhopadhyay (1976,1990):

$$\begin{aligned} &\lim_{d\to 0} \mathrm{E}_{\mu,\sigma}\left(\frac{N}{C}\right) = 1 \quad [\textit{Asymptotic Efficiency or} \\ &\qquad\qquad\qquad\qquad\quad \textit{Asymptotic First-Order Efficiency}]; \\ &\lim_{d\to 0} \mathrm{P}_{\mu,\sigma}\{\mu \in J_N\} = 1 - \alpha \quad [\textit{Asymptotic Consistency}]. \end{aligned} \quad (6.2.41)$$

Hall (1981) and Mukhopadhyay (1990) gave the *asymptotic second-order* results for the three-stage strategy (6.2.39)-(6.2.40). The second-order results are summarized from Hall (1981) and Mukhopadhyay (1990) in the following as $d \to 0$:

$$\begin{aligned}
&\mathrm{E}_{\mu,\sigma}(N - C) = \frac{1+a^2}{2\rho} + o(1);\\
&\frac{N-C}{\sqrt{C}} \text{ converges to } N(0, \frac{2}{\rho}) \text{ in distribution;}\\
&\mathrm{V}_{\mu,\sigma}(N) = \frac{2}{\rho}C + o(C);\\
&\mathrm{P}_{\mu,\sigma}\{\mu \in J_N\} = 1 - \alpha + o(C^{-1}).
\end{aligned} \tag{6.2.42}$$

See also Ghosh and Mukhopadhyay (1981), Mukhopadhyay and Solanky (1994, Chapter 2) and Ghosh et al. (1997, Chapter 6) for many details. From (6.2.42) we can summarize the following conclusion.

The three-stage procedure (6.2.39)- (6.2.40) is *asymptotically consistent* and *asymptotically second-order efficient.*

If d is small, that is when the sample size is large, we can explain how these expansions may guide one in practice. In the case of the first part, we can say that $\mathrm{E}_{\mu,\sigma}(N - C)$ may lie in a close proximity of $\frac{1}{2}\rho^{-1}(1+a^2)$ which is a computable number. But, note that $\frac{1}{2}\rho^{-1}(1+a^2)$ is exactly the same as $q - \frac{2}{\rho}$ that was quoted in (6.2.38). The third part suggests that $\mathrm{V}_{\mu,\sigma}(N)$ may be nearly $\frac{2}{\rho}C$. The normalized stopping variable, namely $\frac{N-C}{\sqrt{C}}$, would have an approximate normal distribution. From the last expression in (6.2.42), we get an idea about the departure of the confidence coefficient from its target $1 - \alpha$.

Right away, we want everyone to realize an important point. The variance of N under three-stage sampling is approximately $\frac{2}{\rho}C$ which is strictly larger than the variance ($\approx 2C$) of N under sequential sampling. This may sound like bad news, but the news in itself should surprise no one. The increased variance under three-stage sampling is a direct result of reducing sampling operations. In practice, one should balance the gain from operational convenience and the added uncertainty from increased variability in N. Unfortunately, there is no one single or "optimal" way to achieve any desired balance. Program number 7 within **Seq06.exe** would help one to explore these remarks further.

Figure 6.2.11 shows an output from the said program for $\alpha = 0.01$, $d = 0.5$, $\sigma = 1, 4$ and $\rho = 0.2, 0.4, 0.6, 0.8$. The program again obtains the values of n_bar -C, P_diff, and var_n/C in the same spirit as it did so in the accelerated sequential case. Again, n_bar -C ought to be compared with $\frac{1}{2}\rho^{-1}(1+a^2)$. Also, ρvar_n values may be checked against $2C$. We note that all P_diff ($= \overline{p} - 0.99$) values are near zero, but P_diff value has turned out negative a number of times. That is, the estimated coverage probability $\overline{p}$ fell under the target ($= 0.95$) a number of times. We saw a similar feature for both accelerated sequential and sequential procedures earlier.

The reader should explore more of these and other features by running program number 7 within **Seq06.exe**. That will truly help in understanding many intricate issues.

```
        Simulated Second-Order Terms of the
          Three-Stage Sequential Procedure
          --------------------------------

        Input values:
                     minimum  maximum  number
              sigma    1.000    4.000       2
              alpha    0.010    0.010       1
              d         0.50     0.50       1
              rho       0.20     0.80       4
              -------------------------------

          Number of simulation replications = 2000
          Initial Sample size (m) = 10

          P_diff  = p_bar-(1-alpha)

sigma = 1.00; alpha = 0.01; d = 0.50; C =  26.54

 rho   n_bar     var_n  n_bar-C  var_n/C   P_diff
0.20   55.46    132.76    28.92   5.0025   0.0100
0.40   38.82     88.65    12.28   3.3401   0.0070
0.60   33.70     86.86     7.16   3.2728   0.0050
0.80   31.30     85.00     4.76   3.2027   0.0035

sigma = 4.00; alpha = 0.01; d = 0.50; C = 424.63

 rho   n_bar     var_n  n_bar-C  var_n/C   P_diff
0.20  442.40   5416.21    17.76  12.7550  -0.0025
0.40  432.01   2880.27     7.37   6.7829  -0.0035
0.60  437.63   3045.13    13.00   7.1712   0.0005
0.80  462.70   8438.33    38.06  19.8720   0.0015
```

Figure 6.2.11. *Second-order terms of the three-stage procedure.*

6.2.6 Two-Stage Procedure with a Known Lower Bound for Variance

We pointed out earlier that the Stein procedure (6.2.6) did not enjoy great asymptotic properties. This was evident from (6.2.14). Mukhopadhyay's (1980) modified two-stage procedure (6.2.23)-(6.2.24) improved upon that situation markedly, and yet this led only to first-order properties. Now, consider the following modification proposed by Mukhopadhyay and Duggan (1997).

We recall that $C = a^2\sigma^2/d^2$ where σ^2 is unknown. But, suppose that we have available a number $\sigma_L(>0)$ and $\sigma > \sigma_L$. In some practical situations, such a positive lower bound σ_L could be readily available.

In this case, observe that $C > a^2\sigma_L^2/d^2$ which is a known number. Hence, we define

$$m \equiv m(d) = \max\left\{m_0, \left\langle \frac{a^2\sigma_L^2}{d^2} \right\rangle + 1\right\}, \tag{6.2.43}$$

and start with $X_1, \ldots, X_m$, a pilot sample of size $m(\geq m_0)$. Here, $m_0(\geq 2)$ is a minimum pilot size that one must have if $\langle a^2\sigma_L^2/d^2 \rangle + 1$ happens to be "small" in the judgement of a practitioner. Next, we define the final sample size:

$$N \equiv N(d) = \max\left\{m, \left\langle \frac{a^2S_m^2}{d^2} \right\rangle + 1\right\}. \tag{6.2.44}$$

This two-stage procedure is implemented in the same way as we did in the case of the Stein procedure (6.2.6). Combining both stages of sampling, the final dataset is composed of N and $X_1, \ldots, X_N$. These would give rise to the sample mean $\overline{X}_N$ and the associated fixed-width interval $J_N = \left[\overline{X}_N \pm d\right]$ for μ.

This procedure is known to enjoy *asymptotic second-order* (see Mukhopadhyay and Duggan, 1997) properties as well as some *higher-order* (see Mukhopadhyay, 1999a) properties. This is so because the pilot size m from (6.2.43) is guaranteed to be smaller than C, and yet both m, C would converge to infinity at the same rate when $d \to 0$. This makes a very substantial difference in the theory.

Computations and Simulations

When we have available a known positive lower bound of σ, the two-stage fixed-width intervals can be constructed using program number 2 within **Seq06.exe**. This is similar to the construction of Stein's two-stage confidence interval explained in Section 6.2.1. Once procedure number 1 is selected from program number 2, one should simply type 'Y' when prompted to answer the question on a lower bound for σ. This will allow one to input a value for σ_L. Then, the program will compute m using σ_L. Figure 6.2.12 shows the output of a simulation study of this procedure. We generated data from a $N(5,4)$ distribution. Here, we assumed that $\sigma_L = 1$. The results show that on an average $124.33(= \overline{n})$ observations were required to construct a 99% confidence interval with width $2d = 1$.

Figure 6.2.13 shows results from a simulation study similar to the one that was reported in Figure 6.2.12. But, in this case we have not given an input regarding σ_L. We decided to use a pilot size $m = 5$. Now, we find that the average sample size $\overline{n} = 352.33$ which is much larger than the average sample size $\overline{n} = 124.33$ shown in Figure 6.2.12. The standard deviation of n shown in Figure 6.2.12 is also significantly smaller than that shown in Figure 6.2.13. On the other hand, the estimated coverage probabilities ($\overline{p}$) and their standard errors are very similar in the two situations!

```
         Fixed-Width Confidence Interval
              Using Two-Stage Procedure
              =========================

Simulation study for  99% confidence interval for
the normal mean. Data were generated from normal
distribution with mean =  5.00 and variance =  4.00

Number of simulation replications        =  1000
Width of the confidence interval (2d) =  1.00

Initial sample size =  27
is obtained using sigma_L =  1.00

     Optimal sample size (C)           =  106.16

    Average sample size (n_bar)        =  124.33
    Std. dev. of n (s)                 =   34.14
    Minimum sample size (min n)        =   44
    Maximum sample size (max n)        =  249

    Coverage probability (p_bar)       = 0.990
    Standard error of p_bar            = 0.003
```

Figure 6.2.12. *Simulation results for the two-stage procedure (6.2.43) - (6.2.44) with a known lower bound σ_L for unknown σ.*

```
         Fixed-Width Confidence Interval
              Using Two-Stage Procedure
              =========================

Simulation study for  99% confidence interval for
the normal mean. Data were generated from normal
distribution with mean =  5.00 and variance =  4.00

Number of simulation replications        =  1000
Width of the confidence interval (2d) =  1.00

Initial sample size =  5

     Optimal sample size (C)           =  106.16

    Average sample size (n_bar)        =  352.33
    Std. dev. of n (s)                 =  255.21
    Minimum sample size (min n)        =    6
    Maximum sample size (max n)        =  1757

    Coverage probability (p_bar)       = 0.993
    Standard error of p_bar            = 0.003
```

Figure 6.2.13. *Simulation results for a two-stage procedure (6.2.6) with unknown σ.*

6.2.7 More Comparisons and Some Conclusions

This section briefly examines the six methodologies that have already been included in our discourses. The methodologies are:

1. Stein's Two-Stage Procedure (STwoSP);
2. Modified Two-Stage Procedure (MSTwoP);
3. Two-Stage Procedure with a Known Lower Bound for the Population Variance (TwoSPLB);
4. Purely Sequential Procedure (PureSP);
5. Accelerated Sequential Procedure (AccSP);
6. Three-Stage Procedure (ThreeSP).

It should be obvious that each methodology has its own set of strengths and weaknesses. Hence, it is not possible to declare a clear winner. We examine some of the advantages and disadvantages of each methodology with the help of an example and simulations.

Example 6.2.2 : Let us consider a simulation study for examining the behavior of the sample size (N) that may be required to construct a 95% fixed-width ($2d = 1$) confidence interval for μ. We successively implemented each procedure by generating observations from a $N(5, 4)$ distribution. We selected or initialized the random number sequence 5 within **Seq06.exe**. In each case, simulations were replicated 2000 times. The findings are summarized in Table 6.2.2.

With $a = 1.96$, the optimal fixed-sample-size C from (6.2.4) became:

$$C = \frac{a^2\sigma^2}{d^2} = \frac{1.96^2 \times 4}{0.5^2} = 61.466.$$

Table 6.2.2 shows that the Stein procedure from Section 6.2.1 has oversampled significantly ($\overline{n} = 81.37$) compared with $C = 61.466$. But it guarantees the required confidence coefficient. We observe that our simulations estimated the coverage probability $\overline{p} = 0.952$ which exceeded the target 0.95.

In a modified two-stage procedure from Section 6.2.3, $\overline{n}$ and $\overline{p}$ will depend on the choice of $\gamma(> 0)$. If we can choose γ in such a way that m is close to C but $m < C$, a modified procedure will produce very attractive results. Since C is unknown, an appropriate choice of γ may not come very easily. The third two-stage procedure from Section 6.2.6 requires the knowledge of $\sigma_L(> 0)$, a lower bound for σ. Again, it is clear that the performance will improve as σ_L increases but stays below σ. These two-stage procedures require only two batches of sampling. This is a great advantage over other multistage and sequential procedures. But, the key plus point in favor of two-stage procedures is this: The confidence coefficient is at least the preassigned target, $1 - \alpha$.

Table 6.2.2: Simulation study for 95% fixed-width confidence intervals

The Procedure	Initializing Parameters	Ave. Sample Size $\overline{n}$	Std. Dev. of n (s)	Coverage Probability $\overline{p}$	# Additional Samples
STwoSP	$m = 10$	81.37	36.09	0.952	-
MSTwoP	$\gamma = 0.1$	78.96	33.67	0.955	-
TwoSPLB	$\sigma_L = 1$	72.60	26.77	0.951	-
PureSP	$m = 10$	59.48	12.42	0.937	4
AccSP	$m = 10$; $\rho = 0.5$	66.11	17.57	0.952	-
ThreeSP	$m = 10$; $\rho = 0.5$	64.35	18.03	0.947	3

The purely sequential procedure introduced in Section 6.2.2 often underestimates C by a small margin and the coverage probability falls below target. Table 6.2.2 shows that it required 4 additional observations on an average beyond N to achieve the preset confidence coefficient 0.95. In general, we have found in numerous situations that a sequential procedure underestimates C by less than 10 observations. Hence, taking a few additional observations beyond N may do the trick! However, in many situations, gathering one observation at-a-time may be costly and time consuming.

An accelerated procedure from Section 6.2.4 overcomes the problem of underestimatig C by a sequential procedure. The required sampling operations may be divided into three main parts. In the first part, we gather a pilot sample in one batch. In the second part, we estimate a fraction of the final sample size by gathering one observation at-a-time in a sequential fashion. In the final step, we gather the remaining observations in one single batch. Note that for this procedure, we have $\overline{n} = 66.11$, which is slightly larger than C and $\overline{p} = 0.952$. These estimated values are very close to the respective target values.

The three-stage procedure from Section 6.2.5 gathers all necessary observations in three single batches. Consider all sequential and multistage methodologies introduced in Sections 6.2.2, 6.2.4 and 6.2.5. Among these, a three-stage procedure is operationally the simplest one to implement when batch sampling is feasible in practice. A key point, however, is this: The confidence coefficient associated with any of these sequential and multistage methodologies is no longer at least $1 - \alpha$. Instead, the confidence coefficient is the preassigned target $1 - \alpha$ only in a limiting sense.

In a practical scenario, if a positive lower bound σ_L for σ is not readily available, we may suggest using a modified two-stage procedure, MSTwoP.

If a suitable σ_L is available, we suggest using the procedure, TwoSPLB. In offering these suggestions, we are guided by two fundamental assumptions: (i) the exact consistency property is deemed important and (ii) batch sampling in two steps is implementable.

On the other hand, if asymptotic (second-order) approximations are good enough to have in a practical scenario, we suggest using a three-stage procedure, ThreeSP, assuming that batch sampling is implementable.

In some situations, for example, in continuous monitoring of a production process, the observations may arrive naturally in a sequential order. Then, one would likely be swayed by a purely sequential procedure, PureSP. In some of these situations, it may be possible to augment purely sequential sampling with batch sampling. For example, in the first phase of a clinical trial, patients may arrive sequentially. But, in the second phase, groups of patients may be available. Then, an accelerated sequential procedure, AccSP, may be a good sampling strategy to implement.

A word of caution may be in order. The suggestions offered here have not taken into account any kind of detailed cost-benefit analysis. Our suggestions are essentially guided by the range of known theoretical properties and only very rudimentary operational considerations of each procedure. Only the true nature of some fieldwork accompanied with serious cost-benefit analysis should be the hallmark to guide an experimenter to select and implement one procedure over another.

6.3 Bounded Risk Point Estimation

Now, we introduce some multistage procedures for bounded risk point estimation of the unknown mean μ of a $N(\mu, \sigma^2)$ distribution. The variance σ^2 is also assumed unknown, $-\infty < \mu < \infty$ and $0 < \sigma < \infty$. Suppose that $X_1, \ldots, X_n$ are i.i.d. random variables from this population, $n \geq 2$. We propose to estimate μ under the following loss function:

$$L_n(\mu, \overline{X}_n) = A\left|\overline{X}_n - \mu\right|^k \text{ with } A > 0 \text{ and } k > 0 \text{ known,} \tag{6.3.1}$$

where $\overline{X}_n = \frac{1}{n}\sum_{i=1}^{n} X_i$.

When we fix $k = 1$ or 2, the loss function corresponds to *absolute error* and *squared error* respectively. The goal is to make the associated risk not to exceed a preassigned number $\omega(> 0)$ for all $0 < \sigma < \infty$. That is, we require:

$$\begin{gathered}\mathrm{E}_{\mu,\sigma}\left\{L_n(\mu, \overline{X}_n)\right\} = \mathrm{E}_{\mu,\sigma}\left[A\left|\overline{X}_n - \mu\right|^k\right] = B\sigma^k n^{-k/2} \leq \omega \\ \text{for all } 0 < \sigma < \infty, \text{ where } B = \frac{2^{k/2}A}{\sqrt{\pi}}\Gamma\left(\frac{1}{2}(k+1)\right).\end{gathered} \tag{6.3.2}$$

But, we cannot achieve this goal with any prefixed sample size n. Refer to Example 2.3.2. The verification of (6.3.2) is left as an exercise.

We note, however, that the risk associated with $\overline{X}_n$ will not exceed

$\omega(>0)$ if and only if

$$n \geq \left(\frac{B}{\omega}\right)^{2/k} \sigma^2 = n^*, \text{ say.} \tag{6.3.3}$$

Let us pretend for a moment that n^* is an integer. This n^* is referred to as the required optimal fixed sample size had σ been known. But, we realize that the magnitude of n^* remains unknown even though its expression looks so simple.

So, we discuss some multistage estimation strategies for μ. We recall the expression of a sample variance, $S_n^2 = \frac{1}{n-1}\sum_{i=1}^{n}\left(X_i - \overline{X}_n\right)^2$.

6.3.1 Two-Stage Procedure

We first discuss sampling in two-stages. We start with pilot observations $X_1, \ldots, X_m$ where $m(> k+1)$ is the pilot size and obtain S_m^2. Define the final sample size as:

$$N \equiv N(A) = \max\left\{m, \left\langle\left(\frac{b_m B}{\omega}\right)^{2/k} S_m^2\right\rangle + 1\right\} \tag{6.3.4}$$

where $b \equiv b_m$ is a fixed positive number that is to be appropriately determined in Theorem 6.3.1. Ghosh and Mukhopadhyay (1976) gave an analogous two-stage procedure.

In the definition (6.3.4) for N, observe that n^* from (6.3.3) is estimated by $\left(\frac{b_m B}{\omega}\right)^{2/k} S_m^2$ based on the pilot observations. In the expression of n^*, σ^2 and B are respectively replaced by S_m^2 and $b_m B$. So, N itself is viewed as an estimator of n^*.

The two-stage sampling scheme (6.3.4) is implemented as follows. If $N = m$, it indicates that we have taken too many observations at pilot stage, and hence we do not need more observations at second stage. But, if $N > m$, t indicates that we have started with too few observations at pilot stage, and hence we sample the difference at second stage to compensate for deficiency. That is, if $N > m$, we take additional observations $X_{m+1}, \ldots, X_N$ at the second stage of sampling. That is,

$$\begin{aligned}&\text{if } N = m, \text{ the final dataset is } X_1, \ldots, X_m\\ &\text{if } N > m, \text{ the final dataset is } X_1, \ldots, X_m, X_{m+1}, \ldots, X_N\end{aligned} \tag{6.3.5}$$

Combining the two situations from (6.3.5), one can see that the final dataset is composed of N and $X_1, \ldots, X_N$ which give rise to an estimator $\overline{X}_N$ for μ.

Now, we observe that for all fixed $n(\geq m)$, the event $\{N = n\}$ depends only on the random variable S_m^2. But, for all fixed $n(\geq m)$, $\overline{X}_n$ and S_m^2 are independent random variables. So, any event defined only through $\overline{X}_n$ is independent of the event $\{N = n\}$ for all fixed $n(\geq m)$. Hence, we can express the risk associated with the estimator $\overline{X}_N$ as follows:

$$\begin{aligned}
&\mathrm{E}_{\mu,\sigma}\left\{L_N(\mu,\overline{X}_N)\right\} \\
&= \mathrm{E}_{\mu,\sigma}\left[A\left|\overline{X}_N-\mu\right|^k\right] \\
&= \textstyle\sum_{n=m}^{\infty} \mathrm{E}_{\mu,\sigma}\left[A\left|\overline{X}_n-\mu\right|^k \cap N=n\right] \\
&= \textstyle\sum_{n=m}^{\infty} \mathrm{E}_{\mu,\sigma}\left[A\left|\overline{X}_n-\mu\right|^k\right]\mathrm{E}_{\mu,\sigma}[I(N=n)] \\
&= \textstyle\sum_{n=m}^{\infty}\left[B\sigma^k n^{-k/2}\right]\mathrm{P}_{\mu,\sigma}(N=n) \\
&= B\sigma^k\mathrm{E}_{\mu,\sigma}\left[N^{-k/2}\right].
\end{aligned} \tag{6.3.6}$$

with B defined in (6.3.2). Now, let us summarize some of the crucial properties of the estimation procedure (6.3.4).

Theorem 6.3.1 : *For the two-stage estimation procedure $(N,\overline{X}_N)$ defined by (6.3.4), for all fixed μ,σ,A and ω, one has the following properties.*

(i) $(b_m B/\omega)^{2/k}\sigma^2 \le \mathrm{E}_{\mu,\sigma}[N] \le m + (b_m B/\omega)^{2/k}\sigma^2$;

(ii) $\overline{X}_N$ *is an unbiased estimator of* μ *with its variance* $\sigma^2\mathrm{E}_{\mu,\sigma}\left[N^{-1}\right]$;

(iii) $\mathrm{E}_{\mu,\sigma}\left\{L_N(\mu,\overline{X}_N)\right\} \le \omega$;

where $b_m = \left\{\frac{1}{2}(m-1)\right\}^{k/2}\Gamma\left(\frac{1}{2}(m-k-1)\right)\left\{\Gamma\left(\frac{1}{2}(m-1)\right)\right\}^{-1}$.

Proof: Observe that the requirement $m > k+1$ ensures that the gamma functions involved in defining b_m are both finite. Clearly, parts (i)-(ii) would follow as in the case of Theorem 6.2.1, parts (i)-(ii). The details are left out as exercise.

Next, from (6.3.4), observe the basic inequality

$$\left(\frac{b_m B}{\omega}\right)^{2/k} S_m^2 \le N \le m + \left(\frac{b_m B}{\omega}\right)^{2/k} S_m^2 \text{ w.p. 1.} \tag{6.3.7}$$

This together with (6.3.6) imply that

$$\begin{aligned}
\mathrm{E}_{\mu,\sigma}\left\{L_N(\mu,\overline{X}_N)\right\} &= B\sigma^k\mathrm{E}_{\mu,\sigma}\left[N^{-k/2}\right] \le \frac{\omega}{b_m}\mathrm{E}_{\mu,\sigma}\left[\left(\frac{\sigma^2}{S_m^2}\right)^{k/2}\right] \\
&= \frac{(m-1)^{k/2}\omega}{b_m}\mathrm{E}_{\mu,\sigma}\left[\left(\frac{\sigma^2}{(m-1)S_m^2}\right)^{k/2}\right] \\
&= \frac{(m-1)^{k/2}\omega}{b_m}\mathrm{E}\left[Y^{-k/2}\right],
\end{aligned} \tag{6.3.8}$$

where Y has a χ^2_{m-1} distribution.

Now, we can express $\mathrm{E}\left[Y^{-k/2}\right]$ as follows:

$$\begin{aligned}\mathrm{E}\left[Y^{-k/2}\right] &= \frac{1}{2^{\frac{1}{2}(m-1)}\Gamma(\frac{1}{2}(m-1))}\int_0^\infty y^{-k/2}e^{-y/2}y^{\frac{1}{2}(m-1)-1}dy \\ &= \frac{1}{2^{\frac{1}{2}(m-1)}\Gamma(\frac{1}{2}(m-1))}\int_0^\infty e^{-y/2}y^{\frac{1}{2}(m-k-1)-1}dy \\ &= \frac{2^{\frac{1}{2}(m-k-1)}\Gamma(\frac{1}{2}(m-k-1))}{2^{\frac{1}{2}(m-1)}\Gamma(\frac{1}{2}(m-1))} \text{ since } m > k+1 \\ &= \frac{\Gamma(\frac{1}{2}(m-k-1))}{2^{k/2}\Gamma(\frac{1}{2}(m-1))}. \end{aligned} \tag{6.3.9}$$

In the step before last, we identified the integral as a gamma integral with $\alpha = \frac{1}{2}(m-k-1)$ and $\beta = 2$. Now, we combine (6.3.8) and (6.3.9) to write

$$\mathrm{E}_{\mu,\sigma}\left\{L_N(\mu, \overline{X}_N)\right\} \le \frac{(m-1)^{k/2}\omega}{b_m}\frac{\Gamma(\frac{1}{2}(m-k-1))}{2^{k/2}\Gamma(\frac{1}{2}(m-1))} = \omega,$$

for all $0 < \sigma < \infty$ if we choose

$$b_m = \left\{\frac{1}{2}(m-1)\right\}^{k/2}\Gamma\left(\frac{1}{2}(m-k-1)\right)\left\{\Gamma\left(\frac{1}{2}(m-1)\right)\right\}^{-1}. \tag{6.3.10}$$

The proof is complete. ∎

Some comments are in order. First, part (iii) shows that the risk-bound ω is met for all unknown $0 < \sigma < \infty$. Next, one should note that b_m defined in (6.3.10) exceeds one for all $k > 0$ and $m > k+1$. A proof of this claim is left as an exercise with some hints. Also, from Theorem 6.3.1, part (i), we note:

$$\mathrm{E}_{\mu,\sigma}(N) \ge \left(\frac{b_m B}{\omega}\right)^{2/k}\sigma^2 > \left(\frac{B}{\omega}\right)^{2/k}\sigma^2 = n^* \text{ since } b_m > 1. \tag{6.3.11}$$

That is, on an average, this procedure requires more observations than the optimal fixed-sample-size n^*. This should not be surprising because n^* is "optimal" when we pretend that σ is known! But, in fact σ is unknown and we end up taking more than n^* observations.

So far we have discussed only the exact properties. But, one may ask: What are the associated large-sample properties? Now, we should think carefully how we may make the random sample size N approach infinity. We can make N approach infinity (w.p. 1) by making ω tend to zero or by letting σ tend to infinity. If ω is made smaller, we obtain a smaller risk and hence N will tend to increase. On the other hand, if σ or A becomes larger, N will tend to increase too. In order to state large-sample properties precisely, we make $A \to \infty$.

From Theorem 6.3.1, part (i), it follows easily that

$$\lim_{A\to\infty} \mathrm{E}_{\mu,\sigma}\left(\frac{N}{n^*}\right) = b_m > 1 \quad \text{[\textit{Asymptotic Inefficiency or Asymptotic First-Order Inefficiency}]} \tag{6.3.12}$$

The property given in (6.3.12) is not as easily disposed of. When $A \to \infty$, we know that $N \to \infty$ w.p. 1 and $n^* \to \infty$. But, even then (6.3.12) indicates that $\mathrm{E}_{\mu,\sigma}(N)$ exceeds to exceed the optimal fixed-sample-size n^*! That is, $\mathrm{E}_{\mu,\sigma}(N)$ and n^* do not agree in the large-sample case. Hence, the latter property is more disturbing than what we found in (6.3.11). We may summarize by saying that the two-stage procedure is *asymptotically (first-order) inefficient.* See Ghosh and Mukhopadhyay (1981).

The performance of the procedure depends very heavily on the choice of m. The procedure remains inefficient, even asymptotically, perhaps because the pilot size m is kept fixed whereas n^* is allowed to grow! Another issue may be that σ^2 is estimated only once and this estimator may be a little unstable if m is chosen too small compared with n^*.

Computations and Simulations

Computations and simulation studies for the bounded risk point estimation (BRPE) problem can be performed by selecting program number 8 from **Seq06.exe** program. This selection will prompt the user to select one of the four subprograms connected with a bounded risk point estimation strategy. See Figure 6.3.1 for further details.

The subprogram number 1 simulates the two-stage bounded risk estimation procedure (6.3.4). This subprogram can also work with live data. One is prompted to select either the absolute loss function or the squared error loss function. After selecting a loss function, the user is required to input values of the parameters such as the risk-bound ω and A. Figure 6.3.1 shows input for a simulation study using BRPE.

The example used the absolute error loss function ($k = 1$) with $A = 2$ and $\omega = 0.5$. Simulation was replicated 2000 times using data generated from a $N(5, 4)$ distribution with $m = 5$. The output of this simulation study is shown in Figure 6.3.2. Here, we have $n^* = 40.74$ which was computed from (6.3.3). The results show the simulated average sample size:

$$\overline{n} = 64.79 > 40.74 = n^*.$$

This empirically validated the result in (6.3.11).

```
*****************************************************
* Number          Program                           *
*  1   Exact value of C for given parameters        *
*  2   Fixed-Width Confidence Interval Estimation   *
*  3   E(N) and Confidence Coefficient for Two-Stage *
*  4   Role Gamma in Modified Two-Stage Procedure   *
*  5   Second Order Terms of Purely Sequential      *
*  6   Second Order Terms of Accelerated Sequential *
*  7   Second Order Terms of Three-Stage Procedure  *
*  8   Bounded Risk Point Estimation                *
*  9   Minimum Risk Point Estimation                *
*  0   Exit                                         *
*****************************************************

     Enter the program number: 8

Do you want to save the results in a file(Y or N)? N

   Select one of the following Procedures:
   ***************************************************
   * Number      Subprogram                          *
   *   1 Estimation using Two-Stage Procedure        *
   *   2 Expected N and Risk in Two-Stage Procedure  *
   *   3 Estimation using Purely Sequential Procedure *
   *   4 Simulated & Approx Risk in Purely Sequential *
   ***************************************************

   Enter subprogram number: 1

   Select one of the following loss functions
   *******************************
   *  Number   Loss Function     *
   *     1     Absolute Error    *
   *     2     Squared Error     *
   *******************************

   Enter loss function number (k): 1

   Enter the following preassigned values
         Risk-bound, w = 0.5
         A (Constant in the loss function) = 2

   Note initial sample size, m > 2(=k+1)
   Enter a value for m: 5

Do you want to change the input parameters?[Y or N] N

   Do you want to simulate data or input real data?
   Type S (Simulation) or D (Real Data): S
   Do you want to store simulated sample sizes?
   Type Y (Yes) or N (No): N
   Number of replications for a simulation? 2000

   Enter a positive integer(<10) to initialize
            the random number sequence: 2

   For the data simulation, input
        mean of the normal distribution: 5
                               variance: 4
```

Figure 6.3.1. *Screen input for bounded risk point estimation.*

```
          Bounded Risk Point Estimation
             Using Two-Stage Procedure
             =========================

Input Parameters:
     Loss function number,    k =  1
     Loss function constant, A =  2.00
     Risk-bound,              w =  0.50
     Initial sample size,     m =   5

Simulation study by generating data from
normal distribution with mean =  5.00
and variance =  4.00

Number of simulation replications =  2000

   Optimal sample size (n_star)      =  40.74
   Average final sample size(n_bar) =  64.79
   Std. dev. of n (s_n)              =  43.39
   Minimum sample size (n_min)       =    5
   Maximum sample size (n_max)       = 261
```

Figure 6.3.2. *Bounded risk point estimation using a two-stage procedure.*

Probability Distribution of the Final Sample Size N

Now, we set out to evaluate the probability distribution of N given by (6.3.4) which involves only σ. Let us denote $Y = (m-1)S_m^2/\sigma^2$ which has a χ^2_{m-1} distribution. One may easily verify that

$$\begin{aligned} \mathrm{P}_\sigma(N = m) &= \mathrm{P}_\sigma\left\{\left\langle \left(\frac{b_m B}{\omega}\right)^{2/k} S_m^2 \right\rangle + 1 \le m\right\} \\ &= \mathrm{P}_\sigma\left\{0 < \left(\frac{b_m B}{\omega}\right)^{2/k} S_m^2 \le m\right\} \\ &= \mathrm{P}_\sigma\left\{0 < Y \le \frac{m(m-1)}{\sigma^2}\left(\frac{\omega}{b_m B}\right)^{2/k}\right\}. \end{aligned}$$

Similarly, for any positive integer r we have:

$$\begin{aligned} \mathrm{P}_\sigma(N = m+r) &= \mathrm{P}_\sigma\left\{\left\langle \left(\frac{b_m B}{\omega}\right)^{2/k} S_m^2 \right\rangle = m + r - 1\right\} \\ &= \mathrm{P}_\sigma\left\{m + r - 1 < \left(\frac{b_m B}{\omega}\right)^{2/k} S_m^2 \le m + r\right\} \\ &= \mathrm{P}_\sigma\left\{\frac{(m+r-1)(m-1)}{\sigma^2}\left(\frac{\omega}{b_m B}\right)^{2/k} < Y \right. \\ &\qquad\qquad \left. \le \frac{(m+r)(m-1)}{\sigma^2}\left(\frac{\omega}{b_m B}\right)^{2/k}\right\}. \end{aligned}$$

Again, let us write:

$$g_m(y) = \left\{2^{(m-1)/2}\Gamma\left(\frac{m-1}{2}\right)\right\}^{-1} \exp[-y/2]y^{(m-3)/2}, y > 0;$$

$$B_r = \frac{(m+r)(m-1)}{\sigma^2}\left(\frac{\omega}{b_m B}\right)^{2/k}, \ k = 0, 1, 2, \ldots.$$

Then,

$$\begin{aligned} &\mathrm{P}_\sigma(N = m) = \mathrm{P}_\sigma\left\{0 < \chi^2_{m-1} \le B_0\right\} = \textstyle\int_0^{B_0} g_m(y)dy; \\ &\mathrm{P}_\sigma(N = m + r) = \mathrm{P}_\sigma\left\{B_{r-1} < \chi^2_{m-1} \le B_r\right\} = \textstyle\int_{B_{r-1}}^{B_r} g_m(y)dy, \end{aligned} \tag{6.3.13}$$

for $r = 1, 2, 3, \ldots$. Now, the exact probability distribution of N can be expressed numerically pretending that σ is known.

Utilizing (6.3.13), we can evaluate the expected value and the variance of N as well as the risk associated with $\overline{X}_N$. Subprogram number 2 under program number 8 within **Seq06.exe** will make these tasks easy. This subprogram computes n^*, $\mathrm{E}(N)$, $\mathrm{V}(N)$, standard deviation of N and the risk, $B\sigma^k \mathrm{E}_{\mu,\sigma}\left[N^{-k/2}\right]$, for both absolute and squared error loss functions. Figure 6.3.3 shows an output from this subprogram when $A = 2$, $\omega = 0.5$, $m = 5$ and $\sigma^2 = 4$.

```
Expected N and Risk of the Bounded Risk
Point Estimation Using Two-Stage Procedure
==========================================

Input Parameters:
     Loss function constant, A =  2.00
     Risk-bound,             w =  0.50
     Initial sample size,    m =  5
     Population Variance,      =  4.0

Loss Function Type      Absolute  Squared
                           Error    Error
 Optimal sample size        40.7     16.0
 Expected value of N        64.5     32.5
 Variance of N            2044.7    508.7
 Std deviation of N         45.2     22.6
 Risk                      0.490    0.416
```

Figure 6.3.3. *Program output for* $\mathrm{E}(N)$ *and* risk.

At this point, one may introduce an appropriate modified two-stage, purely sequential, accelerated sequential or a three-stage procedure in the spirit of Sections 6.2.3-6.2.5. But, for brevity, we move right away to a purely sequential estimation strategy. One may refer to Mukhopadhyay (1985a) and Mukhopadhyay et al. (1987) for other details.

6.3.2 Purely Sequential Procedure

In order to reduce oversampling by the two-stage procedure (6.3.4), we proceed to estimate σ^2 successively in a sequential manner. We begin with a few observations (say, $m \geq 2$) and then we continue to take one additional observation at-a-time, but terminate sampling when we have gathered enough observations. As soon as a sample size exceeds the corresponding estimate of n^*, the sampling is terminated right there, and a point estimator of μ is constructed. This idea was also pursued in Section 6.2.2 along the lines of Anscombe (1952), Ray (1957) and Chow and Robbins (1965).

Recall that $n^* = (B/\omega)^{2/k}\sigma^2$ but σ^2 is unknown. We start with $X_1, \ldots, X_m$, a pilot of size $m(\geq 2)$, and then we take one additional X at-a-time as needed. The stopping time is defined as follows:

$$\begin{gathered} N \equiv N(A) \text{ is the smallest integer } n(\geq m) \\ \text{for which we observe } n \geq (B/\omega)^{2/k} S_n^2. \end{gathered} \tag{6.3.14}$$

This purely sequential procedure is implemented in the same way we did in the case of (6.2.17). Again, $\mathrm{P}_{\mu,\sigma}(N < \infty) = 1$, that is sampling terminates with probability one. Upon termination, we estimate μ by the sample mean, $\overline{X}_N$, obtained from the full dataset $N, X_1, \ldots, X_m, \ldots, X_N$. We leave it as an exercise to prove that $\overline{X}_N$ is an unbiased estimator of μ and its variance is $\sigma^2\mathrm{E}_{\mu,\sigma}(N^{-1})$ for all fixed μ, σ, ω, k and A.

The risk associated with $\overline{X}_N$ can be expressed as:

$$\begin{aligned} \mathrm{E}_{\mu,\sigma}\left\{L_N(\mu, \overline{X}_N)\right\} &= B\sigma^k \mathrm{E}_{\mu,\sigma}\left[N^{-k/2}\right] \\ &= \omega \mathrm{E}_{\mu,\sigma}\left[(N/n^*)^{-k/2}\right]. \end{aligned} \tag{6.3.15}$$

The *asymptotic second-order* results are cumbersome, and these are summarized below. The tools that are essential in their proofs can be borrowed from Woodroofe (1977,1982) and Mukhopadhyay (1988a). See also Ghosh and Mukhopadhyay (1981) and Section 2.4.2 of Mukhopadhyay and Solanky (1994) for many details. Recall the expression of $\eta(p)$ from (6.2.33) with $p = 1$. We have as $A \to \infty$:

$$\begin{gathered} \mathrm{E}_{\mu,\sigma}(N - n^*) = \eta(1) + o(1) \text{ if } m \geq 4; \\ \frac{N - n^*}{\sqrt{n^*}} \text{ converges to } N(0, 2) \text{ in distribution;} \\ \mathrm{V}_{\mu,\sigma}(N) = 2n^* + o(n^*) \text{ if } m \geq 4. \end{gathered} \tag{6.3.16}$$

Under an appropriate condition on m, the pilot size, the risk function from (6.3.15) has the following *second-order* expansion as $A \to \infty$:

$$\mathrm{E}_{\mu,\sigma}\left\{L_N(\mu, \overline{X}_N)\right\} = \omega - \frac{k\omega}{2n^*}\left(\eta(1) - 1 - \frac{1}{2}k\right) + o(n^{*-1}). \tag{6.3.17}$$

In the case of an absolute error loss function (that is, $k = 1$) we require

$m \geq 13$, but in the case of a squared error loss function (that is, $k = 2$) we require $m \geq 14$.

The purely sequential procedure (6.3.14) is *asymptotically second-order efficient* and *second-order risk efficient.*

For large n^*, let us interpret the results from (6.3.16). From the first part,, we can say that $\mathrm{E}_{\mu,\sigma}(N - n^*)$ may be expected to lie in a close proximity of the computable number, $\eta(1) + 1$. The third part says that $\mathrm{V}_{\mu,\sigma}(N)$ should be around $2n^*$. Also, the normalized stopping variable $(N - n^*)/\sqrt{n^*}$ would have an approximate normal distribution.

From the expression in (6.3.17), we get some feelings about the departure of a sequential risk from the risk-bound ω. We observe that $\mathrm{E}_{\mu,\sigma}\left\{L_N(\mu, \overline{X}_N)\right\} - \omega$ may be close to

(i) $-\frac{\omega}{2n^*}\left(\eta(1) - \frac{1}{2}\right)$ in the case of an absolute error loss if $m \geq 13$,

(ii) $-\frac{\omega}{n^*}\left(\eta(1) - 1\right)$ in the case of a squared error loss if $m \geq 14$.

Pretending that σ is known, these departures can be assessed for given ω using subprogram number 4 of program number 8 within **Seq06.exe**. The simulated difference in the program is obtained by subtracting ω from $(\omega/n_s)\sum_{i=1}^{n_s}(n_i/n^*)^{-k/2}$ where n_i is the value of N in the ith simulation and n_s is the number of simulations. This estimates of $\mathrm{E}_{\mu,\sigma}\left\{L_N(\mu, \overline{X}_N)\right\}$ in (6.3.15). The subprogram obtains results for both absolute error ($k = 1$) and the squared error ($k = 2$) loss functions. The simulations are carried out by generating data from a $N(0, \sigma^2)$ distribution with an input for σ^2. This subprogram uses user input, $m(\geq 14)$, ω, σ^2 and A values, however the check for given ω, σ^2 and A gives $n^* > m$. In case $n^* \leq m$, the program prints the n^* values and prompt for different user input for these parameters.

Figure 6.3.4 shows an output from subprogram number 4 with $\omega = 0.5$, $\sigma^2 = 9$ and $m = 15$ and $A = 1, 2, \ldots, 10$. Simulations were replicated 2000 times to obtain the average sample size, $\overline{n}$. From the output in Figure 6.3.4, we feel that the asymptotic and simulated entities are generally comparable for large n^*.

6.3.3 Comparisons and Conclusions

From the results presented in Sections 6.3.1 and 6.3.2, it is clear that two-stage procedure oversampled and purely sequential procedure tended to undersample. To examine the behavior of these two procedures further, we consider the following example.

Example 6.3.1: Consider estimation of μ by generating data form a $N(25, 16)$ distribution. We use both absolute and squared error loss functions. We input the following parameter values in subprogram number 1 and subprogram number 3 from program number 8 within **Seq06.exe**.

```
          Asymptotic and Simulated Risk for Bounded Risk
         Point Estimation Using Purely Sequential Procedure
         ----------------------------------------------------

         Input values:
              Risk-bound,          w = 0.500
              Population variance = 9.00

              Number of Simulation Replications = 2000
              Initial Sample size (m)   = 15
              Computed value of eta(1) = -1.183

                        E(Loss Function)-(Risk-Bound)

            Absolute Error Loss Function     Squared Error Loss Function
    A     n_star Asymptotic Simulated       n_star Asymptotic Simulated
   1.00     22.9     0.0184    0.0168         18.0     0.0606    0.0061
   2.00     91.7     0.0046    0.0074         36.0     0.0303    0.0844
   3.00    206.3     0.0020    0.0028         54.0     0.0202    0.0564
   4.00    366.7     0.0011    0.0021         72.0     0.0152    0.0271
   5.00    573.0     0.0007    0.0009         90.0     0.0121    0.0202
   6.00    825.1     0.0005    0.0006        108.0     0.0101    0.0162
   7.00   1123.0     0.0004    0.0006        126.0     0.0087    0.0154
   8.00   1466.8     0.0003    0.0004        144.0     0.0076    0.0127
   9.00   1856.4     0.0002    0.0002        162.0     0.0067    0.0121
  10.00   2291.8     0.0002    0.0004        180.0     0.0061    0.0122
```

Figure 6.3.4. *Asymptotic and simulated risk for bounded risk point estimation.*

Random number initialization sequence number = 5;
The risk-bound $\omega = 0.4$; $A = 2$;
Simulation replications = 2000.

The summary results obtained from the subprograms, are presented in Table 6.3.1. The entries show that the two-stage procedure significantly oversamples and as expected the purely sequential procedure undersamples by a few observations.

Table 6.3.1: Simulation study for bounded risk point estimation

Procedure:	Two-Stage		Purely Sequential	
Loss Function:	Absolute	Squared	Absolute	Squared
n^*	254.65	80.00	254.65	80.00
$\overline{n}$	286.78	94.26	253.14	78.12
s_n	104.79	34.31	22.76	13.09
$\min(n)$	36	15	160	18
$\max(n)$	815	267	329	120

It is interesting see that

- the minimum number samples used in the simulation is lower and
- the maximum is higher

for the two-stage procedure. We also note larger simulated values of $\mathrm{V}(N)$ in the case of two-stage procedures. This indicates that the purely sequential procedure tend to estimate n^* more tightly than the a two-stage procedure.

Next we examine the two loss functions. Table 6.3.1 shows that both $\overline{n}$ and n^* are higher for the absolute error loss function. First we rewrite $B \equiv B(k)$ where $k = 1(2)$ represents the absolute (squared) error loss function. Now from (6.3.2) we have:

$$B(k) = B = \begin{cases} A\sqrt{2/\pi} & \text{if } k = 1, \\ A & \text{if } k = 2. \end{cases} \tag{6.3.18}$$

Similarly, we rewrite the optimal sample size n^* from (6.3.3) as n_k^*,

$$n_k^* = \left(\frac{B(k)}{\omega}\right)^{2/k} \sigma^2. \tag{6.3.19}$$

Now combining (6.3.18) and (6.3.19), we get

$$\frac{n_1^*}{n_2^*} = \frac{2A}{\pi\omega} = \frac{0.6366A}{\omega}, \; k = 1, 2. \tag{6.3.20}$$

Thus, if $\omega < 0.6366A$, we have $n_1^* > n_2^*$. That is, BRPE under an absolute error loss function would require more observations than under a squared error loss function. Conversely, if $\omega > 0.6366A$, we have $n_1^* < n_2^*$. One may explore these points using the subprograms under program number 8.

Consider the distribution of N given in (6.3.13) for a two-stage procedure. Here the fixed positive number b_m in (6.3.10) is a function of k. Therefore as before, we may rewrite:

$$b_m(k) \equiv b_m = \left\{\frac{1}{2}(m-1)\right\}^{k/2} \Gamma\left(\frac{1}{2}(m-k-1)\right) \left\{\Gamma\left(\frac{1}{2}(m-1)\right)\right\}^{-1}, \tag{6.3.21}$$

for $k = 1, 2$. Next, we may

$$B_r(k) \equiv B_r = \frac{(m+r)(m-1)}{\sigma^2} \left(\frac{\omega}{b_m(k)B(k)}\right)^{2/k}, \; k = 1, 2. \tag{6.3.22}$$

Now, using (6.3.18), (6.3.21) and (6.3.18), one can check that

$$\frac{b_m(2)}{b_m^2(1)} = \frac{\Gamma\left(\frac{1}{2}(m-3)\right)\Gamma\left(\frac{1}{2}(m-1)\right)}{\Gamma^2\left(\frac{1}{2}(m-2)\right)} \tag{6.3.23}$$

and

$$\frac{B_r(1)}{B_r(2)} = \frac{\pi\omega}{2A}\frac{b_m(2)}{b_m^2(1)} = \frac{\pi\omega}{2A}\frac{\Gamma\left(\frac{1}{2}(m-3)\right)\Gamma\left(\frac{1}{2}(m-1)\right)}{\Gamma^2\left(\frac{1}{2}(m-2)\right)}. \tag{6.3.24}$$

Thus, if $m = 5$, (6.3.24) easily reduces to:

$$\frac{B_r(1)}{B_r(2)} = \frac{2\omega}{A}$$

So, the distributions of N under these two loss functions are identical if we choose $\omega = 0.5A$ when $m = 5$. The subprogram number 2 from program number 8 may be used to explore other scenarios pertaining to similar features.

6.4 Minimum Risk Point Estimation

Now, we are about to introduce a minimum risk point estimation problem for μ in a $N(\mu, \sigma^2)$ distribution. A pathbreaking paper on the original formulation was due to Robbins (1959). We assume that μ, σ^2 are both unknown with $-\infty < \mu < \infty, 0 < \sigma < \infty$. Suppose that $X_1, \ldots, X_n, \ldots$ are i.i.d. observations from this distribution where $n \geq 2$. We propose to estimate μ under the following loss function which is composed of a squared error loss due to estimation and the cost due to sampling: Having recorded $X_1, \ldots, X_n$, suppose that the loss in estimating μ by the sample mean, $\overline{X}_n$, is given by

$$L_n(\mu, \overline{X}_n) = A\left(\overline{X}_n - \mu\right)^2 + cn \text{ with } A > 0 \text{ and } c > 0 \text{ known.} \quad (6.4.1)$$

Here, c represents the cost per unit observation and $\left(\overline{X}_n - \mu\right)^2$ represents the loss due to estimation of μ. If n is small, then sampling cost cn will be small but we may expect a large loss due to estimation. On the other hand, if n is large, then the sampling cost cn will be large but we may expect a small loss due to estimation. So, this problem must balance the expenses incurred by sampling and the achieved margin of estimation error.

Next, suppose that c is measured in dollars and the X's are measured in inches. Then, the weight A found in (6.4.1) would have an unit 'dollars per square inch'. In other words, the weight A dollars per square inch should reflect an experimenter's feeling about the likely cost per unit squared error loss due to estimation.

The fixed-sample-size risk function associated with (6.4.1) can be expressed as:

$$R_n(c) \equiv \mathrm{E}_{\mu,\sigma}\left\{L_n(\mu, \overline{X}_n)\right\} = A\sigma^2 n^{-1} + cn. \quad (6.4.2)$$

Our goal is to design estimation strategies that would minimize this risk for all $0 < \sigma < \infty$. Now, let us pretend that n is a continuous variable and examine the nature of the curve, $g(n) \equiv A\sigma^2 n^{-1} + cn$ for $n > 0$. It is a strictly convex curve, but $g(n) \to \infty$ whether $n \downarrow 0$ or $n \uparrow \infty$. In other words, $g(n)$ must attain a minimum at some point within $(0, \infty)$. Consider $\frac{d}{dn}g(n) = -A\sigma^2 n^{-2} + c$, and solve $\frac{d}{dn}g(n) = 0$ to obtain $n \equiv n^*(c) = (A/c)^{1/2}\sigma$. Obviously, $\left.\frac{d^2}{dn^2}g(n)\right|_{n=n^*} > 0$ so that $n \equiv n^*(c)$ minimizes the function $g(n)$.

We note, however, that

$$n \equiv n^*(c) = (A/c)^{1/2}\sigma \quad (6.4.3)$$

may not be a (positive) integer, but we refer to $n^*(c)$ as the optimal fixed-sample-size had σ been known. The minimum fixed-sample-size risk is then given by

$$R_{n^*}(c) \equiv A\sigma^2 n^{*-1} + cn^* = 2cn^*. \tag{6.4.4}$$

Surely, the magnitude of the optimal fixed sample size remains unknown since σ is unknown. In fact, no fixed sample size procedure will minimize the risk (6.4.2) under the loss function (6.4.1), uniformly in σ for all $0 < \sigma < \infty$. Robbins (1959) proposed an appropriate purely sequential procedure.

Starr (1966b) presented a detailed study of the Robbins procedure. Ghosh and Mukhopadhyay (1976) investigated an appropriate two-stage procedure. Mukhopadhyay (1990) gave a three-stage sampling strategy and Mukhopadhyay (1996) and Mukhopadhyay et al. (1987) included accelerated sequential procedures. See also Ghosh and Mukhopadhyay (1981). For brevity, we introduce only the purely sequential procedure and a multistage strategy for this estimation problem.

6.4.1 Purely Sequential Procedure

Recall that $S_n^2 = (n-1)^{-1}\sum_{i=1}^{n}\left(X_i - \overline{X}_n\right)^2, n \geq 2$, stands for a sample variance. The purely sequential stopping rule of Robbins (1959) is summarized as follows.

We start with $X_1, \ldots, X_m$, a pilot of size $m(\geq 2)$, and then we take one additional X at-a-time. The stopping time is defined as:

$$\begin{aligned} &N \equiv N(c) \text{ is the smallest integer } n(\geq m) \text{ for which} \\ &\qquad\qquad \text{we observe } n \geq (A/c)^{1/2} S_n. \end{aligned} \tag{6.4.5}$$

This sequential procedure is implemented in the same way we did in the case of (6.2.17). Again, $\mathrm{P}_{\mu,\sigma}(N < \infty) = 1$, that is sampling terminates with probability one.

Upon termination, we estimate μ by $\overline{X}_N$ obtained from the full dataset $N, X_1, \ldots, X_m, \ldots, X_N$. We leave it as an exercise to prove that $\overline{X}_N$ is an unbiased estimator of μ and its variance is $\sigma^2\mathrm{E}_{\mu,\sigma}(N^{-1})$ for all fixed μ, σ, A, c.

The risk associated with $\overline{X}_N$ can be expressed as:

$$R_N(c) \equiv \mathrm{E}_{\mu,\sigma}\left\{L_N(\mu, \overline{X}_N)\right\} = A\sigma^2\mathrm{E}_{\mu,\sigma}\left[N^{-1}\right] + c\mathrm{E}_{\mu,\sigma}\left[N\right]. \tag{6.4.6}$$

Robbins (1959) defined two measures for comparing the sequential strategy $(N, \overline{X}_N)$ from (6.4.5) with the optimal strategy $(n^*, \overline{X}_{n^*})$ pretending that n^* is known. The notions are:

$$\begin{aligned} &\textit{Risk Efficiency}: \text{RiskEff}(c) \equiv \frac{R_N(c)}{R_{n^*}(c)} = \tfrac{1}{2}\mathrm{E}_{\mu,\sigma}\left[N/n^*\right] + \tfrac{1}{2}\mathrm{E}_{\mu,\sigma}\left[n^*/N\right]; \\ &\textit{Regret}: \text{Regret}(c) \equiv R_N(c) - R_{n^*}(c) = c\mathrm{E}_{\mu,\sigma}\left[\frac{(N-n^*)^2}{N}\right]. \end{aligned} \tag{6.4.7}$$

Robbins (1959) included an algorithm to find the exact probability distribution of N pretending that σ were known, and showed that the risk efficiency and regret came incredibly close to one and zero respectively, even for rather small values of n^*. Starr (1966b) embarked upon a very thorough investigation and showed the following first-order results. For all fixed μ, σ, m and A, as $c \to 0$:

$$\begin{gathered}\mathrm{E}_{\mu,\sigma}\left[N/n^*\right] \to 1 \text{ if } m \geq 2;\ \mathrm{E}_{\mu,\sigma}\left[n^*/N\right] \to 1 \text{ if and only if } m \geq 3; \\ \text{RiskEff}(c) \to 1 + (2\pi)^{-1/2}\sigma^{-1} \text{ if } m = 2; \\ \text{RiskEff}(c) \to 1 \text{ if } m \geq 3 \ [\textit{Asymptotic First-Order Risk Efficiency}]\end{gathered} \tag{6.4.8}$$

See also Mukhopadhyay (1991) for many details.

By the way, in Ghosh and Mukhopadhyay (1976), one will find the following results. For all fixed μ, σ, m, A and c, one has:

$$\begin{gathered}\mathrm{E}_{\mu,\sigma}\left[N\right] \leq n^* + m + 1; \text{ and} \\ \mathrm{E}_{\mu,\sigma}\left[N^2\right] \leq \left(n^* + m + 1\right)^2 \text{ if } m \geq 2.\end{gathered} \tag{6.4.9}$$

Since $N/n^* \xrightarrow{\mathrm{P}} 1$, one may note that Fatou's lemma and (6.4.7) would lead to the following assertions. With fixed μ, σ, m and A, the following first-order asymptotic results hold as $c \to 0$:

$$\begin{gathered}\mathrm{E}_{\mu,\sigma}\left[N/n^*\right] \to 1 \text{ and} \\ \mathrm{E}_{\mu,\sigma}\left[N^2/n^{*2}\right] \to 1 \text{ if } m \geq 2.\end{gathered} \tag{6.4.10}$$

The first-order properties from (6.4.8) may be interpreted as follows: For large n^* values, that is for small c values, we would expect the average sample size $\mathrm{E}_{\mu,\sigma}[N]$ to hover in a close proximity of n^*. One may note from (6.4.7) that the expression of the risk efficiency can be rewritten as

$$\begin{aligned}\text{RiskEff}(c) &\equiv 1 + \frac{1}{2}\left\{\mathrm{E}_{\mu,\sigma}\left[\frac{N}{n^*}\right] + \mathrm{E}_{\mu,\sigma}\left[\frac{n^*}{N}\right] - 2\right\} \\ &= 1 + \frac{1}{2}\mathrm{E}_{\mu,\sigma}\left[\left(\sqrt{N/n^*} - \sqrt{n^*/N}\right)^2\right],\end{aligned} \tag{6.4.11}$$

that is, RiskEff(c), would always exceed one for fixed μ, σ, m, c and A. Hence, for fixed μ, σ, m, c and A, the sequential risk $R_N(c)$ will exceed the optimal fixed sample risk $R_{n^*}(c)$.

But, from the last part of (6.4.8) it should be clear that for large n^* values, we would expect RiskEff(c) to hover in a close proximity of one. That is, for large n^* values we would expect the sequential risk $R_N(c)$ and the optimal fixed sample risk $R_{n^*}(c)$ to stay close to one another when $m \geq 3$.

Starr and Woodroofe (1969) strengthened considerably the asymptotic risk efficiency result from the last step in (6.4.8) by proving the following result for all fixed μ, σ, m and A, as $c \to 0$:

$$\text{Regret}(c) \equiv R_N(c) - R_{n^*}(c) = O(c) \text{ if and only if } m \geq 3. \qquad (6.4.12)$$

This paper did not throw much light on the constant multiple that was involved in the $O(c)$ term. This constant term actually turned out to be $\frac{1}{2}$ which was proved in the remarkable paper of Woodroofe (1977).

The *asymptotic second-order* results are involved. The tools that are essential in their proofs may be borrowed from Woodroofe (1977,1982) and Mukhopadhyay (1988a). See also Mukhopadhyay and Solanky (1994, Section 2.4.2) and Ghosh et al. (1997, Section 7.2) for many details. Let us denote

$$\eta_1(1) = -\frac{1}{2}\sum_{n=1}^{\infty}\frac{1}{n}\text{E}\left[\max\left\{0, \chi_n^2 - 3n\right\}\right] \qquad (6.4.13)$$

From Woodroofe (1977), for all fixed μ, σ, m and A, we summarize the following asymptotic results as $c \to 0$:

$$\begin{aligned} &\text{E}_{\mu,\sigma}(N - n^*) = \eta_1(1) + o(1) \text{ if } m \geq 3; \\ &\frac{N - n^*}{\sqrt{n^*}} \text{ converges to } N(0, \tfrac{1}{2}) \text{ in distribution if } m \geq 2; \\ &\text{V}_{\mu,\sigma}(N) = \tfrac{1}{2}n^* + o(n^*) \text{ if } m \geq 3; \\ &\text{Regret}(c) \equiv R_N(c) - R_{n^*}(c) = \tfrac{1}{2}c + o(c) \text{ if } m \geq 4. \end{aligned} \qquad (6.4.14)$$

The purely sequential procedure (6.4.5) is *asymptotically second-order efficient* and *second-order risk efficient.*

For large n^*, let us interpret the results from (6.4.14). Note that $\eta_1(1)$ from (6.4.13) is a computable number. So, from the first result in (6.4.14), we know on an average how much the sample size N may differ from n^* when n^* is large. We can safely say that $\text{E}_{\mu,\sigma}(N - n^*)$ will be in a close proximity of $\eta_1(1)$. The second part says that the normalized stopping variable, namely $(N - n^*)/\sqrt{n^*}$, would have an approximate normal distribution. The third part says that $\text{V}_{\mu,\sigma}(N)$ should be nearly $\frac{1}{2}n^*$. The last part says that the asymptotic regret amounts to the cost of one-half of one observation. In other words, in the absence of any knowledge about σ, we may end up wasting the price of only one-half of one observation in the process of using the purely sequential strategy (6.4.5). The purely sequential estimation strategy is that efficient!

Computations and Simulations

The minimum risk point estimation (MRPE) would be explained more with the help of hands-on simulations and data analysis using program number 9 within **Seq06.exe** program. This subprogram also obtains the MRPE of μ when one inputs live data.

Example 6.4.1 : Consider a simulation study under MRPE of μ by generating data from a $N(5,4)$ distribution. We fix $m = 5$ and replicate simulations 2000 times using purely sequential procedure with the following loss function:

$$L_n(\mu, \overline{X}_n) = 100\left(\overline{X}_n - \mu\right)^2 + 0.5n.$$

An output from subprogram number 1 in program number 9 is shown in Figure 6.4.1. This example is considered again later.

```
        Minimum Risk Point Estimation
      Using Purely Sequential procedure
      =================================

   Input Parameters:
         Loss function constant, A = 100.00
         Cost of unit sample     c =   0.50
         Initial sample size,    m =    5

  Simulation study by generating data from
  normal distribution with mean =  5.00
  and variance =  4.00

  Number of simulation replications =  2000

  Optimal sample size (n_star)        =     28.28
  Average final sample size (n_bar) =     28.24
  Std. dev. of n (s_n)                =      4.12
  Minimum sample size (n_min)         =      5
  Maximum sample size (n_max)         =     39

  Minimum fixed-sample risk (Rn*)     =     28.28
  Simulated average risk (Rn_bar)     =     27.63
  Simulated average regret            =      0.42
```

Figure 6.4.1. *Minimum risk point estimation using a purely sequential procedure.*

The program number 9 in **Seq06.exe** the 'Minimum fixed-sample risk (Rn*)' is same as $R_{n^*} = 2cn^*$ given in (6.4.4). Note that the 'Simulated average risk (Rn_bar)' and the 'Simulated average regret' are respectively the estimates $R_N(c)$ given in (6.4.6) and Regret(c) in (6.4.7). Let n_i be the value of N and s_i^2 be the estimate of σ^2 in the ith simulation replication for $i = 1, \ldots, n_s$ where n_s is the number of simulations. The program computes these estimators as follows:

$$\text{Simulated average risk (Rn\_bar)} = \frac{1}{n_s}\sum_{i=1}^{n}\left(\frac{As_i^2}{n_i} + cn_i\right)$$

and

$$\text{Simulated average regret} = \frac{c}{n_s}\sum_{i=1}^{n}\left(\frac{(n_i - n^*)^2}{n_i}\right).$$

Note that the sample size N is estimated by $n_i = \sqrt{As_i^2/c}$ to obtained the above estimates.

6.4.2 Parallel Piecewise Sequential Procedure

We have mentioned before that a purely sequential procedure can be operationally inconvenient because of gathering one observation at-a-time until termination. For added convenience, we advocated earlier a three-stage or an accelerated sequential procedure in a number of problems. Such procedures would also preserve crucial second-order properties.

But, in order to expose the readers to another class of procedures, we now introduce a *parallel piecewise sequential* methodology. The original idea was developed by Mukhopadhyay and Sen (1993).

Let $k(\geq 2)$ be a prefixed integer. Consider k independent sequences $X_{ij}, j = 1, \ldots, n_i, \ldots$ and $i = 1, \ldots, k$, of i.i.d. observations from a $N(\mu, \sigma^2)$ distribution.

We may visualize k investigators gathering data from the same population at the same time, but independently of each other. In simulation runs, one may think of vector parallel processing with k arms where each arm simultaneously process same type of job independently of other arms. Finally, the experimenter will combine the data from all k sources and come up with an estimator for μ.

Having recorded $X_{i1}, \ldots, X_{in_i}$ from the ith arm with $n_i \geq 2$, let us denote:

$$\overline{X}_{in_i} = \frac{1}{n_i}\sum_{j=1}^{n_i} X_{ij} \text{ and } S_{in_i}^2 = \frac{1}{n_i - 1}\sum_{j=1}^{n_i}(X_{ij} - \overline{X}_{in_i})^2$$

for the sample mean and the sample variance respectively, $i = 1, \ldots, k$. We also write:

$$\boldsymbol{n}' = (n_1, \ldots, n_k),\ n = \sum_{i=1}^{k} n_i,$$

and denote the pooled estimator of μ:

$$T_k(\boldsymbol{n}) = n^{-1}\sum_{i=1}^{k} n_i\overline{X}_{in_i},$$

which coincides with the customary sample mean from combined data $X_{i1}, \ldots, X_{in_i}, i = 1, \ldots, k$.

Suppose that the loss in estimating μ by $T_k(\boldsymbol{n})$ is given by:

$$L_{\boldsymbol{n}}(\mu, T_k(\boldsymbol{n})) = A\left(T_k(\boldsymbol{n}) - \mu\right)^2 + cn \text{ with } A > 0 \text{ and } c > 0 \text{ known.} \tag{6.4.15}$$

The fixed-sample-size risk function associated with (6.4.15) can be expressed as in (6.4.2). Again, our goal is to design an estimation strategy that would minimize this risk for all $0 < \sigma < \infty$. We recall from (6.4.3) that the optimal fixed sample size would be:

$$n_i \equiv n_i^*(c) = k^{-1}(A/c)^{1/2}\sigma, n \equiv n^*(c) = (A/c)^{1/2}\sigma,$$
$$\text{and } \boldsymbol{n}^* = (n_1^*, \ldots, n_k^*).$$

The associated minimum fixed-sample-size risk is:

$$R_{\boldsymbol{n}^*}(c) \equiv A\sigma^2 n^{*-1} + cn^* = 2cn^*. \tag{6.4.16}$$

Now, on the ith arm, we carry out purely sequential sampling with one observation at-a-time. The stopping variable is defined as follows. We start with $X_{i1}, \ldots, X_{im}$, a pilot of size $m(\geq 2)$, and then we take one additional X at-a-time as needed. The stopping time is defined as:

$$\begin{aligned} &N_i \equiv N_i(c) \text{ is the smallest integer } n(\geq m) \text{ for which} \\ &\text{we observe } n \geq \frac{1}{k}\left(A/c\right)^{1/2} S_{in},\ i = 1, \ldots, k. \end{aligned} \tag{6.4.17}$$

When each arm terminates sampling, we estimate μ by the combined sample mean $T_k(\boldsymbol{n})$, obtained from the full dataset $X_{i1}, \ldots, X_{im}, \ldots, X_{iN_i}$, $i = 1, \ldots, k$. We denote:

$$\boldsymbol{N}' = (N_1, \ldots, N_k),\ N = \sum_{i=1}^{k} N_i,\ \overline{N} = k^{-1}N. \tag{6.4.18}$$

Again, $\mathrm{P}_{\mu,\sigma}(N_i < \infty) = 1$ for $i = 1, \ldots, k$. That is, the parallel piecewise sequential sampling would terminate with probability one. We leave it as an exercise to prove that $T_k(\boldsymbol{N})$ is an unbiased estimator of μ and its variance is $\sigma^2 \mathrm{E}_{\mu,\sigma}(N^{-1})$ for all fixed μ, σ, A and c.

The risk associated with the sample mean $T_k(\boldsymbol{N})$ can be expressed as

$$R_{\boldsymbol{N}}(c) \equiv \mathrm{E}_{\mu,\sigma}\left\{L_{\boldsymbol{N}}\left(\mu, T_k(\boldsymbol{N})\right)\right\} = A\sigma^2 \mathrm{E}_{\mu,\sigma}\left[N^{-1}\right] + c\mathrm{E}_{\mu,\sigma}\left[N\right]. \tag{6.4.19}$$

We have:

$$\begin{aligned} &\textit{Risk Efficiency}: \text{RiskEff}(c) \equiv \frac{R_{\boldsymbol{N}}(c)}{R_{\boldsymbol{n}^*}(c)} = \tfrac{1}{2}\mathrm{E}_{\mu,\sigma}\left[N/n^*\right] + \tfrac{1}{2}\mathrm{E}_{\mu,\sigma}\left[n^*/N\right] \\ &\textit{Regret}: \text{Regret}(c) \equiv R_{\boldsymbol{N}}(c) - R_{\boldsymbol{n}^*}(c) = c\mathrm{E}_{\mu,\sigma}\left[\frac{(N-n^*)^2}{N}\right]. \end{aligned} \tag{6.4.20}$$

Now, we summarize the following *second-order* results from Mukhopadhyay and Sen (1993) as $c \to 0$:

$$\begin{aligned} &\mathrm{E}_{\mu,\sigma}\left(N - n^*\right) = k\eta_1(1) + o(1) \text{ if } m \geq 3; \\ &\frac{N - n^*}{\sqrt{n^*}} \text{ converges to } N(0, \tfrac{1}{2}) \text{ in distribution if } m \geq 2; \\ &\mathrm{V}_{\mu,\sigma}(N) = \tfrac{1}{2}n^* + o(n^*) \text{ if } m \geq 3; \\ &\text{Regret}(c) \equiv R_{\boldsymbol{N}}(c) - R_{\boldsymbol{n}^*}(c) = \tfrac{1}{2}c + o(c) \text{ if } m \geq 4. \end{aligned} \tag{6.4.21}$$

Note that the expansions of the regret function, Regret(c), from (6.4.14) and (6.4.21) are exactly the same. In other words, as far as the regret of a procedure is concerned, the parallel piecewise sequential procedure (6.4.17) fares as well as the purely sequential sampling strategy (6.4.5). Certainly, the results from (6.4.10) and the asymptotic risk efficiency (RiskEff(c) $\to$ 1) continue to hold.

Next, since $N_1, \ldots, N_k$ are i.i.d. random variables, the $\mathrm{V}(N)$ can be unbiasedly estimated by:

$$\Delta = k(k-1)^{-1}\sum_{i=1}^{k}(N_i - \overline{N})^2 \text{ with } \overline{N} = k^{-1}\sum_{i=1}^{k} N_i. \tag{6.4.22}$$

This is an added feature under parallel networking! In the case of the purely sequential estimation procedure (6.4.5), we could not provide an unbiased estimator for the variance of the sample size.

> The parallel piecewise sequential procedure (6.4.17) is *asymptotically second-order efficient* and *second-order risk efficient.*

Observe that a parallel piecewise sequential methodology (6.4.17) will require significantly less time to complete all k runs. Approximately, the run-time of a parallel piecewise sequential strategy will be:

$\frac{1}{k}\times$ the length of time a purely sequential strategy (6.4.5) requires to wrap up the problem.

This translates into significant reductions in operations and logistics. Such interesting features will be explained more with the help of hands-on simulations and data analyses.

The basic idea of piecewise sampling can also be implemented in the context of two-stage, three-stage, and accelerated sequential sampling. Mukhopadhyay and Datta (1994) developed machineries for parallel piecewise multistage methodologies.

Another modification could also be looked at. One has noted that in the original parallel piecewise sequential methodology, we ran *all* k pieces simultaneously. But, as long as the notion of parallel sampling is in place, one does not have to run *all* k pieces. We may decide to run only *some* of the k pieces simultaneously in order to save resources even more. Mukhopadhyay and de Silva (1998a) developed machineries for parallel *partial piecewise* multistage methodologies. The interesting features of such methodologies include (i) tremendous savings due to operational convenience, and (ii) second-order properties comparable to those of three-stage, accelerated sequential, and purely sequential sampling.

Subprogram number 2 of program number 9 carries out minimum risk point estimation of a normal mean using a parallel piecewise procedure. This program would perform simulation studies and it would also work with live data.

6.4.3 Comparisons and Conclusions

Table 6.4.1 presents a summary of the results and compares them in the light of Examples 6.4.1 and 6.4.2. Here, purely sequential results are obtained assuming that there is only one investigator to collect full data. The piecewise procedure used 3 parallel investigators. Since each investigator stars with $m = 5$ observations, $\min(n) = 3 \times 5 = 15$ for a piecewise procedure, but $\min(n) = m = 5$ for the purely sequential procedure. Further, in a piecewise procedure one can compute Δ, an unbiased estimate of $\mathrm{V}(N)$. The results obtained by either procedure are very similar to each other. One can readily see that the final average sample sizes or the estimated risks are in the same ballpark.

Example 6.4.2: Repeat Example 6.4.1 with 3 "parallel" investigators (arms) using subprogram Number 1 from program number 9. The output is shown in Figure 6.4.2.

```
     Minimum Risk Point Estimation Using
     Parallel Piecewise Sequential Procedure
     =======================================

 Input Parameters:
       Number of investigators,k =    3
       Loss function constant, A = 100.00
       Cost of unit sample     c =   0.50
       Initial sample size,    m =    5

 Simulation study by generating data from
 normal distribution with mean =  5.00
 and variance =  4.00

 Number of simulation replications =  2000

 Optimal sample size (n_star)        =       28.28
 Average final sample size (n_bar)   =       27.41
 Std. dev. of n (s_n)                =        4.23
 Minimum sample size (n_min)         =       15
 Maximum sample size (n_max)         =       41

 Minimum fixed-sample risk (Rn*)     =       28.28
 Simulated average risk (Rn_bar)     =       27.56
 Simulated average regret            =        0.38

 Ave investigators variances,(Delta) =       17.89
 Estimate of Var(N) using simulation =       17.85
```

Figure 6.4.2. *Minimum risk point estimation using parallel piecewise sequential procedure.*

A parallel piecewise strategy only demands availability of k equally reliable and efficient arms that may run simultaneously. So, what is the advantage of a parallel piecewise strategy over a purely sequential strategy? Suppose that a purely sequential strategy takes t_S seconds to provide all the information needed to fill the entries in column 2 of Table 6.4.1. Next, suppose that a parallel piecewise strategy takes t_P seconds to provide all the information needed to fill the entries in column 3 of Table 6.4.1. Then, $t_P \approx \frac{1}{k} t_S$, which means that a parallel piecewise strategy would save nearly $100(1 - \frac{1}{k})\%$ of an experiment's total duration compared with a purely sequential experiment! Note, that this important conclusion has nothing to do with the unit of time (for example, seconds or hours or days) or how old or new or fast or slow a batch of k computers may be as long as they are equally reliable and efficient.

Table 6.4.1: Simulation results comparison for minimum risk point estimations
$n^* = 28.28; \quad R_{n^*} = 28.28$

	Purely Sequential Procedure	Parallel Piecewise Procedure
$\overline{n}$	27.70	27.18
s_n	4.51	4.21
Simulate $\mathrm{V}(N)$	20.34	17.76
Δ	-	17.58
min(n)	5	15
max(n)	39	39
$R_{\overline{n}}$	27.08	27.28
Regret	-1.21	-1.01

Example 6.4.3 : Sequential data given below were obtained from a normal population with unknown mean μ and variance σ^2. Obtain minimum risk point estimators for μ using both purely sequential and parallel piecewise sequential procedures under the following loss function:

$$L_n(\mu, \overline{X}_n) = 100\left(\overline{X}_n - \mu\right)^2 + 0.5n.$$

Data*:*

5.3	3.1	3.3	5.7	6.9	2.3	2.8	4.3	3.6	4.9
4.6	3.2	6.2	6.0	3.7	2.3	3.0	4.4	3.4	4.0

Figure 6.4.3 shows results obtained from subprogram number 1 & 2 within program number 9. The information on different aspects of minimum risk point estimation (MRPE) of μ using purely sequential and parallel piecewise sequential procedures are included.

When purely sequential procedure was adopted, it used 20 observations to obtain an estimate (4.15) of μ. A parallel piecewise procedure with 3 investigators obtained the following results: Investigators #1, #2 and #3 terminated sampling with 8, 6 and 5 observations with means 4.212, 4.750 and 3.360 respectively. In summary the procedure used 19 observations to obtain an estimate (4.158) of μ. In addition, the piecewise procedure gave:

$$\text{Estimated V}(N) = 7.00 \text{ and Estimated } R_N(c) = 18.788.$$

```
   Minimum Risk Point Estimation
Using Purely Sequential procedure
=================================

Input Parameters:
     Loss function constant, A = 100.00
     Cost of unit sample     c =   0.50
     Initial sample size,    m =    5

Final Sample size =     20
Sample mean       =     4.15
Sample variance   =     1.77
```

```
Minimum Risk Point Estimation Using
Parallel Piecewise Sequential Procedure
=======================================

Input Parameters:
     Number of investigators,k =     3
     Loss function constant, A = 100.00
     Cost of unit sample     c =   0.50
     Initial sample size,    m =    5

Investigator  sample    sample    sample
   number       mean   variance     size
     1         4.212      2.621        8
     2         4.750      1.487        6
     3         3.360      0.613        5

 Pooled        4.158      1.765       19

   Unbiased Estimate of V(N) = Delta =   7.000
   Estimate Risk, RN(c) =  18.788
```

Figure 6.4.3. *MRPE of μ for normal population data.*

6.5 Some Selected Derivations

In this section, we provide some selected sketches of details. Some of these may sound simple. Others may appear technical. We include them for completeness. The readers may pick and choose as they wish.

6.5.1 Independence of Sample Mean and Sample Variances

Suppose that $X_1, \ldots, X_n$ are i.i.d. random variables from a $N(\mu, \sigma^2)$ distribution having μ, σ^2 both unknown. Recall $\overline{X}_n = n^{-1}\sum_{i=1}^{n} X_i$ and $S_n^2 = (n-1)^{-1}\sum_{i=1}^{n}\left(X_i - \overline{X}_n\right)^2$ for $n \geq m(\geq 2)$. Here, m is a pilot size. In what follows, we argue that $\overline{X}_n$ and $(S_m^2, S_{m+1}^2, \ldots, S_n^2)$ are independently distributed.

For an arbitrary but fixed value of σ^2, we note that (i) $\overline{X}_n$ is a complete and sufficient statistic for μ, and (ii) $(S_m^2, S_{m+1}^2, \ldots, S_n^2)$ is an ancillary statistic for μ.

Then, by Basu's (1955) Theorem, for one fixed value of σ^2 and for all $n \geq m(\geq 2)$, $\overline{X}_n$ and $(S_m^2, S_{m+1}^2, \ldots, S_n^2)$ are independent. But, note that independence between $\overline{X}_n$ and $(S_m^2, S_{m+1}^2, \ldots, S_n^2)$ holds as long as σ^2 is kept fixed. Also, such independence does not have anything to do with a particular value of σ^2 that is fixed.

Hence, $\overline{X}_n$ and $(S_m^2, S_{m+1}^2, \ldots, S_n^2)$ are independent for all fixed $n(\geq m)$ whatever be $(\mu, \sigma^2), -\infty < \mu < \infty, 0 < \sigma^2 < \infty$. ■

Independence between $I(N = n)$ and $\overline{X}_n$

In all multistage and sequential procedures discussed in this chapter, we have noticed that the event $\{N = n\}$ is determined exclusively by the observed values of $(S_m^2, S_{m+1}^2, \ldots, S_n^2)$. In other words, whether or not we stop at $N = n$ depends solely on the observed values of $(S_m^2, S_{m+1}^2, \ldots, S_n^2)$. We may put this in another way: The random variable $I(N = n)$ is a function of $(S_m^2, S_{m+1}^2, \ldots, S_n^2)$ alone. Hence, a function of $\overline{X}_n$ alone, including $\overline{X}_n$ itself, and a function of $(S_m^2, S_{m+1}^2, \ldots, S_n^2)$ alone must be independently distributed for all fixed $n(\geq m)$.

It should be obvious now that $I(N = n)$ and $\overline{X}_n$ would be independently distributed for all fixed $n(\geq m)$. ■

6.5.2 Second-Order Results for Sequential Strategy (6.2.17)

Let us go back to the stopping variable N from (6.2.17). The boundary condition involves S_n^2, but we know that $(n-1)S_n^2/\sigma^2 \sim \chi_{n-1}^2$. Let us think of random variables $W_1, W_2, \ldots$ which are i.i.d. χ_1^2. Now, we rewrite (6.2.17) as:

$$\begin{aligned} N \equiv \;& N(d) \text{ is the smallest integer } n(\geq m) \text{ for which} \\ & \text{we observe } n(n-1) \geq \frac{a^2\sigma^2}{d^2}\frac{(n-1)S_n^2}{\sigma^2}. \end{aligned}$$

This can be further rewritten as:

$$\begin{aligned} & N \equiv N(d) \text{ is the smallest integer } n(\geq m) \text{ for which} \\ & \quad \text{we observe } C^{-1}n(n-1) \geq \textstyle\sum_{i=1}^{n-1} W_i \end{aligned} \tag{6.5.1}$$

since $C = a^2\sigma^2/d^2$. We can express N from (6.2.17) or (6.5.1) or (6.5.2) as:

$$N = T + 1 \text{ w.p. } 1$$

where a new stopping variable T is defined as:

$$\begin{array}{c} T \equiv T(d) \text{ is the smallest integer } n(\geq m-1) \text{ for which} \\ \text{we observe } C^{-1}(n+1)n \geq \sum_{i=1}^{n} W_i \text{ or equivalently} \\ \text{observe } C^{-1}n^2(1+\frac{1}{n}) \geq \sum_{i=1}^{n} W_i. \end{array} \quad (6.5.2)$$

The random variable T is obtained by substituting "$n-1$" with "n" on both sides of the boundary condition in (6.5.2). Note that the setup for T matches with that in (1.1) of Woodroofe (1977) or Mukhopadhyay (1988a). This is summarized in Section A.4 of the Appendix. Match (6.5.2) with (A.4.1) to note:

$$\begin{array}{c} m_0 = m-1, \delta = 2, h^* = C^{-1}, L_n = (n+1)n, L_0 = 1, \\ \theta = 1, \tau^2 = 2, \beta^* = 1 \Rightarrow n_0^* = C \text{ and } q = 2. \end{array} \quad (6.5.3)$$

Also, since $W_1 \sim \chi_1^2$, we have for $u > 0$:

$$\begin{aligned} \mathrm{P}(W_1 \leq u) &= \frac{1}{\sqrt{2\pi}} \int_0^u \exp(-w/2) w^{-1/2} dw \\ &\leq \frac{1}{\sqrt{2\pi}} \int_0^u w^{-1/2} dw = \frac{2}{\sqrt{2\pi}} u^{1/2}. \end{aligned}$$

From (6.5.5) we see that the condition from (A.4.5) holds with $r = \frac{1}{2}$.

Again, since $\sum_{i=1}^{n} W_i \sim \chi_n^2$, from (A.4.3)-(A.4.4), we obtain:

$$\begin{aligned} &\nu = \tfrac{3}{2} - \textstyle\sum_{n=1}^{\infty} \frac{1}{n} \mathrm{E}\left[\max\left(0, \chi_n^2 - 2n\right)\right] \\ &\Rightarrow \kappa = \nu - 3 = -\tfrac{3}{2} - \textstyle\sum_{n=1}^{\infty} \frac{1}{n} \mathrm{E}\left[\max\left(0, \chi_n^2 - 2n\right)\right]. \end{aligned} \quad (6.5.4)$$

One should check easily that our κ in (6.5.4) matches exactly with the expression of $\eta(1)$ from (6.2.33).

The distribution of T or N is affected by σ^2 only. But, we will continue to write $\mathrm{P}_{\mu,\sigma}(\cdot)$ or $\mathrm{E}_{\mu,\sigma}(\cdot)$. Next, let us denote $U^* = (T-C)/C^{1/2}$. Now, we can state some conclusions as $d \to 0$:

$$\begin{array}{rl} \text{(i)} & \mathrm{P}_{\mu,\sigma}(T \leq \frac{1}{2}C) = O(h^{*(m-1)/2}) = O(C^{-(m-1)/2}) \text{ if } m \geq 2; \\ \text{(ii)} & U^* \text{ converges in distribution to } N(0,2) \text{ if } m \geq 2; \text{ and} \\ \text{(iii)} & U^{*2} \text{ is uniformly integrable if } m \geq 4. \end{array} \quad (6.5.5)$$

These conclusions follow directly from Theorem A.4.1 in the Appendix.

Expected Value of N:

Now, Theorem A.4.2 clearly implies with $\omega = 1$:

$$\begin{aligned} &\mathrm{E}_{\mu,\sigma}[T] = C + \kappa + o(1) \text{ if } m-1 > 2, \text{ that is if } m \geq 4 \\ &\Rightarrow \mathrm{E}_{\mu,\sigma}[N] = \mathrm{E}_{\mu,\sigma}[T+1] = C + \kappa + 1 + o(1) \text{ if } m \geq 4. \end{aligned} \quad (6.5.6)$$

The constant term "$\kappa + 1$" from (6.5.6) is exactly the same as the constant "$\eta(1)$" that we had defined in (6.2.33). ■

Variance of N:

Parts (ii) and (iii) from (6.5.5) immediately lead to the following conclusion:

$$\mathrm{V}_{\mu,\sigma}[N] = 2C + o(C), \tag{6.5.7}$$

if $m \geq 4$. ■

Expansion of Confidence Coefficient:

The confidence coefficient associated with J_N is:

$$\begin{array}{c} h(Nd^2/\sigma^2) \equiv \mathrm{E}_{\mu,\sigma}\left[2\Phi\left(\sqrt{N}d/\sigma\right) - 1\right] \text{ where we let} \\ h(x) = 2\Phi\left(\sqrt{x}\right) - 1, x > 0. \end{array} \tag{6.5.8}$$

Since, $\sqrt{N}d/\sigma$ converges to a in probability, we want to expand $h(x)$ around $x = a^2$. A simple application of Taylor expansion gives:

$$\begin{array}{l} \mathrm{E}_{\mu,\sigma}\left[2\Phi\left(\sqrt{N}d/\sigma\right) - 1\right] \\ \quad = h(a^2) + \mathrm{E}_{\mu,\sigma}\left[\frac{Nd^2}{\sigma^2} - a^2\right] h'(a^2) + \frac{1}{2}\mathrm{E}_{\mu,\sigma}\left[\left(\frac{Nd^2}{\sigma^2} - a^2\right)^2 h''(W)\right], \end{array} \tag{6.5.9}$$

where $W \equiv W(d)$ is an appropriate random variable between Nd^2/σ^2 and a^2. Note that

$$h(a^2) = 2\Phi(a) - 1 = 1 - \alpha.$$

Next, we may rewrite (6.5.9) as:

$$\begin{array}{l} \mathrm{E}_{\mu,\sigma}\left[2\Phi\left(\sqrt{N}d/\sigma\right) - 1\right] \\ \quad = h(a^2) + \frac{a^2}{C}h'(a^2)\mathrm{E}_{\mu,\sigma}[N - C] + \frac{1}{2}\frac{a^4}{C}\mathrm{E}_{\mu,\sigma}\left[\frac{(N-C)^2}{C}h''(W)\right] \\ \quad = 1 - \alpha + \frac{a^2}{C}h'(a^2)\left[\eta(1) + o(1)\right] + \frac{1}{2}\frac{a^4}{C}\left[2h''(a^2) + o(1)\right]. \end{array} \tag{6.5.10}$$

Now, in the last step, the middle term follows from (6.5.10) if $m \geq 4$. But, the last term is more complicated because W is also random variable that depends on N. The convergence of $\frac{(N-C)^2}{C}h''(W)$ in distribution is guaranteed, but its uniform integrability may not be. Without its uniform integrability, we cannot claim the convergence of the expectation!

With more analysis, however, one may indeed show that $\left|\frac{(N-C)^2}{C}h''(W)\right|$ is uniformly integrable when $m \geq 7$. Hence, the last step in (6.5.10) holds when $m \geq 7$.

From this, the confidence coefficient expansion shown in (6.2.34) works out since

$$\begin{array}{l} \mathrm{E}_{\mu,\sigma}\left[2\Phi\left(\sqrt{N}d/\sigma\right) - 1\right] \\ \quad = 1 - \alpha + \frac{a^2}{C}\left[h'(a^2)\eta(1) + a^2h''(a^2)\right] + o(C^{-1}), \end{array} \tag{6.5.11}$$

and

$$h^{'}(x) = \frac{1}{\sqrt{2\pi}}\exp(-x/2)x^{-1/2}, h^{''}(x) = -\frac{1}{2\sqrt{2\pi}}\exp(-x/2)x^{-1/2}(1+x^{-1}), \quad (6.5.12)$$

hold.

Let $f(x;1)$ be the p.d.f. of a χ_1^2 random variable, $x > 0$, as in (6.2.33). Then, we have $h'(a^2) = f(a^2;1)$ and $h''(a^2) = -\frac{1}{2}(1 + a^{-2})f(a^2;1)$.

Now, one can show that (6.5.12) leads to the expansion shown in (6.2.34). The intermediate steps are left out as an exercise. ■

6.5.3 Second-Order $\mathrm{E}_{\mu,\sigma}(\boldsymbol{N})$ *and Regret Expansion (6.4.14)*

Let us go back to the stopping variable N from (6.4.5). The boundary condition involves S_n, but we know that $(n-1)S_n^2/\sigma^2 \backsim \chi_{n-1}^2$. As before, let us think of random variables $W_1, W_2, \ldots$ which are i.i.d. χ_1^2. Now, we rewrite (6.4.5) as:

$$\begin{aligned} N \equiv\ & N(c) \text{ is the smallest integer } n(\geq m) \text{ for which} \\ & \text{we observe } n^2(n-1) \geq \frac{A\sigma^2}{c}\frac{(n-1)S_n^2}{\sigma^2}. \end{aligned}$$

This can be further rewritten as

$$\begin{aligned} N \equiv\ & N(c) \text{ is the smallest integer } n(\geq m) \text{ for which} \\ & \text{we observe } n^{*-2}n^2(n-1) \geq \textstyle\sum_{i=1}^{n-1} W_i, \end{aligned} \quad (6.5.13)$$

since $n^* = \sqrt{A/c}\sigma$. We can express N from (6.5.13) as:

$$N = T + 1 \text{ w.p. } 1$$

where a new stopping variable T is defined as:

$$\begin{aligned} T \equiv\ & T(d) \text{ is the smallest integer } n(\geq m-1) \text{ for which} \\ & \text{we observe } n^{*-2}(n+1)^2 n \geq \textstyle\sum_{i=1}^{n} W_i \text{ or equivalently} \\ & \text{observe } n^{*-2}n^3(1 + \frac{2}{n} + o(n^{-1})) \geq \textstyle\sum_{i=1}^{n} W_i. \end{aligned} \quad (6.5.14)$$

The random variable T is obtained by substituting "$n-1$" with "n" on both sides of the boundary condition in (6.5.14). The setup for T matches with that in (1.1) of Woodroofe (1977) or Mukhopadhyay (1988a). This is summarized in Section A.4 of the Appendix. Match (6.5.14) with (A.4.1) to note:

$$\begin{aligned} m_0 = m-1, \delta = 3, h^* = n^{*-2}, L_n = (n+1)^2 n, L_0 = 2, \\ \theta = 1, \tau^2 = 2, \beta^* = \tfrac{1}{2} \Rightarrow n_0^* = n^* \text{ and } q = \tfrac{1}{2}. \end{aligned} \quad (6.5.15)$$

Expansion of Average Sample Size:

Now, since $W_1 \sim \chi_1^2$, we have for $u > 0$:

$$\mathrm{P}(W_1 \leq u) \leq \frac{1}{\sqrt{2\pi}}\int_0^u w^{-1/2}dw = \frac{2}{\sqrt{2\pi}}u^{1/2},$$

as in (6.5.5). So, the condition from (A.4.3) again holds with $r = \frac{1}{2}$.

Again, since $\sum_{i=1}^{n-1} W_i \sim \chi_n^2$, from (A.4.3)-(A.4.4), we obtain:

$$\begin{aligned} &\nu = \tfrac{3}{2} - \textstyle\sum_{n=1}^{\infty} \tfrac{1}{n}\mathrm{E}\left[\max\left(0, \chi_n^2 - 3n\right)\right] \\ &\Rightarrow \kappa = \tfrac{1}{2}\nu - 1 - \tfrac{3}{4} = -1 - \tfrac{1}{2}\textstyle\sum_{n=1}^{\infty} \tfrac{1}{n}\mathrm{E}\left[\max\left(0, \chi_n^2 - 3n\right)\right]. \end{aligned} \quad (6.5.16)$$

Again, the distribution of T or N is affected by σ^2 only. But, we will continue to write $\mathrm{P}_{\mu,\sigma}(\cdot)$ or $\mathrm{E}_{\mu,\sigma}(\cdot)$. Let us denote $U^* = (T - n^*)/n^{*1/2}$. Now, we can state some conclusions as $d \to 0$:

$$\begin{aligned} &\text{(i)} \quad \mathrm{P}_{\mu,\sigma}(T \le \tfrac{1}{2}n^*) = O(h^{*(m-1)/2}) = O(n^{*-(m-1)}) \text{ if } m \ge 2; \\ &\text{(ii)} \quad U^* \text{ converges in distribution to } N(0, \tfrac{1}{2}) \text{ if } m \ge 2; \text{ and} \\ &\text{(iii)} \quad U^{*2} \text{ is uniformly integrable if } m \ge 3. \end{aligned} \quad (6.5.17)$$

These conclusions follow directly from Theorem A.4.1 in Appendix. With some minimal effort, one may verify that the exact same conclusions would hold when T is replaced by N in (6.5.17). Details are left out as an exercise.

Expansion of Expected Value of N:

Now, Theorem A.4.2 clearly implies with $\omega = 1$:

$$\begin{aligned} &\mathrm{E}_{\mu,\sigma}[T] = n^* + \kappa + o(1) \text{ if } m - 1 > 1, \text{ that is if } m \ge 3 \\ &\Rightarrow \mathrm{E}_{\mu,\sigma}[N] = \mathrm{E}_{\mu,\sigma}[T+1] = n^* + \kappa + 1 + o(1) \text{ if } m \ge 3. \end{aligned} \quad (6.5.18)$$

The constant term "$\kappa+1$" from (6.5.18) is exactly the same as the constant "$\eta_1(1)$" that we had defined in (6.4.13). ■

Variance of N:

Parts (ii) and (iii) from (6.5.17) immediately lead to the following conclusion:

$$\mathrm{V}_{\mu,\sigma}[N] = \frac{1}{2}n^* + o(n^*), \quad (6.5.19)$$

if $m \ge 3$. ■

Expansion of Regret:

From (6.4.7), we rewrite:

$$\mathrm{Regret}(c) = c\mathrm{E}_{\mu,\sigma}\left[\frac{(N-n^*)^2}{N}\right] \quad (6.5.20)$$

Combining (ii)-(iii) from (6.5.17), we may claim:

$$\mathrm{E}_{\mu,\sigma}\left[\frac{(N-n^*)^2}{n^*}\right] = \frac{1}{2} + o(1) \quad (6.5.21)$$

if $m \ge 3$.

Now, let us rewrite the regret function from (6.5.20) as follows:

$$\begin{aligned}\tfrac{1}{c}\text{Regret}(c) = \quad & \mathrm{E}_{\mu,\sigma}\left[\frac{(N-n^*)^2}{N}I\left(N > \tfrac{1}{2}n^*\right)\right] \\ & +\mathrm{E}_{\mu,\sigma}\left[\frac{(N-n^*)^2}{N}I\left(N \le \tfrac{1}{2}n^*\right)\right]\end{aligned} \qquad (6.5.22)$$

and observe that

$$\left|\frac{(N-n^*)^2}{N}I\left(N > \frac{1}{2}n^*\right)\right| \le 2\frac{(N-n^*)^2}{n^*} \text{ w.p. 1.}$$

But, $\dfrac{(N-n^*)^2}{n^*}$ is uniformly integrable (see (6.5.17), part (iii)) if $m \ge 3$, and hence so is $\dfrac{(N-n^*)^2}{N}I\left(N > \tfrac{1}{2}n^*\right)$. But, we also know that

$$\frac{(N-n^*)^2}{N}I\left(N > \frac{1}{2}n^*\right) \text{ converges in distribution to } N(0, \frac{1}{2}).$$

Hence, we have:

$$\mathrm{E}_{\mu,\sigma}\left[\frac{(N-n^*)^2}{N}I\left(N > \frac{1}{2}n^*\right)\right] = \frac{1}{2} + o(1), \qquad (6.5.23)$$

if $m \ge 3$.

On the other hand, we may write:

$$\begin{aligned} & \mathrm{E}_{\mu,\sigma}\left[\frac{(N-n^*)^2}{N}I\left(N \le \tfrac{1}{2}n^*\right)\right] \\ & \le \mathrm{E}_{\mu,\sigma}\left[\frac{N^2+n^{*2}}{N}I\left(N \le \tfrac{1}{2}n^*\right)\right] \\ & \le \mathrm{E}_{\mu,\sigma}\left[\frac{\tfrac{1}{4}n^{*2}+n^{*2}}{m}I\left(N \le \tfrac{1}{2}n^*\right)\right] \\ & = \frac{5}{4m}n^{*2}\mathrm{P}_{\mu,\sigma}\left(N \le \tfrac{1}{2}n^*\right) \\ & = O\left(n^{*2}\right)O\left(n^{*-m+1}\right) \text{ by (6.5.17), part (i)} \\ & = o(1) \text{ if } 3-m<0, \text{ that is, if } m \ge 4. \end{aligned} \qquad (6.5.24)$$

At this point one combines (6.5.23)-(6.5.24) and (6.5.20) to claim:

$$\frac{1}{c}\text{Regret}(c) = c\mathrm{E}_{\mu,\sigma}\left[\frac{(N-n^*)^2}{N}\right] = \frac{1}{2}c + o(c),$$

when $m \ge 4$. ■

6.6 Exercises

Exercise 6.2.1: Suppose that $X_1, \ldots, X_n$ are i.i.d. $N(2\mu, 5\sigma^2)$ with $n \geq 2$. Denote

$$U_1 = \frac{X_1 + \cdots + X_n}{\sqrt{n}},$$
$$U_i = \frac{X_1 + \cdots + X_{i-1} - (i-1)X_i}{\sqrt{i(i-1)}}, \ i = 2, 3, \ldots, n.$$

(i) Show that $U_1, \ldots, U_n$ are independent random variables.
(ii) Show that U_1 is distributed as $N(2\sqrt{n}\mu, 5\sigma^2)$.
(iii) Show that $U_2, \ldots, U_n$ are i.i.d. $N(0, 5\sigma^2)$.

Exercise 6.6.2 (Exercise 6.2.1 Continued): Recall the random variables $U_1, \ldots, U_n$. Let $\overline{X}_n = \dfrac{1}{n}\sum_{i=1}^n X_i$ and $S_n^2 = \frac{1}{n-1}\sum_{i=1}^n (X_i - \overline{X}_n)^2$ be respectively the sample mean and the sample variance.

(i) Show that $\overline{X}_n = U_1/\sqrt{n}$ and $(n-1)S_n^2 = \sum_{i=2}^n U_i^2$. Hence, prove that $\overline{X}_n$ and S_n^2 are independent, and that $\dfrac{(n-1)S_n^2}{5\sigma^2}$ has a χ_{n-1}^2 distribution.
(ii) Show that $\overline{X}_n$ and S_k^2 are independent for each $k, 2 \leq k \leq n$.
(iii) Show that $\overline{X}_n$ and $(S_2^2, \ldots, S_n^2)$ are independent.

Exercise 6.2.3: Verify the expression for the variance of $\overline{X}_N$ given in Theorem 6.2.1, part (ii).

Exercise 6.2.4: For the purely sequential procedure (6.2.17), show that $E_{\mu,\sigma}(N) \leq C + m + 1$, for all fixed μ, σ, d and α.

{*Hint*: Since we do not stop at stage $N-1$, we must have $N - 1 < \frac{a^2 S_{N-1}^2}{d^2} + (m-1)$. Thus, $(N-m)(N-2) < \frac{a^2}{d^2}\sum_{i=1}^{N-1}(X_i - \overline{X}_{N-1})^2$ which is smaller than $\frac{a^2}{d^2}\sum_{i=1}^{N-1}(X_i - \mu)^2 < \frac{a^2}{d^2}\sum_{i=1}^{N}(X_i - \mu)^2$. So, we have $N^2 - (m+2)N < (N-m)(N-2) < \frac{a^2}{d^2}\sum_{i=1}^{N}(X_i - \mu)^2$. Now, use Wald's first equation (Theorem 3.5.4) with $U_i = (X_i - \mu)^2$. Hence, we can write: $E_{\mu,\sigma}(N^2) - (m+2)E_{\mu,\sigma}(N) < \frac{a^2}{d^2}E_{\mu,\sigma}(N)\sigma^2 = CE_{\mu,\sigma}(N) \Rightarrow E_{\mu,\sigma}(N) - (m+2) < C$ since $\{E_{\mu,\sigma}(N)\}^2 < E_{\mu,\sigma}(N^2)$.}

Exercise 6.2.5: For the three-stage procedure (6.2.39)-(6.2.40), show that

(i) $\overline{X}_N$ is an unbiased estimator for μ.

(ii) $\mathrm{P}_{\mu,\sigma}\{\mu \in J_N = [\overline{X}_N \pm d]\} = \mathrm{E}_{\mu,\sigma}\left[2\Phi\left(\sqrt{N}d/\sigma\right) - 1\right]$.

(iii) $Q \equiv \sqrt{N}(\overline{X}_N - \mu)/\sigma$ has a standard normal distribution.

Exercise 6.2.6: Show that the probability distribution of the final sample size N involves σ^2 but not μ when N is defined by (6.2.6), (6.2.17), (6.2.23)-(6.2.24), (6.2.35), (6.2.39)-(6.2.40) or (6.2.43)-(6.2.44).

Exercise 6.3.1: Verify (6.3.2).

{*Hint*: Observe that

$$\mathrm{E}_{\mu,\sigma}\left\{L_n(\mu, \overline{X}_n)\right\} = \frac{A\sigma^k}{n^{k/2}}\mathrm{E}_{\mu,\sigma}\left[\left|\sqrt{n}\left(\overline{X}_n - \mu\right)/\sigma\right|^k\right] = \frac{A\sigma^k}{n^{k/2}}\mathrm{E}\left[|Z|^k\right]$$

where Z is a standard normal variable. Next, one may check that $\mathrm{E}\left[|Z|^k\right] = \frac{2^{k/2}}{\sqrt{\pi}}\Gamma\left(\frac{1}{2}(k+1)\right)$.}

Exercise 6.3.2: Verify that

$$b_m = \left\{\frac{1}{2}(m-1)\right\}^{k/2}\Gamma\left(\frac{1}{2}(m-k-1)\right)\left\{\Gamma\left(\frac{1}{2}(m-1)\right)\right\}^{-1}$$

exceeds one for all $k > 0$ and $m > k+1$.

{*Hint*: U be distributed as Gamma(α, β) with $\alpha = \frac{1}{2}(m-1)$ and $\beta = 1$. Then,

$$\begin{aligned}\mathrm{E}[U^{-k/2}] &= \left\{\Gamma\left(\tfrac{1}{2}(m-1)\right)\right\}^{-1}\int_0^\infty e^{-u}u^{\frac{1}{2}(m-k-1)-1}du \\ &= \Gamma\left(\tfrac{1}{2}(m-k-1)\right)\left\{\Gamma\left(\tfrac{1}{2}(m-1)\right)\right\}^{-1},\end{aligned}$$

since $m > k+1 (> 1)$. Now, look at the function $g(u) = u^{-k/2}, u > 0$. It is a strictly convex function. Hence, using Jensen's inequality, we can claim that $\mathrm{E}[U^{-k/2}] > [\mathrm{E}(U)]^{-k/2} = \left(\frac{1}{2}(m-1)\right)^{-k/2}$. In other words, we have $b_m > 1$.}

Exercise 6.3.3: Prove parts (i)-(ii) of Theorem 6.3.1 along the lines of Theorem 6.2.1.

Exercise 6.3.4: For two-stage procedure (6.3.4), show that $Q \equiv \sqrt{N}(\overline{X}_N - \mu)/\sigma$ has a standard normal distribution.

Exercise 6.3.5: In a bounded-risk point estimation problem, propose an appropriate modified two-stage procedure in the spirit of Section 6.2.3. Then, investigate some of its crucial properties.

Exercise 6.3.6: In a bounded-risk point estimation problem, propose an appropriate accelerated sequential procedure in the spirit of Section

6.2.4. Then, investigate some of its crucial properties. See Mukhopadhyay (1985a) and Mukhopadhyay et al. (1987).

Exercise 6.3.7: In a bounded-risk point estimation problem, propose an appropriate three-stage procedure in the spirit of Section 6.2.5. Then investigate some of its crucial properties. See Mukhopadhyay (1985) and Mukhopadhyay et al. (1987).

Exercise 6.3.8: In the purely sequential procedure (6.3.14), show that $\overline{X}_N$ is an unbiased estimator of μ and its variance is $\sigma^2 \mathrm{E}_{\mu,\sigma}(N^{-1})$ for all fixed μ, σ, w, k and A.

Exercise 6.3.9: Show that the probability distribution of the final sample size N involves σ^2 but not μ when N is defined by (6.3.4) or (6.3.14).

Exercise 6.4.1: For the purely sequential procedure (6.4.5), show that $\overline{X}_N$ is an unbiased estimator of μ and its variance is $\sigma^2 \mathrm{E}_{\mu,\sigma}(N^{-1})$ for all fixed μ, σ, c and A.

Exercise 6.4.2: Verify (6.4.6).

Exercise 6.4.3: Verify (6.4.9). See Ghosh and Mukhopadhyay (1976).

Exercise 6.4.4: For the parallel piecewise sequential procedure (6.4.17), show that $T_k(\boldsymbol{N})$ is an unbiased estimator of μ and its variance has the form $\sigma^2 \mathrm{E}_{\mu,\sigma}(N^{-1})$ for all fixed μ, σ, c and A.

Exercise 6.4.5: Verify (6.4.20).

Exercise 6.4.6: Consider the expression of Δ from (6.4.22). Show that Δ is an unbiased estimator of $\mathrm{V}_\sigma(N)$.

Exercise 6.4.7: Show that the probability distribution of the final sample size N involves σ^2 but not μ when N is defined by (6.4.5) or (6.4.17).

CHAPTER 7

Location Estimation: Negative Exponential Distribution

7.1 Introduction

In this chapter, we focus on a negative exponential distribution with the following p.d.f.:

$$f(x;\theta,\sigma) = \begin{cases} \frac{1}{\sigma}\exp\left(-(x-\theta)/\sigma\right) & \text{if } x > \theta \\ 0 & \text{if } x \le \theta. \end{cases} \quad (7.1.1)$$

We denote this distribution by NExpo(θ,σ) and assume that θ,σ are unknown parameters, $-\infty < \theta < \infty$ and $0 < \sigma < \infty$. This distribution also goes by another name, a two-parameter exponential distribution.

The parameter θ, when positive, may be interpreted as a minimum guarantee time or a threshold of the distribution. The parameter σ is called the "scale" of this distribution. This distribution has been used in many reliability and lifetesting experiments to describe, for example, failure times of complex equipment, vacuum tubes, and small electrical components. One is referred to Johnson and Kotz (1970), Bain (1978), Lawless and Singhal (1980), Grubbs (1971) and other sources. It has been recommended as a statistical model in some clinical trials in cancer research studying tumor systems and survival data in animals. One is referred to Zelen (1966). The NExpo(θ,σ) model has also been used in areas of soil science and weed propagation. In these applications, θ may stand for the minimum number of days a weed variety needs to germinate or spread. Zou (1998) mentioned such applications.

Having recorded n independent observations $X_1, \ldots, X_n$, each following a NExpo(θ,σ) distribution (7.1.1), θ is customarily estimated by the *maximum likelihood estimator* (MLE):

$$\widehat{\theta}_{\text{MLE}} \equiv X_{n:1} = \min\{X_1,\ldots,X_n\}. \quad (7.1.2)$$

The scale parameter σ is customarily estimated by the *uniformly minimum variance unbiased estimator* (UMVUE):

$$\widehat{\sigma}_{\text{UMVUE}} \equiv U_n = \frac{1}{n-1}\Sigma_{i=1}^{n}(X_i - X_{n:1}), n \ge 2. \quad (7.1.3)$$

By the way, the MLE of σ is given by

$$\widehat{\sigma}_{\text{MLE}} \equiv V_n = \frac{1}{n}\sum_{i=1}^{n}(X_i - X_{n:1}), n \geq 2. \tag{7.1.4}$$

The mean μ of a NExpo(θ, σ) distribution is given by $\theta + \sigma$. Hence, we have $\widehat{\mu}_{\text{MLE}} \equiv \widehat{\theta}_{\text{MLE}} + \widehat{\sigma}_{\text{MLE}} = \overline{X}_n$, the sample mean. The following properties are well-known:

$$\begin{array}{rl} \text{(i)} & n(X_{n:1} - \theta)/\sigma \text{ is distributed as NExpo}(0,1) \text{ or} \\ & \text{standard exponential;} \\ \text{(ii)} & 2(n-1)U_n/\sigma \text{ is distributed as } \chi^2_{2n-2}, n \geq 2; \text{ and} \\ \text{(iii)} & X_{n:1} \text{ and } (U_2, \ldots, U_n) \text{ are independent, } n \geq 2. \end{array} \tag{7.1.5}$$

One can easily see that $\text{E}_{\theta,\sigma}[X_{n:1}] = \theta + n^{-1}\sigma$, and hence, an *unbiased estimator* of θ would be

$$T_n \equiv X_{n:1} - n^{-1}U_n, n \geq 2. \tag{7.1.6}$$

It is indeed the UMVUE for θ.

Now, let us explain the layout of this chapter. We begin with a fixed-width confidence interval estimation problem for θ using its MLE in Section 7.2. We develop a two-stage procedure (Section 7.2.1) of Ghurye (1958) and a purely sequential strategy (Section 7.2.2) of Mukhopadhyay (1974a). Some of the other multistage sampling techniques are briefly mentioned in Section 7.2.3. Comparisons and conclusions are given in Section 7.2.4.

In Section 7.3, we first develop a purely sequential strategy (Section 7.3.1) for minimum risk point estimation of θ using its MLE, $X_{n:1}$ (Basu, 1971 and Mukhopadhyay, 1974a). Next, minimum risk estimation of θ using its UMVUE, T_n, is briefly mentioned under an improved purely sequential strategy (Section 7.3.2) of Ghosh and Mukhopadhyay (1990). Some of the other multistage sampling techniques are briefly mentioned (Section 7.3.3). Again, comparisons and conclusions are given in Section 7.3.4. Section 7.4 includes selected derivations.

7.2 Fixed-Width Confidence Intervals

Suppose that one has n independent observations $X_1, \ldots, X_n$, each following a NExpo(θ, σ) distribution (7.1.1). Consider the following confidence interval

$$J_n = (X_{n:1} - d, X_{n:1}) \tag{7.2.1}$$

for θ where the interval's width $d(> 0)$ is a preassigned number. Since $X_{n:1} > \theta$ w.p. 1, this fixed-width confidence interval with an upper confidence limit $X_{n:1}$ appears natural.

We also require, however, that the associated confidence coefficient be at least $1-\alpha$ where $0<\alpha<1$ is preassigned. Thus, we must have:

$$\begin{aligned} \mathrm{P}_{\theta,\sigma}\left\{X_{n:1}-d<\theta<X_{n:1}\right\} &= \mathrm{P}_{\theta,\sigma}\left\{0<\frac{n\left(X_{n:1}-\theta\right)}{\sigma}<\frac{nd}{\sigma}\right\} \\ &= 1-\exp(-nd/\sigma), \end{aligned} \tag{7.2.2}$$

since $n\left(X_{n:1}-\theta\right)/\sigma$ is NExpo$(0,1)$. Refer to part (i) in (7.1.5).

So, the confidence coefficient associated with J_n will be at least $1-\alpha$ provided that

$$1-\exp(-nd/\sigma)\geq 1-\alpha.$$

That is, the required sample size

$$n \text{ must be the smallest integer } \geq \frac{a\sigma}{d}=C, \text{ say,} \\ \text{where } a=\log(1/\alpha). \tag{7.2.3}$$

Here, "log" stands for the natural logarithm.

Now, C is referred to as the required *optimal* fixed sample size had σ been known. We note that the magnitude of C remains unknown! So, we must opt for multistage estimation strategies.

Computations and Simulations

The values of C can be easily computed using the program **Seq07.exe** for given values of α, σ and d. **Seq07.exe** can also be used for other computations and simulation studies. Figure 7.2.1 shows the screen input for computation C for $\sigma=10$, $\alpha=0.01$ and $d=0.2,0.4,0.6,0.8,1.0$. The results output from the 'program number 1', which computes C values are given in Figure 7.2.2.

7.2.1 Two-Stage Procedure

Ghurye (1958) proposed the following two-stage procedure along the lines of Stein (1945,1949). We start with pilot observations $X_1,\ldots,X_m$ where $m(\geq 2)$ is the initial sample size. Let us define:

$$N\equiv N(d)=\max\left\{m,\left\langle\frac{b_mU_m}{d}\right\rangle+1\right\}. \tag{7.2.4}$$

Here, b_m is the upper $100\alpha\%$ point of a F-distribution with $2,2m-2$ degrees of freedom.

We implement this two-stage procedure in usual fashion by sampling the difference at the second stage if $N>m$. Then, using final consisting of $N,X_1,\ldots,X_N$ obtained by combining both stages, the combining both stages, the fixed-width confidence interval

$$J_N\equiv\left(X_{N:1}-d,X_{N:1}\right)$$

is proposed for θ. It is easy to see that N is finite w.p. 1 and J_N is a genuine confidence interval.

```
Select a program using the number given
in the left hand side. Details of these
programs are given in Chapter 7.

*****************************************************
* Number          Program                           *
*                                                   *
*  1   Exact value of C for given parameters        *
*  2   Fixed-Width Confidence Interval Estimation   *
*  3   E(N) and Confidence Coefficient for Two-Stage *
*  4   Second-Order Terms of Purely Sequential      *
*  5   Minimum Risk Point Estimation                *
*  0   Exit                                         *
*****************************************************

         Enter the program number:1

Do you want to save the results in a file(Y or N)? y

Note that your default outfile name is "SEQ07.OUT".

Do you want to change the outfile name?
    [Y(yes) or N(no)]: n

Optimal Sample Size (C) for Fixed-Width Confidence
Interval for the Location Parameter (theta) of a
Negative Exponential Population
--------------------------------------------------

This program computes the values of C for the
range of values of d, alpha and sigma in a
negative exponential distribution.

For each of these parameter, enter:
number of values (k) and
value of the parameter if k=1;
minimum and maximum values if k > 1

For scale parameter (sigma):
              number of values = 1
              enter the value  = 10
Do you want to change the above values  (type Y or N): n

For alpha (1- confidence coefficient):
              number of values = 1
              enter the value  = .01
Do you want to change the above values  (type Y or N): n

For d (width):
              number of values = 5
              minimum  value   = .2
              maximum  value   = 1
Do you want to change the above values  (type Y or N): n
```

Figure 7.2.1. *Screen input to compute C for given σ, α and d.*

```
Optimal Sample Size (C) for Fixed-Width Confidence
Interval for the Location Parameter (theta) of a
Negative Exponential Population
--------------------------------------------------

          Input values:
                 minimum   maximum   Number
          Sigma    10.00     10.00        1
          Alpha     0.01      0.01        1
          d         0.20      1.00        5
          --------------------------------

          Sigma      Alpha          d  Computed C
          10.00      0.010       0.20     230.26
          10.00      0.010       0.40     115.13
          10.00      0.010       0.60      76.75
          10.00      0.010       0.80      57.56
          10.00      0.010       1.00      46.05
```

Figure 7.2.2. *Computed values of C: output from 'program number 1'.*

From part (iii) in (7.1.5), it is clear that random variables $I(N = n)$ and $X_{n:1}$ are independent for all fixed $n(\geq m)$. Hence, we express the confidence coefficient associated with J_N as follows:

$$\mathrm{P}_{\theta,\sigma}\left\{X_{N:1} - d < \theta < X_{N:1}\right\} = \mathrm{E}_{\theta,\sigma}\left[1 - e^{-Nd/\sigma}\right]. \tag{7.2.5}$$

By the way, the distribution of N involves σ and not θ. Yet, we may interchangeably write, for example, $\mathrm{E}_{\theta,\sigma}\left[1 - e^{-Nd/\sigma}\right]$ or $\mathrm{E}_{\sigma}\left[1 - e^{-Nd/\sigma}\right]$. Now, we summarize some of the crucial properties of the two-stage strategy. Their proofs are given in Section 7.4.1.

Theorem 7.2.1 : *For the stopping variable N defined by (7.2.4), for all fixed θ, σ, d, m and α, one has the following properties:*

(i) $b_m \sigma d^{-1} \leq \mathrm{E}_{\theta,\sigma}(N) \leq m + b_m \sigma d^{-1}$;

(ii) $Q \equiv N(X_{N:1} - \theta)/\sigma$ *has a standard exponential distribution, that is,* NExpo(0, 1);

(iii) $\mathrm{P}_{\theta,\sigma}\{\theta \in J_N\} \geq 1 - \alpha$ [*Consistency or Exact Consistency*];

where b_m is the upper $100\alpha\%$ point of a F-distribution with $2, 2m-2$ degrees of freedom.

Now, some other comments are in order. One can easily verify that

$$b_m \equiv (m-1)\left\{\alpha^{-1/(m-1)} - 1\right\}. \tag{7.2.6}$$

See Mukhopadhyay (2000, p. 631). We may rewrite (7.2.6) as follows:

$$a \equiv \log(1/\alpha) = (m-1)\log\left(1 + \frac{b_m}{m-1}\right) < (m-1)\left(\frac{b_m}{m-1}\right) \equiv b_m, \tag{7.2.7}$$

since $\log(1+x) < x$ for $x > 0$.

Thus, we have,

$$\mathrm{E}_{\theta,\sigma}(N) \geq b_m \sigma d^{-1} > C. \tag{7.2.8}$$

In other words, on an average, the Ghurye procedure (7.2.4) oversamples compared with the optimal fixed sample size C. This should not be surprising because C is "optimal" when we pretend that σ is known!

From Theorem 7.2.1, part (i), it follows easily that

$$\lim_{d\to 0} \mathrm{E}_{\theta,\sigma}\left(\frac{N}{C}\right) = \frac{b_m}{a} > 1. \tag{7.2.9}$$

The property given in (7.2.9) is not easily disposed of. It shows that $\mathrm{E}_{\theta,\sigma}(N)$ and C do not match in a large-sample case. Hence, the latter property is more disturbing that the Ghurye procedure is *asymptotically (first-order) asymptotically (first-order) inefficient* in the sense of Chow and Robbins (1965) and Ghosh and Mukhopadhyay (1981).

The status of the limiting confidence coefficient is described below:

$$\mathrm{P}_{\theta,\sigma}\{\theta \in J_N\} \to 1-\alpha \text{ as } d \to 0.$$

That is, the two-stage strategy (7.2.4) is *asymptotically consistent.* The proof of this result is left as an exercise with hints.

> The two-stage strategy (7.2.4) is consistent, asymptotically consistent, and asymptotically first-order inefficient.

```
Simulation study for 99% confidence interval for
the location in a Negative Exponential Population.
==================================================

Data were generated from an negative exponential
distribution with location= 10.00 and scale= 5.00

Number of simulation replications    = 10000
Width of the confidence interval (d) =  0.50
Initial sample size (m) =  5

    Optimal sample size (C)          =     46.05

    Average sample size (n_bar)      =     87.59
    Std. dev. of n (s)               =     42.66
    Minimum sample size (min n)      =      7
    Maximum sample size (max n)      =    325

    Coverage probability (p_bar)     = 0.991
    Standard error of p_bar          = 0.001
```

Figure 7.2.3. *Simulation results for two-stage procedure.*

The 'program number 2' in **Seq07.exe** can obtain a simulated fixed-width interval for θ using either a two-stage or a sequential (see Section

7.2.2) strategy. Also, by selecting the option of live data input, the user would obtain a confidence interval using one of the two procedures. Figure 7.2.3 shows results from a simulation study of a 99% confidence interval with $d = 0.5$ using the two-stage strategy. We generated data from NExpo(10, 5) and fixed $m = 5$. The results show that $C = 46.05$, but for the simulated data we have $\overline{n} = 87.59$. One will note that $a = 4.6052$ and $b_m = 8.6491$ so that $b_m/a \approx 1.8781$. But, we have $\overline{n}/C \approx 1.9021$, which is rather close to 1.8781. So, an estimated or simulated oversampling percentage is nearly 90% whereas theoretically one would expect nearly 88%. Also, an estimated coverage probability p_bar is 0.991(> 0.99), which validates the (exact) consistency result in practice.

Probability Distribution of the Final Sample Size N

Since the final sample size N from (7.2.4) is a random variable, it makes sense to determine the exact distribution of N pretending that σ is known. Let us denote $Y = 2(m-1)U_m/\sigma$ which has a χ^2_{2m-2} distribution. Refer to part (ii) in (7.1.5).

One may write:

$$\begin{aligned}\mathrm{P}_\sigma(N = m) &= \mathrm{P}_\sigma\left\{\langle b_m U_m/d\rangle + 1 \le m\right\}\\ &= \mathrm{P}_\sigma\left\{0 < b_m U_m/d \le m\right\}\\ &= \mathrm{P}_\sigma\left\{0 < Y \le \frac{2m(m-1)d}{\sigma b_m}\right\}.\end{aligned}$$

Similarly, for any positive integer k, we may write:

$$\begin{aligned}&\mathrm{P}_\sigma(N = m+k)\\ &= \mathrm{P}_\sigma\left\{\langle b_m U_m/d\rangle = m+k-1\right\}\\ &= \mathrm{P}_\sigma\left\{m+k-1 < b_m U_m/d \le m+k\right\}\\ &= \mathrm{P}_\sigma\left\{\frac{2(m+k-1)(m-1)d}{\sigma b_m} < Y \le \frac{2(m+k)(m-1)d}{\sigma b_m}\right\}.\end{aligned}$$

Let us denote

$$g_m(y) = \left\{2^{(m-1)}\Gamma(m-1)\right\}^{-1}\exp[-y/2]y^{m-2}, y > 0;$$

$$A_k = \frac{2(m+k)(m-1)d}{\sigma b_m},\ k = 0, 1, 2, \ldots.$$

Then, we can rewrite:

$$\begin{aligned}\mathrm{P}_\sigma(N = m) = \mathrm{P}_\sigma\left\{0 < \chi^2_{2m-2} \le A_0\right\} = \textstyle\int_0^{A_0} g_m(y)dy;\\ \mathrm{P}_\sigma(N = m+k) = \mathrm{P}_\sigma\left\{A_{k-1} < \chi^2_{2m-2} \le A_k\right\} = \textstyle\int_{A_{k-1}}^{A_k} g_m(y)dy,\end{aligned} \tag{7.2.10}$$

for $k = 1, 2, 3, \ldots$.

Once an exact distribution of N is found, pretending that σ were known, we can evaluate (i) expected value and variance of N as well as (ii) confidence coefficient associated with $J_N \equiv (X_{N:1} - d, X_{N:1})$. The 'program number 3' within **Seq07.exe** computes exact values of $\text{E}_\sigma(N)$, $\text{V}_\sigma(N)$ and $\text{P}_\sigma(\theta \in J_N)$ given α, d, σ and m.

Figure 7.2.4 shows an output from the 'program number 3'. Here, we obtained $C = 59.9$, $\text{E}_\sigma(N) = 89.7$, $\text{V}_\sigma = 1987.2$ and $\text{P}_\sigma(\theta \in J_N) = 0.951$ given $\alpha = 0.05$, $d = 0.5$, $\sigma = 10$ and $m = 5$. This validates our general sentiment that

$$\text{E}_\sigma(N) > C \text{ and } \text{P}_\sigma(\theta \in J_N) \geq (1 - \alpha),$$

in practice under a two-stage strategy.

```
E(N), Var(N) and the Associated Confidence
Coefficient for the Location Confidence Interval
of a Negative Exponential Population with known
Scale Parameter Using Two-Stage Procedure
============================================

Input Parameters:
    alpha                    =  0.05
    d (interval width)       =  0.50
    Initial sample size (m)  =  5
    Scale parameter (sigma)  =  10.0

Computed Results:
    Optimal sample size, C =       59.9
    Expected value of N    =       89.7
    Variance of N          =     1987.2

    Associated Confidence Coefficient = 0.951
```

Figure 7.2.4. *Exact values of* $\text{E}_\sigma(N)$, $\text{V}_\sigma(N)$ *and confidence coefficient.*

7.2.2 Purely Sequential Procedure

We recall that we had $C = a\sigma/d$ with $a = -\log(\alpha)$. Now, unlike (7.2.9), we want to be able to claim first-order efficiency property, namely

$$\lim_{d \to 0} \text{E}_{\theta,\sigma}(N/C) = 1.$$

Mukhopadhyay (1974a) proposed a purely sequential methodology along the lines of Chow and Robbins (1965).

We start with observations $X_1, \ldots, X_m$ where $m(\geq 2)$ is a pilot size. Next, we proceed sequentially with one additional X at-a-time, but terminate sampling according to the following stopping time:

$$\begin{aligned} &N \equiv N(d) \text{ is the smallest integer } n(\geq m) \text{ for which} \\ &\qquad \text{we observe } n \geq \left(\frac{a}{d}\right) U_n \text{ with } a = -\log(\alpha). \end{aligned} \tag{7.2.11}$$

This purely sequential procedure is implemented in the same way we did in the case of (6.2.17). Again, $P_{\theta,\sigma}(N < \infty) = 1$, that is sampling terminates w.p. 1. Upon termination, we estimate θ by the fixed-width interval:

$$J_N \equiv (X_{N:1} - d, X_{N:1})$$

obtained from the full dataset $N, X_1, \ldots, X_m, \ldots, X_N$.

```
Fixed-Width Confidence Interval
Using Purely Sequential Procedure
=================================

Simulation study for  99% confidence interval for
the location in a negative exponential population.

Data were generated from negative exponential
distribution with location= 10.00 and scale= 5.00

Number of simulation replications     = 10000
Width of the confidence interval (d) =  0.50

Initial sample size =  5

    Optimal sample size (C)           =     46.05

    Average sample size (n_bar)       =     45.75
    Std. dev. of n (s)                =      7.26
    Minimum sample size (min n)       =      5
    Maximum sample size (max n)       =     71

    Coverage probability (p_bar)      = 0.987
    Standard error of p_bar           = 0.001

k  = No. observations added
cp = Coverage probability
se = Standard error of the coverage probability

    k     cp     se
    1  0.988  0.001
    2  0.988  0.001
    3  0.989  0.001
    4  0.990  0.001
```

Figure 7.2.5. *Simulation results for purely sequential procedure.*

Figure 7.2.5 shows some results from simulations using the 'program number 2' within **Seq07.exe** with $\alpha = 0.01$ under sequential sampling. The data were generated from an NExpo(10, 5) distribution. We fixed $m = 5, \alpha = 0.01$ and $d = 0.5$. The results clearly show that sequential procedure (7.2.11) undersampled. We see that we have $\overline{n} = 45.75$ that estimates $E(N)$

but we had $C = 46.05$. Also, we have p_bar $= 0.987$ that fell below the target, 0.99.

This program has an added feature to take additional observations (k) that is supposed to make the estimated coverage probability to hit or just surpass the target, $1 - \alpha$. In our simulations, 4 additional observations beyond stopping via (7.2.11) did it. The last line in Figure 7.2.5 shows this output.

From part (iii) in (7.1.5), it is obvious that the two random variables $I(N = n)$ and $X_{n:1}$ are independent for all fixed $n(\geq m)$. Hence, we express the confidence coefficient associated with J_N as follows:

$$\mathrm{P}_{\theta,\sigma}\left\{X_{N:1} - d < \theta < X_{N:1}\right\} = \mathrm{E}_{\theta,\sigma}\left[1 - e^{-Nd/\sigma}\right]. \tag{7.2.12}$$

The following *first-order*results were obtained by Mukhopadhyay (1974a).

Theorem 7.2.2: *For the stopping variable N defined by (7.2.11), for all fixed θ, σ, m and α, one has the following properties:*

(i) $\mathrm{E}_{\theta,\sigma}(N) \leq C + m + 2$ *with d fixed;*

(ii) $Q \equiv N(X_{N:1} - \theta)/\sigma$ *has a standard exponential, namely* NExpo$(0, 1)$, *distribution with d fixed;*

(iii) $\mathrm{E}_{\theta,\sigma}(N/C) \to 1$ *as $d \to 0$;*

(iv) $\mathrm{P}_{\theta,\sigma}\{\theta \in J_N\} \to 1 - \alpha$ *as $d \to 0$.*

Part (i) gives an idea about the cost of ignorance of the scale parameter σ. This amounts to the cost of $m + 1$ observations at the most! Parts (iii) and (iv) show that the sequential methodology (7.2.11) is both *asymptotically first-order efficient* and *asymptotically consistent.*

The *second-order*results are more involved which were obtained by Swanepoel and van Wyk (1982). They first used an elegant embedding technique that was due to Lombard and Swanepoel (1978), and then employed tools from Woodroofe (1977). Swanepoel and van Wyk's (1982) key results are summarized below.

Recall $h(p)$ from (6.2.33) with $p = 2$. Then, we have as $d \to 0$:

$$\begin{aligned} &\mathrm{E}_{\theta,\sigma}(N - C) = \eta(2) + o(1) \text{ if } m \geq 3; \\ &\frac{N - C}{\sqrt{C}} \text{ converges to } N(0, 1) \text{ in distribution if } m \geq 2; \\ &\mathrm{V}_{\theta,\sigma}(N) = C + o(C) \text{ if } m \geq 3. \end{aligned} \tag{7.2.13}$$

With $m \geq 3$ and $a = -\log(\alpha)$, the associated confidence coefficient (7.2.12) has the following *second-order* expansion as $d \to 0$:

$$\mathrm{P}_{\theta,\sigma}\left\{X_{N:1} - d < \theta < X_{N:1}\right\} = 1 - \alpha + \frac{a\alpha}{C}\left(\eta(2) - \frac{1}{2}a\right) + o(C^{-1}). \tag{7.2.14}$$

See also Ghosh and Mukhopadhyay (1981), Mukhopadhyay (1988a,1995a), and Section 2.4.2 of Mukhopadhyay and Solanky (1994) for many details.

The purely sequential strategy (7.2.11) is *asymptotically second-order efficient.*

For large C, let us interpret the results from (7.2.13). From the first part, we may expect $\text{E}_{\theta,\sigma}(N - C)$ to be in a close proximity of $\eta(2)$, a computable number. The second part says that the normalized stopping variable $(N-C)/\sqrt{C}$ would have an approximate normal distribution. The third part says that $\text{V}_{\theta,\sigma}(N)$ may be nearly C.

From the expression in (7.2.14), we get a feel for the departure of the achieved sequential confidence coefficient from its target, $1-\alpha$. In practice, we may expect

$$\text{P}_{\theta,\sigma}\{X_{N:1} - d < \theta < X_{N:1}\} - (1-\alpha)$$

to be close to

$$\frac{a\alpha}{C}\left(\eta(2) - \tfrac{1}{2}a\right)$$

for large C.

```
Theoretical and Simulated Second-Order Terms
of the Purely Sequential Procedure for the
Location Confidence Interval of a Negative
Exponential Population
============================================

    Input values:
                minimum  maximum  Number
         sigma    10.00    10.00       1
         alpha    0.050    0.050       1
         d         0.20     1.00       5
         -------------------------------

    Number of Simulation Replications =   10000
    Initial Sample size (m) =  5

    P_diff  = p_bar-(1-alpha)
    T_so    = second-order term

eta(2) =-1.2550;  sigma =10.00;  alpha =0.050

   d       C   n_bar  var_n  n_bar-C  var_n/C   P_diff     T_so
0.20  149.79  149.61 153.93    -0.18   1.0276   0.0008  -0.0028
0.40   74.89   74.43  81.08    -0.47   1.0826  -0.0011  -0.0055
0.60   49.93   49.52  57.62    -0.41   1.1540  -0.0066  -0.0083
0.80   37.45   36.84  47.83    -0.61   1.2774  -0.0134  -0.0110
1.00   29.96   29.39  39.39    -0.57   1.3147  -0.0202  -0.0138
```

Figure 7.2.6. *Asymptotic expressions from (7.2.13)-(7.2.14) for sequential procedure.*

The 'program number 4' within **Seq07.exe** will help to examine the second-order terms from (7.2.13) - (7.2.14) for small and moderate sample size, C. This program computes the following second-order term:

$$\text{T\_so} = \frac{a\alpha}{C}(\eta(2) - \tfrac{1}{2}a).$$

From (7.2.13), we see that we should compare $\overline{n} - C$ with $\eta(2)$ where $\overline{n}$ is the simulated estimate of the C. Also, from (7.2.14), we know that we should compare P_diff $= \overline{p} - (1 - \alpha)$ with T_so where $\overline{p}$ is the simulated estimate of the coverage probability.

Figure 7.2.6 shows the output from the 'program number 4' when $\sigma = 10$, $\alpha = 0.05$, $m = 5$ and $d = 0.2, 0.4, 0.6, 0.8, 1$. Simulations were carried out by generating data from NExpo(0, 10). We observe that $\overline{n} - C < 0$ and $|\overline{n} - C| < |\eta(2)| = 1.255$. Also, P_diff values are not outrageously away from T_so values. These approximations will usually work better if one fixes $m = 10, 15$ and $C = 200, 250$.

7.2.3 Other Multistage Procedures

Costanza et al. (1986) developed two-stage fixed-width intervals for the common location parameter of several negative exponential distributions.

Mukhopadhyay (1982a) extended (i) Stein's (1945,1949) two-stage fixed-width interval procedure and (ii) his (Mukhopadhyay, 1980) modified two-stage procedure for a general class of distributions with certain pivotal conditions. In that paper, a modified version of two-stage procedure (7.2.4) was shown to be *consistent*, *asymptotically consistent*, and *asymptotically first-order efficient*.

Now, observe that $\alpha = \exp\{\log \alpha\}$ so that we may rewrite b_m as follows:

$$
\begin{aligned}
b_m &= (m-1)\left\{e^{a/(m-1)} - 1\right\} \\
&= (m-1)\left\{\frac{a}{m-1} + \left(\frac{a}{m-1}\right)^2 + O(m^{-3})\right\} \\
&= a + \frac{a^2}{m-1} + O(m^{-2}) \\
&= a + \frac{a^2}{m} + O(m^{-2}) \text{ with } a = -\log(\alpha).
\end{aligned}
\tag{7.2.15}
$$

Mukhopadhyay (1982a) used the expansion from (7.2.15) to prove that the modified two-stage strategy was not *asymptotically second-order efficient*. See also Swanepoel and van Wyk (1982).

Mukhopadhyay and Mauromoustakos (1987) gave a three-stage fixed-width interval procedure that was both *asymptotically consistent*and *asymptotically second-order efficient*. In the general unified theory paper of Mukhopadhyay and Solanky (1991), they investigated an accelerated sequential fixed-width interval procedure that was also both *asymptotically consistent*and *asymptotically second-order efficient*.

In reliability and lifetest analyses, one may argue that a threshold parameter θ is positive. Then, fixed-width confidence intervals

$$J_N \equiv (X_{N:1} - d, X_{N:1})$$

may not be appropriate because the lower confidence limit $X_{N:1} - d$ may be negative. Mukhopadhyay (1988b) gave an alternative fixed precision interval in a NExpo(θ, σ) distribution when $\theta > 0$.

Mukhopadhyay and Duggan (2000a) included a two-stage fixed-width interval procedure for θ assuming a known positive lower bound σ_L for σ. This procedure enjoys *asymptotic second-order* properties parallel to those summarized in (7.2.13) - (7.2.14).

Mukhopadhyay and Datta (1994) gave general multistage piecewise methodologies by extending the original parallel piecewise methodology of Mukhopadhyay and Sen (1993). Mukhopadhyay (1995a, Section 26.2.6) summarized a piecewise sequential strategy and its *asymptotic second-order* properties.

7.2.4 Comparisons and Conclusions

Let us reconsider the simulation results summarized in Figures 7.2.3 and 7.2.5. The output correspond to 99% confidence intervals for θ using a two-stage strategy and a purely sequential strategy respectively. Table 7.2.1 emphasizes some of the important entities.

Table 7.2.1: Simulation results for a 99% fixed-width confidence interval when $d = 0.5$, $m = 5$, $\theta = 10$, $\sigma = 5$ and number of simulations = 10000

	Purely Sequential	Two-Stage Procedure
C	46.05	46.05
$\overline{n}$	87.59	45.75
s_n	42.66	7.26
$\min(n)$	7	5
$\max(n)$	325	71
$\overline{p}$	0.991	0.987
$\text{se}(\overline{p})$	0.001	0.001

Table 7.2.1 shows clearly that the sequential procedure is more efficient than the two-stage procedure. On an average, two-stage procedure oversampled by a large margin and the estimated coverage probability ($\overline{p}$) exceeded the target, 0.99. On an average, sequential procedure undersampled by slightly and the estimated coverage probability fell below the target, 0.99. The estimated standard error of $\overline{p}$, $\text{se}(\overline{p})$, remain comparable under the two procedures.

However, we should not forget some of the important points mentioned before: A two-stage procedure delivers the consistency or exact consistency property (Theorem 7.2.1, part (iii)). There is no denying that a two-stage strategy will oversample, but oversampling can be cut down by increasing m as much as one can afford. Most importantly, two-stage sampling is operationally the most convenient one to implement. In contrast, all properties associated with a sequential strategy hold only asymptotically. That is, consistency or exact consistency property (Theorem 7.2.1, part (iii)) is completely out of the question under a sequential strategy. There is no denying that a sequential strategy has many attractive asymptotic second-order characteristics, but its implementation may become laborious and operationally costly.

7.3 Minimum Risk Point Estimation

Now, we develop a minimum risk point estimation problem for an unknown location parameter θ in a NExpo(θ, σ) distribution along the lines of Robbins (1959) as introduced in Section 6.4. We continue to assume that the scale parameter σ is unknown with $-\infty < \theta < \infty, 0 < \sigma < \infty$.

Suppose that $X_1, \ldots, X_n, \ldots$ are i.i.d. observations from a NExpo(θ, σ) distribution and we propose to estimate θ under a loss function which is composed of estimation error and cost due to sampling. The basic formulation is borrowed from Basu (1971) and Mukhopadhyay (1974a).

Having recorded $X_1, \ldots, X_n$ with $n \geq 2$,we suppose that the loss in estimating θ by the sample minimum order statistic, $X_{n:1} \equiv \min\{X_1, \ldots, X_n\}$, is given by:

$$\begin{aligned} L_n(\theta, X_{n:1}) = \quad & A \left| X_{n:1} - \theta \right|^k + cn^t \\ & \text{with } A > 0, c > 0, k > 0, \text{ and } t > 0 \text{ known.} \end{aligned} \tag{7.3.1}$$

Observe that $X_{n:1} > \theta$ w.p. 1 so that $\mid X_{n:1} - \theta \mid^k$ coincides with $(X_{n:1} - \theta)^k$.

Here, cn^t represents the cost for gathering n observations and $(X_{n:1} - \theta)^k$ represents the loss due to estimation of θ. If n is small, then the sampling cost cn^t will likely be small but we may expect $(X_{n:1} - \theta)^k$ to be large. On the other hand, if n is large, then the sampling cost cn^t will likely be large but we may expect $(X_{n:1} - \theta)^k$ to be small. The loss function (7.3.1) tries to create a balance between sampling cost and estimation error.

Now, the fixed-sample-size risk function associated with (7.3.1) is expressed as:

$$\begin{aligned} R_n(c) \equiv \mathrm{E}_{\theta,\sigma} \left\{ L_n(\theta, X_{n:1}) \right\} &= \frac{A\sigma^k}{n^k} \mathrm{E}_{\theta,\sigma} \left\{ \left[\frac{n\left(X_{n:1} - \theta\right)}{\sigma} \right]^k \right\} + cn^t \\ &= B\sigma^k (kn^k)^{-1} + cn^t \text{ with} B = Ak^2\Gamma(k), \end{aligned} \tag{7.3.2}$$

since $n\left(X_{n:1} - \theta\right)/\sigma$ is distributed as NExpo$(0, 1)$. Refer to part (i) in (7.1.5). Basu (1971) considered the loss function (7.3.1) where $k = t = 1$. Mukhopadhyay (1974a) worked with the loss function as seen in (7.3.1).

Our goal is to design estimation strategies that would minimize the risk for all $0 < \sigma < \infty$. Let us pretend that n is a continuous variable and examine the nature of the function:

$$g(n) \equiv B\sigma^k(kn^k)^{-1} + cn^t \text{ for } n > 0.$$

It is a strictly convex curve, but $g(n) \to \infty$ whether $n \downarrow 0$ or $n \uparrow \infty$. In other words, $g(n)$ must attain a minimum at some point within $(0, \infty)$. Consider solving $\frac{d}{dn}g(n) = 0$, obtain $n \equiv n^*(c) = (B\sigma^k/(ct))^{1/(t+k)}$, and check that $\left.\frac{d^2}{dn^2}g(n)\right|_{n=n^*} > 0$. Hence, we know that $n \equiv n^*(c)$ minimizes the function $g(n)$.

We note that

$$n \equiv n^*(c) = \left(\frac{B\sigma^k}{ct}\right)^{1/(t+k)} \tag{7.3.3}$$

may not be an integer, and yet we would refer to $n^*(c)$ as an optimal fixed sample size had σ been known.

The minimum fixed-sample-size risk is given by:

$$R_{n^*}(c) \equiv B\sigma^k(kn^{*k})^{-1} + cn^{*t} = c\left(1 + \frac{t}{k}\right)n^{*t}. \tag{7.3.4}$$

Surely, the magnitude of the optimal fixed sample size remains unknown since σ is unknown, but an expression of n^* is known. In fact, no fixed sample size strategy will minimize the risk under the loss function (7.3.1), uniformly in σ for all $0 < \sigma < \infty$. Thus, we introduce two kinds of purely sequential procedures. Other multistage procedures are left out for brevity.

7.3.1 Purely Sequential Procedure

Recall $U_n = (n-1)^{-1}\sum_{i=1}^{n}(X_i - X_{n:1})$ from (7.1.3) that unbiasedly estimates $\sigma, n \geq 2$. The purely sequential stopping rule of Mukhopadhyay (1974a) is summarized as follows:

We start with $X_1, \ldots, X_m$, a pilot of size $m(\geq 2)$, and we proceed with one additional X at-a-time as needed. The stopping time is defined as:

$$\begin{aligned} &N \equiv N(c) \text{ is the smallest integer } n(\geq m) \text{ for which} \\ &\qquad \text{we observe } n \geq \left(\frac{BU_n^k}{ct}\right)^{1/(t+k)}. \end{aligned} \tag{7.3.5}$$

This purely sequential procedure is implemented in the same way we did in the case of (6.2.17) or (6.4.5). Again, $\mathrm{P}_{\theta,\sigma}(N < \infty) = 1$, that is, sampling would terminate w.p. 1.

Upon termination, we estimate θ by the $X_{N:1}$ obtained from full dataset $N, X_1, \ldots, X_m, \ldots, X_N$. We leave it as an exercise to prove that the *bias* and *mean squared error* (MSE) of $X_{N:1}$ are given by:

$$\text{Bias}_{X_{N:1}} \equiv \sigma\mathrm{E}_{\theta,\sigma}\left[N^{-1}\right] \text{ and } \text{MSE}_{X_{N:1}} \equiv 2\sigma^2\mathrm{E}_{\theta,\sigma}\left[N^{-2}\right].$$

The risk associated with $X_{N:1}$ can be expressed as follows:

$$R_N(c) \equiv \mathrm{E}_{\theta,\sigma}\left\{L_N(\theta, X_{N:1})\right\} = B\sigma^k k^{-1}\mathrm{E}_{\theta,\sigma}\left[N^{-k}\right] + c\mathrm{E}_{\theta,\sigma}\left[N^t\right], \tag{7.3.6}$$

since $I(N = n)$ and $X_{n:1}$ are independent for all $n \geq m$. Here, one exploits the last part in (7.1.5).

By the way, the distribution of N depends only on σ and not θ. Yet, we will interchangeably write, for example, $\mathrm{E}_{\theta,\sigma}\left[N^{-k}\right]$ or $\mathrm{E}_{\theta,\sigma}\left[N^t\right]$ instead of $\mathrm{E}_{\sigma}\left[N^{-k}\right]$ or $\mathrm{E}_{\sigma}\left[N^t\right]$.

One pretends, however, that n^* is an integer! Along the lines of Robbins (1959), Basu (1971) and Mukhopadhyay (1974a) that defined the following measures to compare a sequential strategy $(N, X_{N:1})$ from (7.3.5) with a fictitious and optimal strategy $(n^*, X_{n^*:1})$:

$$\begin{aligned} &\textit{Risk Efficiency}\text{: RiskEff}(c) \equiv \frac{R_N(c)}{R_{n^*}(c)} \\ &= \left(1 + \frac{t}{k}\right)^{-1}\left\{\mathrm{E}_{\theta,\sigma}\left[N^t n^{*-t}\right] + \frac{t}{k}\mathrm{E}_{\theta,\sigma}\left[n^{*k}N^{-k}\right]\right\}; \\ &\textit{Regret}\text{:Regret}(c) \equiv R_N(c) - R_{n^*}(c). \end{aligned} \tag{7.3.7}$$

Basu (1971) and Mukhopadhyay (1974a) independently embarked upon investigations and for all fixed θ, σ, m, A, k and t showed the following *first-order* results as $c \to 0$.

Theorem 7.3.1: *For the stopping variable N defined by (7.3.5), for all fixed θ, σ, m, A, k and t, one has following properties as $c \to 0$:*

(i) $\mathrm{E}_{\theta,\sigma}(N) \leq n^* + O(1)$;

(ii) $\mathrm{E}_{\theta,\sigma}\left[N^t/n^{*t}\right] \to 1$;

(iii) $\text{RiskEff}(c) \to \begin{cases} 1 & \text{if } m > 1 + k^2(t+k)^{-1} \\ 1 + \gamma^* & \text{if } m = 1 + k^2(t+k)^{-1} \\ \infty & \text{if } m < 1 + k^2(t+k)^{-1}; \end{cases}$

where γ^* is an appropriate known positive constant that depends only on m, k and t.

The first-order properties from Theorem 7.3.1 may be interpreted as follows: For large n^* values, part (i) asserts that cost of not knowing σ , that is $\mathrm{E}_{\theta,\sigma}(N - n^*)$, is bounded! Part (ii) asserts that we may expect the average sample size $\mathrm{E}_{\theta,\sigma}(N)$ to hover in a close proximity of n^* for large n^* values. From part (iii), we may expect risk efficiency, RiskEff(c), to lie in a close proximity of one if $m > 1 + k^2(t+k)^{-1}$ for large n^* values. In other words, we may expect $R_N(c)$ and $R_{n^*}(c)$ to be close to one another when n^* is large. For example, we may expect $R_N(c), R_{n^*}(c)$ to be close to each other under (i) absolute error loss $(k = 1)$, $t = 1$ and $m \geq 2$ or (ii) squared error loss $(k = 2)$, $t = 1$ and $m \geq 3$.

Mukhopadhyay (1982b) combined some major techniques from Lombard and Swanepoel (1978), Swanepoel and van Wyk (1982), and Starr and Woodroofe (1972) to strengthen considerably the asymptotic risk efficiency result (Theorem 7.3.1, part (iii)). Mukhopadhyay's (1982b,1988a) *second-order* results are summarized as follows when $t = 1$ for all fixed θ, σ and A as $c \to 0$:

$$
\begin{aligned}
&(i) \quad \text{Regret}(c) \equiv R_N(c) - R_{n^*}(c) = O(c) \text{ if and only if } m \geq k+1;\\
&(ii) \quad \text{Regret}(c) \equiv \tfrac{1}{4}c + o(c) \text{ if } k = 1 \text{ (absolute error) and } m \geq 3;\\
&(iii) \quad \text{Regret}(c) \equiv \tfrac{2}{3}c + o(c) \text{ if } k = 2 \text{ (squared error) and } m \geq 5.
\end{aligned}
\tag{7.3.8}
$$

For all fixed θ, σ, A, k and t, Mukhopadhyay Mukhopadhyay (1982b,1988a) also gave the following results as $c \to 0$:

$$
\begin{aligned}
&\mathrm{E}_{\theta,\sigma}(N - n^*) = \eta^* + o(1);\\
&\frac{N - n^*}{\sqrt{n^*}} \text{ converges to } N(0,p) \text{ in distribution;}\\
&\mathrm{V}_{\theta,\sigma}(N) = pn^* + o(n^*);
\end{aligned}
\tag{7.3.9}
$$

with some appropriate numbers η^* and $p(> 0)$ under suitable conditions on m.

For large n^*, let us interpret the results from (7.3.9). Note that η^* is a computable number. So, from the first result in (7.3.9), we should know on an average how much N could differ from n^*. The second part says that the normalized stopping variable would have an approximate normal distribution. The third part says that $\mathrm{V}_{\theta,\sigma}(N)$ should be nearly pn^*. One may refer to Theorems 9-10 in Mukhopadhyay (1988a) for details.

The purely sequential procedure (7.3.5) is *asymptotically second-order efficient* and *second-order risk efficient.*

These and other interesting features will be explained more with the help of hands-on simulations and data analyses.

Computation of minimum risk point estimates for θ can be carried out by the 'program number 5' within **Seq07.exe**. This program gives the choice of using either purely sequential procedure or its improved version from Section 7.3.2.

These programs do not evaluate the second-order terms from (7.3.8)-(7.3.9).

Example 7.3.1 : Figure 7.3.1 shows the output of a simulation study under a purely sequential procedure. We generated data from a NExpo(10, 5) distribution. Other parameter values used in these simulations were $A = 500$, $k = 1$, $c = 0.01$, $t = 2$ and $m = 5$. That is, the loss function used is given by:

$$L_n(\theta, X_{n:1}) = 500\,|X_{n:1} - \theta| + 0.01n^2.$$

The results were obtained by averaging over 10,000 simulations. We see that $\overline{n}(= 50.38)$, the estimate of $\mathrm{E}_{10,5}(N)$, which is very close to $C(= 50)$. The average simulated risk, $\overline{R}_n(= 74.66)$, is also very close to $R_{n^*} = 75$, the minimum fixed-sample-size risk.

```
Minimum Risk Point Estimation for the
Location of a Negative Exponential Distribution
Using Purely Sequential Procedure
=================================

Input Parameters:
    Loss function constant, A = 500.00
    Loss function index,    k =   1.00
    Cost of unit sample     c =   0.01
    Loss function index,    t =   2.00
    Initial sample size,    m =    5

Simulation study by generating data from
negative exponential distribution with
location = 10.00  and scale =  5.00

Number of simulation replications = 10000

    Optimal sample size (n_star)       = 50.00
    Average final sample size (n_bar) = 50.38
    Std. dev. of n (s_n)               =  2.42
    Minimum sample size (n_min)        = 40
    Maximum sample size (n_max)        = 59

    Minimum fixed-sample risk (Rn*)    = 75.00
    Simulated average risk (Rbar_n)    = 74.66
    Regret (Rbar_n - Rn*)              = -0.34
```

Figure 7.3.1. *Minimum risk point estimation using purely sequential procedure.*

7.3.2 Improved Purely Sequential Procedure

Here, we reconsider the same minimum risk point estimation problem for θ in an NExpo(θ, σ) distribution. Now, we use the UMVUE, $T_n (\equiv X_{n:1} - n^{-1} U_n)$, from (7.1.6) to estimate $\theta, n \geq 2$. Ghosh and Mukhopadhyay (1990) developed a sequential minimum risk point estimation methodology under the following loss function:

$$L_n(\theta, T_n) = A\left(T_n - \theta\right)^2 + cn \text{ with } A > 0 \text{ and } c > 0 \text{ known.} \qquad (7.3.10)$$

This amounts to squared error loss due to estimation plus a linear cost of sampling.

In view of the properties mentioned in (7.1.5), the fixed-sample-size risk function $R_n(c)$ can be expressed as:

$$\begin{aligned} &\mathrm{E}_{\theta,\sigma}\left\{L_n(\theta, T_n)\right\} \\ &= A\mathrm{E}_{\theta,\sigma}\left\{(X_{n:1} - \theta)^2\right\} + \frac{A}{n^2}\mathrm{E}_{\theta,\sigma}\left\{U_n^2\right\} \\ &\qquad - \frac{2A}{n}\mathrm{E}_{\theta,\sigma}\left\{(X_{n:1} - \theta)\, U_n\right\} + cn \end{aligned}$$

$$
\begin{aligned}
&= \frac{A\sigma^2}{n^2}\mathrm{E}_{\theta,\sigma}\left\{\left[\frac{n\,(X_{n:1}-\theta)}{\sigma}\right]^2\right\} + \frac{A}{n^2}\left\{\mathrm{V}_{\theta,\sigma}(U_n^2) + \mathrm{E}^2_{\theta,\sigma}(U_n)\right\} \\
&\qquad\qquad - \frac{2A}{n}\mathrm{E}_{\theta,\sigma}\,(X_{n:1}-\theta)\,\mathrm{E}_{\theta,\sigma}(U_n) + cn \\
&= \frac{2A\sigma^2}{n^2} + \frac{A}{n^2}\left\{\frac{\sigma^2}{n-1} + \sigma^2\right\} - \frac{2A\sigma}{n^2}\mathrm{E}_{\theta,\sigma}\left\{\frac{n\,(X_{n:1}-\theta)}{\sigma}\right\}\sigma + cn \\
&= \frac{2A\sigma^2}{n^2} + \frac{A\sigma^2}{n(n-1)} - \frac{2A\sigma^2}{n^2} + cn \\
&= \frac{A\sigma^2}{n(n-1)} + cn = \frac{A\sigma^2}{n^2} + cn + O(n^{-3}).
\end{aligned}
\tag{7.3.11}
$$

Now, our goal is to minimize $R_n(c)$ for all $0 < \sigma < \infty$. Clearly, the optimal fixed sample size, had σ been known, turns out to be:

$$
n \equiv n^{**}(c) = \left(\frac{2A\sigma^2}{c}\right)^{1/3}, \tag{7.3.12}
$$

after neglecting the term $O(n^{-3})$ from (7.3.11). Again, we pretend that n^{**} is an integer, and the minimum fixed-sample-size risk is given by

$$
R_{n^{**}}(c) \equiv \frac{A\sigma^2}{n^{**2}} + cn^{**} + O(n^{**-3}) = \frac{3}{2}cn^{**} + O(c^{3/2}). \tag{7.3.13}
$$

Let us turn around and reexamine these entities in the light of the loss function (7.3.1) where θ was estimated by $X_{n:1}$ by fixing $k = 2, t = 1$. We find $n^*(c) = \left(4A\sigma^2/c\right)^{1/3}$ and $R_{n^*}(c) = \frac{3}{2}cn^*$ from (7.3.3) and (7.3.4) respectively. But one would observe that

$$
\begin{aligned}
n^{**}(c) &= 2^{-1/3}n^*(c) \approx 0.7937n^*(c) \text{ and} \\
R_{n^{**}}(c) &= 2^{-1/3}R_{n^*}(c) \approx 0.7937R_{n^*}(c).
\end{aligned}
\tag{7.3.14}
$$

In other words, under squared error plus linear cost, the present strategy $(n^{**}, T_{n^{**}}, R_{n^{**}})$ would save 20.63% in both *optimal* fixed sample size and *minimum risk* compared with the optimal strategy $(n^*, X_{n^*:1}, R_{n^*})$ from Section 7.3.1. So, one ought to pursue the present strategy $(n^{**}, T_{n^{**}}, R_{n^{**}})$ to define an appropriate sequential strategy. Such a sequential strategy may have both average sample size and sequential risk reduced by nearly 20.63% compared with those for the earlier sequential strategy (7.3.5) in Section 7.3.1. Ghosh and Mukhopadhyay (1990) proceeded along that route.

We recall that $U_n = (n-1)^{-1}\sum_{i=1}^{n}(X_i - X_{n:1})$ from (7.1.3) estimates $\sigma, n \geq 2$. Next, the purely sequential stopping rule of Ghosh and Mukhopadhyay (1990) is summarized.

One starts with $X_1, \ldots, X_m$, a pilot of size $m(\geq 2)$, and takes one additional X at-a-time as needed. The stopping time is defined as:

$$N \equiv N(c) \text{ is the smallest integer } n(\geq m) \text{ for which}$$
$$\text{we observe } n \geq \left(\frac{2AU_n^2}{c}\right)^{1/3}, \qquad (7.3.15)$$

by imitating the expression of n^{**} from (7.3.12). This sequential procedure is implemented in the same way we did in the case of (7.3.5).

Again, $P_{\theta,\sigma}(N < \infty) = 1$, so that sampling terminates w.p. 1. Upon termination, we estimate θ by $T_N \equiv X_{N:1} - N^{-1}U_N$ obtained from full dataset $N, X_1, \ldots, X_m, \ldots, X_N$.

The risk associated T_N can be expressed as:

$$R_N(c) \equiv \mathrm{E}_{\theta,\sigma}\{L_N(\theta, T_N)\} = \mathrm{E}_{\theta,\sigma}\left\{A(T_N - \theta)^2 + cN\right\}. \qquad (7.3.16)$$

But, unlike (7.3.6), $R_N(c)$ cannot be simplified involving moments of N alone. Complications arise because the random variables $I(N = n)$ and T_n are *not* independent for any fixed $n \geq m$.

Given this backdrop, Ghosh and Mukhopadhyay (1990) combined major techniques from Lombard and Swanepoel (1978), Swanepoel and van Wyk (1982), Starr and Woodroofe (1972), and Mukhopadhyay (1982b,1988a) to provide asymptotic *second-order* results for the improved sequential strategy (7.3.15). Their major results are summarized for all fixed θ, σ, and A as $c \to 0$:

$$\begin{aligned} &(i) \quad \mathrm{E}_{\theta,\sigma}(N - n^{**}) = \eta^{**} + o(1) \text{ if } m \geq 2; \\ &(ii) \quad \text{Regret}(c) \equiv R_N(c) - R_{n^{**}}(c) = \tfrac{2}{3}c + o(c) \text{ if } m \geq 5; \end{aligned} \qquad (7.3.17)$$

where

$$\eta^{**} \equiv \frac{1}{6} - \frac{1}{3}\sum_{n=1}^{\infty}\frac{1}{n}\mathrm{E}\left[\max\left(0, \chi_{2n}^2 - 5n\right)\right] = 0.086489. \qquad (7.3.18)$$

For more details, one may refer to Ghosh and Mukhopadhyay (1990).

For large n^{**}, let us interpret the results from (7.3.17). From part (i) in (7.3.17), we should expect N to differ from n^{**} by nearly 0.086489 on an average if $m \geq 2$ when n^{**} is large. Part (ii) in (7.3.17) says that we should expect regret to be close to the cost of two-thirds of one observation if $m \geq 5$ when n^{**} is large.

> The purely sequential procedure (7.3.15) is *asymptotically second-order efficient* and *second-order risk efficient.*

These and other interesting features can be examined with the help of hands-on simulations and data analyses by selecting the option to run improved purely sequential procedure (7.3.15) using the 'program number 5' within **Seq07.exe**.

Example 7.3.2 : We examine this procedure by simulating data from a NExpo(10, 5) distribution under a loss function:

$$L_n(\theta, T_n) = 500\,(T_n - \theta)^2 + 0.01n.$$

Figure 7.3.2 summarizes results from simulations by fixing $m = 5$ and 10,000 replications. From the results shown in this figure, we can feel good knowing that the alternative sequential procedure (7.3.15) formed so well. In this particular case, we have $\overline{n} - n^{**} = 0.15$. But, if we consider the interval

$$\begin{aligned}(\overline{n} - n^{**}) \pm 2 \times s_n/\sqrt{10000} &= 0.15 \pm 2 \times 7.82/\sqrt{10000} \\ &= 0.15 \pm 0.1564 \Rightarrow (-0.0042, 0.3064).\end{aligned}$$

This interval does include the value $\eta^{**} = 0.086489$.

```
Minimum Risk Point Estimation for the
Location of a Negative Exponential Distribution
Using Improved Purely Sequential Procedure
===============================================

Input Parameters:
    Loss function constant, A = 500.00
    Cost of unit sample      c =   0.01
    Initial sample size,     m =    5

Simulation study by generating data from
negative exponential distribution with
location = 10.00  and scale =  5.00

Number of simulation replications = 10000

 Optimal sample size (n_star)        =     135.72
 eta** [see equation (7.3.18)]       =   0.086489

 Average final sample size (n_bar) =     135.87
 Std. dev. of n (s_n)                =       7.82
 Minimum sample size (n_min)         =     107
 Maximum sample size (n_max)         =     165

 Minimum fixed-sample risk (Rn*)     =       2.04
 Simulated average risk (Rbar_n)     =       2.06
 Regret (Rbar_n - Rn*)               =       0.02
```

Figure 7.3.2. *Minimum risk point estimation using improved purely sequential procedure.*

7.3.3 Comparisons and Conclusions

Now, we want to compare performances of the sequential and improved sequential strategies. Hence, we reconsider the simulation studies from Examples 7.3.1 - 7.3.2 one more time. Also, we ran simulations under sequential strategy (7.3.5) with $k = 2, t = 1$ and $m = 5$. Table 7.3.1 summarizes the findings.

In Table 7.3.1, the entries in columns 3 and 4 should be contrasted. We note that

$$\text{(i) } n^*_{\text{column } 4}/n^*_{\text{column } 3} = 135.72/171.00 = 0.79368 \approx 0.7937,$$

and

$$\text{(ii) } R_{n^*\text{column } 4}/R_{n^*\text{column } 3} = 2.04/2.56 = 0.79688 \approx 0.7969.$$

These ratios are close to theoretically expected value 0.7937.

On the other hand, from simulations, we have found that

$$\text{(i) } \overline{n}_{\text{column } 4}/\overline{n}_{\text{column } 3} = 135.93/171.18 = 0.79408,$$

and

$$\text{(ii) } \overline{R}_{n \text{ column } 4}/\overline{R}_{n\text{column } 3} = 2.02/2.56 = 0.78906.$$

In other words, the simulated estimates of the average sample sizes and regrets are extremely close to what one ought to expect these to be theoretically. So, in practice we should expect the improved sequential procedure to beat the purely sequential procedure in all aspects, and by a margin that is accurately suggested by the theory.

Table 7.3.1: Comparison of purely sequential and improved purely sequential procedures with common parameters $A = 500$, $c = 0.01$, $m = 5$, $\theta = 10$, $\sigma = 5$ and number of simulations = 10,000

	Purely Sequential Procedure		Improved Purely Sequential Procedure
	$k = 1, t = 2$	$k = 2, t = 1$	$k = 2, t = 1$
n^*	50.00	171.00	135.72
$\overline{n}$	50.38	171.18	135.93
s_n	2.42	8.86	7.79
$\min(n)$	40	139	105
$\max(n)$	59	202	162
R_{n^*}	75.00	2.56	2.04
$\overline{R}_n$	74.66	2.56	2.02
Regret	-0.34	-0.01	-0.02

Example 7.3.3 : Here is another example to compare the two sequential procedure further. In this example, we use the data provided in Table 7.3.2 which were generated from a NExpo$(15, 5)$ distribution.

Table 7.3.2: Data from a NExpo(15,5) distribution
The sequential data given below should be taken row by row: $x_1 = 24.45, x_2 = 19.71, \ldots$

24.45	19.71	16.23	23.98	17.66	15.66
17.57	25.19	19.84	15.65	17.92	15.27
17.22	18.26	22.55	19.04	29.35	41.25
20.25	27.63	16.25	20.55	20.41	17.29
20.88	16.03	17.49	17.30	22.34	24.36
15.65	25.25	21.07	16.69	15.51	19.71
16.32	16.32	19.59	22.49	29.54	18.69
19.12	18.64	17.10	16.58	17.34	19.65
17.74	20.02	22.48	24.53	15.74	17.45
16.56	17.25	40.43	16.53	15.74	17.74
15.56	23.97	17.63	20.10	27.85	29.09
15.27	19.09	18.32	29.27	19.91	15.51
20.64	24.15	16.58	21.41	17.57	35.68
19.27	16.65	15.52	15.48	31.83	17.19
16.57	23.17	21.34	15.62	17.89	18.02
15.67	16.41	23.06	26.42	20.45	18.73
17.25	17.63	17.39	21.18	16.68	19.99
22.20	15.14	16.15	17.92	16.14	29.65
18.74	21.50	17.48	23.00	20.57	17.27
19.41	24.14	19.34	35.25	19.19	24.71
22.70	17.72	21.01	16.36	15.18	17.42
18.39	25.68	27.59	17.61	19.54	18.39
18.67	16.19	15.92	25.90	16.41	15.09
20.36	24.93	15.72	28.66	23.01	19.97
17.83	16.17	16.17	16.46	18.38	19.50
40.69	16.84	19.74	15.06	22.38	18.09
29.32	15.92	16.57	20.18	17.02	16.58
26.79	17.86	20.36	20.41	15.96	17.84
24.51	22.11	15.39	23.70	18.07	17.96

A summary of the results obtained for data from Table 7.3.2 using the 'program number 5' within **Seq07.exe** is presented in Table 7.3.3. Here, we have first used sequential procedure (7.3.5) when (i) $m = 5, k = 1, t = 2$ and (ii) $m = 5, k = 2, t = 1$. Then, we used improved sequential procedure (7.3.15) with $m = 5$. Purely sequential procedure (7.3.5) estimates θ using the sample minimum, $X_{n:1}$, which is the MLE of θ based on full sequentially observed data $n, x_1, \ldots, x_n$ at the stopped stage $N = n$. The improved purely sequential procedure (7.3.15) estimates θ using UMVUE, T_n given in (7.1.6) based on full sequentially observed data $n, x_1, \ldots, x_n$ at the stopped stage $N = n$. In addition, the program also computes U_n, the UMVUE of σ from (7.1.3), based on full sequentially observed data $n, x_1, \ldots, x_n$ at the stopped stage $N = n$.

One should check that the entries shown in Table 7.3.3 are entirely within reason of what one ought to expect theoretically. We leave this as a simple exercise in data analyses.

Table 7.3.3: Results obtained using the data given in Table 7.3.2
Common parameters: $A = 500$, $c = 0.01$, $m = 5$

	Purely Sequential Procedure		Improved Purely Sequential Procedure
	$k = 1, t = 2$	$k = 2, t = 1$	$k = 2, t = 1$
n	50	174	137
$\widehat{\theta}_{\text{MLE}} = X_{n:1}$	15.27	15.06	15.14
$\widehat{\sigma}_{\text{UMVUE}}$	4.84	5.10	5.05
T_n	—	—	15.10

7.3.4 Other Multistage Procedures

In case of the sequential fixed-width interval procedure (7.2.11), recall the *second-order* expansion of confidence coefficient from (7.2.14). The leading term was $1 - \alpha$, but its second term involved unknown C. This dampened usefulness of such a *second-order* expansion because a departure of confidence coefficient from $1 - \alpha$ could not be readily evaluated. By the way, replacing C with N gave a crude estimate of a departure of confidence coefficient from $1 - \alpha$. Mukhopadhyay and Datta (1995) resolved this important issue by developing a *fine-tuned* sequential fixed-width interval procedure with the confidence coefficient, $1 - \alpha + o(C^{-1})$.

In connection with a minimum risk estimation problem for θ, one may also refer to Chaturvedi and Shukla's (1990) commentaries. Ghosh and Mukhopadhyay (1989) developed sequential minimum risk point estimation methodologies for the percentiles of a negative exponential distribution. They obtained *second-order* results including an expansion for the regret function.

Mukhopadhyay and Hilton (1986) introduced two-stage, modified two-stage, sequential minimum risk and bounded risk estimation methodologies for θ. They included asymptotic *first-order* results and "cost of ignorance" issues. For bounded risk estimation problems, one may also refer to Hilton (1984) and Hamdy et al. (1988). Mukhopadhyay and Solanky (1991) included an accelerated sequential minimum risk estimation strategy for θ along with *second-order* results.

Along the lines of Mukhopadhyay (1985a) and Mukhopadhyay et al. (1987), a three-stage minimum risk estimator of θ under weighted squared error loss plus linear cost was developed by Hamdy et al. (1988) and Hamdy (1988). They gave *second-order* analyses of regret function.

7.4 Selected Derivations

In this section, we highlight some parts of crucial derivations.

7.4.1 Proof of Theorem 7.2.1

Proof of Part (i):

From (7.2.4), first observe a basic inequality:

$$b_m U_m/d \le N \le m + b_m U_m/d \text{ w.p. 1.} \tag{7.4.1}$$

Now, we take expectations throughout (7.4.1) and write:

$$b_m d^{-1}\mathrm{E}_{\theta,\sigma}[U_m] \le N \le m + b_m d^{-1}\mathrm{E}_{\theta,\sigma}[U_m]. \tag{7.4.2}$$

But, $\mathrm{E}_{\theta,\sigma}[U_m] = \sigma$. Hence, we have part (i) ■

Proof of Part (ii):

The random variable Q is positive since $X_{n:1} > \theta$ w.p. 1 for all fixed n. Since $X_{n:1}$ and $I(N = n)$ are independent for all $n \ge m$, for all $x > 0$, we write:

$$\begin{aligned}
&\mathrm{P}_{\theta,\sigma}\{0 < Q \le x\} \\
&= \textstyle\sum_{n\ge m} \mathrm{P}_{\theta,\sigma}\left\{\frac{N(X_{N:1}-\theta)}{\sigma} \le x \cap N = n\right\} \\
&= \textstyle\sum_{n\ gem} \mathrm{P}_{\theta,\sigma}\left\{\frac{n(X_{n:1}-\theta)}{\sigma} \le x \cap N = n\right\} \\
&= \textstyle\sum_{n\ge m} \mathrm{P}_{\theta,\sigma}\left\{\frac{n(X_{n:1}-\theta)}{\sigma} \le x\right\} \mathrm{P}_{\theta,\sigma}\{N = n\} \\
&\qquad \text{since } X_{n:1} \text{ and } I(N = n) \text{ are independent} \qquad (7.4.3) \\
&= \textstyle\sum_{n\ge m}\{1 - \exp(-x)\}\mathrm{P}_{\theta,\sigma}\{N = n\} \\
&\qquad \text{since } \frac{n(X_{n:1}-\theta)}{\sigma} \text{ is standard exponential} \\
&= \{1 - \exp(-x)\} \textstyle\sum_{n\ge m} \mathrm{P}_{\theta,\sigma}\{N = n\} \\
&= \{1 - \exp(-x)\}\mathrm{P}_{\theta,\sigma}\{N < \infty\} \\
&= 1 - \exp(-x).
\end{aligned}$$

Now, the distribution function $\mathrm{P}_{\theta,\sigma}\{Q \le x\}$ is zero when $x \le 0$, and it is $1 - \exp(-x)$ when $x > 0$. The result follows. ■

Proof of Part (iii):

First, one will verify (7.2.5) in the same way we began (7.4.3). Then, we use the lower bound from (7.4.1) to write

$$\begin{aligned} &\exp(Nd/\sigma) \geq \exp\left(b_m d^{-1} U_m d/\sigma\right) = \exp(b_m U_m/\sigma) \text{ w.p. } 1 \\ &\quad \Rightarrow 1 - \exp(-Nd/\sigma) \geq 1 - \exp(-b_m U_m/\sigma) \text{ w.p. } 1. \end{aligned} \tag{7.4.4}$$

Now, let Y be a standard exponential random variable distributed independently of U_m. Then, from (7.2.5) we have:

$$\begin{aligned} &\mathrm{P}_{\theta,\sigma}\{\theta \in J_N\} \\ &\quad = \mathrm{E}_{\theta,\sigma}\left[1 - \exp(-Nd/\sigma)\right] \\ &\quad \geq \mathrm{E}_{\theta,\sigma}\left[1 - \exp(-b_m U_m/\sigma)\right] \text{ using (7.4.4)} \\ &\quad = \mathrm{E}_{\theta,\sigma}\left[\mathrm{P}_{\theta,\sigma}\{0 < Y \leq b_m U_m/\sigma \mid U_m\}\right] \\ &\quad = \mathrm{E}_{\theta,\sigma}\left[\mathrm{P}_{\theta,\sigma}\left\{0 < \frac{Y}{U_m/\sigma} \leq b_m \mid U_m\right\}\right]. \end{aligned} \tag{7.4.5}$$

But, conditionally given U_m, the random variable $\dfrac{Y}{U_m/\sigma}$ has a $F_{2,2m-2}$ distribution. Hence, we have:

$$\begin{aligned} &\mathrm{P}_{\theta,\sigma}\{\theta \in J_N\} \\ &\quad \geq \mathrm{E}_{\theta,\sigma}\left[\mathrm{P}\{0 < F_{2,2m-2} \leq b_m\}\right] \\ &\quad = \mathrm{P}\{0 < F_{2,2m-2} \leq b_m\} \\ &\quad = 1 - \alpha, \end{aligned} \tag{7.4.6}$$

by the choice of b_m. ■

7.4.2 Proof of Theorem 7.2.2

Proof of Part (i):

From (7.2.11), first observe a basic inequality:

$$aU_N/d \leq N \leq m + aU_{N-1}/d \text{ w.p. } 1. \tag{7.4.7}$$

Assume for a moment that $\mathrm{E}_{\theta,\sigma}[N] < \infty$. Now, we work from the right-hand side of (7.4.7) to write:

$$\begin{aligned} &N \leq m + \frac{a}{d}\textstyle\sum_{i=1}^{N-1}(X_i - X_{N-1:1}) \text{ w.p. } 1 \\ &\Rightarrow (N-m)(N-2) \leq \frac{a}{d}\textstyle\sum_{i=1}^{N-1}(X_i - \theta) \leq \frac{a}{d}\sum_{i=1}^{N}(X_i - \theta) \text{ w.p. } 1 \\ &\quad \Rightarrow N^2 - (m+2)N \leq \frac{a}{d}\textstyle\sum_{i=1}^{N}(X_i - \theta) \text{ w.p. } 1. \end{aligned} \tag{7.4.8}$$

Now, we take expectations throughout the last step in (7.4.8) and use

Wald's first equation (Theorem 3.5.4) to obtain:

$$\begin{aligned}
\mathrm{E}_{\theta,\sigma}[N^2] - (m+2)\mathrm{E}_{\theta,\sigma}[N] &\le \frac{a}{d}\mathrm{E}_{\theta,\sigma}\left[\textstyle\sum_{i=1}^{N}(X_i-\theta)\right] = \frac{a}{d}\sigma\mathrm{E}_{\theta,\sigma}[N] \\
\Rightarrow \mathrm{E}^2_{\theta,\sigma}[N] - (m+2)\mathrm{E}_{\theta,\sigma}[N] &\le C\mathrm{E}_{\theta,\sigma}[N] \text{ since } C = \frac{a}{d}\sigma \\
\Rightarrow \mathrm{E}_{\theta,\sigma}[N] - (m+2) &\le C.
\end{aligned} \tag{7.4.9}$$

That is, $\mathrm{E}_{\theta,\sigma}[N] \le C + m + 2$.

Next, if $\mathrm{E}_{\theta,\sigma}[N]$ is not finite, then one may employ a truncated stopping variable, namely $N_k = \min\{k, N\}, k = 1, 2, \ldots$. From (7.4.9), we can claim that $\mathrm{E}_{\theta,\sigma}[N_k] \le C + m + 2$ for each fixed k. Now, an application of monotone convergence theorem will complete the proof. ■

Proof of Part (ii):

Its proof is very similar to that of Theorem 7.2.1, part (ii).

Proof of Part (iii):

First, observe that $N \equiv N(d) \to \infty$ w.p. 1 as $d \to 0$. We may rewrite the basic inequality from (7.4.7) as follows:

$$\frac{U_N}{\sigma} \le \frac{N}{C} \le \frac{m}{C} + \frac{U_{N-1}}{\sigma}. \tag{7.4.10}$$

But, as $d \to 0$, we must have $m/C \to 0$ and both $U_N/\sigma, U_{N-1}/\sigma$ converge to 1 w.p. 1. Hence, we can claim that $N/C \to 1$ w.p. 1 as $d \to 0$. Now, Fatou's Lemma will apply, and we can claim:

$$\liminf_{d\to 0} \mathrm{E}_{\theta,\sigma}[N/C] \ge \mathrm{E}_{\theta,\sigma}[\liminf_{d\to 0} N/C] = \mathrm{E}[1] = 1.$$

Also, using part (i), we have $\limsup_{d\to 0} \mathrm{E}_{\theta,\sigma}[N/C] \le 1$. Hence, part (iii) follows. ■

Proof of Part (iv):

One may proceed along the hints given in Exercise 7.2.3 and write:

$$\begin{aligned}
\lim_{d\to 0} \mathrm{P}_{\theta,\sigma}\{\mu \in J_N\} &= \mathrm{E}_{\theta,\sigma}\left[1 - \exp\left(-\lim_{d\to 0}(Nd/\sigma)\right)\right] \\
&= \mathrm{E}_{\theta,\sigma}\left[1 - \exp\left(-a\right)\right] = 1 - \alpha,
\end{aligned} \tag{7.4.11}$$

since clearly $\lim_{d\to 0}(Nd/\sigma) = a$ because $\lim_{d\to 0}(N/C) = 1$ w.p. 1. In the first line of (7.4.11), note that the limit and the expectation could be interchanged since the function $1 - \exp(-x)$ is bounded for $x > 0$, and hence dominated convergence theorem applies. ■

7.5 Exercises

Exercise 7.1.1: Verify (7.1.5).

Exercise 7.1.2: Show that T_n from (7.1.6) is the UMVUE for θ.

Exercise 7.2.1: For the two-stage procedure (7.2.4), show that

$$\mathrm{E}_{\theta,\sigma}\left[X_{N:1}\right] = \mu + \sigma \mathrm{E}_{\theta,\sigma}\left[N^{-1}\right].$$

Exercise 7.2.2: Prove that

$$\lim_{d \to 0} \mathrm{P}_{\theta,\sigma}\{\mu \in J_N\} = 1 - \alpha,$$

for the two-stage procedure (7.2.4). {*Hints*: Note that $\mathrm{P}_{\theta,\sigma}\{\mu \in J_N\} = \mathrm{E}_{\theta,\sigma}\left[1 - \exp(-Nd/\sigma)\right]$. But, observe that $Nd/\sigma \xrightarrow{\mathrm{P}} b_m U_m/\sigma$ as $d \to 0$. Also, $g(x) = 1 - \exp(-x), x > 0$, is bounded. Thus, by the dominated convergence theorem, one has:

$$\lim_{d \to 0} \mathrm{P}_{\theta,\sigma}\{\mu \in J_N\} = \mathrm{E}_{\theta,\sigma}\left[1 - \exp\left(-\lim_{d \to 0}(Nd/\sigma)\right)\right].$$

Suppose that Y is a standard exponential variable that is independent of U_m. Now,

$$\begin{aligned}\mathrm{E}_{\theta,\sigma}\left[1 - \exp(-b_m U_m/\sigma)\right] &= \mathrm{E}_{\theta,\sigma}\left[\mathrm{P}_{\theta,\sigma}\left\{0 < Y < b_m U_m/\sigma \mid U_m\right\}\right] \\ &= \mathrm{P}\left\{0 < F_{2,2m-2} < b_m\right\} = 1 - \alpha.\end{aligned}$$

This property assures us that for a large sample size, the associated confidence coefficient will be in the vicinity of the target, $1 - \alpha$.}

Exercise 7.2.3: Prove (7.2.6). { *Hints*: The number b_m is chosen in such a way that $\mathrm{P}\left\{0 < F_{2,2m-2} < b_m\right\} = 1 - \alpha$. Suppose that Y is a standard exponential variable that is independent of U_m. Also, let W be a χ^2_{2m-2} random variable. Then, $Y\left(b_m U_m/\sigma\right)^{-1}$ has a $F_{2,2m-2}$ distribution. Now,

$$\begin{aligned}&\mathrm{P}\left\{0 < F_{2,2m-2} < b_m\right\} \\ &= \mathrm{E}_{\theta,\sigma}\left[\mathrm{P}_{\theta,\sigma}\left\{0 < Y < b_m U_m/\sigma \mid U_m\right\}\right] \\ &= \mathrm{E}_{\theta,\sigma}\left[1 - \exp(-b_m U_m/\sigma)\right] \\ &= 1 - \mathrm{E}_{\theta,\sigma}\left[\exp(-b_m U_m/\sigma)\right] \\ &= 1 - \mathrm{E}_{\theta,\sigma}\left[\exp\left(-\{b_m/(2m-2)\}W\right)\right] \\ &= 1 - \left(1 + \frac{2b_m}{2m-2}\right)^{-(m-1)},\end{aligned}$$

using the moment generating function of W.

Thus, we have:

$$\left(1 + \frac{2b_m}{2m-2}\right)^{-(m-1)} = \alpha,$$

which can be easily rewritten in the form of (7.2.6).}

Exercise 7.2.4: For the purely sequential procedure (7.2.11), show that

$$\mathrm{E}_{\theta,\sigma}\left[X_{N:1}\right]=\theta+\sigma\mathrm{E}_{\theta,\sigma}\left[N^{-1}\right].$$

Exercise 7.3.1: For the purely sequential procedure (7.3.5), show that

$$\text{Bias}_{X_{N:1}}\equiv\sigma\mathrm{E}_{\theta,\sigma}\left[N^{-1}\right]\text{ and }\text{MSE}_{X_{N:1}}\equiv 2\sigma^2\mathrm{E}_{\theta,\sigma}\left[N^{-2}\right].$$

Exercise 7.3.2: For the purely sequential procedure (7.3.5) with $t=k=1$, prove parts (i) and (ii) of Theorem 7.3.1.

Exercise 7.3.3: For the purely sequential procedure (7.3.15), prove that

$$\lim_{d\to 0} N/n^{**}=1\text{ w.p. }1.$$

CHAPTER 8

Point Estimation of Mean of an Exponential Population

8.1 Introduction

In this chapter, we focus on an exponential distribution, referred to as Expo(θ), having the following probability density function:

$$f(x;\theta) = \begin{cases} \dfrac{1}{\theta}\exp(-x/\theta) & \text{if } x > 0 \\ 0 & \text{if } x \le 0. \end{cases} \tag{8.1.1}$$

Here, θ is the unknown mean of this distribution, $0 < \theta < \infty$.

This distribution has been used in many reliability and lifetesting experiments to describe, for example, a failure time of complex equipment, vacuum tubes and small electrical components. In these contexts, θ is interpreted as the *mean time to failure* (MTTF). One is referred to Johnson and Kotz (1970), Bain (1978), Lawless and Singhal (1980), Grubbs (1971), Basu (1991) and other sources. This distribution has also been used extensively to model survival times, especially under random censoring. See, for example, Aras (1987,1989), Gardiner and Susarla (1983,1984,1991) and Gardiner et al. (1986).

For overviews on reliability theory, one may refer to Lomnicki (1973), Barlow and Proschan (1975) and Ansell and Phillips (1989). The articles of Tong (1977), Brown (1977), Beg and Singh (1979) and Beg (1980) will provide added perspectives. A volume edited by Balakrishnan and Basu (1995) presented a wide spectrum of statistical methodologies with exponential distributions.

Suppose that we have recorded n independent observations $X_1, \ldots, X_n$, each following Expo(θ) distribution from (8.1.1). Then, the MTTF parameter θ is customarily estimated by the *maximum likelihood estimator* (MLE). The MLE coincides with the *uniformly minimum variance unbiased estimator* (UMVUE), namely the sample mean $\overline{X}_n \equiv n^{-1}\sum_{i=1}^{n} X_i$.

Now, let us mention the layout of this chapter. We begin with a purely sequential strategy for a minimum risk point estimation problem for θ in Section 8.2. The loss function consists of squared estimation error by means of $\overline{X}_n$ plus a linear cost of sampling. This situation is very different from other sequential methodologies discussed in Chapters 6 and 7. One will see that we cannot express the associated sequential risk involving moments of N, the final sample size, alone. In spite of this difficulty, Starr and Woodroofe (1972) and Woodroofe (1977) came up with elegant *second-order* results. Next, in Section 8.3, we introduce a new two-stage bounded risk point estimation procedure for θ. Mukhopadhyay and Pepe (2006) have shown that a preassigned *risk-bound* can be met exactly through a properly designed Stein-type two-stage strategy. This is a striking result given that one cannot express the associated two-stage risk involving moments of N alone. Section 8.3.2 gives an improvement due to Zacks and Mukhopadhyay (2006a). Section 8.4 is devoted to computer related work and data analyses. Some of the other multistage methodologies are briefly mentioned in Section 8.5. We mention very briefly analogous developments in (i) a *two-parameter exponential family* in Section 8.5.1, (ii) a *natural exponential family* (NEF) with a *power variance function* (PVF) in Section 8.5.2, and (iii) *reliability* and *time-sequential* problems in Sections 8.5.3 and 8.5.4 respectively. Selected derivations are included in Section 8.6.

One may view this exposition largely as a reunion of reliability estimation and sequential analysis in the light of the seminal papers of Epstein and Sobel (1953,1954,1955) and Sobel (1956). We hope that this effort will energize investigations under models that are more sophisticated than (8.1.1).

8.2 Minimum Risk Estimation

Having recorded n independent observations $X_1, \ldots, X_n$, each following an Expo(θ) distribution (8.1.1), we suppose that the loss in estimating θ by $\overline{X}_n$ is given by:

$$L_n(\theta, \overline{X}_n) = A\left(\overline{X}_n - \theta\right)^2 + cn \text{ with } A > 0, c > 0 \text{ known.} \tag{8.2.1}$$

Here, "cn" represents the cost of gathering n observations and $\left(\overline{X}_n - \theta\right)^2$ represents a loss due to estimation of θ by $\overline{X}_n$. If n is small, cn will be small but we may expect $\left(\overline{X}_n - \theta\right)^2$ to be large. But, if n is large, the sampling cost cn will be large, and we may expect $\left(\overline{X}_n - \theta\right)^2$ to be small.

This estimation problem must try to balance expenses incurred due to sampling and the achieved estimation error. Suppose that c is measured in dollars and the X's are measured in hours. Then, the weight A found in (8.2.1) would have an unit 'dollar per hour'. The weight A expressed in 'dollar per hour' would reflect an experimenter's feelings about the cost per unit estimation error.

Starr and Woodroofe (1972) used the loss function (8.2.1). Now, the fixed-sample-size risk function associated with (8.2.1) is expressed as follows:

$$\begin{aligned} R_n(c) &\equiv \mathrm{E}_\theta\left[L_n(\theta, \overline{X}_n)\right] \\ &= A\mathrm{E}_\theta\left[\left(\overline{X}_n - \theta\right)^2\right] + cn \\ &= A\theta^2 n^{-1} + cn. \end{aligned} \tag{8.2.2}$$

Our goal is to design a purely sequential strategy that would minimize this risk for all $0 < \theta < \infty$. We may pretend that n is a continuous variable and examine the nature of the curve,

$$g(n) \equiv A\theta^2 n^{-1} + cn \text{ for } n > 0.$$

It is a strictly convex curve, but $g(n) \to \infty$ whether $n \downarrow 0$ or $n \uparrow \infty$. Hence, $g(n)$ must attain a minimum at some point inside $(0, \infty)$. Consider solving $\frac{d}{dn}g(n) = 0$ to obtain $n \equiv n^*(c) = (A\theta^2/c)^{1/2}$ and one can check that $\frac{d^2}{dn^2}g(n)\Big|_{n=n^*} > 0$ so that $n \equiv n^*(c)$ does minimize the function $g(n)$.

We note, however, that

$$n \equiv n^*(c) = \left(\frac{A}{c}\right)^{1/2} \theta \tag{8.2.3}$$

may not be a (positive) integer, but we continue to refer to $n^*(c)$ as the optimal fixed sample size had θ been known. The minimum fixed-sample-size risk is given by

$$R_{n^*}(c) \equiv A\theta^2 n^{*-1} + cn^* = 2cn^*. \tag{8.2.4}$$

Surely, the magnitude of the optimal fixed sample size n^* remains unknown since θ is unknown. In fact, no fixed sample size procedure will minimize the risk (8.2.2) under the loss function (8.2.1), uniformly in θ, $0 < \theta < \infty$. Thus, we introduce a sequential procedure that was originally developed by Starr and Woodroofe (1972).

8.2.1 Purely Sequential Procedure

Starr and Woodroofe's (1972) sequential stopping rule may be summarized as follows. We start with $X_1, \cdots, X_m$, a pilot of size $m(\geq 1)$, and we proceed with one additional X at-a-time as needed. The stopping time is defined as follows:

$$\begin{aligned} &N \equiv N(c) \text{ is the smallest integer } n(\geq m) \text{ for which} \\ &\qquad \text{we observe } n \geq (A/c)^{1/2}\, \overline{X}_n. \end{aligned} \tag{8.2.5}$$

This purely sequential procedure is implemented in the same way we did in the case of (6.2.5). Again, $\mathrm{P}_\theta(N < \infty) = 1$, that is, the process would terminate with probability one. This is verified in Section 8.6.1.

Upon termination, we estimate θ by the sample mean, $\overline{X}_N$, obtained from full dataset $N, X_1, \ldots, X_m, \ldots, X_N$. Now, the associated risk can be expressed as follows:

$$R_N(c) \equiv \mathrm{E}_\theta\left[L_N(\theta, \overline{X}_N)\right] = A\mathrm{E}_\theta\left[\left(\overline{X}_N - \theta\right)^2\right] + c\mathrm{E}_\theta\left[N\right]. \tag{8.2.6}$$

Let us emphasize a crucial difference between $\mathrm{E}_\theta\left[\left(\overline{X}_N - \theta\right)^2\right]$ from (8.2.6) and the analogous expressions found in, for example, (6.3.6), (6.3.15), (6.4.6), and (7.3.6). We may try to rewrite $\mathrm{E}_\theta\left[\left(\overline{X}_N - \theta\right)^2\right]$ as:

$$\begin{aligned} &\textstyle\sum_{n=m}^{\infty} \mathrm{E}_\theta\left[\left(\overline{X}_N - \theta\right)^2 \cap N = n\right] \\ &\quad = \textstyle\sum_{n=m}^{\infty} \mathrm{E}_\theta\left[\left(\overline{X}_N - \theta\right)^2 \mid N = n\right] \mathrm{P}_\theta[N = n] \\ &\quad = \textstyle\sum_{n=m}^{\infty} \mathrm{E}_\theta\left[\left(\overline{X}_n - \theta\right)^2 \mid N = n\right] \mathrm{P}_\theta[N = n]. \end{aligned} \tag{8.2.7}$$

At this point, we may focus on the event $\{N = n\}$ which is equivalent to the event:

$$\left\{k < (A/c)^{1/2}\overline{X}_k, k = m, \ldots, n-1, \text{ but } n \geq (A/c)^{1/2}\overline{X}_n\right\}.$$

Hence, the two random variables $\left(\overline{X}_n - \theta\right)^2$ and $I(N = n)$ ought to be dependent. That is, we cannot reduce $\mathrm{E}_\theta\left[\left(\overline{X}_n - \theta\right)^2 \mid N = n\right]$ to simply $\mathrm{E}_\theta\left[\left(\overline{X}_n - \theta\right)^2\right]$. A net effect of this is that unfortunately we cannot hope to rewrite the last expression from (8.2.7) involving appropriate moment(s) of N alone. In other words, in order to evaluate the sequential risk function $R_N(c)$, one must diligently work with (8.2.6) "as is". Starr and Woodroofe (1972) and Woodroofe (1977) did just that, and did so extremely elegantly.

Along the lines of Robbins (1959), one may define the following standard efficiency measures:

$$\begin{aligned} &\textit{RiskEfficiency}: \ \mathrm{RiskEff}(c) \equiv \frac{R_N(c)}{R_{n^*}(c)} \\ &\textit{Regret}: \ \mathrm{Regret}(c) \equiv R_N(c) - R_{n^*}(c). \end{aligned} \tag{8.2.8}$$

A procedure is called *asymptotically efficient* or *asymptotically first-order efficient* if $\lim_{c\to 0} \mathrm{E}_\theta\left[N/n^*\right] = 1$. A procedure is called *asymptotically second-order efficient* if $\lim_{c\to 0} \mathrm{E}_\theta\left[N - n^*\right]$ is finite.

Some preliminary results may be summarized now. For the stopping variable N defined by (8.2.5), for all fixed θ, m and A, the following properties hold:

$$\begin{aligned} &\text{(i)} \quad \mathrm{E}_\theta(N) \leq n^* + m + 1 \text{ for all fixed } c > 0; \\ &\text{(ii)} \quad \mathrm{E}_\theta\left[N/n^*\right] \to 1 \text{ as } c \to 0. \end{aligned} \tag{8.2.9}$$

These results are verified in Section 8.6.2.

A procedure is called *asymptotically risk-efficient* or *asymptotically first-order risk-efficient* if $\lim_{c\to 0}\text{RiskEff}(c) = 1$. A procedure is called *asymptotically second-order risk-efficient* if $\lim_{c\to 0} c^{-1}\,\text{Regret}(c)$ is finite.

Since $\overline{X}_n$ and $I(N=n)$ are dependent for all $n \ge m$, one encounters formidable technical difficulties in the present problem unlike anything that we faced in Chapters 6 and 7. Next, we state some extraordinary asymptotic *second-order* results.

Theorem 8.2.1: *For the estimation strategy $(N, \overline{X}_N)$ defined by (8.2.5), for all fixed θ, m and A, the following properties hold as $c \to 0$:*

(i) $\text{Regret}(c) = O(c)$ *if and only if* $m \ge 2$;
(ii) $\text{Regret}(c) = 3c + o(c)$ *if* $m \ge 3$.

Starr and Woodroofe (1972) proved part (i) which says that $\{R_N(c) - R_{n^*}(c)\}/c$ is a finite number for sufficiently small c or equivalently for large n^*. That is, as we implement the purely sequential procedure (8.2.5), the difference between the achieved risk $R_N(c)$ and $R_{n^*}(c)$ may be equivalent to the cost of "k" observations with some finite k. But, this number k remains unknown in part (i).

With more painstaking analysis, Woodroofe (1977) showed that the regret amounts to the cost of 3 observations. This is the crucial difference between the results in parts (i) and (ii). From a technical perspective, one should note the sufficient conditions on the pilot size, m, in the two parts. We supply some of the crucial steps in the proof of part (ii) in Section 8.6.3.

Obviously, part (i) follows from part (ii) when $m \ge 3$. Next, observe that $R_{n^*}(c) \equiv 2cn^* = 2\,(Ac)^{1/2}$. Now, part (i) implies that

$$\frac{\text{Regret}(c)}{R_{n^*}(c)} = \frac{O(c)}{2\,(Ac)^{1/2}} = O(c^{1/2}) \to 0 \text{ as } c \to 0, \tag{8.2.10}$$

if $m \ge 2$. But, note that

$$\text{RiskEff}(c) \equiv \frac{R_N(c)}{R_{n^*}(c)} = \frac{\text{Regret}(c)}{R_{n^*}(c)} + 1.$$

Thus, from (8.2.10), we can claim the following *asymptotic risk efficiency* result for the purely sequential methodology (8.2.5):

$$\text{RiskEff}(c) \to 1 \text{ as } c \to 0 \text{ if } m \ge 2. \tag{8.2.11}$$

The sequential procedure (8.2.5) is *asymptotically second-order efficient* and *asymptotically second-order risk-efficient.*

8.3 Bounded Risk Estimation

Suppose that $X_1, X_2, \ldots, X_n$ are i.i.d. observations from an Expo(θ) distribution (8.1.1). Let the loss function in estimating θ by $\overline{X}_n = n^{-1}\sum_{i=1}^{n} X_i$ be given by

$$L_n(\theta, \overline{X}_n) = A\left(\overline{X}_n - \theta\right)^2, A(>0) \text{ is known.} \tag{8.3.1}$$

Given a preassigned number $\omega(>0)$, the *risk-bound*, our goal is to make the associated risk,

$$\mathrm{E}_\theta\left[L_n(\theta, \overline{X}_n)\right] \equiv A\mathrm{E}_\theta\left[\left(\overline{X}_n - \theta\right)^2\right] \le \omega \text{ for all fixed } \theta > 0. \tag{8.3.2}$$

Then, with $n(\ge 1)$ fixed, we must have

$$\mathrm{E}_\theta\left[L_n(\theta, \overline{X}_n)\right] \le \omega \Rightarrow n \ge A\theta^2/\omega = n^*, \text{ say.} \tag{8.3.3}$$

We note that the magnitude of the optimal fixed-sample-size n^* remains unknown since θ is unknown. We tacitly disregard the fact that n^* may not be an integer! It is important to note that there does not exist any fixed-sample-size procedure for this bounded risk problem. For a proof of this result, one should refer to Takada (1986).

One may be tempted to implement a sampling procedure along the line of Birnbaum and Healy (1960). The Birnbaum-Healy methodology will involve initial observations $X_1, X_2, \ldots, X_m$ with a pilot size $m(\ge 3)$. From this data, one would obtain $\overline{X}_m$ and estimate the optimal fixed-sample-size n^* by:

$$Q \equiv Q(\omega) = \left\langle K\overline{X}_m^2/\omega\right\rangle + 1. \tag{8.3.4}$$

Here, $K \equiv K_m(>0)$ is an appropriate expression to be determined involving only A and m. In the second stage, one would gather a *new* set of Q observations $X_{m+1}, X_{m+2}, \ldots, X_{m+Q}$ with a sample mean:

$$\overline{X}_Q^* \equiv Q^{-1}\sum_{i=1}^{Q} X_{m+i}. \tag{8.3.5}$$

Then, $\overline{X}_Q^*$ is an unbiased estimator of θ. Under the Birnbaum-Healy strategy, one will have the following risk associated with $\overline{X}_Q^*$:

$$\begin{aligned} A\mathrm{E}_\theta\left[\left(\overline{X}_Q^* - \theta\right)^2\right] &= A\theta^2\mathrm{E}_\theta\left[Q^{-1}\right] \le A\theta^2 wK^{-1}\mathrm{E}_\theta\left[\overline{X}_m^{-2}\right] \\ &= 4AwK^{-1}m^2\mathrm{E}\left[U^{-2}\right], \end{aligned} \tag{8.3.6}$$

where $U \equiv 2\theta^{-1}\sum_{i=1}^{m} X_i$. Obviously, U is a Chi-square random variable with degree of freedom $2m$ and one can easily verify that

$$\mathrm{E}\left[U^{-k}\right] = \left\{2^k\Gamma(m)\right\}^{-1}\Gamma(m-k) \text{ if } m > k. \tag{8.3.7}$$

That is, we have $\mathrm{E}\left[U^{-2}\right] = \frac{1}{4}(m-1)^{-1}(m-2)^{-1}$ provided that $m \ge 3$.

Hence, if $m \geq 3$, it immediately follows from (8.3.6)-(8.3.7) that

$$AE_\theta\left[\left(\overline{X}^*_Q - \theta\right)^2\right] \leq \omega \text{ for all fixed } \theta > 0 \text{ if one uses} \quad K = Am^2\{(m-1)(m-2)\}^{-1}. \tag{8.3.8}$$

A major advantage one rips from Birnbaum-Healy strategy (8.3.4) - (8.3.5) happens to be the following *crucial* fact: The final estimator $\overline{X}^*_q$ of θ obtained from the observations $X_{m+1}, X_{m+2}, \ldots, X_{m+q}$ *alone* is distributed independently of a random variable $I(Q = q)$ for all fixed $q \geq m$.

In contrast, a major criticism against the Birnbaum-Healy technique is that the initial m observations are entirely disregarded while proposing to finally estimate θ by $\overline{X}^*_Q$. This is undesirable from a practical point of view. Samuel (1966) proposed to broaden the Birnbaum-Healy technique. Incidentally, one may note that Graybill and Connell's (1964) two-stage fixed-width confidence interval procedure for estimating σ^2 in a $N(\mu, \sigma^2)$ population where both μ and σ^2 were unknown was also proposed along similar lines of Birnbaum and Healy (1960).

Kubokawa (1989) combined $\overline{X}_m$ and $\overline{X}^*_Q$ linearly to propose a class of final estimators of θ such that their associated risks fell under that of $\overline{X}^*_Q$. Thus, Kubokawa (1989) improved upon the original Birnbaum-Healy estimator $\overline{X}^*_Q$. But, the following facts remain: (1) The Birnbaum-Healy sampling technique (8.3.4) does not constitute a *genuine* two-stage sampling design in the sense of Stein (1945,1949); (2) For moderately large Q, Kubokawa's (1989) improved estimators would practically coincide with the Birnbaum-Healy estimator $\overline{X}^*_Q$ because then the weight given to $\overline{X}_m$ is nearly zero; and (3) Under a genuine two-stage sampling design, Kubokawa's (1989) estimators may not be relevant. One may raise similar concerns about Samuel's (1966) modifications.

In Section 8.3.1, we introduce the new two-stage bounded risk point estimation methodology of Mukhopadhyay and Pepe (2006). Two important points ought to be noted to appreciate this methodology. First, it is a genuine two-stage strategy in the sense of Stein (1945,1949), and the second point is that the associated risk

$$AE_\theta\left[\left(\overline{X}_N - \theta\right)^2\right] \leq \omega,$$

so that the risk-bound is exactly ω.

This is a striking result given that $I(N = n)$ and $\overline{X}_n$ are dependent for all n. In the literature, there is no other result quite like this. Section 8.3.2 gives an improvement due to Zacks and Mukhopadhyay (2006a) over the two-stage methodology (8.3.9).

8.3.1 A Genuine Two-Stage Procedure

In the new two-stage methodology (8.3.9) of Mukhopadhyay and Pepe (2006), one will observe that the final estimator of θ is indeed the *average* of *all* observations from stages 1 and 2 *combined.* In other words, the proposed final estimator of θ and the final sample size are no longer independent!

In certain kinds of bounded risk estimation problems, we have noticed that a genuine two-stage sample size and the final estimator turned out to be independent! In such instances, the analyses became more tractable even though the final estimator was constructed by using *all* observations from stages 1 and 2 *combined.* One will find some notable examples from Ghosh and Mukhopadhyay (1976), Mukhopadhyay and Hilton (1986), and Section 6.3.

Let $X_1, X_2, \ldots, X_m$ be the pilot observations where $m(\geq 3)$. Now, we define the following two-stage stopping variable:

$$N \equiv N(\omega, B) = \max\left\{m, \left\langle B\overline{X}_m^2/\omega\right\rangle + 1\right\}, \qquad (8.3.9)$$

where $B \equiv B_m(A)\,(> 0)$ is a number that will be determined appropriately.

Now, given B, if we observe $N = m$, we do not require any more observations at second stage.

But, if $N > m$, we sample the difference at second stage by gathering additional observations $X_{m+1}, \ldots, X_N$.

Then, based on the *combined* set of observations $N, X_1, X_2, \ldots, X_N$ from stages 1 and 2, we estimate the MTTF θ by the sample mean, $\overline{X}_N$. This is what we call a *genuine* two-stage methodology in the sense of Stein (1945,1949) discussed in Section 6.2.1.

The risk associated with $\overline{X}_N$ is obviously given by

$$\mathrm{E}_\theta\left[L_N(\theta, \overline{X}_N)\right] \equiv A\mathrm{E}_\theta\left[\left(\overline{X}_N - \theta\right)^2\right]. \qquad (8.3.10)$$

Again, we emphasize that unlike the simplification of $\mathrm{E}_\theta\left[(\overline{X}_Q^* - \theta)^2\right]$ given in (8.3.6), we cannot express $\mathrm{E}_\theta\left[L_N(\theta, \overline{X}_N)\right]$ as $A\theta^2\mathrm{E}_\theta\left[N^{-1}\right]$ in the present situation. If we could, then $B \equiv B_m$ would be replaced by K that was identified easily in (8.3.8).

Mukhopadhyay and Pepe's main result is summarized below. A sketch of its proof is given in Section 8.6.4.

Theorem 8.3.1 : *For the two-stage procedure $(N, \overline{X}_N)$ defined by (8.3.9) with its associated risk (8.3.10), for all fixed θ, m and A, one can conclude that $\mathrm{E}_\theta\left[L_N(\theta, \overline{X}_N)\right] \leq \omega$ provided that*

$$B \equiv B_m \equiv B_m(A) = \frac{2m\,(m+1)\,A}{(m-1)\,(m-2)} \qquad (8.3.11)$$

and $m \geq 3$.

In the Birnbaum-Healy procedure (8.3.4), one collects data of size $m+Q$, but uses only the sample mean $\overline{X}_Q^*$ to estimate the MTTF, θ. Note that Q is the estimator of n^* from (8.3.3). The expected sample size, $m + \mathrm{E}_\theta[Q]$, will then be approximately $m + K\theta^2/\omega$, but the pilot observations are thrown away from any further consideration! To the contrary, through two-stage procedure (8.3.9), Mukhopadhyay and Pepe (2006) recommend collecting data of size N and utilize the fully combined sample mean $\overline{X}_N$ to estimate θ. Here, N estimates n^*. The corresponding expected sample size, $\mathrm{E}_\theta[N]$, will be approximately $B\theta^2/\omega$. This may be summarized as follows:

$$\begin{aligned} \mathrm{E}_\theta\left[(m+Q)/n^*\right] &\approx mn^{*-1} + \frac{m^2}{(m-1)(m-2)} \\ \text{and } \mathrm{E}_\theta\left[N/n^*\right] &\approx \frac{2m(m+1)}{(m-1)(m-2)}. \end{aligned} \tag{8.3.12}$$

For large m, n^* such that $m = o(n^*)$, we can expect $\mathrm{E}_\theta\left[(m+Q)/n^*\right]$ to be close to one. But, $\mathrm{E}_\theta\left[N/n^*\right]$ would be close to two. In all fairness, we add that the stopping variables Q and N are not comparable because the strategy (8.3.4) may not even be called "two-stage"! The next Section 8.3.2 gives an improvement due to Zacks and Mukhopadhyay (2006a) over the two-stage methodology (8.3.9).

8.3.2 An Improved Two-Stage Procedure

Since the goal is to achieve an *exact* risk-bound ω where the final sample size and the terminal estimator of θ are dependent, we would generally expect to record more than n^* observations. Mukhopadhyay and Pepe's (2006) analyses revealed that on an average N turned out to be nearly twice the magnitude of n^*. They also noted that the risk associated with $\overline{X}_N$ was significantly smaller than the risk-bound ω because of considerable oversampling compared with n^*.

Zacks and Mukhopadhyay (2006a) came up with a remarkably more efficient two-stage strategy compared with Mukhopadhyay and Pepe's (2006) strategy (8.3.9). Zacks and Mukhopadhyay (2006a) developed methods for exact computation of the risk function and the distribution of N to arrive at an optimal choice of "B". They found that B used in (8.3.9) ought to be replaced by:

$$B^* \equiv B_m^* \equiv B_m^*(A) = 0.565B_m(A) \text{ with } B_m(A) \text{ from (8.3.11).} \tag{8.3.13}$$

Such modified two-stage methodology maintained $\mathrm{E}_\theta[N]$ in a very close proximity of n^*, but at the same time achieved a risk that was only a shade under ω.

8.4 Data Analyses and Conclusions

The computer program **Seq08.exe** implements the methodologies introduced in Sections 8.2.1 and 8.3.2 for the point estimation of the mean of an exponential distribution. Figure 8.4.1 shows the data input screen from this program which can be used for either simulation studies or with live data.

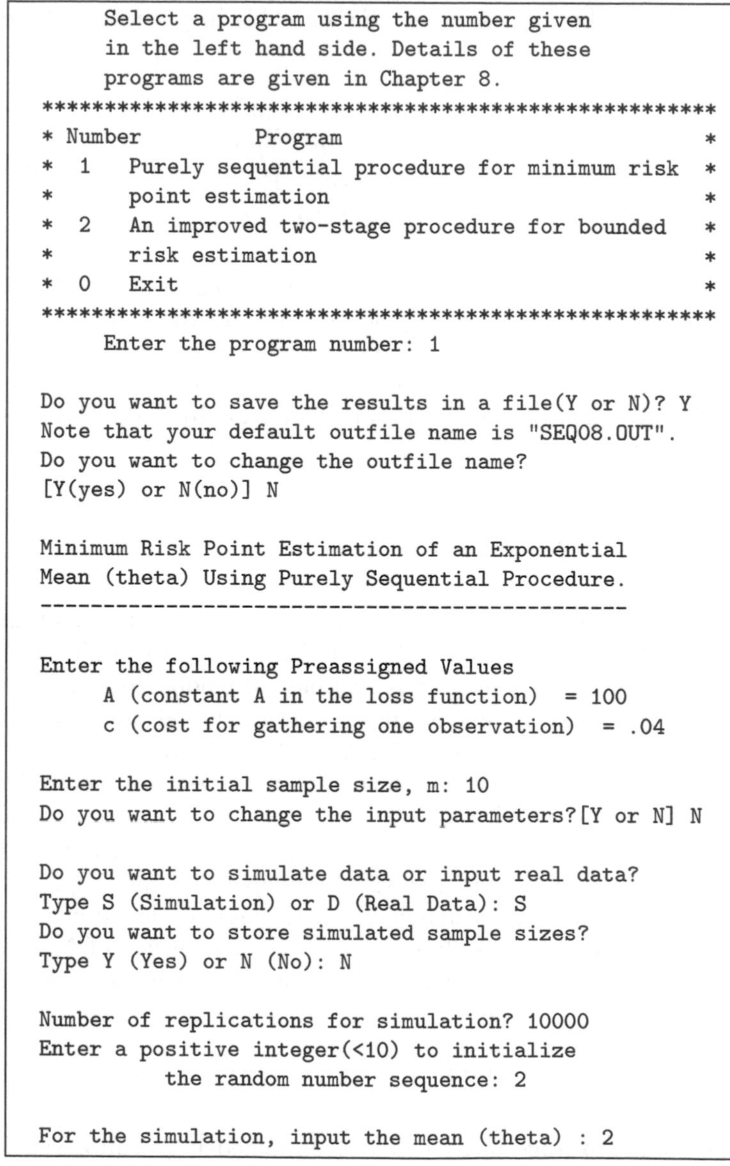

```
     Select a program using the number given
     in the left hand side. Details of these
     programs are given in Chapter 8.
*******************************************************
* Number        Program                               *
*  1   Purely sequential procedure for minimum risk   *
*      point estimation                               *
*  2   An improved two-stage procedure for bounded    *
*      risk estimation                                *
*  0   Exit                                           *
*******************************************************
     Enter the program number: 1

Do you want to save the results in a file(Y or N)? Y
Note that your default outfile name is "SEQ08.OUT".
Do you want to change the outfile name?
[Y(yes) or N(no)] N

Minimum Risk Point Estimation of an Exponential
Mean (theta) Using Purely Sequential Procedure.
-----------------------------------------------


Enter the following Preassigned Values
     A (constant A in the loss function)  = 100
     c (cost for gathering one observation)  = .04

Enter the initial sample size, m: 10
Do you want to change the input parameters?[Y or N] N

Do you want to simulate data or input real data?
Type S (Simulation) or D (Real Data): S
Do you want to store simulated sample sizes?
Type Y (Yes) or N (No): N

Number of replications for simulation? 10000
Enter a positive integer(<10) to initialize
          the random number sequence: 2

For the simulation, input the mean (theta) : 2
```

Figure 8.4.1. *Data input screen of **Seq08.exe** program.*

```
Minimum Risk Point Estimation of an Exponential Mean
Simulation Study of  the Purely Sequential Procedure
====================================================

Input Parameters:   A =  100.00
                    c =    0.04
                theta =    2.00
Initial Sample size   =      10
Number of Simulations =    10000

      Optimal sample size (n_star)        = 100.00
      Average final sample size (n_bar) =  99.87
      Std. dev. of n (s_n)                =  10.11
      Minimum sample size (n_min)         =  53
      Maximum sample size (n_max)         = 134
      Minimum fixed-sample risk (Rn*)     =   8.00
      Simulated average risk (Rbar_n)     =   7.93
      Simulated Risk Efficiency           =   0.99
      Regret (Rbar_n - Rn*)               =  -0.07
      Theta_hat (simulated average)       = 1.9825
      Std Error of theta_hat              = 0.2018
```

Figure 8.4.2. *Output for the minimum risk point estimation subprogram 1 within* ***Seq08.exe***.

```
Bounded Risk Point Estimation of an Exponential Mean
    Simulation Study of  the Two-Stage Procedure
    ============================================

Input Parameters:   A =  100.00
                omega =    4.00
                theta =    2.00
Initial Sample size   =      10
         e (epsilon) =   0.5650
Number of Simulations =    10000

      Optimal sample size (n_star)        =    100.00
      Average final sample size (n_bar) =    191.42
      Std. dev. of n (s_n)                =    125.31
      Minimum sample size (n_min)         =     10
      Maximum sample size (n_max)         =   1168
      Simulated average risk (Rbar_n)     =      2.93
      Simulated Risk Efficiency           =      0.73
      Regret (Rbar_n - Rn*)               =     -1.07
      Theta_hat (simulated average)       =    1.9695
      Std Error of theta_hat              =    0.1905
```

Figure 8.4.3. *Output for the bounded risk point estimation subprogram 2 within* ***Seq08.exe***.

Figures 8.4.2 and 8.4.3 respectively show the output screens from simulation studies using 'program number: 1' and 'program number: 2' respectively within **Seq08.exe**. These refer to minimum risk and bounded risk point estimation problems respectively.

These were obtained by generating data from an Expo(2) distribution and replicating simulations 10,000 times. Here, we fixed $A = 100$, $c = 0.04$ and $\omega = 4$ so that we had $n^* = 100$ for both methodologies. Also, $m = 10$ and random number initialization 2 were used in either methodology. The program output is summarized in Table 8.4.1. The symbol "e" in Figure 8.4.2 is the value of ε in

$$B^* \equiv B_m^*(A) = \varepsilon B_m. \tag{8.4.1}$$

The above equation is equivalent to (8.3.13) with $\varepsilon = 0.565$. Zacks and Mukhopadhyay (2006a) showed that B^* was optimal when $\varepsilon = 0.565$.

Table 8.4.1: Simulated results summary for minimum and bounded risk estimation

	Minimum Risk Estimation	Bounded Risk Estimation
$\widehat{\theta}$	1.9825	1.9695
Std Error of $\widehat{\theta}$	0.2018	0.1905
n^*	100	100
$\overline{n}$	99.87	191.42
s_N	10.11	125.31
$n_{\min}$	53	10
$n_{\max}$	134	1168
R_{n^*}	8	4
$\overline{R}_N$	7.93	2.93
Simulated RiskEff	0.99	0.73
Simulated Regret	−0.07	−1.07

In these simulation studies we have used both the sequential minimum risk procedure and two-stage bounded risk procedure. As expected, the sequential procedure has undersampled ($\overline{n} < n^*$) and the two-stage procedure has oversampled ($\overline{n} > n^*$). Note that $\overline{n}$ is a simulated estimate of $E_\theta[N]$, thus for minimum risk point estimation, one would expect

$$\overline{n} - n^* \approx -D = -0.255$$

from (8.6.10). The simulation study gave $\overline{n} - n^* = 99.87 - 100 = -0.13$. They are in the same ballpark! Also note that

$$
\begin{aligned}
(\overline{n} - n^{**}) \pm 2 \times s_n/\sqrt{10000} &= -0.13 \pm 2 \times 7.93/\sqrt{10000} \\
&= -0.13 \pm 0.1586 \Rightarrow (-0.2886, 0.0286).
\end{aligned}
$$

This interval does include the value $-D$.

In the bounded risk problem, average simulated risk was 2.93 whereas the risk-bound was set with $\omega = 4$. Also, average simulated sample size $\overline{n} = 191.42$ came considerably over the target $n^* = 100$. The magnitude of oversampling can be reduced significantly by choosing $m = 20$ or 30 instead of $m = 10$.

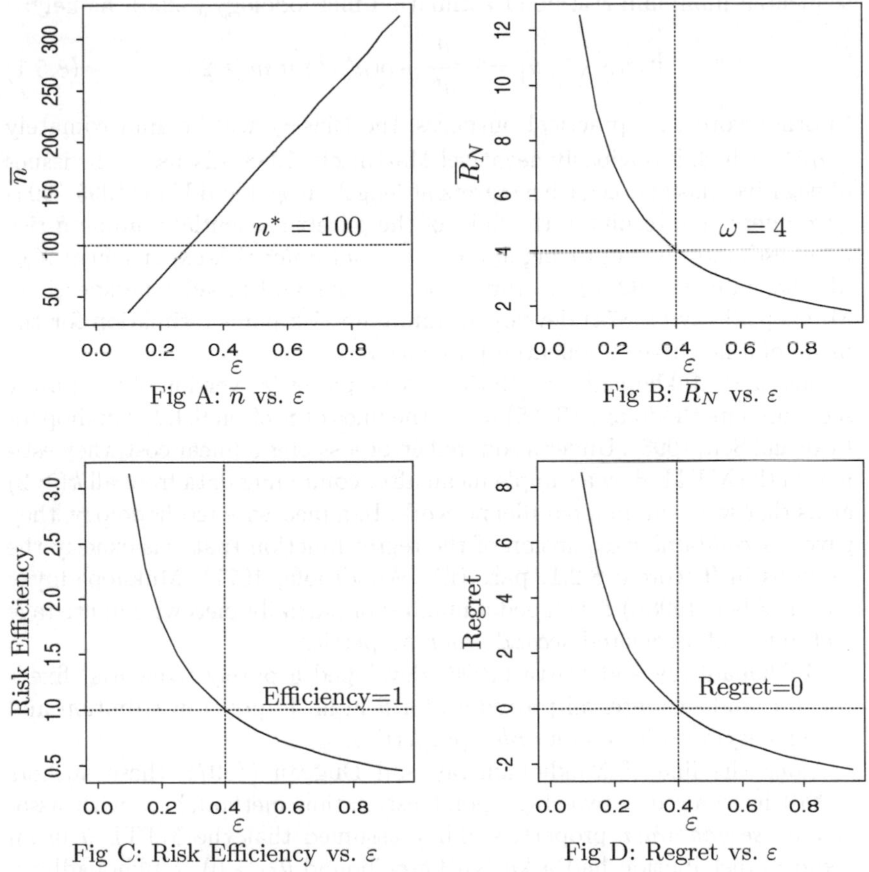

Figure 8.4.4. *Simulated results of the bound risk estimation for different* ε.

Next, we ran the simulation study for the bounded risk problem with various choices of $\varepsilon, 0 < \varepsilon < 1$. We replicated simulations 10,000 times by generating data from an Expo(2) distribution. We fixed $A = 100$, $\omega = 4$ and $m = 10$. The results are graphically presented in Figure 8.4.4.

From Fig A in Figure 8.4.4 we see that the average sample size ($\overline{n}$) increases with ε. When $\varepsilon = 0.3$, this figure gives the optimal sample size, $\overline{n} \approx 100 = n^*$. However, Fig B, Fig C and Fig D together seem to indicate that an "optimal" choice of ε may be 0.4. By considering further simulations with increasing n^*, one can show that the optimal $\varepsilon \approx 0.565$ for large n^*.

8.5 Other Selected Multistage Procedures

Mukhopadhyay and Chattopadhyay (1991) obtained a *second-order* expression of the (negative) bias for the final estimator $\overline{X}_N$ under the purely sequential minimum risk point estimation methodology (8.2.5), namely:

$$\mathrm{E}_\theta\left[\overline{X}_N - \theta\right] = -\frac{\theta}{n^*} + o(n^{*-1}) \text{ if } m \geq 2. \tag{8.5.1}$$

In other words, for practical purposes, the $\text{Bias}_{\overline{X}_N}$ will be approximately $-\theta n^{*-1}$ which is obviously negative! Martinsek (1988) discussed the issues of negative bias and negative regrets at length. Isogai and Uno (1993,1994) gave interesting results in the light of the purely sequential minimum risk point estimator by improving upon the *second-order* risk-expansion of $\overline{X}_N$. Mukhopadhyay (1994) gave further extensions and broader perspectives. Mukhopadhyay (1987a) developed minimum risk point estimation for the mean of a negative exponential population.

Bose and Mukhopadhyay (1995a) gave a piecewise version of the purely sequential methodology (8.2.5) along the lines of Section 6.4.2 (Mukhopadhyay and Sen, 1993). Under a squared error loss plus a linear cost, they estimated the MTTF θ by a sample mean after combining data from all $k(\geq 2)$ arms that were run in a parallel network. In a piecewise methodology, they gave a *second-order* expansion of the regret function that was exactly the same as in Theorem 8.2.1, part (ii) (Woodroofe, 1977). Mukhopadhyay and de Silva (1998b) developed a number of partially piecewise multistage methods and associated *second-order* properties.

Mukhopadhyay and Datta (1996) developed a purely sequential fixed-width confidence interval procedure for θ in an Expo(θ) distribution and gave its asymptotic *second-order* properties.

Along the line of Mukhopadhyay and Duggan (1997), these authors (1996) introduced a two-stage point estimation methodology with associated *second-order* properties. They assumed that the MTTF θ in an Expo(θ) distribution had a known lower bound $\theta_L(> 0)$. Mukhopadhyay and Duggan (2000b) investigated bias-corrected forms of estimators of θ, including one depending on the stopping variable alone, under a sequential strategy (8.2.5). As far as we know, Mukhopadhyay and Duggan (2000b,2001) were first to propose the following estimator of θ,

$$\widehat{\theta}_N \equiv \left(\frac{c}{A}\right)^{1/2} N \tag{8.5.2}$$

and its bias-corrected form based on a sequential strategy (8.2.5). This estimator performs remarkably well compared with $\overline{X}_N$.

In a one-parameter exponential family, Woodroofe (1987) discussed a three-stage procedure for estimating the mean in order to achieve asymptotic *local minimax regret*, a concept that was defined earlier (Woodroofe, 1985).

Estimation of the variance σ^2 in a normal distribution, when μ was unknown, was treated by Starr and Woodroofe (1972). A sequential minimum risk point estimation of the scale parameter σ in a negative exponential distribution, NExp(θ, σ) from (7.1.1), when θ is unknown, requires careful handling of additional layers of difficulties. These were investigated by Mukhopadhyay and Ekwo (1987). One is also referred to Govindarajulu (1985) and Govindarajulu and Sarkar (1991) for some closely related studies.

8.5.1 Two-Parameter Exponential Family

Bar-Lev and Reiser (1982) gave a two-parameter exponential subfamily which admitted an *uniformly most powerful unbiased* (UMPU) test based on a single test statistic. In this exponential subfamily, Bose and Boukai (1993a) formulated a minimum risk point estimation problem for a mean parameter μ_2 with $\overline{T}_{2n}$ under weighted squared error plus a linear cost as the loss function. They proposed a sequential methodology along the lines of Robbins (1959). But, unlike what we saw in the case of (8.2.5), the Bose-Boukai stopping time N enjoyed the following characteristic: The random variables $I(N = n)$ and $\overline{T}_{2n}$ were independent for all $n \geq m$. Along the lines of Chapters 6 and 7, they provided various *second-order* expansions, including that for the regret function.

Bose and Mukhopadhyay (1994a) developed an accelerated version of the sequential methodology of Bose and Boukai (1993a). Then, Bose and Mukhopadhyay (1994b) developed sequential estimation methods via piecewise stopping numbers in a two-parameter exponential family of distributions along the lines of Section 6.4.2 (Mukhopadhyay and Sen, 1993). These investigations were aimed at *second-order* analyses.

In another collaboration, Bose and Mukhopadhyay (1995b) investigated a sequential interval estimation method with proportional closeness via piecewise stopping times in a two-parameter exponential family of distributions. Again, various *second-order* analyses were highlighted.

8.5.2 NEF-PVF Family

Let $\mathcal{F} = \{F_\theta : \theta \in \Theta\}$ be the class of *natural exponential family* (NEF) of distributions. That is, $\mathcal{F}$ is minimal NEF of order 1 whose members are of the form:

$$F_\theta(dx) = \exp\{\theta x + a(\theta)\}\Delta(dx), \theta \in \Theta, \tag{8.5.3}$$

where Δ is a sigma-finite measure on the Borel sets of the real line $\Re$ and the parameter space Θ consists of all $\theta \in \Re$ for which $\int \exp(\theta x)\Delta(dx)$ is finite. It is well-known that F_θ has finite moments of all order. See Barndorff-Nielsen (1978). For $\theta \in \Theta$, we let $\mu \equiv \mu(\theta) = -da(\theta)/d\theta$ and $\Omega \equiv \mu(\text{int}\Theta)$ respectively denote the mean of F_θ and the mean's parameter space. We also write $V(\mu)$ for the variance function for (8.5.3). Let us assume that the members of the NEF $\mathcal{F}$ have a *power variance function* (PVF) so that we may write:

$$V(\mu) = \alpha\mu^\gamma, \mu \in \Omega \tag{8.5.4}$$

with some known constant $\alpha \neq 0$ and γ.

For basic understanding, however, one may effectively bypass these jargons. We ask our readers to primarily focus on Poisson (θ) and Normal $(0, \theta)$ distributions, $\theta > 0$. For a comprehensive discussion, one may refer to Morris (1982,1983) and Bar-Lev and Enis (1986).

Bose and Boukai (1993b) discussed a minimum risk sequential estimation method for μ, the mean of NEF-PVF distributions (8.5.3 - 8.5.4), with a sample mean. They worked under the squared error plus linear cost as a loss function. The highlight was the derivation of a *second-order* expansion of their regret function along the line of Theorem 8.2.1, part (ii).

Bose and Mukhopadhyay (1995c) developed an accelerated version of the sequential estimation methodology for the mean of NEF-PVF distributions. The Bose-Mukhopadhyay acceleration turned out to be operationally more efficient than the Bose-Boukai methodology, but at the same time the accelerated technique had a comparable *second-order* regret expansion.

8.5.3 Reliability Problems

Sequential methodologies for reliability problems have a long history. A marriage between reliability estimation and sequential analysis took place in the seminal papers of Epstein and Sobel (1953,1954,1955) and Sobel (1956). Basu (1991) gave an excellent review of this field.

Mukhopadhyay et al. (1997) developed *second-order* properties in the context of estimating a reliability function, $\rho(t) \equiv \exp(-t/\theta), t > 0$, in an Expo$(\theta)$ distribution (8.1.1). But, they did so only after the purely sequential sampling from (8.2.5) terminated first. In other words, given the data $(N, X_1, \ldots, X_N)$ generated by the stopping rule (8.2.5), Mukhopadhyay et al. (1997) discussed various ways to estimate $\rho(t)$ and investigated *second-order* characteristics.

In a preliminary report, Mukhopadhyay and Cicconetti (2000) expanded this investigation to include several types of estimators of $\rho(t)$. Later, Cicconetti (2002) and Mukhopadhyay and Cicconetti (2005) gave a comprehensive treatment of both one- and two-sample problems involving system reliabilities. Again, they addressed these problems under some previously run multistage strategies.

Under a number of strategies, Zacks and Mukhopadhyay (2006b) have developed some fast and unique style of computing exact risks for a variety of reliability estimators. In that light, they compared a number of natural estimators and some bias-corrected estimators for $\rho(t)$.

8.5.4 Time-Sequential Problems

These methodologies cover a wide variety of applications including quality control, survival analysis, and clinical trials. For, example, Sen's (1980) formulation has direct relevance in quality control and lifetesing experiments. Whereas Gardiner et al.'s (1986) piece tackled a particularly difficult problem in sequential clinical trials where patients or subjects tend to drop off from a trial randomly at different times.

A first time-sequential methodology was introduced by Sen (1980). Clever generalizations of the problem from Section 8.2 in the presence of random censoring were put forth by Gardiner and Susarla (1983,1984) as well as in Aras (1987,1989). Gardiner et al. (1986) gave important directions under random withdrawals. For other references, one may refer to Gardiner and Susarla (1991). Mukhopadhyay (1995b) developed an asymptotic *second-order* expansion of the regret function associated with Sen's (1980) time-sequential purely sequential methodology.

8.6 Some Selected Derivations

In this section, we include some crucial derivations. Parts of these derivations are simple while others may appear hard. We have included both kinds to cater to as many readers as possible with different prerequisites. In a first pass, some readers may avoid all proofs here, especially the hard ones, without compromising their understanding of the basic statistical import.

8.6.1 Proof of Termination of N from (8.2.5)

We want to show that $\mathrm{P}_\theta(N < \infty) = 1$. We first write $\mathrm{P}_\theta(N = \infty) = \lim_{n\to\infty}\mathrm{P}_\theta(N > n)$, and note that the event $\{N > n\}$ implies that we did not stop with n observations. Hence, at the stage when we have $n(\geq m)$ observations, we must have all the events $k < (A/c)^{1/2}\,\overline{X}_k$ satisfied, $k = m, \ldots, n$. Thus, we can write:

$$\begin{aligned} \mathrm{P}_\theta(N > n) &= \mathrm{P}_\theta\left\{k < (A/c)^{1/2}\,\overline{X}_k, k = m, \ldots, n\right\} \\ &\leq \mathrm{P}_\theta\left\{n < (A/c)^{1/2}\,\overline{X}_n\right\} \\ &= \mathrm{P}_\theta\left\{\overline{X}_n - \theta > n\,(A/c)^{-1/2} - \theta\right\}. \end{aligned} \tag{8.6.1}$$

Now, we may consider n sufficiently large, that is, when $n\,(A/c)^{-1/2} > \theta$. Thus, from (8.6.1), we obtain:

$$\begin{aligned}
&\mathrm{P}_\theta(N > n) \\
&\quad \le \mathrm{P}_\theta\left\{\left|\overline{X}_n - \theta\right| > n\,(A/c)^{-1/2} - \theta\right\} \text{ since } n\,(A/c)^{-1/2} - \theta > 0 \\
&\quad \le \mathrm{E}_\theta\left\{\left|\overline{X}_n - \theta\right|^2\right\}\left(n\,(A/c)^{-1/2} - \theta\right)^{-2} \text{ by Markov's inequality} \\
&\quad = \mathrm{V}_\theta\left(\overline{X}_n\right)\left(n\,(A/c)^{-1/2} - \theta\right)^{-2}.
\end{aligned} \tag{8.6.2}$$

But, $\mathrm{V}_\theta\left(\overline{X}_n\right) = \theta^2 n^{-1}$. Hence, from the last step in (8.6.2), we have:

$$0 \le \lim_{n\to\infty} \mathrm{P}_\theta(N > n) \le \lim_{n\to\infty} \theta^2 n^{-1}\left(n\,(A/c)^{-1/2} - \theta\right)^{-2} = 0,$$

so that the proof of termination is complete. ■

8.6.2 Proof of (8.2.9)

Proof of Part (i):

From (8.2.5), first observe the basic inequality:

$$(A/c)^{1/2}\,\overline{X}_N \le N \le m + (A/c)^{1/2}\,\overline{X}_{N-1} \text{ w.p. 1.} \tag{8.6.3}$$

Assume for the time being that $\mathrm{E}_\theta[N] < \infty$. Now, we work from the right hand side of (8.6.3) to write:

$$\begin{aligned}
& N \le m + (A/c)^{1/2}\,\frac{1}{N-1}\sum_{i=1}^{N-1} X_i \\
\Rightarrow\ & (N-m)(N-1) \le (A/c)^{1/2}\sum_{i=1}^{N-1} X_i \le (A/c)^{1/2}\sum_{i=1}^{N} X_i \\
\Rightarrow\ & N^2 - (m+1)N \le (A/c)^{1/2}\sum_{i=1}^{N} X_i
\end{aligned} \tag{8.6.4}$$

each step holding w.p. 1.

Now, we take expectations throughout the last step in (8.6.4) and use Wald's first equation (Theorem 3.5.4) to write:

$$\begin{aligned}
& \mathrm{E}_\theta[N^2] - (m+1)\mathrm{E}_\theta[N] \le (A/c)^{1/2}\,\mathrm{E}_\theta\left[\sum_{i=1}^{N} X_i\right] \\
& \qquad = (A/c)^{1/2}\,\theta\mathrm{E}_\theta[N] \\
\Rightarrow\ & \mathrm{E}_\theta^2[N] - (m+1)\mathrm{E}_\theta[N] \le n^*\mathrm{E}_\theta[N] \text{ since } n^* = (A/c)^{1/2}\,\theta \\
\Rightarrow\ & \mathrm{E}_\theta[N] - (m+1) \le n^*,
\end{aligned} \tag{8.6.5}$$

that is, $\mathrm{E}_\theta[N] \le n^* + m + 1$.

If $\mathrm{E}_\theta[N]$ is not finite, one may use a truncated stopping variable $N_k = \min\{k, N\}, k = 1, 2, \ldots$ and claim that $\mathrm{E}_\theta[N_k] \le n^* + m + 1$ for each fixed k. An application of the monotone convergence theorem will complete the proof. ■

Proof of Part (ii):

From (8.2.5), observe that $N \to \infty$ w.p. 1 as $c \to 0$. Next, we divide throughout the basic inequality (8.6.3) by n^* and observe:

$$\begin{aligned} \theta^{-1}\overline{X}_N \le N/n^* \le mn^{*-1} + \theta^{-1}\overline{X}_{N-1} \text{ w.p. } 1 \\ \Rightarrow \lim_{c\to 0} N/n^* = 1 \text{ w.p. } 1. \end{aligned} \quad (8.6.6)$$

Now, Fatou's lemma implies that

$$\liminf_{c\to 0} \mathrm{E}_\theta[N/n^*] \ge \mathrm{E}_\theta[\liminf_{c\to 0} N/n^*] = \mathrm{E}_\theta[1] = 1.$$

Next, from part (i) we can claim that

$$\limsup_{c\to 0} \mathrm{E}_\theta[N/n^*] \le \limsup_{c\to 0}(n^* + m + 1)n^{*-1} = 1.$$

Hence, $\lim_{c\to 0} \mathrm{E}_\theta[N/n^*]$ exists and this limit is one. ■

8.6.3 Proof of Theorem 8.2.1 Part (ii)

This is an adaptation of Woodroofe's (1977) original proof. See also Ghosh et al (1977, pp. 240-242). We offer only those key steps that may help in understanding a basic 'road map' of this complex proof.

We do not explicitly mention any requisite condition on "m". We leave those details out as an exercise.

Observe that the stopping variable from (8.2.5) can be equivalently expressed as:

$$\begin{aligned} &N \equiv N(c) \text{ is the smallest integer } n(\ge m) \text{ for which} \\ &\text{we observe } n^2/n^* \ge S_n \text{ with } S_n = \textstyle\sum_{i=1}^n Y_i, \end{aligned} \quad (8.6.7)$$

where Y's are i.i.d. standard exponential random variables. That is, the Y's have a common p.d.f. (8.1.1) with $\theta = 1$. This representation is the same as in Woodroofe (1977,1982). This also coincides with the representation in (2.4) of Mukhopadhyay (1988a) with $\delta = 2, L_0 = 0, \beta^* = 1, h^* = 1/n^*, p = 1, b = 1, \mathrm{E}[Y] \equiv \theta = 1$, and $\mathrm{V}[Y] \equiv \tau^2 = 1$. The excess over the boundary is:

$$\mathcal{R} \equiv \mathcal{R}_c = N^2 n^{*-1} - S_N \text{ with } S_N = \sum_{i=1}^{N} Y_i.$$

From Ghosh and Mukhopadhyay's (1975) theorem, we claim:

$$\frac{S_N - N}{\sqrt{N}} \text{ converges to } N(0,1) \text{ in distribution as } c \to 0. \quad (8.6.8)$$

One may also refer to Mukhopadhyay and Solanky (1994, Section 2.4) or Ghosh et al. (1997, Exercise 2.7.4).

Next, let us denote

$$\nu = 1 - D \text{ with } D = \frac{1}{2}\sum_{n=1}^{\infty}\frac{1}{n}\mathrm{E}\left[\max\left(0, \chi_n^2 - 4n\right)\right]. \quad (8.6.9)$$

Now, from Woodroofe (1977) or Mukhopadhyay (1988a, Theorem 3), we can conclude the following second-order expansion of $\text{E}_\theta[N]$:

$$\text{E}_\theta[N] = n^* - D + o(1) \text{ with } D = 0.254965. \tag{8.6.10}$$

Let us rewrite $A\text{E}_\theta\left[\left(\overline{X}_N - \theta\right)^2\right]$ as:

$$\begin{aligned}
&A\theta^2\text{E}_\theta\left[\frac{\left(S_N - N\right)^2}{N^2}\right] \\
&= A\theta^2\text{E}_\theta\left[\frac{\left(S_N - N\right)^2}{n^{*2}}\right] + \frac{A\theta^2}{n^{*2}}\text{E}_\theta\left[\left(\frac{n^{*2}}{N^2} - 1\right)\left(S_N - N\right)^2\right] \qquad (8.6.11)\\
&= A\theta^2\text{E}_\theta[I] + \frac{A\theta^2}{n^{*2}}\text{E}_\theta[J], \text{ say.}
\end{aligned}$$

We may use Wald's first equation (Theorem 3.5.4) and (8.6.10) to claim:

$$\text{E}_\theta[I] = \frac{\text{E}_\theta[N]}{n^{*2}} = \frac{1}{n^*} - \frac{D}{n^{*2}} + o(c), \tag{8.6.12}$$

since $n^{*-2} = c/(A\theta^2)$ which implies that $n^{*-2}o(1) = o(c)$.

In view of the second term on the right-hand side of (8.6.11), we need to expand $\text{E}_\theta[J]$ up to the order $o(1)$. We can write:

$$\begin{aligned}
&\text{E}_\theta[J] \\
&= \text{E}_\theta\left[\left(1 - \frac{N^2}{n^{*2}}\right)\left(S_N - N\right)^2\right] + \text{E}_\theta\left[\left(1 - \frac{N^2}{n^{*2}}\right)^2 \frac{n^{*2}}{N^2}\left(S_N - N\right)^2\right] \\
&= \text{E}_\theta[J_1] + \text{E}_\theta[J_2], \text{ say.}
\end{aligned} \tag{8.6.13}$$

Now, one rewrites $\text{E}_\theta[J_2]$ as follows:

$$\begin{aligned}
&\text{E}_\theta\left[\left(1 + \frac{N}{n^*}\right)^2 \frac{n^{*2}}{N^2}\left(1 - \frac{N}{n^*}\right)^2\left(S_N - N\right)^2\right] \\
&= \text{E}_\theta\left[\left(1 + \frac{N}{n^*}\right)^2 \frac{n^{*2}}{N^2}\left(1 - \frac{\mathcal{R}}{N} - \frac{S_N}{N}\right)^2\left(S_N - N\right)^2\right] \\
&= \text{E}_\theta\left[\left(1 + \frac{N}{n^*}\right)^2 \frac{n^{*2}}{N^2}\left(\frac{\mathcal{R}}{N} + \frac{S_N - N}{N}\right)^2\left(S_N - N\right)^2\right] \qquad (8.6.14)\\
&= \text{E}_\theta\left[\left(1 + \frac{N}{n^*}\right)^2 \frac{n^{*2}}{N^2}\left(\mathcal{R}^2\frac{\left(S_N - N\right)^2}{N^2} + \frac{\left(S_N - N\right)^4}{N^2}\right.\right. \\
&\qquad\qquad \left.\left. + \frac{2\mathcal{R}\left(S_N - N\right)^3}{N^2}\right)^2\right] \\
&= (1+1)^2(3) + o(1) \; = 12 + o(1).
\end{aligned}$$

In (8.6.14), we have used (i) the fourth moment (= 3) of a standard normal variable in view of (8.6.8); (ii) the fact that the overshoot $\mathcal{R}$ is dominated and $\mathrm{E}_\theta(\mathcal{R}) = 1 - D + o(1)$ from Woodroofe (1977); (iii) certain positive moments of $\frac{n^*}{N}$ which have finite expectations; and (iv) other techniques including repeated applications of Hölder's inequality.

Next, we address $\mathrm{E}_\theta[J_1]$ by expressing it as follows:

$$\begin{aligned}
&\mathrm{E}_\theta\left[\left(1+\frac{N}{n^*}\right)\left(1-\frac{N}{n^*}\right)(S_N-N)^2\right] \\
&= \mathrm{E}_\theta\left[-\left(1+\frac{N}{n^*}\right)\left(\frac{\mathcal{R}}{N}+\frac{S_N-N}{N}\right)(S_N-N)^2\right] \\
&= \mathrm{E}_\theta\left[-\left(1+\frac{N}{n^*}\right)\frac{\mathcal{R}\,(S_N-N)^2}{N}\right] - \mathrm{E}_\theta\left[\left(1+\frac{N}{n^*}\right)\frac{(S_N-N)^3}{N}\right] \\
&= -\mathrm{E}_\theta[J_{11}] - \mathrm{E}_\theta[J_{12}], \text{ say.}
\end{aligned} \tag{8.6.15}$$

But, we can claim that

$$\mathrm{E}_\theta[J_{11}] = 2(1-D)(1) + o(1) = 2 - 2D + o(1), \tag{8.6.16}$$

where we used (i) the second moment (=1) of a standard normal variable in view of (8.6.8); and (ii) the fact that certain positive moments of overshoot $\mathcal{R}$ are dominated and $\mathrm{E}_\theta(\mathcal{R}) = 1-D+o(1)$ from Woodroofe (1977, Lemma 2.1).

Next, we move to handle $\mathrm{E}_\theta[J_{12}]$ as follows:

$$\begin{aligned}
&\mathrm{E}_\theta\left[\left(1+\frac{N}{n^*}\right)\frac{(S_N-N)^3}{N}\right] \\
&= \mathrm{E}_\theta\left[2\frac{(S_N-N)^3}{n^*}\right] + \mathrm{E}_\theta\left[\frac{n^*-N}{n^*N}(S_N-N)^3\right] \\
&= 2\mathrm{E}_\theta[J_{121}] + \mathrm{E}_\theta[J_{122}], \text{ say.}
\end{aligned} \tag{8.6.17}$$

Now, we write $\mathrm{E}_\theta[J_{121}]$ as:

$$\begin{aligned}
&\mathrm{E}_\theta\left[\frac{(S_N-N)^3}{n^*}\right] = \frac{\mathrm{E}_\theta\left[2N + 3N(S_N-N)\right]}{n^*} \\
&= 2\left[1+\frac{D}{n^*}+o(n^{*-1})\right] + 3\mathrm{E}_\theta\left\{\left[\frac{\mathcal{R}}{N}+\frac{S_N}{N}-1+1\right](S_N-N)\right\} \\
&= 2+o(1)+3\mathrm{E}_\theta\left\{\left[\frac{\mathcal{R}\,(S_N-N)}{N}+\frac{(S_N-N)^2}{N}+0\right](S_N-N)\right\} \\
&= 2+o(1)+3(1)+o(1) \;= 5+o(1).
\end{aligned} \tag{8.6.18}$$

In the second step of (8.6.18), we used an expression of the third moment of $S_N - N$ from Chow et al. (1965). One may also refer to Theorem 2.4.6 in Ghosh et al. (1997, pp. 28-31).

Next, we express $\mathrm{E}_\theta[J_{122}]$ as:

$$\begin{aligned}
&\mathrm{E}_\theta\left[\frac{n^*-N}{n^*N}(S_N-N)^3\right] \\
&= \mathrm{E}_\theta\left[\frac{1}{N}\left(1-\frac{N}{n^*}\right)(S_N-N)^3\right] \\
&= \mathrm{E}_\theta\left[\frac{1}{N}\left(1-\frac{\mathcal{R}}{N}-\frac{S_N}{N}\right)(S_N-N)^3\right] \\
&= -\mathrm{E}_\theta\left[\frac{1}{N}\left(\frac{\mathcal{R}}{N}+\frac{S_N-N}{N}\right)(S_N-N)^3\right] \\
&= -\mathrm{E}_\theta\left[\mathcal{R}\frac{(S_N-N)^3}{N^2}+\frac{(S_N-N)^4}{N^2}\right] \\
&= -3+o(1).
\end{aligned} \tag{8.6.19}$$

In the last step of (8.6.19), we used (i) the fourth moment (= 3) of a standard normal variable in view of (8.6.8); (ii) the fact that the overshoot $\mathcal{R}$ is dominated; (iii) certain positive moments of $\dfrac{n^*}{N}$ which have finite expectations; and (iv) other techniques including repeated applications of Hölder's inequality.

Hence, by combining (8.6.15)-(8.6.19) we obtain:

$$\begin{aligned}
\mathrm{E}_\theta[J_1] &= -2+2D+o(1)-2\mathrm{E}_\theta[J_{121}]-\mathrm{E}_\theta[J_{122}] \\
&= -2+2D+o(1)-2[5+o(1)]-[-3+o(1)] \\
&= -9+2D+o(1).
\end{aligned} \tag{8.6.20}$$

Thus, from (8.6.13)-(8.6.14) and (8.6.20) we find:

$$\begin{aligned}
\mathrm{E}_\theta[J] &= \mathrm{E}_\theta[J_1]+\mathrm{E}_\theta[J_2] \\
&= -9+2D+o(1)+12+o(1) \\
&= 3+2D+o(1).
\end{aligned} \tag{8.6.21}$$

At this point, by combining (8.6.10), (8.6.11), (8.6.12), and (8.6.21), the regret function can be expressed as:

$$
\begin{aligned}
&A\mathrm{E}_\theta\left[\left(\overline{X}_N-\theta\right)^2\right]+c\mathrm{E}_\theta[N]-2cn^*\\
&=A\theta^2\mathrm{E}_\theta[I]+\frac{A\theta^2}{n^{*2}}\mathrm{E}_\theta[J]+c[n^*-D+o(1)]-2cn^*\\
&=cn^{*2}\left[\frac{1}{n^*}-\frac{D}{n^{*2}}+o(c)\right]+c[3+2D+o(1)]\\
&\quad+c[n^*-D+o(1)]-2cn^*\\
&=3c+o(c).
\end{aligned}
\tag{8.6.22}
$$

The proof is now complete. ■

8.6.4 Proof of Theorem 8.3.1

This proof is adapted from Mukhopadhyay and Pepe (2006). Let $\overline{X}_m$ be the average of $X_1, X_2, \ldots, X_m$ and $\overline{X}^{**}$ be the average of $X_{m+1}, X_{m+2}, \ldots, X_N$. We will derive a bound for $\mathrm{E}_\theta\left[L_N(\theta, \overline{X}_N)\right]$ so that B can be determined explicitly involving A and m. Observe that $\mathrm{E}_\theta\left[\left(\overline{X}_N-\theta\right)^2\right]$ can be expressed as:

$$
\begin{aligned}
&\mathrm{E}_\theta\left[\frac{1}{N^2}\left\{m\left(\overline{X}_m-\theta\right)+(N-m)\left(\overline{X}^{**}-\theta\right)\right\}^2\right]\\
&=\mathrm{E}_\theta\left[\frac{m^2}{N^2}\left(\overline{X}_m-\theta\right)^2\right]+\mathrm{E}_\theta\left[\frac{(N-m)^2}{N^2}\left(\overline{X}^{**}-\theta\right)^2\right]\\
&\quad+\mathrm{E}_\theta\left[\frac{2m(N-m)}{N^2}\left(\overline{X}_m-\theta\right)\left(\overline{X}^{**}-\theta\right)\right].
\end{aligned}
\tag{8.6.23}
$$

Now, let us use a conditional argument to simplify each expectation in the last step above. We may write:

$$
\begin{aligned}
\mathrm{E}_\theta\left[\left(\overline{X}_N-\theta\right)^2\right]&=\mathrm{E}_\theta\left[\frac{m^2}{N^2}\left(\overline{X}_m-\theta\right)^2\right]+\theta^2\mathrm{E}_\theta\left[\frac{N-m}{N^2}\right]\\
&=I+J,\text{ say.}
\end{aligned}
\tag{8.6.24}
$$

Next, we note the following inequalities:

$$
\begin{gathered}
N\ge B\overline{X}_m^2\omega^{-1}\text{ w.p. 1 }\Rightarrow N^{-1}\le\omega B^{-1}\overline{X}_m^{-2}\text{ w.p. 1, and}\\
N\ge m\text{ w.p. 1 }\Rightarrow mN^{-1}\le 1\text{ w.p. 1}\Rightarrow m^2N^{-2}\le mN^{-1}\text{ w.p. 1}
\end{gathered}
\tag{8.6.25}
$$

First, we use the inequalities from (8.6.25) to derive the following bounds on I and J:

$$\begin{aligned} I &= \mathrm{E}_\theta\left[\frac{m^2}{N^2}\left(\overline{X}_m-\theta\right)^2\right] \le \mathrm{E}_\theta\left[\frac{m}{N}\left(\overline{X}_m-\theta\right)^2\right] \\ &\le m\omega B^{-1}\mathrm{E}_\theta\left[\frac{\left(\overline{X}_m-\theta\right)^2}{\overline{X}_m^2}\right] = mwB^{-1}\mathrm{E}_\theta\left[1-\frac{2\theta}{\overline{X}_m}+\frac{\theta^2}{\overline{X}_m^2}\right] \\ &= m\omega B^{-1}I_1, \text{ say.} \end{aligned} \tag{8.6.26}$$

$$\begin{aligned} J &= \theta^2\mathrm{E}_\theta\left[\frac{N-m}{N^2}\right] = \mathrm{E}_\theta\left[\left(1-\frac{m}{N}\right)\frac{\theta^2}{N}\right] \\ &= \mathrm{E}_\theta\left[\left(1-\frac{m}{N}\right)\frac{\theta^2}{N}I\left(N=m\right)\right] + \mathrm{E}_\theta\left[\left(1-\frac{m}{N}\right)\frac{\theta^2}{N}I\left(N>m\right)\right] \\ &= \mathrm{E}_\theta\left[\left(1-\frac{m}{N}\right)\frac{\theta^2}{N}I\left(N>m\right)\right], \end{aligned} \tag{8.6.27}$$

since $\mathrm{E}_\theta\left[\left(1-\frac{m}{N}\right)\frac{\theta^2}{N}I\left(N=m\right)\right]=0$. From (8.6.27), we can obviously write:

$$J \le \mathrm{E}_\theta\left[\theta^2N^{-1}\right] \le \omega B^{-1}\mathrm{E}_\theta\left[\theta^2\overline{X}_m^{-2}\right] = \omega B^{-1}J_1, \text{ say.} \tag{8.6.28}$$

Next, combining (8.6.24), (8.6.26) and (8.6.28), we can claim:

$$\begin{aligned} &\mathrm{E}_\theta\left[L_N(\theta,\overline{X}_N)\right] = A\mathrm{E}_\theta\left[\left(\overline{X}_N-\theta\right)^2\right] \\ &= A[I+J] \le A\omega B^{-1}\left(mI_1+J_1\right) \\ &= A\omega B^{-1}\left\{m\mathrm{E}_\theta\left[1-\frac{2\theta}{\overline{X}_m}+\frac{\theta^2}{\overline{X}_m^2}\right]+\mathrm{E}_\theta\left[\frac{\theta^2}{\overline{X}_m^2}\right]\right\}. \end{aligned} \tag{8.6.29}$$

Let us denote $U=2\theta^{-1}\sum_{i=1}^m X_i$ or equivalently $\theta\overline{X}_m^{-1}=2mU^{-1}$, where U is distributed as a Chi-square random variable with the degree of freedom $2m$ whatever be θ. Then, from (8.6.29) we obtain

$$\begin{aligned} \mathrm{E}_\theta\left[L_N(\theta,\overline{X}_N)\right] &\le \frac{m\omega A}{B}\mathrm{E}\left[1-\frac{4m}{U}+\frac{4m^2}{U^2}\right]+\frac{\omega A}{B}4m^2\mathrm{E}\left[\frac{1}{U^2}\right] \\ &= \frac{m\omega A}{B}\mathrm{E}\left[1-\frac{4m}{U}+4m\left(m+1\right)\frac{1}{U^2}\right]. \end{aligned} \tag{8.6.30}$$

Now, one can easily verify the following expressions: $\mathrm{E}\left[U^{-1}\right]=\frac{1}{2}(m-1)^{-1}$ if $m>1$ and $\mathrm{E}\left[U^{-2}\right]=\frac{1}{4}(m-1)^{-1}(m-2)^{-1}$ if $m>2$. Therefore, from (8.6.30), we can claim that

$$
\begin{aligned}
&\mathrm{E}_\theta\left[L_N(\theta,\overline{X}_N)\right] \\
&\le \frac{m\omega A}{B}\left\{1-4m\frac{1}{2(m-1)}+4m(m+1)\frac{1}{4(m-1)(m-2)}\right\} \\
&= \frac{m\omega A}{B}\left\{1-\frac{2m}{m-1}+\frac{m(m+1)}{(m-1)(m-2)}\right\} \\
&= \frac{2m(m+1)A\omega}{(m-1)(m-2)B} \text{ for } m\ge 3.
\end{aligned}
\tag{8.6.31}
$$

Hence, in order to achieve our goal of risk-bound ω, that is to claim $\mathrm{E}_\theta\left[L_N(\theta,\overline{X}_N)\right]\le\omega$, we would require:

$$
\frac{2m(m+1)A}{(m-1)(m-2)B}=1\Rightarrow B\equiv B_m(A)=\frac{2m(m+1)A}{(m-1)(m-2)} \text{ for } m\ge 3.
\tag{8.6.32}
$$

Thus, the appropriate two-stage stopping variable is given by N defined in (8.2.5) with $B\equiv B_m(A)$ determined from (8.6.32). ■

8.7 Exercises

Exercise 8.2.1: For the stopping variable N from (8.2.5), use part (i) from (8.2.9) to show that $\mathrm{E}_\theta[N]$ is finite for all fixed θ, m, A and c.

Exercise 8.2.2: For all fixed θ, m, A and c, for the stopping variable N from (8.2.5), show that

(i) $\mathrm{E}_\theta\left[N^2\right]$ is finite.

(ii) $\mathrm{E}_\theta\left[N^2\right]\le(n^*+m+1)^2$.

Hence, show that $\lim_{c\to 0}\mathrm{E}_\theta\left[N^2/n^{*2}\right]=1$ for all fixed θ, m and A.

Exercise 8.2.3: Suppose that $X_1,\ldots,X_n,\ldots$ is a sequence of i.i.d. $N(0,\sigma^2)$ random variables, $0<\sigma^2<\infty$. We may like to estimate σ^2 by its MLE, namely $T_n\equiv n^{-1}\sum_{i=1}^n X_i^2$. Write down a set up similar to that in Section 8.2 that will allow one to treat a minimum risk point estimation problem for σ^2. Try the loss function $L_n(\sigma^2,T_n)=A\left(T_n-\sigma^2\right)^2+cn$ with $A>0, c>0$.

(i) Determine the optimal fixed sample size, n^*.

(ii) Propose a natural purely sequential stopping rule along the line of (8.2.5).

(iii) Show that $\mathrm{E}_\theta[N]$ is finite.

(iv) Show that $\mathrm{E}_\theta[N] \le n^* + O(1)$.
(v) Show that $\mathrm{E}_\theta[N/n^*] \to 1$ as $c \to 0$.

Exercise 8.2.4: Is it possible to reduce the problem from Exercise 8.2.3 to the basic problem from Section 8.2? If so, (i) show explicitly how to do so, and (ii) write down a set of results similar to those found in Theorem 8.2.1.

Exercise 8.3.1: In the two-stage procedure (8.3.9), clearly θ^2 in the expression for n^* was replaced by $\overline{X}_m^2$. But, suppose that we now replace $\overline{X}_m^2$ by the best unbiased estimator of θ^2, namely $m(m+1)^{-1}\overline{X}_m^2$. For the corresponding two-stage methodology, state and prove a result similar to that found in Theorem 8.3.1.

Exercise 8.3.2: Suppose that $X_1, \ldots, X_n, \ldots$ is a sequence of i.i.d. $N(0, \sigma^2)$ random variables, $0 < \sigma^2 < \infty$. We may like to estimate σ^2 by its MLE, namely $T_n \equiv n^{-1}\sum_{i=1}^n X_i^2$. Write down a set up similar to that in Section 8.3 that will allow one to treat a bounded risk point estimation problem for σ^2. Try the loss function $L_n(\sigma^2, T_n) = A\left(T_n - \sigma^2\right)^2$ with $A > 0$.

(i) Determine the optimal fixed sample size, n^*.
(ii) Propose a natural two-stage stopping rule along the line of (8.3.9).
(iii) For the corresponding two-stage methodology, state and prove a result similar to that found in Theorem 8.3.1.

Exercise 8.3.3: Is it possible to reduce the problem from Exercise 8.3.2 to the basic problem from Section 8.3? If so, (i) show explicitly how to do so, and (ii) write down a result similar to that found in Theorem 8.3.1.

Exercise 8.4.1: The following data were obtained from an exponential population with unknown mean θ.

11.1	11.9	16.0	10.2	11.9	13.9	14.5	26.0	10.9	10.5
17.2	17.3	10.8	10.0	13.5	15.0	25.0	11.7	26.7	12.1
13.9	15.2	11.7	10.0	14.1	20.1	11.8	30.9	12.6	10.7

Obtain the minimum risk point estimation of θ. Use program number 1 in **Seq08.exe** with initial sample size, $m = 10$ and the loss function

$$L_n(\theta, \overline{X}_n) = \left(\overline{X}_n - \theta\right)^2 + 0.25n.$$

Exercise 8.4.2: Obtain the bounded risk point estimate of θ with risk not exceeding 5, from the data given in **Example 8.4.1** with an initial sample sample size $m = 10$ and the loss function

$$L_n(\theta, \overline{X}_n) = 0.5\left(\overline{X}_n - \theta\right)^2.$$

Use $\varepsilon = 0.5$.

Exercise 8.4.3: Use bounded risk point estimation program in **Seq08.exe** to show that $\varepsilon = 0.4227$ is an optimal choice for the following simulation studies.

Consider estimation of exponential mean, θ by generating random sample from Expo(2) using random number initialization 2. Replicate each simulation 10,000 times with initial sample size, $m = 10$ and parameters A and ω given below:

(i) $A = 50$, $\omega = 1$.
(ii) $A = 100$, $\omega = 2$.
(iii) $A = 150$, $\omega = 3$.
(iv) $A = 200$, $\omega = 4$.
(v) $A = 300$, $\omega = 6$.

CHAPTER 9

Fixed-Width Intervals from MLEs

9.1 Introduction

So far, we have discussed a large number of multistage methodologies of fixed-width confidence intervals for a parameter in a variety of continuous probability distributions. But, so far, we have hardly said anything about constructing fixed-width intervals for a parameter in a discrete probability distribution. That is going to change now.

We will focus on probability distributions involving only a single unknown parameter θ. We will assume that we can find a *maximum likelihood estimator* (MLE), $\widehat{\theta}_n \equiv \widehat{\theta}_n(X_1, \ldots, X_n)$, based on i.i.d. observations $X_1, \ldots, X_n$ from a one-parameter family of distributions. Under mild regularity conditions, Khan's (1969) general sequential approach will apply. In Section 9.2, we provide an elementary introduction to Khan's (1969) general sequential strategy. That would allow us to construct fixed-width confidence intervals for θ.

In Section 9.3, we include an accelerated version of the sequential fixed-width interval strategy for θ. Operationally, an accelerated sequential procedure for θ will be more convenient to apply. Section 9.4 will provide examples including binomial and Poisson distributions. We also discuss a not-so-standard example from a continuous distribution.

All sequential and accelerated sequential fixed-width interval strategies will satisfy desirable *first-order* asymptotic properties. Section 9.5 is devoted to data analyses and concluding remarks. Selected derivations are presented in Section 9.6.

9.2 General Sequential Approach

Under mild regularity conditions, Khan (1969) gave a general method to sequentially construct fixed-width confidence intervals based on a MLE of θ. This general approach was inspired by the seminal work of Chow and Robbins (1965). In this section, we include an elementary exposition of Khan's (1969) sequential strategy in the case of a one-parameter family of distributions.

Suppose that we have available a sequence of independent observations following a common distribution with its probability mass (discrete case) or probability density (continuous case) function $f(x; \theta)$ that is indexed by a single parameter θ. The unknown parameter θ belongs to a parameter space Θ which is a subinterval of real line, for example, $(-\infty, \infty)$ or $(0, \infty)$.

We take it for granted that having recorded $X_1, \ldots, X_n$, one can determine a unique MLE of θ, denoted by

$$\widehat{\theta}_{n,\text{MLE}} \equiv \widehat{\theta}_n = \widehat{\theta}_n(X_1, \ldots, X_n). \tag{9.2.1}$$

Let us assume that the central limit theorem holds for $\widehat{\theta}_n$:

$$\begin{aligned} \sqrt{n}\left(\widehat{\theta}_{n,\text{MLE}} - \theta\right) &\text{ converges in distribution to} \\ N\left(0, 1/\mathcal{I}_X(\theta)\right) &\text{ as } n \to \infty. \end{aligned} \tag{9.2.2}$$

Here, $\mathcal{I}_X(\theta)$ stands for *Fisher-information* about θ in a single observation X, that is,

$$\mathcal{I}_X(\theta) \equiv \mathrm{E}_\theta\left[\left(\frac{\partial}{\partial\theta}\log f(X;\theta)\right)^2\right]. \tag{9.2.3}$$

We obviously assume that $0 < \mathcal{I}_X(\theta) < \infty$ for all $\theta \in \Theta$.

We assume regularity conditions that will allow us to (i) determine $\widehat{\theta}_n$ uniquely for every fixed n, (ii) conclude the limit theorem stated in (9.2.2), and (iii) guarantee that $\mathcal{I}_X(\theta)$ is positive and finite. A typical set of sufficient conditions can be found in other sources. One can get away with less stringent sufficient conditions, but such refinements are beyond the scope of this book.

Having recorded $X_1, \ldots, X_n$, we proceed to construct a confidence interval J for θ with a confidence coefficient approximately matching $1 - \alpha$ where $0 < \alpha < 1$ is preassigned. Now, we also demand that the width of J must be $2d$ where $d (> 0)$ is a preassigned small number. One surely realizes that words such as "small" or "large" ought to be interpreted in the context of a practical problem on hand.

Let us propose a confidence interval centered at $\widehat{\theta}_{n,\text{MLE}}$:

$$J_n \equiv \left[\widehat{\theta}_{n,\text{MLE}} - d, \widehat{\theta}_{n,\text{MLE}} + d\right] \text{ for } \theta. \tag{9.2.4}$$

This confidence interval has a preassigned width $2d$. Now, when will J_n also have the desired approximate coverage probability $1 - \alpha$? A simple answer is that n must be chosen appropriately, but how? We can express the confidence coefficient associated with J_n from (9.2.4) as follows:

$$\begin{aligned} &\mathrm{P}_\theta\left\{\theta \in J_n\right\} \\ &= \mathrm{P}_\theta\left\{\widehat{\theta}_{n,\text{MLE}} - d \le \theta \le \widehat{\theta}_{n,\text{MLE}} + d\right\} \\ &= \mathrm{P}_\theta\left\{\left|\widehat{\theta}_{n,\text{MLE}} - \theta\right| \le d\right\} \\ &= \mathrm{P}_\theta\left\{\sqrt{n}\sqrt{\mathcal{I}_X(\theta)}\left|\widehat{\theta}_{n,\text{MLE}} - \theta\right| \le \sqrt{n}d\sqrt{\mathcal{I}_X(\theta)}\right\} \\ &\approx 2\Phi\left(\sqrt{n}d\sqrt{\mathcal{I}_X(\theta)}\right) - 1, \text{ for large } n, \end{aligned} \tag{9.2.5}$$

in view of (9.2.2).

Clearly, there is a difference between the last step here and what we had in the last step of (6.2.2). We have a similar expressions for confidence coefficient, however, now it holds only approximately! Actually, the confidence coefficient $2\Phi\left(\sqrt{n}d\sqrt{\mathcal{I}_X(\theta)}\right)-1$ holds asymptotically, as $n\to\infty$.

Now, we know that $\Phi\left(z_{\alpha/2}\right)=1-\frac{1}{2}\alpha$ so that the approximate confidence coefficient in (9.2.5) will be at least $1-\alpha$ provided that $\sqrt{n}d\sqrt{\mathcal{I}_X(\theta)}\ge z_{\alpha/2}$. In other words, the required sample size n must be:

$$\text{the smallest integer such that } n\ge\frac{z^2_{\alpha/2}}{d^2\mathcal{I}_X(\theta)}\equiv C,\text{ say.} \tag{9.2.6}$$

Recall that we used the notation "a" instead of $z_{\alpha/2}$ in (6.2.3).

The expression C is called an *optimal* fixed sample size had $\mathcal{I}_X(\theta)$ been known. In situations where $\mathcal{I}_X(\theta)$ is free from θ, the magnitude of C will be known. In such situations, fixed-sample-size fixed-width interval methodologies for θ would easily follow.

However, in many circumstances, $\mathcal{I}_X(\theta)$ will involve θ and the magnitude of C will remain unknown. Thus, we are interested in situations where $\mathcal{I}_X(\theta)$ would involve θ and we must come up with appropriate sequential or multistage estimation strategies.

The purely sequential stopping rule of Khan (1969) is summarized as follows. We start with $X_1,\ldots,X_m$, a pilot sample of size $m(\ge 2)$, and proceed with an additional observation X at-a-time. The stopping time is defined as:

$$\begin{aligned}&N\equiv N(d)\text{ is the smallest integer } n(\ge m)\text{ for which}\\&\quad\text{we observe } n\ge\frac{z^2_{\alpha/2}}{d^2}\left\{\mathcal{I}_X^{-1}(\widehat{\theta}_{n,\text{MLE}})+n^{-1}\right\}.\end{aligned} \tag{9.2.7}$$

This sequential strategy is implemented in the same way we did, for example, in the case of (6.2.17). We note that the expression $z^2_{\alpha/2}\{\mathcal{I}_X^{-1}(\widehat{\theta}_{n,\text{MLE}})+n^{-1}\}/d^2$ found in the boundary condition of (9.2.7) clearly estimates C from (9.2.6) with n observations on hand. We terminate sampling as soon a sample size n is at least as large as an estimate of C at the time.

One may verify that $\mathrm{P}_\theta(N<\infty)=1$, that is, sampling would terminate w.p. 1 when $\mathcal{I}_X(\theta)$ is positive and finite. Upon termination, we estimate θ by the fixed-width interval:

$$J_N\equiv\left[\widehat{\theta}_{N,\text{MLE}}-d,\widehat{\theta}_{N,\text{MLE}}+d\right], \tag{9.2.8}$$

in view of (9.2.4). Here, $\widehat{\theta}_{N,\text{MLE}}$ is obtained from full dataset N, X_1, …, X_m, …, X_N.

The confidence coefficient associated with J_N is given by:

$$\mathrm{P}_\theta\left\{\widehat{\theta}_{N,\text{MLE}}-d\le\theta\le\widehat{\theta}_{N,\text{MLE}}+d\right\}. \tag{9.2.9}$$

Let us denote $\sigma^2 \equiv \sigma_\theta^2 = 1/\mathcal{I}_X(\theta)$. Unfortunately, the confidence coefficient from (9.2.9) cannot be simplified in general to the expression

$$\mathrm{E}_\theta\left[2\Phi(\sqrt{N}d/\sigma) - 1\right]$$

that was found in (6.2.8).

We need to clarify one other point. In (9.2.7), the expression $\mathcal{I}_X^{-1}(\widehat{\theta}_{n,\text{MLE}}) + n^{-1}$ was used in the boundary condition to estimate $\mathcal{I}_X^{-1}(\theta)$. Why? Well, intuitively speaking, $\mathcal{I}_X^{-1}(\theta)$ would represent the variance of $\sqrt{n}\widehat{\theta}_n$ for large n. It is conceivable that $\mathcal{I}_X^{-1}(\widehat{\theta}_{n,\text{MLE}})$ may be zero with a positive probability, especially in the case of a discrete distribution. Instead of using $\mathcal{I}_X^{-1}(\widehat{\theta}_{n,\text{MLE}})$, we use the estimator $\mathcal{I}_X^{-1}(\widehat{\theta}_{n,\text{MLE}}) + n^{-1}$ in the boundary condition so that the stopping rule (9.2.7) may not quit too early. In the case of continuous distributions, we often use simply $\mathcal{I}_X^{-1}(\widehat{\theta}_{n,\text{MLE}})$ in the boundary condition.

Now, the crucial results of Khan (1969) may be summarized as follows for many kinds of $f(x;\theta)$, discrete or continuous. We sketch its proofs in Section 9.6.1.

Theorem 9.2.1: *Assume that* (a) $\mathcal{I}_X(\theta)$ *is positive and finite, and* (b) $\mathrm{E}_\theta[\sup_{n\geq m}\mathcal{I}_X^{-1}(\widehat{\theta}_{n,\text{MLE}})]$ *is finite. Then, for the purely sequential fixed-width confidence interval procedure (9.2.7), for all fixed f, θ, m and α, one has the following properties as $d \to 0$:*

(i) $N/C \to 1$ *in probability;*
(ii) $\mathrm{E}_\theta[N/C] \to 1$ [*Asymptotic First-Order Efficiency*];
(iii) $\mathrm{P}_\theta\left\{\widehat{\theta}_{N,\text{MLE}} - d \leq \theta \leq \widehat{\theta}_{N,\text{MLE}} + d\right\} \to 1-\alpha$ [*Asymptotic Consistency*];

where $C \equiv z_{\alpha/2}^2\mathcal{I}_X^{-1}(\theta)/d^2$ *from (9.2.6).*

The properties (i) and (ii) make us feel comfortable knowing that on an average, the final sample size N required by Khan's stopping rule will hover around C for all fixed distributions f under consideration. Part (ii) also implies that this strategy is asymptotically *first-order* efficient. Part (iii) seems to imply that on an average, the sequential coverage probability will hover around the set goal, namely $1 - \alpha$. That is, this procedure is asymptotically consistent. These asymptotic properties hold when d is small or equivalently when C is large.

We should point out that part (iii) of Theorem 9.2.1 relies heavily on the following fundamental result of Anscombe (1952):

Anscombe's random central limit theorem for MLE: For the stopping variable $N \equiv N(d)$ from (9.2.7), $N^{1/2}\left(\widehat{\theta}_{N,\text{MLE}} - \theta\right) \to N(0, 1/\mathcal{I}_X(\theta))$ in distribution as $d \to 0$ when $\mathcal{I}_X(\theta)$ is positive and finite. (9.2.10)

9.3 General Accelerated Sequential Approach

The sequential strategy proposed in (9.2.7) was designed to continue with one additional observation at-a-time until termination. That process may continue for a long time. But, from a practical consideration, one may want to reduce the number of sampling operations. One may achieve that goal by proceeding sequentially part of the way and augment with sampling in a single batch. Such a methodology will be operationally more convenient than a sequential strategy from (9.2.7).

Hence, we introduce an accelerated sequential technique in the light of Mukhopadhyay (1996). This particular methodology in the present context has not appeared elsewhere.

We first choose and fix a number ρ, $0 < \rho < 1$. Now, we begin sequential sampling with $X_1, \ldots, X_m$, a pilot of size $m(\geq 2)$. We proceed with one additional observation X at-a-time. The stopping time is defined as follows:

$$t \equiv t(d) \text{ is the smallest integer } n(\geq m) \text{ for which}$$
$$\text{we observe } n \geq \rho \frac{z_{\alpha/2}^2}{d^2} \left\{ \mathcal{I}_X^{-1}(\widehat{\theta}_{n,\text{MLE}}) + n^{-1} \right\}, \tag{9.3.1}$$

along the line of (9.2.7). One may verify that $P_\theta(t < \infty) = 1$, that is, this part of sequential sampling would terminate w.p. 1 when $\mathcal{I}_X(\theta)$ is positive and finite.

There is a difference between what we had in (9.2.7) and what we have now. In (9.2.7), the stopping time N estimated C sequentially. But, in (9.3.1), the stopping time t estimates ρC, a fraction of C. So, with the help of (9.3.1), we proceed to gather approximately ρC observations at termination.

Having obtained t from (9.3.1), we define:

$$N \equiv N(d) = \left\langle \frac{t}{\rho} \right\rangle + 1. \tag{9.3.2}$$

Note that N from (9.3.2) estimates C. Thus, we gather additional $(N - t)$ observations $X_{t+1}, \ldots, X_N$ in *one* batch. The final data would consist of $N, X_1, \ldots, X_t, X_{t+1}, \ldots, X_N$.

Next, we estimate θ by the fixed-width interval:

$$J_N \equiv \left[\widehat{\theta}_{N,\text{MLE}} - d, \widehat{\theta}_{N,\text{MLE}} + d \right] \tag{9.3.3}$$

in view of (9.2.8) where $\widehat{\theta}_{N,\text{MLE}} \equiv \widehat{\theta}_N$, obtained from the full dataset $N, X_1, \ldots, X_m, \ldots, X_N$.

The associated confidence coefficient $P_\theta \{\theta \in J_N\}$ is given by:

$$P_\theta \left\{ \widehat{\theta}_{N,\text{MLE}} - d \leq \theta \leq \widehat{\theta}_{N,\text{MLE}} + d \right\}. \tag{9.3.4}$$

It should be clear that this accelerated sequential methodology will save approximately $100(1 - \rho)\%$ sampling operations compared with that required by Khan's (1969) sequential methodology (9.2.7). For example, with

$\rho = 0.6$, the accelerated sequential methodology (9.3.1) - (9.3.2) would save nearly 40% sampling operations compared with its sequential counterpart (9.2.7). Next, some crucial results are stated. We sketch some proofs in Section 9.6.2.

Theorem 9.3.1: *Assume that* (a) $\mathcal{I}_X(\theta)$ *is positive and finite, and* (b) $\mathrm{E}_\theta\left[\sup_{n\geq m}\mathcal{I}_X^{-1}(\widehat{\theta}_{n,\mathrm{MLE}})\right]$ *is finite. Then, for the accelerated sequential fixed-width confidence interval procedure (9.3.1) - (9.3.2), for all fixed* f, θ, m *and* α, *one has the following properties as* $d \to 0$*:*

(i) $N/C \to 1$ *in probability;*

(ii) $\mathrm{E}_\theta[N/C] \to 1$ [*Asymptotic First-Order Efficiency*];

(iii) $\mathrm{P}_\theta\left\{\widehat{\theta}_{N,\mathrm{MLE}} - d \leq \theta \leq \widehat{\theta}_{N,\mathrm{MLE}} + d\right\} \to 1-\alpha$ [*Asymptotic Consistency*];

where $C \equiv z_{\alpha/2}^2\mathcal{I}_X^{-1}(\theta)/d^2$ *from (9.2.6).*

The results stated here are identical to those stated in Theorem 9.2.1. The interpretations of these results remain the same as we had explained after Theorem 9.2.1.

If ρ is chosen close to zero, an accelerated procedure will behave more like a two-stage strategy. On the other hand, if ρ is chosen close to one, an accelerated procedure will behave more like a full-blown sequential strategy. That is, if ρ is chosen close to one, an accelerated procedure will not lead to sizable operational simplicity compared to its sequential counterpart. Normally, one uses $\rho = 0.4, 0.5$ or 0.6. In many kinds of problems, researchers have found that the choice $\rho = 0.5$ has worked very well.

In practice, an accelerated sequential strategy may tend to oversample slightly compared with its sequential counterpart. Also, an accelerated sequential stopping time is expected to have slightly more variability than its sequential counterpart. On the other hand, the coverage probability of a fixed-width interval will be nearly same whether an accelerated sequential or a sequential strategy is employed.

Under this backdrop, we should remind everyone why an accelerated sequential strategy is considered important at all. An accelerated sequential strategy would always be operationally much more convenient in practical implementation than its sequential counterpart.

9.4 Examples

Suppose that we want to obtain a fixed-width confidence interval for θ in a $N(\theta, \sigma^2)$ distribution where θ is unknown but σ is assumed known, $-\infty < \theta < \infty, 0 < \sigma < \infty$. Now, one may check that $\mathcal{I}_X(\theta) = 1/\sigma^2$ and $\widehat{\theta}_{n,\mathrm{MLE}} \equiv \overline{X}_n$, the sample mean. Since $\mathcal{I}_X(\theta)$ does *not* involve θ, a fixed-sample-size fixed-width confidence interval for θ can be easily constructed.

Let the fixed sample size n be the smallest integer $\geq C \equiv z_{\alpha/2}^2\mathcal{I}_X^{-1}(\theta)/d^2 = z_{\alpha/2}^2\sigma^2/d^2$ and we propose the $100(1-\alpha)\%$ confidence interval $J_n =$

$[\overline{X}_n - d, \overline{X}_n + d]$ for θ. Obviously, no sequential or multistage strategy would be called for.

In the following examples, however, one will notice that Fisher-information in a single observation, $\mathcal{I}_X(\theta)$, would involve θ.

9.4.1 Bernoulli Example

Suppose that $X_1, \ldots, X_n, \ldots$ are i.i.d. Bernoulli(θ) with unknown θ, $0 < \theta < 1$. Here, the parameter θ is a probability of "success". Clearly, $\widehat{\theta}_{n,\text{MLE}} \equiv \overline{X}_n$, the sample mean or equivalently the sample proportion of successes. One may check that $\mathcal{I}_X(\theta) = (\theta(1-\theta))^{-1}$.

Sequential Procedure:

We start with $X_1, \ldots, X_m$, a pilot of size $m(\geq 2)$, and proceed with one additional observation X at-a-time. The sequential stopping time from (9.2.7) would become:

$$\begin{aligned}&N \equiv N(d) \text{ is the smallest integer } n(\geq m) \text{ for which}\\ &\text{we observe } n > \frac{z^2_{\alpha/2}}{d^2}\left\{\overline{X}_n(1-\overline{X}_n) + n^{-1}\right\}.\end{aligned} \tag{9.4.1}$$

Upon termination, we estimate θ by the fixed-width interval:

$$J_N \equiv [\overline{X}_N - d, \overline{X}_N + d] \tag{9.4.2}$$

in view of (9.2.8).

Accelerated Sequential Procedure:

We first choose and fix a number ρ, $0 < n\rho < 1$. Now, we begin the sequential part with $X_1, \ldots, X_m$, a pilot of size $m(\geq 2)$. Then, we proceed with one additional observation X at-a-time. The stopping time is:

$$\begin{aligned}&t \equiv t(d) \text{ is the smallest integer } n(\geq m) \text{ for which}\\ &\text{we observe } n \geq \rho\frac{z^2_{\alpha/2}}{d^2}\left\{\overline{X}_n(1-\overline{X}_n) + n^{-1}\right\},\end{aligned} \tag{9.4.3}$$

along the line of (9.3.1).

Having obtained t from (9.4.3), we define:

$$N \equiv N(d) = \left\langle \frac{t}{\rho} \right\rangle + 1, \tag{9.4.4}$$

along the line of (9.3.2).

Then, we gather additional $(N-t)$ observations $X_{t+1}, \ldots, X_N$ in *one* batch. The final dataset would consist of N, X_1, ..., X_t, X_{t+1}, ..., X_N. Next, we estimate θ by the fixed-width interval:

$$J_N \equiv [\overline{X}_N - d, \overline{X}_N + d] \tag{9.4.5}$$

in view of (9.3.3).

It should be clear that this accelerated sequential methodology (9.4.3) - (9.4.4) saves approximately $100(1-\rho)\%$ sampling operations compared with that required by Khan's (1969) purely sequential methodology (9.4.1).

9.4.2 Poisson Example

Suppose that $X_1, \ldots, X_n, \ldots$ are i.i.d. Poisson(θ) with unknown θ, $0 < \theta < \infty$. Here, the parameter θ is the mean of the probability distribution. Clearly, $\widehat{\theta}_{n,\text{MLE}} \equiv \overline{X}_n$, the sample mean. One may check that $\mathcal{I}_X(\theta) = \theta^{-1}$.

Sequential Procedure:

We start with $X_1, \ldots, X_m$, a pilot of size $m(\geq 2)$, and proceed with one additional observation X at-a-time. The sequential stopping time from (9.2.7) would become:

$$N \equiv N(d) \text{ is the smallest integer } n(\geq m) \text{ for which we observe } n \geq \frac{z_{\alpha/2}^2}{d^2}\left\{\overline{X}_n + n^{-1}\right\}. \quad (9.4.6)$$

Upon termination, we estimate θ by the fixed-width interval:

$$J_N \equiv \left[\overline{X}_N - d, \overline{X}_N + d\right] \quad (9.4.7)$$

in view of (9.2.8).

Accelerated Sequential Procedure:

We first choose and fix a number ρ, $0 < \rho < 1$. Now, we begin purely sequential part with $X_1, \ldots, X_m$, a pilot of size $m(\geq 2)$. Then, we proceed with one additional observation X at-a-time. The stopping time is:

$$t \equiv t(d) \text{ is the smallest integer } n(\geq m) \text{ for which we observe } n \geq \rho\frac{z_{\alpha/2}^2}{d^2}\left\{\overline{X}_n + n^{-1}\right\}, \quad (9.4.8)$$

along the line of (9.3.1).

Having obtained t from (9.4.8), let us define:

$$N \equiv N(d) = \left\langle \frac{t}{\rho} \right\rangle + 1, \quad (9.4.9)$$

along the line of (9.3.2). Then, we gather additional $(N-t)$ observations $X_{t+1}, \ldots, X_N$ in *one* batch. The final dataset would consist of N, X_1, ..., X_t, X_{t+1}, ..., X_N. Next, we estimate θ by the fixed-width interval:

$$J_N \equiv \left[\overline{X}_N - d, \overline{X}_N + d\right] \quad (9.4.10)$$

in view of (9.3.3).

Again, this accelerated sequential methodology (9.4.8) - (9.4.9) saves approximately $100(1-\rho)\%$ operations compared with that required by Khan's (1969) sequential methodology (9.4.6).

9.4.3 Exponential Example

Suppose that $X_1, \ldots, X_n, \ldots$ are i.i.d. random variables with an exponential distribution, abbreviated as Expo(θ), having unknown mean θ, $0 < \theta < \infty$. Clearly, $\widehat{\theta}_{n,\text{MLE}} \equiv \overline{X}_n$, the sample mean. One may check that $\mathcal{I}_X(\theta) = \theta^{-2}$.

Sequential Procedure:

We start with $X_1, \ldots, X_m$, a pilot of size $m(\geq 2)$, and proceed with one additional observation X at-a-time. The sequential stopping time from (9.2.7) would become:

$$N\ equivN(d) \text{ is the smallest integer } n(\geq m) \text{ for which we observe } n \geq \frac{z_{\alpha/2}^2 \overline{X}_n^2}{d^2}. \tag{9.4.11}$$

Note that we have used $\overline{X}_n^2$ instead of $\overline{X}_n^2 + n^{-1}$ in defining the stopping time N. We have done so because $\overline{X}_n^2$ has a continuous probability distribution. However, there will be no harm if one continues to use $\overline{X}_n^2 + n^{-1}$ in the boundary condition.

Upon termination, we estimate θ by the fixed-width interval:

$$J_N \equiv \left[\overline{X}_N - d, \overline{X}_N + d\right] \tag{9.4.12}$$

in view of (9.2.8).

Accelerated Sequential Procedure:

We first choose and fix a number ρ, $0 < \rho < 1$. Now, we begin the sequential part with $X_1, \ldots, X_m$, a pilot of size $m(\geq 2)$. Then, we proceed with one additional observation X at-a-time. The stopping time is:

$$t \equiv t(d) \text{ is the smallest integer } n(\geq m) \text{ for which we observe } n \geq \rho \frac{z_{\alpha/2}^2 \overline{X}_n^2}{d^2}, \tag{9.4.13}$$

along the line of (9.3.1).

Having obtained t from (9.4.13), we define:

$$N \equiv N(d) = \left\langle \frac{t}{\rho} \right\rangle + 1, \tag{9.4.14}$$

along the line of (9.3.2).

Then, we gather additional $(N - t)$ observations $X_{t+1}, \ldots, X_N$ in *one* batch. The final dataset would consist of $N, X_1, \ldots, X_t, X_{t+1}, \ldots, X_N$. Next, we estimate θ by the fixed-width interval:

$$J_N \equiv \left[\overline{X}_N - d, \overline{X}_N + d\right] \tag{9.4.15}$$

in view of (9.3.3). Again, this accelerated sequential estimation methodology (9.4.13) - (9.4.14) saves approximately $100(1-\rho)\%$ sampling operations compared with that required by Khan's (1969) sequential methodology (9.4.11).

9.4.4 Special Normal Example

In Section 9.4.2, we introduced a problem of estimating θ in a Poisson(θ) distribution, $\theta > 0$. But, a Poisson(θ) distribution is approximated well in practice with a $N(\theta, \theta)$ distribution when θ is large. Hence, we consider estimating θ in a $N(\theta, \theta)$ distribution, $\theta > 0$. Mukhopadhyay and Cicconetti (2004a) developed both sequential and two-stage bounded risk point estimation problems for θ with practical applications and analysis of data.

Suppose that $X_1, \ldots, X_n, \ldots$ are i.i.d. random variables with a $N(\theta, \theta)$ distribution involving an unknown parameter θ, $0 < \theta < \infty$. We leave it as an exercise to verify that

$$\widehat{\theta}_{n,\text{MLE}} \equiv \widehat{\theta}_n = -\frac{1}{2} + \sqrt{\frac{1}{n}\sum_{i=1}^{n} X_i^2 + \frac{1}{4}}. \tag{9.4.16}$$

We also leave it as an exercise to check that

$$\mathcal{I}_X(\theta) = \frac{2\theta + 1}{2\theta^2}.$$

Sequential Procedure:

We start with $X_1, \ldots, X_m$, a pilot of size $m(\geq 2)$, and proceed with one additional observation X at-a-time. The sequential stopping time from (9.2.7) would become:

$$\begin{aligned} &N \equiv N(d) \text{ is the smallest integer } n(\geq m) \text{ for which we observe} \\ &n \geq \frac{z_{\alpha/2}^2}{d^2} \frac{2\widehat{\theta}_n^2}{(2\widehat{\theta}_n + 1)} \text{ with } \widehat{\theta}_n \text{ from (9.4.16).} \end{aligned} \tag{9.4.17}$$

Note that we have used $2\widehat{\theta}_n^2/(2\widehat{\theta}_n+1)$ instead of $2\widehat{\theta}_n^2/(2\widehat{\theta}_n+1)+n^{-1}$ in defining the boundary condition for N. We have done so because $2\widehat{\theta}_n^2/(2\widehat{\theta}_n+1)$ has a continuous probability distribution. However, there will be no harm if one continues to use $2\widehat{\theta}_n^2/(2\widehat{\theta}_n + 1) + n^{-1}$ in the boundary condition.

Upon termination, we estimate θ by the fixed-width interval:

$$J_N \equiv \left[\widehat{\theta}_N - d, \widehat{\theta}_N + d\right] \text{ where } \widehat{\theta}_N = -\frac{1}{2} + \sqrt{\frac{1}{N}\sum_{i=1}^{N} X_i^2 + \frac{1}{4}}, \tag{9.4.18}$$

in view of (9.2.8).

Accelerated Sequential Procedure:

We first choose and fix a number ρ, $0 < \rho < 1$. Now, we begin sequential sampling part with $X_1, \ldots, X_m$, a pilot of size $m(\geq 2)$. Then, we proceed with one additional observation X at-a-time. The stopping time is defined as:

$$t \equiv t(d) \text{ is the smallest integer } n(\geq m) \text{ for which}$$
$$n \geq \rho \frac{z_{\alpha/2}^2}{d^2} \frac{2\widehat{\theta}_n^2}{(2\widehat{\theta}_n + 1)} \text{ with } \widehat{\theta}_n \text{ from (9.4.16),} \tag{9.4.19}$$

along the line of (9.3.1).

Having obtained t from (9.4.19), we define:

$$N \equiv N(d) = \left\langle \frac{t}{\rho} \right\rangle + 1, \tag{9.4.20}$$

along the line of (9.3.2).

Then, we gather additional $(N - t)$ observations X_{t+1}, …, X_N in *one* batch. The final dataset would consist of N, X_1, …, X_t, X_{t+1}, …, X_N. Next, we estimate θ by the fixed-width interval:

$$J_N \equiv \left[\widehat{\theta}_N - d, \widehat{\theta}_N + d\right] \text{ where } \widehat{\theta}_N = -\frac{1}{2} + \sqrt{\frac{1}{N}\sum_{i=1}^{N} X_i^2 + \frac{1}{4}}, \tag{9.4.21}$$

in view of (9.3.3).

Again, this accelerated sequential methodology (9.4.19) - (9.4.20) saves approximately $100(1 - \rho)\%$ sampling operations compared with that required by Khan's (1969) sequential methodology (9.4.17).

9.5 Data Analyses and Conclusions

In this section, we explain computations and simulation studies for the four examples mentioned in Section 9.4. The program **Seq09.exe** will help in this regard. This program allows one to use either live data input or to carry out simulations by implementing one of the methodologies of one's choice.

Figure 9.5.1 shows an example of an input screen from **Seq09.exe**. It shows the input for simulations to obtain a 95% confidence interval for a Bernoulli "success" probability θ with width $2d = 0.05, m = 10$. The data were generated from Bernoulli($\theta = 0.6$) distribution using the random number sequence 2.

The fixed-width confidence intervals were obtained first using a sequential procedure and then using an accelerated sequential procedure with $\rho = 0.5$. Simulations were replicated 5000 times in either problem. The corresponding output are shown in Figures 9.5.2 and 9.5.3 respectively. As expected, the accelerated procedure has oversampled slightly with no significant difference in coverage probability.

```
Select a program using the number given
in the left hand side. Details of these
programs are given in Chapter 9.
***************************************
* Number         Program               *
*  1   The Bernoulli Case              *
*  2   The Poisson Case                *
*  3   The Exponential Case            *
*  4   A Special Normal Case           *
*  0   Exit                            *
 ***************************************
     Enter the Program Number: 1
Enter 1 for Purely sequential or
      2 for Accelerated sequential procedure
Enter 1 or 2: 1

Do you want to save the results in a file(Y or N)?Y
Note that your default outfile name is "SEQ09.OUT".
Do you want to change the outfile name (Y or N)? N

Fixed-Width Confidence Interval from MLE
       Purely Sequential Procedure.
       ----------------------------


Enter the following Preassigned Values
  alpha (for 100(1-alpha)% confidence interval) = .05
  d (half the width of the confidence interval) = 0.025

Enter the initial sample size, m: 10
Do you want to change the input parameters?[Y or N] N

Do you want to simulate data or input real data?
Type S (Simulation) or D (Real Data): S
Do you want to store simulated sample sizes (Y or N)? N
Number of simulation replications? 5000
Enter a positive integer(<10) to initialize
          the random number sequence: 2
For the simulation, input theta : 0.6
```

Figure 9.5.1. *Data input screen from **Seq09.exe** program.*

Next, we examine the role of our choice of ρ in the performance of an accelerated procedure with the help of simulations. We do so in the case of all four examples mentioned in Section 9.4 by successively fixing $\alpha = 0.05, \rho = 0.05(0.05)1.0$ and $d = 0.05$. Observe that the choice $\rho = 1.0$ will correspond to a purely sequential strategy.

The results for Bernoulli(θ) and Poisson(θ) examples are summarized in Table 9.5.1. Such results in the case of Expo(θ) and Normal(θ, θ) examples are summarized in Table 9.5.2. In these simulation studies, we generated data from an appropriate distribution on a case by case basis after fixing a θ-value that would lead to an optimal fixed-sample-size nearly $370(= C)$.

```
Fixed-Width Confidence Interval from MLE
      Purely Sequential Procedure.
      ----------------------------


The Bernoulli Case
==================
Simulation study using following input parameters:
                 alpha =  0.05
                     d =  0.025
Initial sample size, m =  10
                 theta =  0.60
Number of simulations  =  5000

Optimal sample size (n_star)      = 1475.12
Average final sample size (n_bar) = 1478.33
Std. dev. of n (s_n)              =   15.85
Minimum sample size (n_min)       = 1383
Maximum sample size (n_max)       = 1526
Coverage probability              =  0.9500
```

Figure 9.5.2. *Simulation results for Bernoulli success probability using a sequential procedure.*

```
   Fixed-Width Confidence Interval from MLE
Accelerated Sequential Procedure With rho=0.500
-----------------------------------------------


The Bernoulli Case
==================
Simulation study using following input parameters:
                 alpha =     0.05
                     d =    0.025
Initial sample size, m =       10
                 theta =    0.60
Number of simulations  =     5000

Optimal sample size (n_star)      = 1475.12
Average final sample size (n_bar) = 1481.97
Std. dev. of n (s_n)              =   22.36
Minimum sample size (n_min)       = 1384
Maximum sample size (n_max)       = 1540
Coverage probability              =     0.9490
```

Figure 9.5.3. *Simulation results for Bernoulli success probability using an accelerated sequential procedure.*

Table 9.5.1: Simulated results for accelerated sequential 95% confidence intervals for Bernoulli and Poisson means θ with $d = 0.05$, $m = 10$ and 5000 simulations

	Bernoulli Case			Poisson Case		
	$\theta = 0.595$, $C = 370.3$			$\theta = 0.241$, $C = 370.3$		
ρ	$\overline{n}$	s_n	$\overline{p}$	$\overline{n}$	s_n	$\overline{p}$
0.05	431.4	43.1	0.966	425.4	134.3	0.950
0.10	403.4	28.2	0.960	384.8	120.0	0.935
0.15	392.8	22.7	0.957	374.4	104.1	0.933
0.20	387.8	18.2	0.955	372.6	94.8	0.930
0.25	383.9	16.1	0.954	370.5	85.2	0.939
0.30	381.9	14.8	0.955	370.0	78.1	0.940
0.35	380.8	13.4	0.957	371.4	71.0	0.938
0.40	379.2	12.0	0.949	369.8	66.0	0.943
0.45	378.6	11.4	0.951	370.8	62.4	0.940
0.50	377.2	10.9	0.952	369.5	59.0	0.937
0.55	377.2	10.3	0.951	370.4	55.5	0.945
0.60	376.4	9.8	0.954	370.7	53.5	0.942
0.65	376.1	9.6	0.951	370.4	51.0	0.942
0.70	375.8	9.1	0.949	371.3	48.7	0.943
0.75	375.2	8.8	0.949	371.2	46.5	0.946
0.80	375.0	8.6	0.954	371.0	44.7	0.943
0.85	374.9	8.2	0.954	370.8	43.7	0.945
0.90	374.7	7.8	0.953	371.0	41.9	0.946
0.95	374.4	7.7	0.951	371.2	41.1	0.944
1.00	373.8	7.4	0.955	370.4	39.2	0.946

In these tables, we present the estimated average sample size $\overline{n}$ and the estimated standard deviation s_n for n, plus the estimated coverage probability $\overline{p}$. The results tend to indicate that $\overline{p}$ values stay in a close proximity of its target, 0.95 and this behavior does not appear to change drastically whatever be the choice of ρ. However, $\overline{n}$ and s_n are clearly affected by the choice of ρ. In Bernoulli and Poisson examples, oversampling appears rather serious when ρ is chosen small (less than 0.3). In exponential and normal examples, undersampling is rather pronounced when ρ is chosen small (less than 0.3). For all four distributions on hand, overall it appears however that a choice $\rho = 0.5$ or 0.6 may be expected to work well.

Many more types of interesting data analysis may be contemplated. One should feel free to explore what may be a "good" choice for ρ if one or more among θ, d, m or α are changed. One may even argue about ways one could perhaps define a "good" choice of ρ.

Table 9.5.2: Simulated results for accelerated sequential 95% confidence intervals for exponential and normal means θ with $d = 0.05$, $m = 10$ and 5000 simulations

	Exponential Case $\theta = 0.491$, $C = 370.4$			Normal Case $\theta = 0.488$, $C = 370.4$		
ρ	$\overline{n}$	s_n	$\overline{p}$	$\overline{n}$	s_n	$\overline{p}$
0.05	356.4	143.4	0.924	351.3	117.2	0.923
0.10	346.1	126.5	0.913	344.6	107.5	0.916
0.15	350.0	110.9	0.922	352.3	91.8	0.924
0.20	355.9	95.8	0.927	356.8	76.4	0.929
0.25	361.3	83.1	0.938	361.2	67.6	0.938
0.30	364.3	75.6	0.937	363.1	58.9	0.936
0.35	365.2	68.1	0.940	365.5	53.3	0.939
0.40	367.4	63.8	0.941	365.5	49.9	0.940
0.45	367.2	59.7	0.943	366.8	46.3	0.943
0.50	367.4	55.5	0.947	366.3	43.4	0.942
0.55	367.6	52.3	0.943	366.5	42.0	0.941
0.60	368.6	50.6	0.944	367.1	39.5	0.941
0.65	369.4	47.9	0.945	367.5	37.8	0.939
0.70	369.5	46.1	0.945	367.5	35.9	0.935
0.75	369.6	44.4	0.946	368.0	34.8	0.939
0.80	369.3	42.7	0.946	368.1	33.3	0.943
0.85	369.2	41.5	0.946	368.3	33.1	0.939
0.90	369.5	40.9	0.942	368.8	31.1	0.945
0.95	369.9	39.6	0.946	368.8	30.7	0.944
1.00	369.4	38.4	0.945	368.5	29.3	0.948

9.6 Some Selected Derivations

In this section, we highlight some parts of crucial derivations for completeness. One should note that we interchangeably write $\widehat{\theta}_{N,\text{MLE}}$ or $\widehat{\theta}_N$ for an estimator of θ.

9.6.1 Proof of Theorem 9.2.1

Proof of Part (i):

From (9.2.7), we first observe a basic inequality (w.p. 1):

$$\frac{z_{\alpha/2}^2}{d^2}\left\{\mathcal{I}_X^{-1}(\widehat{\theta}_N) + N^{-1}\right\} \le N \le m + \frac{z_{\alpha/2}^2}{d^2}\left\{\mathcal{I}_X^{-1}(\widehat{\theta}_{N-1}) + (N-1)^{-1}\right\}. \tag{9.6.1}$$

Next, we divide throughout (9.6.1) by C and write (w.p. 1):

$$\frac{\left\{\mathcal{I}_X^{-1}(\widehat{\theta}_N)+N^{-1}\right\}}{\mathcal{I}_X^{-1}(\theta)} \le \frac{N}{C} \le \frac{m}{C} + \frac{\left\{\mathcal{I}_X^{-1}(\widehat{\theta}_{N-1})+(N-1)^{-1}\right\}}{\mathcal{I}_X^{-1}(\theta)}. \quad (9.6.2)$$

Now, observe that $N \to \infty$ w.p. 1 and $C \to \infty$ as $d \to 0$. Also, $\widehat{\theta}_N \to \theta$ in probability, and both $\mathcal{I}_X(\widehat{\theta}_{N,\text{MLE}})$ and $\mathcal{I}_X(\widehat{\theta}_{N-1,\text{MLE}})$ converge to $\mathcal{I}_X(\theta)$ in probability as $d \to 0$. Thus, taking limits throughout (9.6.2), we complete the proof of part (i). ■

Proof of Part (ii):

We denote $U = \sup_{n\ge m}\mathcal{I}_X^{-1}(\widehat{\theta}_{n,\text{MLE}})$. Now, for $0 < d < d_0$, we can rewrite the right-hand side of (9.6.1) as follows:

$$Nd^2 \le md_0^2 + z_{\alpha/2}^2(U+1) \text{ w.p. } 1. \quad (9.6.3)$$

This shows that N/C is bounded by an integrable function since $\text{E}_\theta[U]$ is assumed finite. At this point, we apply dominated convergence theorem and combine with part (i) to claim that
$\lim_{d\to 0}\text{E}_\theta[N/C] = 1$. ■

Proof of Part (iii):

Recall that the confidence interval for θ is $J_N = [\widehat{\theta}_{N,\text{MLE}} - d, \widehat{\theta}_{N,\text{MLE}} + d]$ from (9.2.8). Now, from (9.2.9), with $\sigma^2 = 1/\mathcal{I}_X(\theta)$, we may write:

$$\begin{aligned} &\text{P}_\theta\left\{\widehat{\theta}_{N,\text{MLE}} - d \le \theta \le \widehat{\theta}_{N,\text{MLE}} + d\right\} \\ &= \text{P}_\theta\left\{\left|\widehat{\theta}_{N,\text{MLE}} - \theta\right| \le d\right\} \\ &= \text{P}_\theta\left\{\frac{\sqrt{C}\left|\widehat{\theta}_{N,\text{MLE}} - \theta\right|}{\sigma} \le \frac{\sqrt{C}d}{\sigma}\right\} \\ &= \text{P}_\theta\left\{\frac{\sqrt{C}\left|\widehat{\theta}_{N,\text{MLE}} - \theta\right|}{\sigma} \le z_{\alpha/2}\right\}. \end{aligned} \quad (9.6.4)$$

Let Z denote a standard normal random variable. We use Anscombe's (1952) random central limit theorem from (9.2.10), Slutsky's theorem, and part (i) to claim:

$$\lim_{d\to 0}\text{P}_\theta\left\{\sqrt{C}\left|\widehat{\theta}_N - \theta\right|/\sigma \le z_{\alpha/2}\right\} = \text{P}\{|Z| \le z_{\alpha/2}\} = 1-\alpha. \quad (9.6.5)$$

Now, the proof is complete. ■

9.6.2 Proof of Theorem 9.3.1

Proof of Part (i):

Note that the stopping time t from (9.3.1) looks and behaves exactly like N from (9.2.7) provided that $1/\mathcal{I}_X(\theta)$ is replaced by $\rho/\mathcal{I}_X(\theta)$. Hence, from Theorem 9.2.1, part (i) we can immediately conclude:

$$\lim_{d\to 0} \frac{t}{\rho C} = 1 \text{ in probability.} \tag{9.6.6}$$

Next, from (9.3.2), we may write:

$$\frac{t}{\rho} \le N \le \frac{t}{\rho} + 1 \text{ w.p. } 1 \tag{9.6.7}$$

which implies that

$$\frac{t}{\rho C} \le \frac{N}{C} \le \frac{t}{\rho C} + \frac{1}{C}. \tag{9.6.8}$$

In view of (9.6.6) and the fact that $C \to \infty$ as $d \to 0$, we conclude part (i) from (9.6.8). ■

Proof of Part (ii):

Using Theorem 9.2.1, part (ii) we immediately conclude:

$$\lim_{d\to 0} \mathrm{E}_\theta\left[\frac{t}{\rho C}\right] = 1. \tag{9.6.9}$$

Next, from (9.6.8) we obtain:

$$\mathrm{E}_\theta\left[\frac{t}{\rho C}\right] \le \mathrm{E}_\theta\left[\frac{N}{C}\right] \le \mathrm{E}_\theta\left[\frac{t}{\rho C}\right] + \frac{1}{C}. \tag{9.6.10}$$

Now, in view of (9.6.9) and the fact that $C \to \infty$ as $d \to 0$, we conclude part (ii) from (9.6.10). ■

Proof of Part (iii):

The steps remain exactly the same as in the proof of Theorem 9.2.1, part (iii). ■

9.7 Exercises

Exercise 9.2.1: Suppose that $X_1, \ldots, X_n, \ldots$ are i.i.d. random variables having a Bernoulli(θ) distribution with unknown θ, $0 < \theta < 1$. Show that $\widehat{\theta}_{n,\text{MLE}} \equiv \overline{X}_n$, the sample mean or equivalently a sample proportion of successes. Also, verify that $\mathcal{I}_X(\theta) = \theta^{-1}(1-\theta)^{-1}$.

Exercise 9.2.2: Suppose that $X_1, \ldots, X_n, \ldots$ are i.i.d. random variables having a Poisson(θ) distribution with unknown θ, $0 < \theta < \infty$. Show that $\widehat{\theta}_{n,\text{MLE}} \equiv \overline{X}_n$, the sample mean. Also, verify that $\mathcal{I}_X(\theta) = \theta^{-1}$.

Exercise 9.2.3: Suppose that $X_1, \ldots, X_n, \ldots$ are i.i.d. random variables having an Expo(θ) distribution with an unknown mean θ, $0 < \theta < \infty$. Show that $\widehat{\theta}_{n,\text{MLE}} \equiv \overline{X}_n$, the sample mean. Also, verify that $\mathcal{I}_X(\theta) = \theta^{-2}$.

Exercise 9.2.4: Suppose that $X_1, \ldots, X_n, \ldots$ are i.i.d. random variables having a $N(\theta, \theta)$ distribution involving an unknown parameter θ, $0 < \theta < \infty$. Show that

$$\widehat{\theta}_{n,\text{MLE}} \equiv \widehat{\theta}_n = \sqrt{\frac{1}{n}\sum_{i=1}^{n} X_i^2 + \frac{1}{4}} - \frac{1}{2}.$$

Prove or disprove that $\widehat{\theta}_{n,\text{MLE}}$ positive w.p. 1. Also, verify that

$$\mathcal{I}_X(\theta) = \frac{2\theta + 1}{2\theta^2}.$$

Exercise 9.2.5: Reconsider the fixed-width confidence interval problem from Section 9.2 and recall that $C = z_{\alpha/2}^2 \mathcal{I}_X^{-1}(\theta)/d^2$. We propose the following (modified) two-stage procedure. Let us define a pilot size:

$$m \equiv m(d) = \max\left\{m_0, \left\langle \left(\frac{z_{\alpha/2}^2}{d^2}\right)^{2/(1+\gamma)} \right\rangle + 1\right\},$$

where $m_0 (\geq 2)$ and $\gamma (> 0)$ are fixed. One will first record pilot observations $X_1, \ldots, X_m$ and determine $\widehat{\theta}_{m,\text{MLE}} \equiv \widehat{\theta}_m$. Next, one would define the final sample size:

$$R \equiv R(d) = \max\left\{m, \left\langle \frac{z_{\alpha/2}^2}{d^2} \left\{\mathcal{I}_X^{-1}(\widehat{\theta}_m) + m^{-1}\right\} \right\rangle + 1\right\}.$$

We would sample the difference at the second stage if $R > m$. Upon termination, we consider the fixed-width interval

$$J_R \equiv \left[\widehat{\theta}_{R,\text{MLE}} - d, \widehat{\theta}_{R,\text{MLE}} + d\right]$$

for θ based on full dataset $R, X_1, \ldots, X_R$.

Assuming (a) $\mathcal{I}_X(\theta)$ is positive and finite and
(b) $\text{E}_\theta\left[\sup_{n \geq m} \mathcal{I}_X^{-1}(\widehat{\theta}_n)\right]$ is finite,

for all fixed f, θ, and α, prove the following results as $d \to 0$:

(i) $R/C \to 1$ in probability.

(ii) $\text{E}_\theta[R/C] \to 1$.

(iii) $\text{P}_\theta\left\{\widehat{\theta}_{R,\text{MLE}} - d \leq \theta \leq \widehat{\theta}_{R,\text{MLE}} + d\right\} \to 1 - \alpha$.

Exercise 9.2.6: Suppose that $X_1, \ldots, X_n, \ldots$ are i.i.d. random variables with the following gamma p.d.f.:

$$f(x;\theta) = \begin{cases} \{\theta^a \Gamma(a)\}^{-1} x^{a-1} \exp(-x/\theta) & \text{if } x > 0 \\ 0 & \text{otherwise.} \end{cases}$$

This distribution involves an unknown parameter θ, $0 < \theta < \infty$. We assume that a is positive and known. Find the expressions of $\widehat{\theta}_{n,\text{MLE}}$ and $\mathcal{I}_X(\theta)$. Then, apply the sequential methodology from Section 9.2 to construct an approximately $100(1-\alpha)\%$ fixed-width confidence interval for θ.

Exercise 9.2.7: Suppose that $X_1, \ldots, X_n, \ldots$ are i.i.d. random variables having a $N(\theta, a\theta)$ distribution involving an unknown parameter θ, $0 < \theta < \infty$. Assume that a is positive and known. Find the expressions of $\widehat{\theta}_{n,\text{MLE}}$ and $\mathcal{I}_X(\theta)$. Then, apply the sequential methodology from Section 9.2 to construct an approximately $100(1-\alpha)\%$ fixed-width confidence interval for θ.

Exercise 9.2.8: Suppose that $X_1, \ldots, X_n, \ldots$ are i.i.d. random variables having a common $N(\theta, a\theta^2)$ distribution involving an unknown parameter $\theta, 0 < \theta < \infty$. Assume that a is positive and known. Find the expressions of $\widehat{\theta}_{n,\text{MLE}}$ and $\mathcal{I}_X(\theta)$. Then, apply the sequential methodology from Section 9.2 to construct an approximately $100(1-\alpha)\%$ fixed-width confidence interval for θ.

Exercise 9.2.9: Suppose that $X_1, \ldots, X_n, \ldots$ are i.i.d. random variables with the following geometric p.m.f.:

$$f(x;\theta) = \begin{cases} \theta(1-\theta)^{x-1} & \text{if } x = 1, 2, 3, \ldots \\ 0 & \text{otherwise.} \end{cases}$$

This distribution involves an unknown parameter θ, $0 < \theta < 1$. Find the expressions of $\widehat{\theta}_{n,\text{MLE}}$ and $\mathcal{I}_X(\theta)$. Then, apply the sequential methodology from Section 9.2 to construct an approximately $100(1-\alpha)\%$ fixed-width confidence interval for θ.

Exercise 9.2.10: Suppose that $X_1, \ldots, X_n, \ldots$ are i.i.d. random variables having a $N(0, \theta)$ distribution with an unknown parameter θ, $0 < \theta < \infty$. Find the expressions of $\widehat{\theta}_{n,\text{MLE}}$ and $\mathcal{I}_X(\theta)$. Then, apply the sequential methodology from Section 9.2 to construct an approximately $100(1-\alpha)\%$ fixed-width confidence interval for θ.

Exercise 9.2.11: Suppose that $X_1, \ldots, X_n, \ldots$ are i.i.d. random variables with the following p.m.f.:

x values :	2	4
Probability :	2θ	$1-2\theta$

This distribution involves an unknown parameter θ, $0 < \theta < \frac{1}{2}$. Find the expressions of $\widehat{\theta}_{n,\text{MLE}}$ and $\mathcal{I}_X(\theta)$. Then, apply the sequential methodology from Section 9.2 to construct an approximately $100(1-\alpha)\%$ fixed-width confidence interval for θ.

Exercise 9.3.1 (Exercise 9.2.6 Continued): Suppose that $X_1, \ldots, X_n, \ldots$ are i.i.d. random variables with a common gamma distribution as given. Write down the accelerated sequential methodology from Section 9.3. Explain how it would be implemented in order to construct an approximately $100(1-\alpha)\%$ fixed-width confidence interval for θ.

Exercise 9.3.2 (Exercise 9.2.7 Continued): Suppose that $X_1, \ldots, X_n, \ldots$ are i.i.d. random variables with a common $N(\theta, a\theta)$ distribution as given. Write down the accelerated sequential methodology from Section 9.3. Explain how it would be implemented in order to construct an approximately $100(1-\alpha)\%$ fixed-width confidence interval for θ.

Exercise 9.3.3 (Exercise 9.2.8 Continued): Suppose that $X_1, \ldots, X_n, \ldots$ are i.i.d. random variables with a common $N(\theta, a\theta^2)$ distribution as given. Write down the accelerated sequential methodology from Section 9.3. Explain how it would be implemented in order to construct an approximately $100(1-\alpha)\%$ fixed-width confidence interval for θ.

Exercise 9.3.4 (Exercise 9.2.9 Continued): Suppose that $X_1, \ldots, X_n, \ldots$ are i.i.d. random variables with a common geometric distribution as given. Write down the accelerated sequential methodology from Section 9.3. Explain how it would be implemented in order to construct an approximately $100(1-\alpha)\%$ fixed-width confidence interval for θ.

Exercise 9.3.5 (Exercise 9.2.10 Continued): Suppose that $X_1, \ldots, X_n, \ldots$ are i.i.d. random variables with a common $N(0, \theta)$ distribution as given. Write down the accelerated sequential methodology from Section 9.3. Explain how it would be implemented in order to construct an approximately $100(1-\alpha)\%$ fixed-width confidence interval for θ.

Exercise 9.3.6: Suppose that $X_1, \ldots, X_n, \ldots$ are i.i.d. random variables with the following common p.m.f.:

x values :	0.25	0.5
Probability :	3θ	$1-3\theta$

This distribution involves an unknown parameter θ, $0 < \theta < \frac{1}{3}$. Find the expressions of $\widehat{\theta}_{n,\text{MLE}}$ and $\mathcal{I}_X(\theta)$. Then, apply the accelerated sequential methodology from Section 9.3 to construct an approximately $100(1-\alpha)\%$ fixed-width confidence interval for θ. Explain how this methodology would be implemented.

Exercise 9.5.1: It is believed that a coin used in a game is biased and $\mathrm{P}(\text{Head})$ may be 0.4. A player would like to know how many times (N) the coin ought to be flipped to estimate $\theta(=\mathrm{P}(\text{Head}))$ within ± 0.025 accuracy with 95% confidence. Estimate $\mathrm{E}(N)$ by simulating a coin flipping experiment using:

(i) a purely sequential procedure; and

(ii) an accelerated sequential procedure with $\rho = 0.5$.

Use 1000 simulations and $m = 10$. Compare the results from two simulation studies.

Exercise 9.5.2: An entomologist looked at a tree that was heavily infested with pests. Each sampling unit consists of a leaf and an observation X is the number of white spots on it. The distribution of X was postulated as Poisson(θ) where $\theta(>0)$ was believed to be large. Consider the following dataset on X from 70 randomly selected leaves:

14	26	28	24	36	19	33	39	22	30
27	15	20	19	32	26	33	34	34	28
25	18	15	19	23	26	31	32	29	25
23	23	23	19	18	27	24	28	26	32
29	29	25	22	28	31	23	19	30	21
25	22	31	27	24	31	30	30	27	21
31	29	16	29	25	28	23	30	25	29

One will start with $x_1 = 14$ and proceed either along the rows or along the columns. Use Program 2 in **Seq09.exe** with $m = 10$ on this live data to obtain a purely sequential 90% confidence interval for θ with width $4(=2d)$ by proceeding first along the rows. Is there enough data here to terminate sampling? What is the final confidence interval?

Next, repeat the process by proceeding along columns starting with $x_1 = 14$.

Exercise 9.5.3 (Exercise 9.5.2 Continued): On given data, apply the methodologies from this chapter assuming an approximate $N(\theta, \theta)$ distribution for X, $\theta > 0$. One will start with $x_1 = 14$ and again proceed either along the rows or along the columns. First, proceed along the rows using Program 4 in **Seq09.exe** with $m = 5$ on this live data to obtain a 90% confidence interval for θ with width $2(=2d)$. Implement

(i) the purely sequential procedure;

(ii) the accelerated sequential procedure with $\rho = 0.5$.

Is there enough data here to terminate sampling under either methodology? What is the final confidence interval in each case?

Next, repeat the process by proceeding along columns starting with $x_1 = 14$.

CHAPTER 10

Distribution-Free Methods in Estimation

10.1 Introduction

In Chapters 6-9, we have introduced a number of multistage methodologies for fixed-width confidence interval, minimum risk point estimation, and bounded risk point estimation problems. These addressed estimation problems for an appropriate mean, location, variance or scale parameter in a number of distributions including a normal, negative exponential, exponential, binomial or Poisson. It should be no surprise that the proposed methodologies so far have relied heavily upon crucial characteristics of the distributions under consideration. In this chapter, we go distribution-free or nonparametric under a banner of estimation. One may recall that Chapter 5 included glimpses of selected nonparametric sequential tests.

Suppose that we have available a sequence of independent observations $X_1, \ldots, X_n, \ldots$ from a population having a distribution function F, that is, $F(x) \equiv \mathrm{P}(X \leq x), x \in I\!R$ where $I\!R$ is the real line. Other than some moment conditions on F, we assume very little else, especially in Sections 10.2 and 10.3. We cite more specifics regarding F as needed while addressing certain problems.

The unknown F may correspond to either a discrete or a continuous distribution under consideration. We denote:

$$\begin{aligned} &\text{Mean:} && \mu \equiv \mu(F) = \mathrm{E}_F[X] = \textstyle\int_{-\infty}^{\infty} x dF(x); \\ &\text{Variance:} && \sigma^2 \equiv \sigma^2(F) = \mathrm{E}_F[(X-\mu)^2] = \textstyle\int_{-\infty}^{\infty} (x-\mu)^2 dF(x). \end{aligned} \tag{10.1.1}$$

In Section 10.2, we first introduce a fixed-width confidence interval problem for the mean of a population having an unknown F. The fundamental purely sequential procedure of Chow and Robbins (1965) is discussed in Section 10.2.1. Mukhopadhyay's (1980) modified two-stage procedure is briefly mentioned in Section 10.2.2.

Then, we discuss a minimum risk point estimation problem for the mean of an unknown F in Section 10.3. Mukhopadhyay (1978) developed the first purely sequential methodology when F was unknown under an assumption that $\sigma^2 > \sigma_L^2$ with $\sigma_L^2 (> 0)$ known. In Section 10.3.1, we introduce Mukhopadhyay's (1978) purely sequential methodology. Under the same assumption, in Section 10.3.2, we have included a new two-stage strategy along the lines of Mukhopadhyay and Duggan (1997). In Section 10.3.3, we summarize Ghosh and Mukhopadhyay's (1979) fundamental purely sequential methodology assuming that σ^2 is completely unknown. They showed asymptotic risk efficiency property of their procedure under an appropriate moment condition on F. Chow and Yu (1981) reinvented an analogous sequential methodology and proved asymptotic risk efficiency property under a less restrictive moment assumption.

In Sections 10.2 and 10.3, we work under the following standing assumption:

> **Assumption:** Throughout Sections 10.2-10.3, the variance σ^2 is assumed positive and finite. The mean μ is obviously finite too.

Section 10.4 introduces an important purely sequential bounded length confidence interval strategy from Geertsema (1970) for the center of symmetry, θ. We assume that the unknown distribution function F is twice differentiable. This nonparametric methodology parallels that of a traditional sign test (Sections 5.4-5.5).

Section 10.5 is devoted to data analyses and concluding remarks. Selected multistage methodologies are briefly mentioned in Section 10.6. Some derivations are presented in Section 10.7.

We note that the area of sequential nonparametric estimation is very broad in its coverage. One may simply read our treatise as an introduction or a gateway to this vast area of research and practice. For a more thorough review, one may refer to Sen (1981,1985) and Ghosh et al. (1997, Chapters 9-10).

10.2 Fixed-Width Confidence Intervals for the Mean

Having recorded $X_1, \ldots, X_n$, let $\overline{X}_n$ and S_n^2 denote the customary sample mean and the sample variance respectively, $n \geq 2$. We construct a confidence interval J for μ having a confidence coefficient that is approximately $1 - \alpha$ where $0 < \alpha < 1$ is preassigned. We also require that the width of J must be $2d$ where $d(> 0)$ is preassigned.

We propose the confidence interval

$$J_n \equiv \left[\overline{X}_n - d, \overline{X}_n + d\right] \tag{10.2.1}$$

for μ. This confidence interval has a preassigned width $2d$. Now, when will the interval J_n will also have a desired approximate confidence coefficient $1 - \alpha$? The simple answer is that n must be chosen appropriately.

Let us express the confidence coefficient associated with J_n as follows:

$$\begin{aligned}
\mathrm{P}_F\{\mu \in J_n\} &= \mathrm{P}_F\{\overline{X}_n - d \le \mu \le \overline{X}_n + d\} \\
&= \mathrm{P}_F\{|\overline{X}_n - \mu| \le d\} \\
&= \mathrm{P}_F\left\{\frac{\sqrt{n}\,|\overline{X}_n - \mu|}{\sigma} \le \frac{\sqrt{n}d}{\sigma}\right\} \\
&\approx 2\Phi\left(\frac{\sqrt{n}d}{\sigma}\right) - 1, \text{ for large } n.
\end{aligned} \tag{10.2.2}$$

There is a clear difference between the last step here and what we had in the last of (6.2.2). We now have the same confidence coefficient that holds asymptotically as $n \to \infty$. This follows from the *central limit theorem* (CLT) which says that $\sqrt{n}(\overline{X}_n - \mu)/\sigma$ converges in distribution to a standard normal random variable as $n \to \infty$.

Now, we know that $\Phi\left(z_{\alpha/2}\right) = 1 - \frac{1}{2}\alpha$ so that the approximate confidence coefficient in (10.2.2) will be at least $1 - \alpha$ provided that $\sqrt{n}d/\sigma \ge z_{\alpha/2}$. In other words, the required fixed-sample-size n must be:

$$\text{the smallest integer such that } n \ge a^2\sigma^2/d^2 = C, \text{ say, with } a = z_{\alpha/2}. \tag{10.2.3}$$

We refer to C as a required *optimal* fixed sample size had σ^2 been known, and this expression is identical with that found in (6.2.4). But, σ^2 is unknown, and hence the magnitude of C remains unknown.

10.2.1 Purely Sequential Procedure

The purely sequential stopping rule of Chow and Robbins (1965) follows. We start with $X_1, \ldots, X_m$, a pilot of size $m(\ge 2)$, and proceed with one additional observation X at-a-time. The stopping time is:

$$\begin{aligned}
&N \equiv N(d) \text{ is the smallest integer } n(\ge m) \text{ for which} \\
&\qquad \text{we observe } n \ge \frac{a^2\left(S_n^2 + n^{-1}\right)}{d^2}.
\end{aligned} \tag{10.2.4}$$

In the boundary condition of (10.2.4), note that we have replaced σ^2 in the expression of C with $S_n^2 + n^{-1}$ instead of S_n^2. If F corresponds to a discrete distribution, with n fixed, the sample variance S_n^2 may be zero with a positive probability. On the other hand, we can claim that $S_n^2 + n^{-1} > 0$ w.p. 1. and yet $S_n^2 + n^{-1}$ continues to be a consistent estimator of σ^2. So, the expression $a^2\left(S_n^2 + n^{-1}\right)/d^2$ that is found in the boundary condition of (10.2.4) will always be a positive estimator of C, whatever be F or n.

The sequential procedure (10.2.4) is implemented in the same way as in the case of (7.2.11). We stop sampling as soon as a sample size n is at least as large as the estimate of C at the time.

Again, $\mathrm{P}_F(N < \infty) = 1$, that is, sampling would terminate w.p. 1.

Upon termination, we estimate μ by the fixed-width interval:

$$J_N \equiv \left[\overline{X}_N - d, \overline{X}_N + d\right] \qquad (10.2.5)$$

where $\overline{X}_N \equiv N^{-1}\sum_{i=1}^{N} X_i$, a sample mean obtained from full dataset $N, X_1, \ldots, X_m, \ldots, X_N$.

The associated confidence coefficient $\mathrm{P}_F\{\mu \in J_N\}$ is given by:

$$\mathrm{P}_F\left\{\overline{X}_N - d \le \mu \le \overline{X}_N + d\right\}. \qquad (10.2.6)$$

But, unfortunately this probability cannot be simplified to an expression such as $\mathrm{E}_F\left[2\Phi\left(\sqrt{N}d/\sigma\right) - 1\right]$ that was found in (6.2.8). One can see readily why that is the case. Observe that the event $\{N = n\}$ depends exclusively on $(S_m^2, \ldots, S_n^2)$, but $\overline{X}_n$ and $(S_m^2, \ldots, S_n^2)$ are dependent random variables for all fixed $n \ge m$. Hence, $I(N = n)$ and $\overline{X}_n$ are dependent so that $\mathrm{P}_F\left\{\overline{X}_n - d \le \mu \le \overline{X}_n + d \mid N = n\right\}$ cannot be replaced by $\mathrm{P}_F\left\{\overline{X}_n - d \le \mu \le \overline{X}_n + d\right\}$ for any fixed $n(\ge m)$.

It is well-known that $\overline{X}_n$ and $(S_m^2, \ldots, S_n^2)$ are independent random variables if and only if F corresponds to a normal distribution, which was handled in Chapter 6. That is, (10.2.6) will reduce to $\mathrm{E}_F\left[2\Phi\left(\sqrt{N}d/\sigma\right) - 1\right]$ if and only if F corresponds to a normal distribution.

If F happens to be *continuous*, one may use S_n^2 instead of $S_n^2 + n^{-1}$ in (10.2.4). Then, for all fixed *continuous* F and μ, σ^2, m, d and α, one can prove:

$$\mathrm{E}_F(N) \le C + m + 2. \qquad (10.2.7)$$

This result explains the magnitude of the cost of ignorance of σ^2. Since F is unknown, we start with a pilot of size m. But, as soon as we start sampling with m pilot observations, (10.2.7) indicates that on an average the number of observations beyond pilot will not exceed $C + 2$. That is, as soon as one starts sampling according to (10.2.4), one would fair nearly as well as one would under the known σ^2 scenario if F is continuous. A proof of (10.2.7) is given in Section 10.7.1. A similar conclusion also holds when F is discrete.

Now, the fundamental results of Chow and Robbins (1965) may be stated as follows for all F, discrete or continuous.

Theorem 10.2.1 : *For purely sequential fixed-width confidence interval procedure (10.2.4), for all fixed $F, \mu, \sigma^2, \sigma_L^2, m$ and α, one has the following properties as $d \to 0$:*

(i) $N/C \to 1$ *in probability*;

(ii) $\mathrm{E}_F[N/C] \to 1$ *[Asymptotic Efficiency or Asymptotic First-Order Efficiency]*;

(iii) $\mathrm{P}_F\left\{\overline{X}_N - d \le \mu \le \overline{X}_N + d\right\} \to 1 - \alpha$ *[Asymptotic Consistency]*;

with $C \equiv a^2\sigma^2/d^2$ from (10.2.3).

The properties (i) and (ii) make us feel comfortable knowing that on an average, final sample size N required by Chow-Robbins's stopping rule will hover around C for all fixed population distributions F having $0 < \sigma^2 < \infty$. Part (ii) shows that this procedure is *asymptotically first-order efficient.* Part (iii) implies that on an average, sequential confidence coefficient will hover around target, $1 - \alpha$. That is, this procedure is *asymptotically consistent.*

We must add that part (iii) of Theorem 10.2.1 relies heavily on a fundamental result that is known as the *random central limit theorem* due to Anscombe (1952) which states:

Anscombe's Random Central Limit Theorem (CLT):
$$N^{1/2}(\overline{X}_N - \mu)/\sigma \to N(0,1) \text{ in distribution as } d \to 0. \tag{10.2.8}$$

Chow-Robbins purely sequential fixed-width confidence intervals methodology influenced the course of subsequent research in nonparametric sequential methodologies very heavily.

We sketch a proof of Theorem 10.2.1 in Section 10.7.2 when F is assumed *continuous*. In a discrete case, the proof is similar.

10.2.2 Modified Two-Stage Procedure

Following the stopping rule (10.2.4) by taking one observation at-a-time beyond the pilot may be operationally inconvenient. So, it is prudent to seek an appropriate two-stage estimation methodology. Mukhopadhyay (1980) set out to do just that.

Observe that N from (10.2.4) may be stated by replacing $S_n^2 + n^{-1}$ with $S_n^2 + n^{-\gamma}$ in the boundary condition where $\gamma(> 0)$ is a fixed number. Theorem 10.2.1 would still hold, but note that

$$\begin{aligned} N &\geq \frac{a^2\left(S_N^2 + N^{-\gamma}\right)}{d^2} \geq \frac{a^2 N^{-\gamma}}{d^2} \text{ w.p. } 1 \\ \Rightarrow N^{1+\gamma} &\geq \frac{a^2}{d^2} \text{ w.p. } 1 \Rightarrow N \geq \left(\frac{a}{d}\right)^{2/(1+\gamma)} \text{ w.p. } 1. \end{aligned} \tag{10.2.9}$$

Since N will be at least $(a/d)^{2/(1+\gamma)}$ anyway, Mukhopadhyay (1980) defined a pilot size as:

$$m \equiv m(d) = \max\left\{m_0, \left\langle \left(\frac{a}{d}\right)^{2/(1+\gamma)} \right\rangle + 1\right\}. \tag{10.2.10}$$

where $m_0(\geq 2)$ is fixed. See also Mukhopadhyay (1982a).

One will record the pilot observations $X_1, \ldots, X_m$ and determine S_m^2. Next, in the spirit of Stein (1945,1949), we define a final sample size:

$$N \equiv N(d) = \max\left\{m, \left\langle \frac{a^2 S_m^2}{d^2} \right\rangle + 1\right\}. \tag{10.2.11}$$

We would sample the difference at the second stage if $N > m$.

Finally, we estimate μ by the fixed-width interval:

$$J_N \equiv \left[\overline{X}_N - d, \overline{X}_N + d\right],$$

where $\overline{X}_N$ is obtained from full dataset $N, X_1, \ldots, X_m, \ldots, X_N$.

From (10.2.10), notice that $m(d) \uparrow \infty$ as $d \downarrow 0$. Also, for small d, the pilot size $m \approx (a/d)^{2/(1+\gamma)}$ so that $m/C \to 0$ as $d \to 0$. That is, as d becomes small, m becomes large, but m tends to stay small compared with C.

In (10.2.11), note that we have not used $S_m^2 + m^{-\gamma}$ to estimate σ^2 unlike plugging $S_n^2 + n^{-1}$ to estimate σ^2 in the boundary condition of (10.2.4). We have used only S_m^2 to estimate σ^2 in (10.2.11) because m is expected to be large as d becomes small.

> The Chow-Robbins purely sequential confidence intervals from (10.2.5) and the Mukhopadhyay modified two-stage confidence intervals (10.2.10) - (10.2.11) have exactly the same asymptotic first-order properties.

Now, the following theorem states some crucial asymptotic first-order properties of this two-stage estimation methodology for all kinds of F, discrete or continuous..

Theorem 10.2.2 : *For the two-stage fixed-width confidence interval procedure (10.2.10) - (10.2.11), for all fixed F, μ, σ^2, γ and α, one has the following properties as $d \to 0$:*

(i) $N/C \to 1$ *in probability;*

(ii) $\mathrm{E}_F[N/C] \to 1$ [*Asymptotic Efficiency or Asymptotic First-Order Efficiency*];

(iii) $\mathrm{P}_F\left\{\overline{X}_N - d \le \mu \le \overline{X}_N + d\right\} \to 1 - \alpha$ [*Asymptotic Consistency*];

with $C \equiv a^2\sigma^2/d^2$ from (10.2.3).

Now, these results and those given in Theorem 10.2.1 look identical! In other words, the two-stage and the sequential strategies share the exact same asymptotic *first-order* properties. But, the two-stage strategy is operationally much more convenient because of sampling in at most two batches.

> Operationally, the Mukhopadhyay modified two-stage confidence intervals (10.2.10) - (10.2.11) wins over Chow-Robbins purely sequential confidence intervals from (10.2.5) hands down!

We sketch a proof of Theorem 10.2.2 in Section 10.7.3 along the line of Mukhopadhyay (1980,1982a).

Asymptotic second-order characteristics of Chow-Robbins purely sequential confidence intervals give this methodology a distinctive edge over the Mukhopadhyay two-stage strategy.

10.3 Minimum Risk Point Estimation for the Mean

Having recorded n independent observations $X_1, \ldots, X_n$ with a common but unknown distribution function F, we suppose that the loss in estimating μ by $\overline{X}_n$ is:

$$L_n(\mu, \overline{X}_n) = A\left(\overline{X}_n - \mu\right)^2 + cn \text{ with } A > 0, c > 0 \text{ known.} \tag{10.3.1}$$

As usual, the term cn represents cost for gathering n observations and $\left(\overline{X}_n - \mu\right)^2$ represents the loss due to estimation of μ by $\overline{X}_n$.

Now, the fixed-sample-size risk with (10.3.1) can be expressed as:

$$\begin{aligned} R_n(c) &\equiv \mathrm{E}_F\left[L_n(\mu, \overline{X}_n)\right] \\ &= A\mathrm{E}_F\left[\left(\overline{X}_n - \mu\right)^2\right] + cn = A\sigma^2 n^{-1} + cn. \end{aligned} \tag{10.3.2}$$

Our goal is to design estimation strategies that will minimize this risk for all F having $0 < \sigma^2 < \infty$.

Along the line of (8.2.2), $R_n(c)$ is approximately minimized when

$$n \text{ is the smallest integer } \geq \left(\frac{A}{c}\right)^{1/2} \sigma. \tag{10.3.3}$$

We note that

$$n^* \equiv n^*(c) = \left(\frac{A}{c}\right)^{1/2} \sigma, \tag{10.3.4}$$

may not be a (positive) integer, but we continue to refer to n^* as the optimal fixed sample size had σ^2 been known. The minimum fixed-sample-size risk is:

$$R_{n^*}(c) \equiv A\sigma^2 n^{*-1} + cn^* = 2cn^*. \tag{10.3.5}$$

Surely, the magnitude of the optimal fixed sample size n^* remains unknown since σ^2 is unknown. In fact, no fixed sample size procedure will minimize the risk (10.3.2) under the loss function (10.3.1), uniformly in F. Thus, we introduce some multistage estimation methodologies.

In Section 10.3.1, we introduce Mukhopadhyay's (1978) asymptotically *first-order* risk efficient purely sequential strategy. We believe that this is the first distribution-free minimum risk sequential point estimation methodology for μ. It was assumed a'priori that $\sigma^2 > \sigma_L^2$ with $\sigma_L^2(> 0)$ known. In Section 10.3.2, we discuss a new asymptotically *first-order* risk efficient two-stage estimation strategy along the lines of Mukhopadhyay and Duggan (1997) under the same assumption. In Section 10.3.3, we introduce Ghosh and Mukhopadhyay's (1979) fundamental purely sequential methodology assuming that F and σ^2 are completely unknown. The

Ghosh-Mukhopadhyay strategy is asymptotically *first-order* risk efficient under appropriate moment conditions. We believe that this is the first distribution-free minimum risk sequential point estimation strategy for μ without assuming any a'priori knowledge about σ^2.

10.3.1 Purely Sequential Procedure: Known Positive Lower Bound for Variance

If one knows that $\sigma^2 > \sigma_L^2$, then surely one has:

$$n^*(c) \equiv (A/c)^{1/2} \sigma > (A/c)^{1/2} \sigma_L,$$

with $\sigma_L^2 (> 0)$ known. Thus, one should start with a pilot of size m that is essentially $(A/c)^{1/2} \sigma_L$.

Hence, Mukhopadhyay (1978) defined a pilot sample size:

$$m \equiv m(c) = \max\left\{ m_0, \left\langle \left(\frac{A}{c}\right)^{1/2} \sigma_L \right\rangle + 1 \right\}. \tag{10.3.6}$$

where $m_0(\geq 2)$ is fixed. Now, we start with $X_1, \ldots, X_m$, a pilot of size m, and proceed with one additional observation X at-a-time. The stopping time is::

$$\begin{aligned} &N \equiv N(c) \text{ is the smallest integer } n(\geq m) \text{ for which} \\ &\qquad \text{we observe } n \geq \left(\frac{A}{c}\right)^{1/2} S_n. \end{aligned} \tag{10.3.7}$$

This sequential procedure is implemented in the same way we did in the case of (8.2.5). Again, $\mathrm{P}_F(N < \infty) = 1$, that is, sampling via (10.3.7) terminates w.p. 1. Upon termination, we estimate μ by $\overline{X}_N \equiv N^{-1} \sum_{i=1}^{N} X_i$, a sample mean obtained from full dataset $N, X_1, \ldots, X_m, \ldots, X_N$.

Notice that $m(c) \uparrow \infty$ as $c \downarrow 0$. Also, for small c, the pilot size $m \approx (A/c)^{1/2} \sigma_L$ so that $m/n^* \approx \sigma_L/\sigma$ which lies between zero and one. As c becomes small, it is clear that m becomes large at the same rate as n^*. But, in comparison with n^*, the pilot size m stays small.

The risk associated with $\overline{X}_N$ may be written as:

$$R_N(c) \equiv \mathrm{E}_F\left[L_N(\mu, \overline{X}_N)\right] = A\mathrm{E}_F\left[\left(\overline{X}_N - \mu\right)^2\right] + c\mathrm{E}_F[N]. \tag{10.3.8}$$

Unfortunately, we cannot simplify it in a form that involves moments of N alone. This happens since the random variables $I(N = n)$ and $\overline{X}_n$ are dependent for all $n \geq m$. But, along the lines of Robbins (1959), we may define the risk efficiency and regret measures:

$$\begin{aligned} &\textit{RiskEfficiency} : \text{RiskEff}(c) \equiv \frac{R_N(c)}{R_{n^*}}(c) \\ &\textit{Regret} : \text{Regret}(c) \equiv R_N(c) - R_{n^*}(c). \end{aligned} \tag{10.3.9}$$

In some ways, the present situation may resemble what we had in Sections 8.2 and 8.3. But, there is a major difference here. We remind readers that in Sections 8.2 and 8.3, we exploited many intricate properties of an Expo(θ) distribution. But, now the distribution function F remains unknown to us!

The following theorem states some crucial asymptotic *first-order* properties of this sequential estimation methodology.

Theorem 10.3.1 : *For a purely sequential minimum risk point estimation strategy (10.3.6) - (10.3.7), for all fixed $F, \mu, \sigma^2, \sigma_L^2$ and A one has the following properties as $c \to 0$:*

(i) $N/n^* \to 1$ *in probability;*

(ii) $\mathrm{E}_F[N/n^*] \to 1$ [*Asymptotic First-Order Efficiency*];

(iii) $\text{RiskEff}(c) \equiv \dfrac{R_N(c)}{R_{n^*}(c)} \to 1$ [*Asymptotic First-Order Risk Efficiency*];

with $n^* \equiv (A/c)^{1/2}\,\sigma$ from (10.3.4).

The properties (i) and (ii) would make us feel that on an average, the final sample size N from stopping rule (10.3.7) will hover around n^* for fixed F having $0 < \sigma_L^2 < \sigma^2 < \infty$. Part (ii) implies that this procedure is asymptotically *first-order* efficient. Part (iii) seems to imply that the achieved sequential risk may be expected to be nearly the optimal fixed-sample-size risk. It means that this strategy is asymptotically *first-order* risk efficient. These asymptotic properties hold when c is small, that is, when n^* is large. A sketchy proof of this theorem is given in Section 10.7.4.

10.3.2 Two-Stage Procedure: Known Positive Lower Bound for Variance

Gathering one observation at-a-time beyond a pilot and following along a stopping rule (10.3.7) may be operationally inconvenient. Hence, it is prudent to seek an appropriate two-stage estimation strategy when $\sigma^2 > \sigma_L^2$ and $\sigma_L^2 > 0$ is known. Again, we consider pilot size:

$$m \equiv m(c) = \max\left\{m_0, \left\langle \left(\frac{A}{c}\right)^{1/2} \sigma_L \right\rangle + 1\right\}. \tag{10.3.10}$$

as in (10.3.6). Then, we opt for a new two-stage strategy proposed in the light of Mukhopadhyay and Duggan (1997).

One will first record pilot observations $X_1, \ldots, X_m$ and determine S_m^2, the sample variance. Then, in the spirit of Stein (1945,1949) and Mukhopadhyay and Duggan (1997), one would define:

$$N \equiv N(c) = \max\left\{m, \left\langle \left(\frac{A}{c}\right)^{1/2} S_m \right\rangle + 1\right\}. \tag{10.3.11}$$

We would sample the difference at the second stage if $N > m$. Finally, we would estimate μ by $\overline{X}_N \equiv N^{-1}\sum_{i=1}^{N} X_i$, obtained from full dataset $N, X_1, \ldots, X_m, \ldots, X_N$.

The following theorem states some crucial asymptotic first-order properties of this two-stage estimation methodology.

Theorem 10.3.2 : *For a two-stage minimum risk point estimation procedure (10.3.10) - (10.3.11), for all fixed $F, \mu, \sigma^2, \sigma_L^2$ and A, one has the following properties as $c \to 0$:*

(i) $N/n^* \to 1$ *in probability;*

(ii) $\mathrm{E}_F[N/n^*] \to 1$ [*Asymptotic First-Order Efficiency*];

(iii) $\text{RiskEff}(c) \equiv \dfrac{R_N(c)}{R_{n^*}(c)} \to 1$ [*Asymptotic First-Order Risk Efficiency*];

with $n^* \equiv (A/c)^{1/2}\sigma$ from (10.3.4).

Now, these results and those given in Theorem 10.3.1 are clearly identical! In other words, two-stage and sequential methodologies share the exact same asymptotic first-order properties. But, the two-stage procedure is operationally much more convenient because of sampling at most two batches.

We sketch a proof of Theorem 10.3.2 in Section 10.7.5.

10.3.3 Purely Sequential Procedure: Arbitrary Variance

Ghosh and Mukhopadhyay (1979) developed the first purely sequential minimum risk methodology for point estimation of μ without any assumption about σ^2 like that in Sections 10.3.1-10.3.2. This is a general scenario where F is assumed unknown and $\sigma^2 \equiv \sigma^2(F)$ is assumed positive and finite. Recall previous notation $\overline{X}_n$ and S_n^2, for $n \geq 2$.

Let us focus on a loss function from (10.3.1) and an optimal fixed sample size $n^*(c) \equiv (A/c)^{1/2}\sigma$ from (10.3.4). Now, we introduce the fundamental Ghosh-Mukhopadhyay purely sequential estimation strategy.

One will start $X_1, \ldots, X_m$, a pilot of size $m(\geq 2)$, and proceed with one additional observation X at-a-time. The stopping time is:

$$N \equiv N(c) \text{ is the smallest integer } n(\geq m) \text{ for which we observe } n \geq \left(\frac{A}{c}\right)^{1/2}(S_n + n^{-\gamma}), \tag{10.3.12}$$

with some fixed number γ, $0 < \gamma < \infty$.

This sequential procedure is implemented in the same way we did in the case of (10.2.4). Again, $\mathrm{P}_F(N < \infty) = 1$. Upon termination, we estimate μ by $\overline{X}_N \equiv N^{-1}\sum_{i=1}^{N} X_i$ obtained from full dataset $N, X_1, \ldots, X_m, \ldots, X_N$.

We express the risk associated with $\overline{X}_N$ as:

$$R_N(c) \equiv \mathrm{E}_F\left[L_N(\mu, \overline{X}_N)\right] = A\mathrm{E}_F\left[\left(\overline{X}_N - \mu\right)^2\right] + c\mathrm{E}_F[N]. \tag{10.3.13}$$

Note that $R_N(c)$ cannot be simplified involving some moments of N alone. This happens because the random variables $I(N = n)$ and $\overline{X}_n$ are dependent for all $n \geq m$.

The Ghosh-Mukhopadhyay (1979) purely sequential non-parametric point estimation methodology (10.3.2) is the first of its kind. They proved the fundamental asymptotic risk efficiency property for their methodology.

Now, the following theorem states some of the crucial asymptotic *first-order* properties of this sequential estimation strategy.

Theorem 10.3.3 : *For the purely sequential minimum risk point estimation procedure (10.3.12), for all fixed $F, \mu, \sigma^2, \gamma, m$ and A, one has the following properties as $c \to 0$:*

(i) $N/n^* \to 1$ *in probability;*

(ii) $\mathrm{E}_F[N/n^*] \to 1$ [*Asymptotic First-Order Efficiency*];

(iii) $\text{RiskEff}(c) \equiv \dfrac{R_N(c)}{R_{n^*}(c)} \to 1$ [*Asymptotic First-Order Risk Efficiency*];

with $n^* \equiv (A/c)^{1/2}\sigma$ from (10.3.4). No additional moment condition is required for parts (i) and (ii) to hold. Part (iii) holds when $\mathrm{E}_F\left[|X|^8\right]$ is finite.

Again, part (iii) seems to imply that the sequential risk may be expected to be near the optimal fixed-sample-size risk. It means that this procedure is asymptotically *first-order* risk efficient. These asymptotic properties hold when c is small, that is when n^* is large. One is referred to Ghosh and Mukhopadhyay (1979) for technical details.

Note that the sequential methodology (10.3.12) may be alternatively stated as follows: Let

$$m \equiv m(c) = \max\left\{m_0, \left\langle \left(\frac{A}{c}\right)^{1/(2+2\gamma)} \right\rangle + 1\right\}. \tag{10.3.14}$$

where $m_0(\geq 2)$ is fixed. We start with $X_1, \ldots, X_m$, a pilot sample of size m, and then proceed with one additional observation X at-a-time. The stopping time is defined as follows:

$$\begin{aligned} &N\ equivN(c) \text{ is the smallest integer } n(\geq m) \text{ for which} \\ &\text{we observe } n \geq \left(\frac{A}{c}\right)^{1/2} \sqrt{\frac{1}{n}\textstyle\sum_{i=1}^{n}(X_i - \overline{X}_n)^2}. \end{aligned} \tag{10.3.15}$$

Upon termination, we estimate μ by $\overline{X}_N \equiv N^{-1}\sum_{i=1}^{N} X_i$ obtained from full dataset $N, X_1, \ldots, X_m, \ldots, X_N$.

The asymptotic *first-order* risk efficiency property of the purely sequential estimation strategy (10.3.14) - (10.3.15) was developed by Chow and

Yu (1981) under a less restrictive moment assumption. Chow and Yu (1981) showed that the finiteness of $\mathrm{E}_F\left[|X|^{4p}\right]$ for some $p > 1$ would suffice for the validity of Theorem 10.3.3, part (iii). Chow and Martinsek (1982) elegantly proved an appropriate boundedness property for the associated regret function when $\mathrm{E}_F\left[|X|^{6p}\right] < \infty$ for some $p > 1$.

Next, under an assumption that F is continuous and $\mathrm{E}_F\left[|X|^{6p}\right] < \infty$ for some $p > 1$, Martinsek (1983) gave the following *second-order* expansion of the regret function associated with the sequential strategy (10.3.14)-(10.3.15):

$$\begin{aligned} &\mathrm{Regret}(c) \equiv R_N(c) - R_{n^*}(c) \\ &= \left[2 - \tfrac{3}{4}\sigma^{-4}\mathrm{V}_F\left\{(X-\mu)^2\right\} + 2\sigma^{-6}\left\{\mathrm{E}_F\left((X-\mu)^3\right)\right\}^2\right] c + o(c), \end{aligned} \tag{10.3.16}$$

with $1 \le \gamma < \infty$. This *second-order* expansion of Regret(c) may be rewritten as follows:

$$\left[2 - \frac{3}{4}\left\{\mu_4\sigma^{-4} - 1\right\} + 2\mu_3^2\sigma^{-6}\right] c + o(c), \tag{10.3.17}$$

where $\mu_3 \equiv \mathrm{E}_F[(X-\mu)^3]$ and $\mu_4 \equiv \mathrm{E}_F[(X-\mu)^4]$. One may also refer to Sen (1981,1985), Martinsek (1988), Aras and Woodroofe (1993), and Ghosh et al. (1997, Chapter 9) for many technical details.

10.4 Bounded Length Confidence Interval for the Median

This bounded length confidence interval methodology is adapted from Geertsema (1970). Suppose that we have available a sequence $X_1, \ldots, X_n$, $\ldots$ of independent observations from a population with distribution function $F(x-\theta)$. We assume that $F(x)$ is symmetric around $x = 0$. That is, the parameter θ is a unique median for the distribution $F(x-\theta)$. We suppose that the parameter space for θ is the whole real line, $\mathbb{R}$.

(i) $F(x)$ is twice differentiable in a neighborhood $\mathcal{N}$ of zero and
(ii) $F''(x)$ is bounded inside $\mathcal{N}$.

Geertsema's (1970) distribution-free methodology was driven by a sign test for a null hypothesis, $H_0 : \theta = 0$. Geertsema considered the statistic $\sum_{i=1}^{n} I(X_i > 0)$, which is the number of positive X's among $X_1, \ldots, X_n$. Let us define two sequences of positive integers as follows:

$$\begin{aligned} b(n) &= \max\left\{1, \left\langle \tfrac{1}{2}\left(n - z_{\alpha/2}n^{1/2} - 1\right)\right\rangle\right\}, \text{ and} \\ a(n) &= n - b(n) + 1, \end{aligned} \tag{10.4.1}$$

where $z_{\alpha/2}$ is the upper $50\alpha\%$ point of a standard normal distribution. Here, $0 < \alpha < 1$ is preassigned.

We look at the order statistics $X_{n:1} \le X_{n:2} \le \ldots \le X_{n:n}$ obtained from $X_1, \ldots, X_n$ and focus on two special order statistics $X_{n:b(n)}, X_{n:a(n)}$ with

$b(n), a(n)$ from 10.4.1). Now, we estimate θ by a confidence interval:

$$J_n = \left(X_{n:b(n)}, X_{n:a(n)}\right). \tag{10.4.2}$$

The associated confidence coefficient $\mathrm{P}_F\left\{\theta \in J_n\right\}$ has the following limiting property:

$$\lim_{n\to\infty} \mathrm{P}_F\left\{X_{n:b(n)} < \theta < X_{n:a(n)}\right\} = 1-\alpha. \tag{10.4.3}$$

This follows from Bahadur's (1966) representation of quantiles in large samples.

10.4.1 Purely Sequential Procedure

Apart from an approximate preassigned confidence coefficient, we have another important requirement to satisfy. The width of the confidence interval J_n from (10.4.2) must be smaller or equal $2d$ where $d(>0)$ is preassigned. Clearly, the confidence coefficient ought to be approximately $1-\alpha$ too where $0<\alpha<1$ is preassigned.

Geertsema (1970) developed a purely sequential strategy. One will note that a similar bounded length confidence interval procedure was also introduced earlier by Farrell (1966).

We will start with pilot observations $X_1, \ldots, X_m$ of size m. We continue by gathering one additional observation X at-a-time according to the stopping time:

$$\begin{array}{c} N \equiv N(d) \text{ is the smallest integer } n(\geq m) \text{ for which} \\ \text{we observe } X_{n:a(n)} - X_{n:b(n)} \leq 2d. \end{array} \tag{10.4.4}$$

Geertsema (1970) showed that $\mathrm{P}_F(N<\infty)=1$ and $\mathrm{E}_F(N)<\infty$ for all fixed m, d, F and α. Hence, based on full dataset $N, X_1, \ldots, X_m, \ldots, X_N$, a natural bounded length interval

$$J_N = \left(X_{N:b(N)}, X_{N:a(N)}\right) \tag{10.4.5}$$

is constructed for the population median, θ.

The Geertsema (1970) purely sequential bounded length confidence intervals methodology (10.4.4) was based on the sign-test.

Now, the fundamental results of Geertsema (1970) are stated as follows for all kinds of F satisfying the standing assumptions.

Theorem 10.4.1 : *Consider a purely sequential bounded length confidence interval procedure (10.4.4) - (10.4.5) for θ. Then, for all fixed (i) F satisfying the standing assumptions and (ii) θ, m and α, one has the following properties as $d \to 0$:*

(i) $N/C^* \to 1$ *in probability;*

(ii) $\mathrm{E}_F[N/C^*] \to 1$ [*Asymptotic First-Order Efficiency*];

(iii) $\mathrm{P}_F\left\{X_{N:b(N)} < \theta < X_{N:a(N)}\right\} \to 1-\alpha$ [*Asymptotic Consistency*];

where $C^* = z^2_{\alpha/2}/\left\{4d^2 f^2(0)\right\}$.

For details, one should refer to Geertsema's (1970) original paper. For a broader review, one may refer to Ghosh et al. (1997, Chapter 10).

The properties (i) and (ii) in Theorem 3.4.1 make us feel comfortable knowing that on an average, the final sample size N will hover around C^* for all fixed distributions F satisfying the standing assumptions. Part (ii) also implies that this procedure is *asymptotically first-order efficient.* Part (iii) indicates that on an average, sequential confidence coefficient will hover around target, $1-\alpha$. That is, this strategy is *asymptotically consistent.* These asymptotic properties hold when d is small, that is when C^* is large.

10.5 Data Analyses and Conclusions

The methodologies described in this chapter can be implemented by program **Seq10.exe**. The protocol for executing this program is similar to programs described previously. The program is setup for computations using either live data or simulations.

An example of screen input is shown in Figure 10.5.1. One could use either a sequential procedure (10.2.1) or a two-stage procedure (10.2.10) - (10.2.11) by selecting program number: 1 or 2 respectively. Figure 10.5.1 shows input for simulations of a 90% sequential bounded length confidence interval for the mean μ of a mixture of normal and double exponential distributions. Simulations were replicated 5000 times with half-width 0.1 $(= d)$. The output from simulations are shown in Figure 10.5.2.

The fixed-width confidence intervals from Section 10.2 and minimum risk point estimators from Section 10.3 may be simulated using one from a list of nine probability distributions shown in Figure 10.5.1. One may note, however, that the choice of distributions, mixture or not, may vary in one available methodology to the next.

Details of these distributions are given in an Appendix. In particular, a gamma random variate is generated using a shape parameter $\alpha = \mu^2/\sigma^2$ and a scale parameter $\beta = \sigma^2/\mu$ where $\mu(>0)$ and σ^2 are respectively the mean and variance of the gamma distribution.

A random variate with a mixture of two selected distributions is generated using the following expression of the distribution function (d.f.) with associated mixture proportion $p, 0 < p < 1$. That is,

$$\text{Mixture D.F.} = p(\text{First D.F.}) + (1-p)(\text{Second D.F.}).$$

One has a number of choices for a mixture distribution and a mixing proportion p. Mixtures of normal, double exponential, and gamma distributions may be used within the program **Seq10.exe** to simulate a bounded length confidence interval from Section 10.4 for a median.

Theorem 10.5.1 : *Suppose that we define:*

$$f(x) = pf_1(x) + (1-p)f_2(x), 0 < p < 1 \tag{10.5.1}$$

where $f_1(.), f_2(.)$ are both p.m.f.'s or p.d.f.'s with identical sample spaces $\mathcal{X}$. Also, suppose that $f_i(.)$ has its mean and variance $\mu_i, \sigma_i^2, i = 1, 2$. Then,

(i) *$f(x)$ is a p.m.f. or p.d.f. with its sample spaces $\mathcal{X}$;*
(ii) *$f(x)$ has its mean, $\mu \equiv \mu(F) = p\mu_1 + (1-p)\mu_2$;*
(iii) *$f(x)$ has its variance,*

$$\sigma^2 \equiv \sigma^2(F) = p\left(\sigma_1^2 + \mu_1^2\right) + (1-p)\left(\sigma_2^2 + \mu_2^2\right) - \mu^2.$$

This is left as an exercise. As a special case, suppose that we have a mixture distribution (10.5.1) where the mixing distributions have the same mean, that is $\mu_1 = \mu_2 = \nu$. Then, part (i) of Example 10.5.1 implies that the mean of the mixture distribution is also ν. Also, part (ii) of Example 10.5.1 implies that the variance of the mixture distribution is $\sigma^2 = p\left(\sigma_1^2 + \nu^2\right) + (1-p)\left(\sigma_2^2 + \nu^2\right) - \nu^2 = p\sigma_1^2 + (1-p)\sigma_2^2$.

Example 10.5.1 : Suppose that we have a mixture of distribution 1 that is normal with $\mu_1 = 10$, $\sigma_1^2 = 16$ and distribution 2 that is double exponential with $\mu_2 = 10$, $\sigma_2^2 = 4$ with $p = 0.4$. To make things very specific, we write:

$$f_1(x) = \frac{1}{\sigma_1\sqrt{2\pi}} \exp\left(-\frac{(x-10)^2}{2\sigma_1^2}\right),$$
$$f_2(x) = \frac{1}{\sigma_2\sqrt{2}} \exp\left(-\frac{\sqrt{2}\,|x-10|}{\sigma_2}\right),$$
$$f(x) = 0.4f_1(x) + 0.6f_2(x), -\infty < x < \infty.$$

That is, $\mu \equiv \mu(F) = 10$ and $\sigma^2 \equiv \sigma^2(F) = 0.4 \times 16 + 0.6 \times 4 = 8.8$. For a 90% Chow-Robbins confidence interval from Section 10.2.1 with fixed half-width 0.2 ($= d$), the optimal fixed sample size C from (10.2.3) reduces to

$$(1.645)^2 \times 8.8/(0.2)^2 = 595.33. \tag{10.5.2}$$

Example 10.5.2 : (Example 10.5.1 Continued): Suppose that we have the same mixture distribution as in Example 10.5.1. Clearly, the mixture p.d.f. f is symmetric around $x \equiv \theta = 10$ and f satisfies all standing assumptions made in Section 10.4. Now, let us pretend using the Geertsema procedure (10.4.4) - (10.4.5). For a 90% Geertsema confidence interval from Section 10.4.1 with fixed half-width 0.2 ($= d$), the optimal fixed sample size from Theorem 10.4.1 will reduce to:

$$C^* = (1.645)^2/4 \times (0.2)^2 \times \left(0.4 \times \tfrac{1}{4\sqrt{2\pi}} + 0.6 \times \tfrac{1}{2\sqrt{2}}\right)^2 = 266.26. \tag{10.5.3}$$

```
*********************************************************
* Number         Program                                *
*  1   Fixed-Width Confidence Interval for the Mean     *
*  2   Minimum Risk Point Estimation for the Mean       *
*  3   Bounded Length Confidence Interval for the Median*
*  0   Exit                                             *
*********************************************************
  Enter the program number: 1

  Enter 1 for Purely Sequential or
        2 for Two-Stage Procedure
  Enter 1 or 2: 1

  Do you want to save the results in a file(Y or N)?Y
  Note that your default outfile name is "SEQ10.OUT".
  Do you want to change the outfile name (Y or N)? n

  Enter the following Preassigned Values
      alpha (for 100(1-alpha)% confidence interval) = .1
      d (half the width of the confidence interval) = .2

  Enter the initial sample size, m: 10
  Do you want to change the input parameters?[Y or N] n
  Do you want to simulate data or input real data?
  Type S (Simulation) or D (Real Data): S
  Do you want to store simulated sample sizes (Y or N)? N

  Number of simulation replications? 5000
  Enter a positive integer(<10) to initialize
            the random number sequence: 2

  Select one of the following distributions:
  ****************************************************
  *     ID#  DISTRIBUTION     Parameter1 Parameter2   *
  *      1   Normal            mean        variance   *
  *      2   Gamma             mean        variance   *
  *      3   Exponential       mean                   *
  *      4   Poisson           mean                   *
  *      5   Bernoulli         mean=P(Success)        *
  *      6   Negative Binomial mean               k   *
  *      7   Mixture of Two Normal Distributions      *
  *      8   Mixture of Normal and Double Exponential *
  *      9   Mixture of Two Gamma Distributions       *
  ****************************************************
  Input distribution  ID#: 8

  The common mean of the two distributions = 10
  Variance of the first normal distribution = 16
  Variance of the double exponential distribution = 4

  p=probability of the mixture distribution,
  p*(Distribution 1) + (1-p)*(Distribution 2)
  Enter p = 0.4
```

Figure 10.5.1. *Screen input for **Seq10.exe**.*

But, when we simulate the Geertsema procedure (10.4.4) - (10.4.5), we do not use any specific knowledge about f or its two components f_1, f_2.

```
Fixed-Width Confidence Interval for the Mean
     Using Purely Sequential Procedure.
     ----------------------------------

Data for the simulation were generated from:
Normal and double exponential mixture distribution with
      Common mean =    10.00
      Variance 1  =    16.00
      Variance 2  =     4.00
      Probability =    0.400

Simulation study using following input parameters:
                 alpha =    0.100
                     d =    0.200
Initial sample size, m =     10
Number of simulations  =   5000

Optimal sample size (C or C* or n*)  =     595.33
Average final sample size (n_bar)    =     593.29
Std. dev. of n (s_n)                 =      47.20
Minimum sample size (n_min)          =     411
Maximum sample size (n_max)          =     737
Coverage probability                 =     0.9038
```

Figure 10.5.2. *Simulation results of a 90% confidence interval for the mean of a normal and double exponential mixture distribution using purely sequential procedure.*

In Figure 10.5.2, the estimated $\overline{n}(= 593.29)$ value was calculated from 5000 replications and it came very close to $C = 595.33$. We verified this number in (10.5.2). Note that the C-value is also shown in Figure 10.5.2. Admittedly, the Chow-Robbins sequential procedure (10.2.4) has undersampled slightly, but the estimated coverage probability $\overline{p}$ has held up nicely at 0.9038, which is close to the target 0.90.

Observe that the pilot size m for purely sequential procedures in Sections 10.2.1 and 10.4 is an input set by the experimenter. But, in the case of the two procedures from Sections 10.3.1 and 10.3.2, one inputs a value for the known lower bound for variance, σ_L^2. A pilot size m is calculated accordingly.

The two-stage procedure from Section 10.2.2 and the sequential procedure from Section 10.3.3 require one to input a value for $\gamma(> 0)$. However, the second-order term for Regret(c) seen from (10.3.16) - (10.3.17) for the sequential procedure is valid if $\gamma \geq 1$. Thus, the program **Seq10.exe** will reject if one inputs a value of γ that is smaller than 1 for this procedure.

10.6 Other Selected Multistage Procedures

In the mid 1960's and immediately afterward, we saw many contributions in designing sequential nonparametric strategies for a variety of inference problems. The literature on sequential nonparametric strategies for estimation and testing problems grew fast at the time. A big chunk of this research was largely influenced by the pathbreaking paper of Chow and Robbins (1965).

Gleser (1965), Srivastava (1967,1971), Khan (1969), Nadas (1969), Ghosh and Sen (1971,1972,1973), Sproule (1969,1974) immediately come to mind. Gleser (1965) and Srivastava (1967,1971), among others, dealt with regression problems. Khan (1969) extended Chow-Robbins theory for sequential confidence intervals based on MLEs. Chapter 9 briefly touched upon that topic. Ghosh-Sen papers dealt with regression and simultaneous confidence estimation problems. Sproule (1969,1974) extended the Chow-Robbins theory to U-statistics which by itself has led to important research later.

More recent contributions are included in Ghosh (1980), Ghosh and Dasgupta (1980), Mukhopadhyay and Moreno (1991a,b), Mukhopadhyay and Vik (1985,1988a,b), Vik and Mukhopadhyay (1988), Ghosh et al. (1997, Chapters 9 and 10) and elsewhere.

Many papers from that period proved essential analogs of Theorem 10.2.1 for all sorts of statistical problems. By all means those proofs were not easy at all, but the results were certainly predictable. Mukhopadhyay's (1980) paper was different since it broke some of that earlier trend. More importantly, it questioned the superiority and importance of a sequential strategy over a properly designed two-stage methodology. After all there is no visible difference between Theorems 10.2.1-10.2.2! This inquiry led Ghosh and Mukhopadhyay (1981) to come up with the concept of *asymptotic second-order efficiency.*

Geertsema's (1970) fundamental methodological contribution had its own followers and there were many. Sen's (1981) invariance principles have allowed generations of researchers to handle very involved nonparametric sequential problems.

The area of nonparametric minimum risk point estimation of a mean took off from the papers of Mukhopadhyay (1978) and Ghosh and Mukhopadhyay (1979). These were followed by excellent contributions by others including Chow and Yu (1981), Chow and Martinsek (1982), and Martinsek (1983). Sen and Ghosh (1981) elegantly extended this line of research for estimating the mean of a U-statistic. Ghosh and Sen (1984) expanded their 1981 results for estimating the mean of a generalized U-statistic.

The area of nonparametric sequential estimation has been a very vast area of vigorous growth. The books by Sen (1981,1985) and Ghosh et al. (1997) will clearly testify to this. In this section, we have merely given a snapshot of some of the selected developments.

10.7 Some Selected Derivations

10.7.1 Proof of (10.2.7)

For brevity, we assume that F is continuous. In this case, $S_n^2 + n^{-1}$ in the boundary condition of (10.2.4) is simply replaced by S_n^2. Now, from (10.2.4), we observe a basic inequality:

$$\frac{a^2}{d^2}S_N^2 \le N \le m + \frac{a^2}{d^2}S_{N-1}^2 \text{ w.p. } 1. \tag{10.7.1}$$

One can easily verify that

$$\sum_{i=1}^{k}(x_i-\mu)^2 = \sum_{i=1}^{k}(x_i-\overline{x}_k)^2 + k(\overline{x}_k-\mu)^2 \tag{10.7.2}$$

for any set of numbers $x_1,\ldots,x_k$ and a positive integer k. So, we may express the right-hand side of (10.7.1) as:

$$\begin{aligned}(N-m) &\le \frac{a^2}{d^2(N-2)}\textstyle\sum_{i=1}^{N-1}(X_i-\overline{X}_{N-1})^2\\ \Rightarrow (N-m)(N-2) &\le \frac{a^2}{d^2}\textstyle\sum_{i=1}^{N-1}(X_i-\mu)^2 \le \frac{a^2}{d^2}\sum_{i=1}^{N}(X_i-\mu)^2,\end{aligned} \tag{10.7.3}$$

w.p. 1. But, we have:

$$(N-m)(N-2) \ge N^2-(m+2)N \text{ and } \mathrm{E}_F[N^2] \ge \mathrm{E}_F^2[N]. \tag{10.7.4}$$

Assume for the time being that $\mathrm{E}_F[N] < \infty$. Thus, we use Wald's first equation (Theorem 3.5.4) to claim:

$$\mathrm{E}_F\left[\sum_{i=1}^{N}(X_i-\mu)^2\right] = \mathrm{E}_F[N]\mathrm{E}_F\left[(X_1-\mu)^2\right] = \mathrm{E}_F[N]\sigma^2. \tag{10.7.5}$$

Now, we take expectations throughout the last step in (10.7.3) and combine with (10.7.4) - (10.7.5) to write:

$$\begin{aligned}\mathrm{E}_F^2[N]-(m+2)\mathrm{E}_F[N] \le \frac{a^2}{d^2}\sigma^2\mathrm{E}_F[N] = C\mathrm{E}_F[N] \text{ since } C = \frac{a^2\sigma^2}{d^2}\\ \Rightarrow \mathrm{E}_F[N]-(m+2) \le C,\end{aligned} \tag{10.7.6}$$

that is, $\mathrm{E}_F[N] \le C+m+2$.

Now, if $\mathrm{E}_F[N]$ is not finite, then one may use a truncated stopping variable

$$N_k = \min\{k,N\}, k=1,2,\ldots$$

and claim immediately that $\mathrm{E}_F[N_k] \le C+m+2$ for each fixed k.

Then, an application of the monotone convergence theorem will complete the proof.

This also shows that $\mathrm{E}_F[N]$ is indeed finite. ■

10.7.2 Proof of Theorem 10.2.1

Again, for brevity, we assume that F is continuous. In this case, $S_n^2 + n^{-1}$ in the boundary condition of (10.2.4) is simply replaced by S_n^2.

Suppose that we choose $0 < d_1 < d_2 < \infty$. Observe that $N(d_2)$ is the first integer $n(\geq m)$ such that $n \geq a^2 S_n^2 d_2^{-2}$ w.p. 1. Also, $N(d_1)$ is the first integer $n(\geq m)$ such that $n \geq a^2 S_n^2 d_1^{-2}$ w.p. 1, but $a^2 S_n^2 d_1^{-2} > a^2 S_n^2 d_2^{-2}$. So, $N(d_1)$ satisfies the boundary condition required by $N(d_2)$. Hence, $N(d_1)$ cannot be smaller than $N(d_2)$ w.p. 1. In other words, $N(d)$ is a non-increasing function of $d > 0$. Actually, $N(d) \uparrow \infty$ w.p. 1 as $d \downarrow 0$.

Proof of Part (i):

Recall basic inequality from (10.7.1):

$$\frac{a^2}{d^2} S_N^2 \leq N \leq m + \frac{a^2}{d^2} S_{N-1}^2 \text{ w.p. } 1. \tag{10.7.7}$$

We have $\lim_{d\to 0} S_N^2 = \lim_{d\to 0} S_{N-1}^2 = \sigma^2$ since $\lim_{d\to 0} N = \infty$ w.p. 1. Now, we divide throughout (10.7.7) by C and take limits as $d \to 0$ to write:

$$1 = \lim_{d\to 0} \frac{S_N^2}{\sigma^2} \leq \lim_{d\to 0} \frac{N}{C} \leq \lim_{d\to 0} \frac{m}{C} + \lim_{d\to 0} \frac{S_{N-1}^2}{\sigma^2} = \lim_{d\to 0} \frac{m}{C} + 1. \tag{10.7.8}$$

But, $\lim_{d\to 0} mC^{-1} = 0$ since m is fixed and $C \to \infty$ as $d \to 0$. So, part (i) follows immediately from (10.7.8). ■

Proof of Part (ii):

This proof may be constructed along the lines of Section 8.6.2. Details are left out as exercise. ■

Proof of Part (iii):

Recall the confidence interval $J_N \equiv \left[\overline{X}_N - d, \overline{X}_N + d\right]$ for μ. Now, rehashing (10.2.2), we write:

$$\mathrm{P}_F\{\mu \in J_N\} = \mathrm{P}_F\left\{\frac{\sqrt{C}\left|\overline{X}_N - \mu\right|}{\sigma} \leq a\right\}. \tag{10.7.9}$$

At this point, we use part (i), Anscombe's CLT from (10.2.8), and Slutsky's theorem to claim:

$$\lim_{d\to 0} \mathrm{P}_F\left\{\sqrt{C}\left|\overline{X}_N - \mu\right|/\sigma \leq a\right\} = 1 - \alpha.$$

Now, the proof is complete. ■

10.7.3 Proof of Theorem 10.2.2

For simplicity, let us pretend that $d \to 0$ in such a way that $(a/d)^{2/(1+\gamma)}$ may be regarded an integer, so that $m \equiv m(d) = (a/d)^{2/(1+\gamma)}$.

Proof of Part (i):

Now, we have basic inequality:

$$\frac{a^2}{d^2}S_m^2 \le N \le m + \frac{a^2}{d^2}S_m^2 \text{ w.p. } 1. \tag{10.7.10}$$

Clearly, $m(d) \to \infty$ and $m(d)/C \to 0$ as $d \to 0$. Thus, we have $\lim_{d\to 0} S_m^2 = \sigma^2$ w.p. 1. Now, we divide throughout (10.7.10) by C and take limits as $d \to 0$ to write w.p. 1:

$$1 = \lim_{d\to 0} \frac{S_m^2}{\sigma^2} \le \lim_{d\to 0} \frac{N}{C} \le \lim_{d\to 0} \frac{m}{C} + \lim_{d\to 0} \frac{S_m^2}{\sigma^2} = 0 + 1 = 1. \tag{10.7.11}$$

So, part (i) follows immediately from (10.7.11). ■

Proof of Part (ii):

We take expectations throughout (10.7.10) and write:

$$\frac{a^2\sigma^2}{d^2} \le \mathrm{E}_F[N] \le m + \frac{a^2\sigma^2}{d^2}, \tag{10.7.12}$$

Then, we divide throughout (10.7.12) by C and take limits as $d \to 0$ to write:

$$1 \le \lim_{d\to 0} \mathrm{E}_F\left[\frac{N}{C}\right] \le \lim_{d\to 0} \frac{m}{C} + 1 = 0 + 1 = 1.$$

Part (ii) follows immediately. ■

Proof of Part (iii):

In view of part (i), this proof remains same as our proof of Theorem 10.2.1, part (iii). ■

10.7.4 Proof of Theorem 10.3.1

For small c, we expect $m(c) \approx (A/c)^{1/2}\,\sigma_L$. For simplicity, we pretend that c converges to zero in such a way that $(A/c)^{1/2}\,\sigma_L$ may be regarded an integer so that $m \equiv m(c) = (A/c)^{1/2}\,\sigma_L$.

Proof of Part (i):

Now, we have a basic inequality:

$$\left(\frac{A}{c}\right)^{1/2} S_N \le N \le mI(N = m) + \left(\frac{A}{c}\right)^{1/2} S_{N-1} \text{ w.p. } 1. \tag{10.7.13}$$

Clearly, $m(c) \to \infty$ and $m(c)/n^* \to \sigma_L/\sigma$ as $c \to 0$. Also, we have $\lim_{d\to 0} S_N = \lim_{d\to 0} S_{N-1} = \sigma$ w.p. 1. Now, divide throughout (10.7.10) by n^*, take limits as $d \to 0$, and write w.p. 1:

$$1 = \lim_{c\to 0} \frac{S_N}{\sigma} \le \lim_{c\to 0} \frac{N}{n^*} \le \lim_{c\to 0} \frac{m}{n^*} I(N=m) + \lim_{c\to 0} \frac{S_{N-1}}{\sigma}$$
$$= \lim_{c\to 0} \frac{m}{n^*} I(N=m) + 1. \tag{10.7.14}$$

So, part (i) will follow immediately from (10.7.14) if we show:

$$\lim_{c\to 0} \frac{m}{n^*} I(N=m) = 0 \text{ in probability.} \tag{10.7.15}$$

But, for any fixed $\varepsilon > 0$, we may write:

$$\begin{aligned} &\mathrm{P}_F \{I(N=m) > \varepsilon\} \\ &\le \varepsilon^{-1} \mathrm{E}_F [I(N=m)] \text{ by Markov inequality} \\ &= \varepsilon^{-1} \mathrm{P}_F (N=m) \\ &\le \varepsilon^{-1} \mathrm{P}_F \left\{ \left(\frac{A}{c}\right)^{1/2} \sigma_L \ge \left(\frac{A}{c}\right)^{1/2} S_m \right\} \\ &= \varepsilon^{-1} \mathrm{P}_F (S_m \le \sigma_L). \end{aligned} \tag{10.7.16}$$

Next, $\mathrm{P}_F (S_m \le \sigma_L)$ is expressed as

$$\begin{aligned} &\mathrm{P}_F \{S_m^2 \le \sigma_L^2\} \\ &\le \mathrm{P}_F \{|S_m^2 - \sigma^2| \ge \sigma^2 - \sigma_L^2\} \text{ since } \sigma_L^2 - \sigma^2 < 0 \\ &\le (\sigma^2 - \sigma_L^2)^{-2} \mathrm{E}_F [S_m^2 - \sigma^2]^2 \text{ by Markov inequality} \\ &= O(m^{-1}), \text{ using Lemma 10.2.3 from} \\ &\qquad \text{Ghosh et al. (1997, pp. 275-276).} \end{aligned} \tag{10.7.17}$$

Now, combining (10.7.16) - (10.7.17), the claim made in (10.7.15) is fully justified. ∎

Proof of Part (ii):

Let $U^2 = \sup_{k\ge 1} k^{-1} \sum_{i=1}^{k} (X_i - \mu)^2$. Now, from the right hand side of (10.7.13), we can write w.p. 1:

$$\begin{aligned} &c^{1/2} N \le c^{1/2} m + A^{1/2} \left\{ (N-2)^{-1} \textstyle\sum_{i=1}^{N-1} (X_i - \mu)^2 \right\}^{1/2} \\ &\Rightarrow c^{1/2} N \le A^{1/2} \sigma_L + (3A)^{1/2} \left\{ (N-1)^{-1} \textstyle\sum_{i=1}^{N-1} (X_i - \mu)^2 \right\}^{1/2} \\ &\Rightarrow c^{1/2} N \le A^{1/2} \sigma_L + (3A)^{1/2} U, \end{aligned} \tag{10.7.18}$$

for sufficiently small c so that $m \equiv m(c) \ge 3$ and $(N-1)(N-2)^{-1} \le 3$.

From the last step in (10.7.18), we note that the right-hand side does not involve c. But, U is integrable. Thus, dominated convergence theorem combined with part (i) leads to part (ii).

The last step from (10.7.18) also proves that $\mathrm{E}_F[N]$ is indeed finite. ■

Proof of Part (iii):

Let us denote $Y_{N,c} \equiv \frac{1}{n^*}\left(\sum_{i=1}^N X_i - N\mu\right)^2$. We combine (10.3.8) - (10.3.9) and $R_{n^*}(c)$ to express RiskEff(c) as:

$$\begin{aligned}
&R_N(c)/R_{n^*}(c)\\
&= \frac{A}{2cn^*}\mathrm{E}_F\left[\frac{1}{N^2}\left(\sum_{i=1}^N X_i - N\mu\right)^2\right] + \frac{1}{2}\mathrm{E}_F\left[\frac{N}{n^*}\right]\\
&= \frac{n^*}{\sigma^2}\mathrm{E}_F\left[\frac{1}{N^2}\left(\sum_{i=1}^N X_i - N\mu\right)^2\right] + \frac{1}{2}\mathrm{E}_F\left[\frac{N}{n^*}\right]\\
&= \frac{1}{2\sigma^2}\mathrm{E}_F\left[Y_{N,c}\right] + \frac{1}{2\sigma^2}\mathrm{E}_F\left[\left(\frac{n^{*2}}{N^2} - 1\right)Y_{N,c}\right] + \frac{1}{2}\mathrm{E}_F\left[\frac{N}{n^*}\right]\\
&= \mathrm{E}_F\left[I_1\right] + \mathrm{E}_F\left[I_2\right] + \mathrm{E}_F\left[I_3\right], \text{ say.}
\end{aligned} \tag{10.7.19}$$

From part (ii), we clearly have:

$$\lim_{c\to 0}\mathrm{E}_F\left[I_3\right] = \lim_{c\to 0}\frac{1}{2}\mathrm{E}_F\left[N/n^*\right] = \frac{1}{2}. \tag{10.7.20}$$

Next, use Wald's second equation (Theorem 3.5.5) to claim:

$$\mathrm{E}_F\left[Y_{N,c}\right] = \sigma^2\mathrm{E}_F[N/n^*] \text{ which converges to } \sigma^2 \text{ as } c \to 0,$$

via part (ii). Thus, one has:

$$\lim_{c\to 0}\mathrm{E}_F\left[I_1\right] = \lim_{c\to 0}\frac{1}{2\sigma^2}\mathrm{E}_F\left[Y_{N,c}\right] = \frac{1}{2}. \tag{10.7.21}$$

Next, in view of Anscombe's CLT from (10.2.8) and Slutsky's theorem, observe that

$$\left(\sum_{i=1}^N X_i - N\mu\right)/n^{*1/2} \text{ converges to } N(0,\sigma^2) \text{ in distribution as } c \to 0,$$

$$\Rightarrow Y_{N,c} \text{ converges to } \sigma^2\chi_1^2 \text{ in distribution as } c \to 0.$$

We have argued in (10.7.21) that $\mathrm{E}_F\left[Y_{N,c}\right]$ converges to the expected value of the limiting distribution of $Y_{N,c}$, namely $\sigma^2\chi_1^2$. This implies:

$Y_{N,c}$ is uniformly integrable!

Recall that $N \geq (A/c)^{1/2}\sigma_L$ so that $N/n^* \geq \sigma_L/\sigma$ w.p. 1. Thus, we have $n^{*2}/N^2 \leq \sigma^2/\sigma_L^2$ and hence, we have w.p. 1:

$$\begin{aligned}
&-\frac{1}{2\sigma^2}Y_{N,c} \leq I_2 \leq \frac{1}{2\sigma^2}\left(\frac{\sigma^2}{\sigma_L^2} - 1\right)Y_{N,c}\\
&\Rightarrow |I_2| \leq \max\left\{\frac{1}{2\sigma^2}, \frac{1}{2\sigma^2}\left(\frac{\sigma^2}{\sigma_L^2} - 1\right)\right\}Y_{N,c}
\end{aligned} \tag{10.7.22}$$

But, since $Y_{N,c}$ is uniformly integrable, the last step in (10.7.22) implies that $|I_2|$ is uniformly integrable too.

In view of part (i), however, we know that I_2 converges to zero in probability as $c \to 0$. Thus, we claim:

$$\lim_{c\to 0} \mathrm{E}_F\left[I_3\right] = 0. \tag{10.7.23}$$

The proof is complete once we combine (10.7.19) - (10.7.21) and (10.7.23). ■

10.7.5 Proof of Theorem 10.3.2

For small c, we expect $m(c) \approx (A/c)^{1/2}\,\sigma_L$. Thus, for simplicity, we again assume that c converges to zero in such a way that $(A/c)^{1/2}\,\sigma_L$ may be regarded an integer and $m \equiv m(c) = (A/c)^{1/2}\,\sigma_L$.

Proof of Part (i):

First, we use basic inequality:

$$\left(\frac{A}{c}\right)^{1/2} S_m \le N \le mI(N=m) + \left(\frac{A}{c}\right)^{1/2} S_m \text{ w.p. 1.} \tag{10.7.24}$$

Clearly, $m(c) \to \infty$ and $m(c)/n^* \to \sigma_L/\sigma$ as $c \to 0$. Also, we have $\lim_{c\to 0} S_m = \sigma$ w.p. 1. Now, we divide throughout (10.7.24) by n^* and take limits as $c \to 0$ to write w.p. 1:

$$\begin{aligned} 1 = \lim\nolimits_{c\to 0} \frac{S_m}{\sigma} \le \lim\nolimits_{c\to 0} \frac{N}{n^*} &\le \lim\nolimits_{c\to 0} \frac{m}{n^*} I(N=m) + \lim\nolimits_{c\to 0} \frac{S_m}{\sigma} \\ &= \lim\nolimits_{c\to 0} \frac{m}{n^*} I(N=m) + 1. \end{aligned} \tag{10.7.25}$$

But, as in the case of (10.7.15), one can again claim that

$$\lim_{c\to 0} \frac{m}{n^*} I(N=m) = 0 \text{ in probability.}$$

So, part (i) follows immediately from (10.7.25). ■

Proof of Part (ii):

In view of part (i) and Fatou's lemma, we have:

$$\liminf\nolimits_{c\to 0} \mathrm{E}_F\left[N/n^*\right] \ge 1. \tag{10.7.26}$$

Next, we rewrite the right-hand side of from (10.7.24) as:

$$
\begin{aligned}
N^2 &\le m^2 I(N=m) + \left(\frac{A}{c}\right) S_m^2 \text{ w.p. } 1 \\
\Rightarrow \mathrm{E}_F\left[N^2\right] &\le m^2 \mathrm{P}_F\{N=m\} + \left(\frac{A}{c}\right) \mathrm{E}_F\left[S_m^2\right] \\
\Rightarrow \mathrm{E}_F^2[N] &\le m^2 \mathrm{P}_F\{N=m\} + \left(\frac{A}{c}\right)\sigma^2 \\
\Rightarrow \mathrm{E}_F^2\left[\frac{N}{n^*}\right] &\le \frac{m^2}{n^{*2}} \mathrm{P}_F\{N=m\} + 1 \\
\Rightarrow \limsup_{c\to 0} \mathrm{E}_F\left[N/n^*\right] &\le 1, \text{ in view of (10.7.17).}
\end{aligned}
\tag{10.7.27}
$$

Now, one combines (10.7.26) with the last step in (10.7.27), and the proof of part (ii) is complete.

This derivation also shows that $\mathrm{E}_F[N]$ is indeed finite. ■

Proof of Part (iii):

This proof remains same as in the case of Theorem 10.3.1, part (iii). We omit further details. ■

10.8 Exercises

Exercise 10.2.1: Suppose that $X_1, X_2, \ldots$ are independent exponentially distributed with a common p.d.f.

$$
f(x;\theta) = \begin{cases} \dfrac{1}{\theta}\exp(-x/\theta) & \text{if } x > 0 \\ 0 & \text{otherwise} \end{cases}
$$

where $\theta > 0$ is an unknown parameter. Consider a fixed-width confidence interval $J_n \equiv \left[\overline{X}_n - d, \overline{X}_n + d\right]$ from (10.2.1) for θ with a preassigned confidence coefficient $1-\alpha$, $0<\alpha<1, d>0$.

(i) Obtain the expression of C from (10.2.3) in this special case.

(ii) Propose Chow-Robbins stopping rule (10.2.4) and the associated interval $J_N \equiv \left[\overline{X}_N - d, \overline{X}_N + d\right]$ for the mean.

(iii) Prove that $\mathrm{P}_\theta\{\theta \in J_N\} \ge 1 - d^{-2}\mathrm{E}_\theta\left[\left(\overline{X}_N - \theta\right)^2\right]$.

(iv) Use part (iii) to investigate the achieved confidence coefficient in this special case. What can be said about $\lim_{d\to 0} \mathrm{P}_\theta\{\theta \in J_N\}$?

Exercise 10.2.2: Let $X_1, X_2, \ldots$ be independent exponentially distributed with a common p.d.f.

$$
f(x;\theta) = \begin{cases} \dfrac{1}{\theta}\exp(-x/\theta) & \text{if } x > 0 \\ 0 & \text{otherwise} \end{cases}
$$

with $\theta > 0$. Consider the following stopping rule:

$$R \equiv R(d) \text{ is the smallest integer } r(\geq m) \text{ for which we observe } r \geq a^2\overline{X}_r^2/d^2.$$

instead of (10.2.4). Upon termination, consider the fixed-width interval $J_R \equiv [\overline{X}_R - d, \overline{X}_R + d]$ for the mean of the distribution. Let $C = a^2\theta^2/d^2$. Now, prove the following results as $d \to 0$:

(i) $R/C \to 1$ in probability.

(ii) $\mathrm{E}_\theta[R/C] \to 1$.

(iii) $\mathrm{P}_\theta\{\overline{X}_R - d \leq \mu \leq \overline{X}_R + d\} \to 1 - \alpha$.

Exercise 10.2.3: Let $\phi(x) = (\sqrt{2\pi})^{-1}\exp(-x^2/2), -\infty < x < \infty$, be a standard normal density function. Suppose that $X_1, X_2, \ldots$ are independently distributed with a common mixture p.d.f.

$$f(x;\theta_1,\tau_1,\theta_2,\tau_2) = \frac{p}{\tau_1}\phi\left(\frac{x-\theta_1}{\tau_1}\right) + \frac{1-p}{\tau_2}\phi\left(\frac{x-\theta_2}{\tau_2}\right)$$

for $-\infty < x, \theta_1, \theta_2 < \infty, 0 < \tau_1, \tau_2 < \infty$ and $0 < p < 1$. Suppose that $p, \theta_1, \theta_2, \tau_1, \tau_2$ are all unknown. Consider a fixed-width confidence interval $J_n \equiv [\overline{X}_n - d, \overline{X}_n + d]$ from (10.2.1) for the mean of the mixture distribution with preassigned confidence coefficient $1-\alpha$, $0 < \alpha < 1$, $d > 0$.

(i) Obtain expressions of μ and σ^2 for this distribution.

(ii) Obtain an expression of C from (10.2.3) in this special case.

(iii) Propose Chow-Robbins stopping rule (10.2.4).

(iv) Prove that $\mathrm{P}_\theta\{\theta \in J_N\} \geq 1 - d^{-2}\mathrm{E}_\theta\left[(\overline{X}_N - \theta)^2\right]$.

Exercise 10.2.4: Let $X_1, X_2, \ldots$ be independent exponentially distributed with a common p.d.f.

$$f(x;\theta) = \begin{cases} \dfrac{1}{\theta}\exp(-x/\theta) & \text{if } x > 0 \\ 0 & \text{otherwise} \end{cases}$$

with $\theta > 0$. Suppose that we propose the following two-stage procedure. Let

$$m \equiv m(d) = \max\left\{m_0, \left\langle \left(\frac{a}{d}\right)^{2/(1+\gamma)} \right\rangle + 1\right\},$$

where $m_0(\geq 2)$ and $\gamma(> 0)$ are fixed. One will first record pilot observations $X_1, \ldots, X_m$ and determine $\overline{X}_m$. Next, one would define

$$R \equiv R(d) = \max\left\{m, \left\langle \frac{a^2\overline{X}_m^2}{d^2} \right\rangle + 1\right\},$$

and sample the difference at second stage if $R > m$. Upon termination, consider a fixed-width interval $J_R \equiv \left[\overline{X}_R - d, \overline{X}_R + d\right]$ for θ. Let $C = a^2\theta^2/d^2$. Now, prove the following results as $d \to 0$:

(i) $R/C \to 1$ in probability.
(ii) $\mathrm{E}_\theta[R/C] \to 1$.
(iii) $\mathrm{P}_\theta\left\{\overline{X}_R - d \le \mu \le \overline{X}_R + d\right\} \to 1 - \alpha$.

Exercise 10.2.5: Consider the two-stage procedure (10.2.10)-(10.2.11) and associated basic inequality (10.7.10). Combine techniques from Section 10.7.3 to obtain suitable upper and lower bounds for the variance of N, namely $\mathrm{V}_F[N]$. What does $\mathrm{V}_F[N]$ converge to when $d \to 0$?

Exercise 10.2.6: Let $X_1, X_2, \ldots$ be i.i.d. with an unknown distribution function F as in Section 10.2. Assume that F is continuous. Consider the following two-stage procedure. Let

$$m \equiv m(A) = \max\left\{m_0, \left\langle A^{1/(1+\gamma)}\right\rangle + 1\right\},$$

where $m_0(\ge 2)$ and $\gamma(> 0)$ are fixed. One will first record pilot observations $X_1, \ldots, X_m$ and determine S_m^2. Next, one would define

$$N \equiv N(A) = \max\left\{m, \langle AS_m\rangle + 1\right\},$$

and sample the difference at second stage if $N > m$. Find an appropriate expression of C and prove the following results as $A \to \infty$:

(i) $N/C \to 1$ in probability.
(ii) $\mathrm{E}_F[N] \le C + O(1)$.
(iii) $\mathrm{E}_F[N/C] \to 1$.

Exercise 10.2.7: Consider two-stage procedure (10.2.10) - (10.2.11) and the basic inequality (10.7.10). Combine the techniques from Section 10.7.3 to obtain suitable upper and lower bounds for the variance of N, namely $\mathrm{V}_F[N]$. What does $\mathrm{V}_F[N]$ converge to when $d \to 0$?

Exercise 10.3.1: Express the second-order correction term, that is

$$\left[2 - \frac{3}{4}\left\{\mu_4\sigma^{-4} - 1\right\} + 2\mu_3^2\sigma^{-6}\right] c$$

from (10.3.17) explicitly involving the population parameters in the following cases. Explain the role of population parameters in increasing or decreasing this correction term for associated regret.

Let $\phi(x) = (\sqrt{2\pi})^{-1}\exp(-x^2/2), -\infty < x < \infty$. Now, consider the following population densities:

(i) $f(x) = \frac{1}{\tau}\phi\left(\frac{x-\theta}{\tau}\right)$ for $-\infty < x, \theta < \infty, 0 < \tau < \infty$.

(ii) $f(x; \theta_1, \tau_1, \theta_2, \tau_2) = \frac{p}{\tau_1}\phi\left(\frac{x-\theta_1}{\tau_1}\right) + \frac{1-p}{\tau_2}\phi\left(\frac{x-\theta_2}{\tau_2}\right)$ for $-\infty < x, \theta_1, \theta_2 < \infty, 0 < \tau_1, \tau_2 < \infty$ and $0 < p < 1$.

(iii) $f(x; \tau_1, \tau_2) = \frac{p}{\tau_1}\exp(-x/\tau_1) + \frac{1-p}{\tau_2}\exp(-x/\tau_2)$ for $0 < x, \tau_1, \tau_2 < \infty$ and $0 < p < 1$.

Exercise 10.3.2: Consider the two-stage procedure (10.3.10)-(10.3.11) and the associated basic inequality (10.7.24). Combine techniques from Section 10.7.5 to obtain suitable upper and lower bounds for the variance of N, namely $\text{V}_F[N]$. What does $\text{V}_F[N]$ converge to when $c \to 0$?

Exercise 10.4.1: Let $\phi(x) = (\sqrt{2\pi})^{-1}\exp(-x^2/2), -\infty < x < \infty$. Suppose that $X_1, X_2, \ldots$ are independently distributed with a common p.d.f.

$$f(x; \theta, \tau) = \frac{1}{\tau}\phi\left(\frac{x-\theta}{\tau}\right) \text{ for } -\infty < x < \infty$$

where $-\infty < \theta < \infty$, $0 < \tau < \infty$ are unknown parameters. Consider a bounded length ($2d$) confidence interval $J_n \equiv \left(X_{n:b(n)}, X_{n:a(n)}\right)$ from (10.4.2) for the median of this distribution with a preassigned confidence coefficient $1-\alpha$, $0 < \alpha < 1$, $d > 0$.

(i) What is the median of the distribution in this special case?
(ii) Obtain an expression of C^* from Theorem 10.4.1.
(iii) Is Geertsema's stopping time N from (10.4.5) relevant here?
(iv) What can be said about the asymptotic behaviors of $\text{E}_{\theta,\tau}[N]$ and $\text{P}_{\theta,\tau}\{\theta \in J_N\}$ as $d \to 0$?

Exercise 10.4.2: Let $\phi(x) = (2\pi)^{-1}\exp(-x^2/2), -\infty < x < \infty$. Suppose that $X_1, X_2, \ldots$ are independently distributed with a common p.d.f.

$$f(x; \theta, \tau) = \frac{1}{\tau}\phi\left(\frac{x-\theta}{\tau}\right) \text{ for } -\infty < x < \infty$$

where $-\infty < \theta < \infty$, $0 < \tau < \infty$ are unknown parameters. Consider a fixed-width ($2d$) confidence interval $J_n \equiv \left(\overline{X}_n - d, \overline{X}_n + d\right)$ for the median of this distribution with a preassigned confidence coefficient $1-\alpha, 0 < \alpha < 1$, $d > 0$. Would Chow-Robbins sequential procedure be applicable in this case? If so, would Chow-Robbins and Geertsema sequential estimation procedures compare favorably with each other? If not, which one performs worse? Would that be expected and why so?

Exercise 10.4.3: Let $\phi(x) = (2\pi)^{-1}\exp(-x^2/2), -\infty < x < \infty$. Suppose that $X_1, X_2, \ldots$ are independently distributed with a common mixture p.d.f.

$$f(x;\theta,\tau_1,\tau_2) = \frac{p}{\tau_1}\phi\left(\frac{x-\theta}{\tau_1}\right) + \frac{1-p}{\tau_2}\phi\left(\frac{x-\theta}{\tau_2}\right)$$

for $-\infty < x, \theta < \infty$, $0 < \tau_1, \tau_2 < \infty$ and $0 < p < 1$. Consider a bounded length $(2d)$ confidence interval $J_n \equiv \left(X_{n:b(n)}, X_{n:a(n)}\right)$ from (10.4.2) for the median of this distribution with a preassigned confidence coefficient $1-\alpha$, $0 < \alpha < 1$, $d > 0$.

(i) What is the median of the distribution in this special case?

(ii) Obtain the expression of C^* from Theorem 10.4.1 in this special case.

(iii) Is Geertsema's stopping time N from (10.4.5) relevant here?

(iv) What can be said about the asymptotic behaviors of $\text{E}_{\theta,\tau}[N]$ and $\text{P}_{\theta,\tau}\{\theta \in J_N\}$ as $d \to 0$?

Exercise 10.4.4: Suppose that $X_1, X_2, \ldots$ are independently distributed with a common Cauchy p.d.f.

$$f(x;\theta,\tau) = \frac{\tau}{\pi}\left\{\tau^2 + (x-\theta)^2\right\}^{-1} \text{ for } -\infty < x < \infty$$

where $-\infty < \theta < \infty$, $0 < \tau < \infty$ are unknown parameters. Consider a bounded length $(2d)$ confidence interval $J_n \equiv \left(X_{n:b(n)}, X_{n:a(n)}\right)$ from (10.4.2) for the median of this distribution with a preassigned confidence coefficient $1-\alpha$, $0 < \alpha < 1$, $d > 0$.

(i) What is the median of the distribution in this special case?

(ii) Obtain the expression of C^* from Theorem 10.4.1

(iii) Is Geertsema's stopping time N from (10.4.5) relevant here?

(iv) What can be said about the asymptotic behaviors of $\text{E}_{\theta,\tau}[N]$ and $\text{P}_{\theta,\tau}\{\theta \in J_N\}$ as $d \to 0$?

Exercise 10.4.5: Suppose that $X_1, X_2, \ldots$ are independently distributed with a common mixture Cauchy p.d.f.:

$$f(x;\theta,\tau_1,\tau_2) = \frac{p\tau_1}{\pi}\left\{\tau_1^2 + (x-\theta)^2\right\}^{-1} + \frac{(1-p)\tau_2}{\pi}\left\{\tau_2^2 + (x-\theta)^2\right\}^{-1}$$

for $-\infty < x, \theta < \infty$, $0 < \tau_1$, $\tau_2 < \infty$ and $0 < p < 1$. Consider a bounded length $(2d)$ confidence interval $J_n \equiv \left(X_{n:b(n)}, X_{n:a(n)}\right)$ from (10.4.2) for the median of this distribution with a preassigned confidence coefficient $1-\alpha$, $0 < \alpha < 1$, $d > 0$.

(i) What is the median of the distribution in this special case?

(ii) Obtain the expression of C^* from Theorem 10.4.1 in this special case.

(iii) Is Geertsema's stopping time N from (10.4.5) relevant here?

(iv) What can be said about the asymptotic behaviors of $\mathrm{E}_{\theta,\tau_1,\tau_2}[N]$ and $\mathrm{P}_{\theta,\tau_1,\tau_2}\{\theta \in J_N\}$ as $d \to 0$?

Exercise 10.4.6: Let $\phi(x) = (\sqrt{2\pi})^{-1}\exp(-x^2/2)$, $-\infty < x < \infty$. Suppose that $X_1, X_2, \ldots$ are independently distributed with a common normal-Cauchy mixture p.d.f.:

$$f(x;\theta,\tau_1,\tau_2) = \frac{p}{\tau_1}\phi\left(\frac{x-\theta}{\tau_1}\right) + \frac{(1-p)\tau_2}{\pi}\left\{\tau_2^2 + (x-\theta)^2\right\}^{-1}$$

for $-\infty < x, \theta < \infty$, $0 < \tau_1, \tau_2 < \infty$ and $0 < p < 1$. Consider a bounded length $(2d)$ confidence interval $J_n \equiv \left(X_{n:b(n)}, X_{n:a(n)}\right)$ from (10.4.2) for the median of this distribution with a preassigned confidence coefficient $1-\alpha$, $0 < \alpha < 1$, $d > 0$.

(i) What is the median of the distribution in this special case?
(ii) Obtain the expression of C^* from Theorem 10.4.1 in this special case.
(iii) Is Geertsema's stopping time N from (10.4.5) relevant here?
(iv) What can be said about the asymptotic behaviors of θ, τ_1, τ_2 and $\mathrm{P}_{\theta,\tau_1,\tau_2}\{\theta \in J_N\}$ as $d \to 0$?

Exercise 10.5.1 (Exercise 10.2.1 Continued): Use program **Seq10.exe** to simulate J_N by generating random variates from

$$f(x;\theta) = \begin{cases} \frac{1}{5}\exp(-x/5) & \text{if } x > 0 \\ 0 & \text{otherwise.} \end{cases}$$

Carry out 5000 replications in a simulation study with fixed $\alpha = 0.1$, $d = 0.5$ and $m = 5$. Use program output to

(i) verify the expression of C obtained in Exercise 10.2.1.
(ii) show that the inequality in part (iii) of Exercise 10.2.1 is valid for all simulated values of N. Use coverage probability to estimate

$$\mathrm{P}_\theta\{\theta \in J_N\}.$$

Exercise 10.5.2 (Exercise 10.2.3 Continued): Given $\theta_1 = \theta_2 = 10$, $\tau_1 = 2$, $\tau_2 = 3$ and $p = 0.4$, carry out 5000 replications to study simulated fixed-width confidence intervals $J_n \equiv \left[\overline{X}_n - 0.25, \overline{X}_n + 0.25\right]$. Compare results with those obtained in parts (i), (ii) and (iv) in Exercise 10.2.3.

Exercise 10.5.3: Generate $X_1, X_2, \ldots$ from a gamma distribution with its p.d.f.:

$$f(x;\theta) = \begin{cases} \frac{1}{\Gamma(2.5)4^{2.5}}x^{1.5}\exp(-x/4) & \text{if } x > 0 \\ 0 & \text{otherwise} \end{cases}$$

to simulate 95% fixed-width confidence intervals $J_n \equiv \left[\overline{X}_n - 0.5, \overline{X}_n + 0.5\right]$ for μ. Here, μ is the mean of the gamma distribution. Pretend that its shape and scale parameters are unknown. Replicate simulations 1000 times using a two-stage procedure with $\gamma = 0.25, 0.75$ and 1.00. Explain the results.

Exercise 10.5.4 : Let $X_1, X_2, \ldots$ be independent identically distributed random observations with a common normal mixture p.d.f.:

$$f(x;\theta,\tau_1,\tau_2) = \frac{p}{\tau_1}\phi\left(\frac{x-\theta}{\tau_1}\right) + \frac{1-p}{\tau_2}\phi\left(\frac{x-\theta}{4}\right)$$

where $-\infty < x, \theta < \infty$, $0 < \tau_1, \tau_2 < \infty$ and $0 < p < 1$. Simulate a sequential minimum risk point estimator for θ from Section 10.3.3 based on a loss function

$$L_n\left(\theta, \overline{X}_n\right) = 10\left(\overline{X}_n - \theta\right)^2 + 0.1n$$

using the program **Seq10.exe**. Generate $X_1, X_2, \ldots$ from a mixture normal p.d.f. with $\theta = 10$, $\tau_1 = 3$, $\tau_2 = 4$ and $p = 0.4$. Obtain output from 1000 replications of a simulation study with constant $\gamma = 1$ and $\gamma = 1.5$. Compare the two sets of results.

Exercise 10.5.5 (Exercise 10.4.6 Continued): Use the computer program **Seq10.exe** to simulate a 95% confidence interval for the median of the mixture normal-Cauchy p.d.f.:

$$f(x;\theta,\tau_1,\tau_2) = \frac{p}{\tau_1}\phi\left(\frac{x-\theta}{\tau_1}\right) + \frac{(1-p)\tau_2}{\pi}\left\{\tau_2^2 + (x-\theta)^2\right\}^{-1}$$

for $-\infty < x, \theta < \infty$, $0 < \tau_1, \tau_2 < \infty$ and $0 < p < 1$. Fix $\theta = 10$, $\tau_1 = 3$, $\tau_2 = 4$ and $p = 0.4$ to generate $X_1, X_2, \ldots$ data. Comment on the results.

CHAPTER 11

Multivariate Normal Mean Vector Estimation

11.1 Introduction

In this chapter, we treat estimation problems for an unknown mean vector $\boldsymbol{\mu}_{p\times 1}$ exclusively in a p-dimensional normal distribution. We assume that the distribution has a positive definite (p.d.) *variance-covariance* or *dispersion* matrix $\boldsymbol{\Sigma}_{p\times p}$. Customarily, we denote such a distribution by $N_p(\boldsymbol{\mu}, \boldsymbol{\Sigma})$.

We focus on both fixed-size confidence region estimators and minimum risk point estimators for $\boldsymbol{\mu}$. In some problems, we will assume a certain structure for the dispersion matrix, $\boldsymbol{\Sigma}$. For example, in Section 11.2, we let $\boldsymbol{\Sigma} = \sigma^2 H_{p\times p}$ where $0 < \sigma^2 < \infty$ is assumed unknown and H is a known p.d. matrix. But, in Sections 11.3 - 11.4, we assume that $\boldsymbol{\Sigma}$ is completely unspecified and p.d.

Historically, two-stage and purely sequential methodologies were first developed for constructing fixed-size confidence regions for a mean vector and regression parameters. These generally followed the spirits of Stein (1945,1949) and Chow and Robbins (1965). Then, the methodologies and the kinds of problems under consideration took off in many directions. One may refer to a large collection of papers including Healy (1956), Chatterjee (1959a,b,1962a,b,1990), Gleser (1965), Srivastava (1967,1971), Khan (1968), Ghosh and Sen (1971,1972,1973), Mukhopadhyay and Abid (1986a,b), Mukhopadhyay and Al-Mousawi (1986), Datta and Mukhopadhyay (1997), Mukhopadhyay (1999b), and Mukhopadhyay and Aoshima (1998). The review papers of Chatterjee (1991), Mukhopadhyay (1991), Sinha (1991) and the books by Sen (1981) and Ghosh et al. (1997) may provide a broader view of this field.

Before approaching a minimum risk point estimation problem for $\boldsymbol{\mu}$ in a general multivariate normal distribution, Mukhopadhyay (1975) and Sinha and Mukhopadhyay (1976) considered a bivariate scenario. In a sequel, Ghosh et al. (1976) introduced a purely sequential methodology in the most general case, that is when $\boldsymbol{\Sigma}_{p\times p}$ was arbitrary p.d. and p was arbitrary. They proved asymptotic first-order efficiency and asymptotic risk efficiency properties. Woodroofe (1977) developed asymptotic *second-order* properties for the purely sequential strategy of Ghosh et al. (1976).

On the other hand, Wang (1980) had developed a purely sequential minimum risk point estimator for $\boldsymbol{\mu}$ in a multivariate normal distribution when $\boldsymbol{\Sigma} = \sigma^2 H_{p\times p}$. Wang (1980) assumed that $0 < \sigma^2 < \infty$ was unknown, but H was a known p.d. matrix.

In Section 11.2, we introduce a fixed-size ellipsoidal confidence region for $\boldsymbol{\mu}$ when $\boldsymbol{\Sigma} = \sigma^2 H_{p\times p}$ where $0 < \sigma^2 < \infty$ is assumed unknown, but H is a known p.d. matrix. The material on two-stage (Section 11.2.1), modified two-stage (Section 11.2.2), purely sequential (Section 11.2.3), and three-stage (Section 11.2.4) methodologies are adapted from Mukhopadhyay and Al-Mousawi (1986).

Section 11.3 introduces a fixed-size spherical confidence region for $\boldsymbol{\mu}$ when $\boldsymbol{\Sigma}$ is assumed p.d., but otherwise completely unknown. The classical two-stage methodology (Section 11.3.1) is adapted from Healy (1956) and Chatterjee (1959a,b). The two-stage strategy has the exact consistency property. The purely sequential methodology (Section 11.3.2) is adapted from Srivastava (1967). The sequential strategy has both asymptotic consistency and first-order efficiency properties.

We address minimum risk point estimation problems for $\boldsymbol{\mu}$ in Section 11.4 when $\boldsymbol{\Sigma}$ is assumed p.d., but otherwise completely unknown. We have included the classical purely sequential estimation methodology (Section 11.4.1) due to Ghosh et al. (1976). Its asymptotic *second-order* properties are summarized from Woodroofe (1977,1981).

Section 11.5 is devoted to data analyses and some concluding remarks. These are followed by a brief overview of other multistage strategies in Section 11.6. Some selected derivations are given in Section 11.7.

11.2 Fixed-Size Confidence Region: $\Sigma = \sigma^2 H$

Suppose that we have available a sequence of independent observations $\boldsymbol{X}_1, \ldots, \boldsymbol{X}_n, \ldots$ from a $N_p(\boldsymbol{\mu}, \boldsymbol{\Sigma})$ distribution where $\boldsymbol{\Sigma} = \sigma^2 H_{p\times p}$. We assume that $0 < \sigma^2 < \infty$ is unknown, but H is a known p.d. matrix. A problem of constructing a fixed-size confidence region for $\boldsymbol{\mu}$ is formulated as follows:

We are given two preassigned numbers $0 < d < \infty$ and $0 < \alpha < 1$. Having recorded $n(\geq 2)$ observations $\boldsymbol{X}_1, \ldots, \boldsymbol{X}_n$, we propose an ellipsoidal confidence region:

$$\mathcal{R}_n = \left\{\boldsymbol{\omega}_{p\times 1} \in \Re^p : (\overline{\boldsymbol{X}}_n - \boldsymbol{\omega})' H^{-1} (\overline{\boldsymbol{X}}_n - \boldsymbol{\omega}) \leq d^2\right\}, \tag{11.2.1}$$

for $\boldsymbol{\mu}$. In (11.2.3) and elsewhere, we use customary notation:

$$\begin{aligned} \overline{\boldsymbol{X}}_n &= n^{-1} \textstyle\sum_{i=1}^n \boldsymbol{X}_i, \text{ and} \\ S_n^2 &= (pn-p)^{-1} \textstyle\sum_{i=1}^n (\boldsymbol{X}_i - \overline{\boldsymbol{X}}_n)' H^{-1} (\boldsymbol{X}_i - \overline{\boldsymbol{X}}_n). \end{aligned} \tag{11.2.2}$$

Obviously, $\overline{\boldsymbol{X}}_n$ and S_n^2 respectively estimate $\boldsymbol{\mu}$ and σ^2 unbiasedly and they are the UMVUEs. The confidence region $\mathcal{R}_n$ is said to have a fixed size because its maximum diameter is known and fixed in advance. The volume of $\mathcal{R}_n$ can be made as small as one pleases by choosing an appropriately small number d. We have explained this idea more in Section 11.7.1.

We require that the confidence coefficient be at least $1 - \alpha$. Now, the confidence coefficient associated with $\mathcal{R}_n$ is:

$$\begin{aligned} &\mathrm{P}_{\boldsymbol{\mu},\sigma}\{\boldsymbol{\mu} \in \mathcal{R}_n\} \\ &= \mathrm{P}_{\boldsymbol{\mu},\sigma}\{(\overline{\boldsymbol{X}}_n - \boldsymbol{\mu})' H^{-1} (\overline{\boldsymbol{X}}_n - \boldsymbol{\mu}) \le d^2\} \\ &= F\left(\frac{nd^2}{\sigma^2}\right) \text{ with } F(x) \equiv F(x;p) = \mathrm{P}(U \le x), \\ &\qquad\qquad U \sim \chi^2_p, x > 0. \end{aligned} \tag{11.2.3}$$

For simplicity, we sometimes drop the index p from $F(x;p)$.

The last step in (11.2.3) holds because of part (i) from (11.2.4), where we have summarized two important distributional results:

$$\begin{aligned} &\text{(i)} \quad n(\overline{\boldsymbol{X}}_n - \boldsymbol{\mu})' H^{-1} (\overline{\boldsymbol{X}}_n - \boldsymbol{\mu})/\sigma^2 \sim \chi^2_p \text{ and} \\ &\text{(ii)} \quad p(n-1)S^2_n/\sigma^2 \sim \chi^2_{p(n-1)}, \text{ under } \boldsymbol{\mu} \text{ and } \boldsymbol{\Sigma} = \sigma^2 H. \end{aligned} \tag{11.2.4}$$

We have sketched its proof in Section 11.7.2.

Let $a(>0)$ be such that $F(a) = 1-\alpha$, that is $a(>0)$ is the upper $100\alpha\%$ of a χ^2_p distribution. Since we require that $F\left(nd^2/\sigma^2\right) \ge F(a) = 1-\alpha$, we must have $nd^2/\sigma^2 \ge a$. So, we determine

$$n \text{ as be the smallest integer } \ge \frac{a\sigma^2}{d^2} = C, \text{ say.} \tag{11.2.5}$$

We refer to C as the optimal fixed sample size had σ^2 been known. But, the magnitude of C remains unknown and no fixed-sample-size methodology would help. By the way, this multivariate problem reduces to the problem of fixed-width confidence intervals discussed in Chapter 6 when $p = 1$.

The present problem was thoroughly investigated by Mukhopadhyay and Al-Mousawi (1986). Our presentation is adapted from their paper.

It is understood that a multistage or sequential methodology would come up with a final sample size $N \equiv N(d)$ and we would propose the confidence region:

$$\mathcal{R}_N = \left\{\boldsymbol{\omega}_{p\times 1} \in I\!R^p : (\overline{\boldsymbol{X}}_N - \boldsymbol{\omega})' H^{-1} (\overline{\boldsymbol{X}}_N - \boldsymbol{\omega}) \le d^2\right\} \tag{11.2.6}$$

for $\boldsymbol{\mu}$ based on full dataset $N, \boldsymbol{X}_1, \boldsymbol{X}_2, \ldots, \boldsymbol{X}_N$.

In the kinds of methodologies summarized here, one will notice that for all fixed $n(\ge m)$, the random variable $I(N = n)$ would be determined by $(S^2_m, S^2_{m+1}, \ldots, S^2_n)$. Thus, $I(N = n)$ and $\overline{\boldsymbol{X}}_n$ would be independent random variables since $\overline{\boldsymbol{X}}_n$ and $\left(S^2_m, S^2_{m+1}, \ldots, S^2_n\right)$ are independent, for all fixed $n(\ge m)$.

Thus, for the methodologies reported here, we will express the confidence coefficient associated with $\mathcal{R}_N$ as:

$$\begin{aligned} & \mathrm{P}_{\boldsymbol{\mu},\sigma}\{\boldsymbol{\mu} \in \mathcal{R}_N\} \\ & \quad = \mathrm{P}_{\boldsymbol{\mu},\sigma}\{(\overline{\boldsymbol{X}}_N - \boldsymbol{\mu})' H^{-1}(\overline{\boldsymbol{X}}_N - \boldsymbol{\mu}) \le d^2\} \\ & \quad = \mathrm{E}_{\boldsymbol{\mu},\sigma}\left[F\left(\frac{Nd^2}{\sigma^2}\right)\right] \text{ with} \\ & \qquad\qquad F(x) = \mathrm{P}\left(U \le x\right), U \sim \chi^2_p, x > 0. \end{aligned} \tag{11.2.7}$$

We sketch a proof of this result in Section 11.7.3. Next, we summarize successively some of the crucial multistage estimation procedures.

11.2.1 Two-Stage Procedure

Motivations for a two-stage procedure came from Stein (1945,1949). Let $\boldsymbol{X}_1, \boldsymbol{X}_2, \ldots, \boldsymbol{X}_m$ be pilot observations where $m(\ge 2)$ is the pilot size. We obtain S^2_m from (11.2.2). Now, suppose that $F_{p,pm-p,\alpha}$ denotes an upper $100\alpha\%$ point of a $F_{p,pm-p}$ distribution and write:

$$b \equiv b_{p,m,\alpha} = pF_{p,pm-p,\alpha}. \tag{11.2.8}$$

Next, we define the following two-stage stopping variable:

$$N \equiv N(d) = \max\left\{m, \left\langle \frac{bS^2_m}{d^2}\right\rangle + 1\right\}, \tag{11.2.9}$$

where $b \equiv b_{p,m,\alpha}(> 0)$ comes from (11.2.8).

At this point, if $N = m$, we do not gather any observation at the second stage.

But, if $N > m$, then we sample the difference at the second stage by gathering additional observations $\boldsymbol{X}_{m+1}, \ldots, \boldsymbol{X}_N$.

Then, based on *combined* data $N, \boldsymbol{X}_1, \boldsymbol{X}_2, \ldots, \boldsymbol{X}_N$ from both stages 1 and 2, we estimate $\boldsymbol{\mu}$ by $\mathcal{R}_N$ from (11.2.6). It is easy to see that N is finite w.p. 1. Thus, $\mathcal{R}_N$ is a genuine confidence region estimator.

Note that the associated confidence coefficient is given by the expression from (11.2.7). The following results were obtained by Mukhopadhyay and Al-Mousawi (1986).

Theorem 11.2.1 : *For the stopping variable N defined by (11.2.9), for all fixed $\boldsymbol{\mu}, \sigma, p, m$ and α, one has the following properties:*

(i) $\mathrm{P}_{\boldsymbol{\mu},\sigma}\{\boldsymbol{\mu} \in \mathcal{R}_N\} \ge 1 - \alpha$ *for all fixed d [Consistency or Exact Consistency]*;

(ii) $\lim_{d\to 0} \mathrm{E}_{\boldsymbol{\mu},\sigma}[N/C] = ba^{-1}$;

(iii) $\lim_{d\to 0} \mathrm{V}_{\boldsymbol{\mu},\sigma}[N]\left\{\frac{1}{2}p(m-1)\left(b\sigma^2/d^2\right)^{-2}\right\} = 1$;

(iv) $\lim_{d\to 0} \mathrm{P}_{\boldsymbol{\mu},\sigma}\{\boldsymbol{\mu} \in \mathcal{R}_N\} = 1 - \alpha$ *[Asymptotic Consistency]*;

where $b \equiv b_{p,m,\alpha}$ is taken from (11.2.8), $\mathcal{R}_N$ came from (11.2.6), and $C = a\sigma^2/d^2$ was defined in (11.2.5).

Two-stage strategy (11.2.9) is consistent, but it is asymptotically inefficient.

In part (ii), the limiting ratio ba^{-1} would exceed one in practice whatever be p, m and α. We had referred to this as asymptotic *first-order inefficiency* property in Chapter 6. A proof of part (i) is sketched in Section 11.7.4.

11.2.2 Modified Two-Stage Procedure

Motivated by the results from Mukhopadhyay (1980,1982a), we choose and fix a number $\gamma, 0 < \gamma < \infty$. Mukhopadhyay and Al-Mousawi (1986) defined a pilot size:

$$m \equiv m(d) = \max\left\{m_0, \left\langle \left(\frac{a}{d^2}\right)^{1/(1+\gamma)} \right\rangle + 1\right\}, \qquad (11.2.10)$$

where $m_0 (\geq 2)$ is fixed.

Now, one will record pilot observations $\boldsymbol{X}_1, \ldots, \boldsymbol{X}_m$ and determine S_m^2 from (11.2.2). Next, define a two-stage stopping variable:

$$N \equiv N(d) = \max\left\{m, \left\langle \frac{bS_m^2}{d^2} \right\rangle + 1\right\}, \qquad (11.2.11)$$

where the number $b \equiv b_{p,m,\alpha}$ comes from (11.2.8).

At this point, one samples the difference at second stage by gathering additional observations $\boldsymbol{X}_{m+1}, \ldots, \boldsymbol{X}_N$ if $N > m$. Then, based on *combined* data $N, \boldsymbol{X}_1, \boldsymbol{X}_2, \ldots, \boldsymbol{X}_N$ from both stages 1 and 2, we estimate $\boldsymbol{\mu}$ by $\mathcal{R}_N$ from (11.2.6). Again, N is finite w.p. 1, and thus $\mathcal{R}_N$ is a genuine confidence region estimator.

Note that the associated confidence coefficient is given by (11.2.7). The following results were obtained by Mukhopadhyay and Al-Mousawi (1986).

Theorem 11.2.2: *For the stopping variable N defined by (11.2.11), for all fixed $\boldsymbol{\mu}, \sigma, p, m$ and α, one has the following properties:*

(i) $\mathrm{P}_{\boldsymbol{\mu},\sigma}\{\boldsymbol{\mu} \in \mathcal{R}_N\} \geq 1 - \alpha$ *for all fixed d [Consistency or Exact Consistency]*;

(ii) $\lim_{d\to 0} \mathrm{E}_{\boldsymbol{\mu},\sigma}[N/C] = 1$ *[Asymptotic First-Order Efficiency]*;

(iii) $\lim_{d\to 0} \mathrm{V}_{\boldsymbol{\mu},\sigma}[N]\left\{\frac{1}{2}p(m-1)\left(b\sigma^2/d^2\right)^{-2}\right\} = 1$;

(iv) $\lim_{d\to 0} \mathrm{P}_{\boldsymbol{\mu},\sigma}\{\boldsymbol{\mu} \in \mathcal{R}_N\} = 1 - \alpha$ *[Asymptotic Consistency]*;

where $b \equiv b_{p,m,\alpha}$ is taken from (11.2.8), $\mathcal{R}_N$ came from (11.2.6), and $C = a\sigma^2/d^2$ was defined in (11.2.5).

Modified two-stage strategy (11.2.10)-(11.2.11) is consistent and asymptotically first-order efficient.

We sketch a proof of part (ii) in Section 11.7.5. In connection with part (ii), we add that although this two-stage procedure is asymptotically *first-order* efficient, we expect $\lim_{d\to 0} \mathrm{E}_{\boldsymbol{\mu},\sigma}[N - C]$ to explode for all practical purposes. That is, this procedure would fail to be second-order efficient. We noted similar circumstances in Chapters 6-9.

Two asymptotic second-order efficient estimation strategies are summarized next. See also Mukhopadhyay (1999b).

11.2.3 Purely Sequential Procedure

Mukhopadhyay and Al-Mousawi (1986) defined the following purely sequential methodology. One would start with pilot observations $\boldsymbol{X}_1, \ldots, \boldsymbol{X}_m$ where $m(\geq 2)$. Then, one will proceed sequentially with one additional X at-a-time and terminate sampling according to a stopping time:

$$
\begin{aligned}
&N \equiv N(d) \text{ is the smallest integer } n(\geq m) \text{ for which we observe} \\
&\qquad n \geq \left(\frac{a}{d^2}\right) S_n^2 \text{ with } F(a;p) \equiv \mathrm{P}\left(\chi_p^2 \leq a\right) = 1 - \alpha.
\end{aligned} \tag{11.2.12}
$$

This sequential strategy is implemented as usual and again the sampling would terminate w.p. 1. That is, $\mathcal{R}_N$ from (11.2.6) will be a genuine confidence region estimator for $\boldsymbol{\mu}$.

Note that the associated confidence coefficient is again given by (11.2.7). The following *first-order* results were obtained by Mukhopadhyay and Al-Mousawi (1986).

Theorem 11.2.3: *For the stopping variable N defined by (11.2.12), for all fixed $\boldsymbol{\mu}, \sigma, p, m$ and α, one has the following properties:*

(i) $\mathrm{E}_{\boldsymbol{\mu},\sigma}[N] \leq C + m + 2$ *for all fixed d;*
(ii) $\lim_{d\to 0} \mathrm{E}_{\boldsymbol{\mu},\sigma}[N/C] = 1$ *[Asymptotic First-Order Efficiency];*
(iii) $\lim_{d\to 0} \mathrm{P}_{\boldsymbol{\mu},\sigma}\{\boldsymbol{\mu} \in \mathcal{R}_N\} = 1 - \alpha$ *[Asymptotic Consistency];*

where $\mathcal{R}_N$ came from (11.2.6) and $C = a\sigma^2/d^2$ was defined in (11.2.5).

We sketch proofs of parts (i) and (ii) in Section 11.7.6. The following *second-order* results were also obtained by Mukhopadhyay and Al-Mousawi (1986).

One should recall the expressions of $\eta(p)$ and $f(x;p)$ from (6.2.33). it should be clear that $f(x;p)$ is the p.d.f. of a χ_p^2 random variable, that is:

$$
\begin{aligned}
&f(x;p) \equiv dF(x;p)/dx, &&f'(x;p) \equiv df(x;p)/dx, x > 0; \\
&f(a;p) = dF(x;p)/dx|_{x=a}, &&f'(a;p) = df(x;p)/dx|_{x=a}.
\end{aligned}
$$

Theorem 11.2.4: *For the stopping variable N defined by (11.2.12), for all fixed $\boldsymbol{\mu}, \sigma, p, m$ and α, one has the following properties as $d \to 0$:*

(i) $\mathrm{E}_{\boldsymbol{\mu},\sigma}[N] = C + \eta(p) + o(1)$ *if* (a) $m \geq 4$ *for* $p = 1$, (b) $m \geq 3$ *for* $p = 2$, *and* (c) $m \geq 2$ *for* $p \geq 3$;

(ii) $\mathrm{P}_{\boldsymbol{\mu},\sigma}\{\boldsymbol{\mu} \in \mathcal{R}_N\} = 1 - \alpha + \dfrac{d^2}{\sigma^2}\left\{\eta(p)f(a;p) + \dfrac{a}{p}f'(a;p)\right\} + o(d^2)$ *if* (a) $m \geq 7$ *for* $p = 1$, (b) $m \geq 3$ *for* $p = 2, 3$, *and* (c) $m \geq 2$ *for* $p \geq 4$;

(iii) $\left(\dfrac{p}{2C}\right)^{1/2}(N - C)$ *converges to* $N(0,1)$ *in distribution;*

where $\mathcal{R}_N$ *came from (11.2.6),* $C = a\sigma^2/d^2$ *was defined in (11.2.5), and* $\eta(p), f(x;p)$ *came from (6.2.33).*

11.2.4 Three-Stage Procedure

In some situations, it may be inconvenient to use one-by-one sequential sampling. It will be useful in practice to have a methodology terminating quickly with a final sample size comparing favorably on an average with that under sequential sampling. That is, we may prefer to preserve *second-order* properties with an operational edge. A three-stage strategy fits the bill.

Sequential strategy (11.2.12) and three-stage strategy (11.2.13)-(11.2.14) are asymptotically consistent and asymptotically second-order efficient. Neither is consistent.

This methodology starts with a pilot of size m and with the help of pilot data, one would estimate ρC, a fraction of C. Up to this step, the methodology resembles two-stage sampling. Next, using combined data from first and second stages, one will estimate C, and sample the difference as needed at the third stage. Using combined data of size N from all three stages, one will estimate $\boldsymbol{\mu}$ by $\mathcal{R}_N$ from (11.2.6). Recall that $C = a\sigma^2/d^2$ where σ^2 is unknown.

To be specific, Mukhopadhyay and Al-Mousawi (1986) gave the following three-stage strategy. One would choose and fix $0 < \rho < 1$. One would start with pilot observations $\boldsymbol{X}_1, \ldots, \boldsymbol{X}_m$ where $m(\geq 2)$ and determine S_m^2 from (11.2.2). Define:

$$T \equiv T(d) = \max\left\{m, \left\langle \rho\frac{aS_m^2}{d^2}\right\rangle + 1\right\}, \tag{11.2.13}$$

which estimates ρC, a fraction of C.

We sample the difference at second stage if $T > m$ and denote the combined set of data from the first and second stages, $\boldsymbol{X}_1, \ldots, \boldsymbol{X}_T$. At this point, one may be tempted to estimate σ^2 and C by S_T^2 and $aS_T^2 d^{-2}$ respectively.

More precisely, we define:

$$N \equiv N(d) = \max\left\{T, \left\langle \frac{aS_T^2}{d^2} + \varepsilon \right\rangle + 1\right\}$$
$$\text{where } \varepsilon = \left\{2(p\rho)^{-1} - \tfrac{1}{2}\right\} - a(p\rho)^{-1}\frac{f'(a;p)}{f(a;p)}. \tag{11.2.14}$$

We sample the difference at the third stage if $N > T$ and denote the combined set of data from all three stages, $\boldsymbol{X}_1, \ldots, \boldsymbol{X}_m, \ldots, \boldsymbol{X}_T, \ldots, \boldsymbol{X}_N$.

Now, $\mathcal{R}_N$ is a genuine confidence region estimator. So, upon termination, we estimate $\boldsymbol{\mu}$ by $\mathcal{R}_N$ from (11.2.6).

Note that the associated confidence coefficient is given by (11.2.7). The following *second-order* results were obtained by Mukhopadhyay and Al-Mousawi (1986).

Theorem 11.2.5: *For the stopping variable N defined by (11.2.13) - (11.2.14), for all fixed $\boldsymbol{\mu}, \sigma, p, m$ and α, one has the following properties as $d \to 0$:*

(i) $\mathrm{E}_{\boldsymbol{\mu},\sigma}[N] = C - a(p\rho)^{-1}\dfrac{f'(a;p)}{f(a;p)} + o(1)$;

(ii) $\mathrm{P}_{\boldsymbol{\mu},\sigma}\{\boldsymbol{\mu} \in \mathcal{R}_N\} = 1 - \alpha + o(d^2)$;

(iii) $\left(\dfrac{p\rho}{2C}\right)^{1/2}(N - C)$ *converges to* $N(0,1)$ *in distribution;*

where $\mathcal{R}_N$ came from (11.2.6), $C = a\sigma^2/d^2$ was defined in (11.2.5), and $f(x;p)$ came from (6.2.33).

Comparing the results from part (iii) in Theorems 11.2.4 and 11.2.5, we notice that $\mathrm{V}_{\boldsymbol{\mu},\sigma}[N_{\text{sequential}}] \approx \frac{2}{p}C$ whereas $\mathrm{V}_{\boldsymbol{\mu},\sigma}[N_{\text{three-stage}}] \approx \frac{2}{p\rho}C$ since $(N - C)^2/C$ is uniformly integrable in both cases. That is, even though both $N_{\text{sequential}}$ and $N_{\text{three-stage}}$ estimates the same optimal fixed-sample-size C, the three-stage strategy (11.2.13)-(11.2.14) is more variable since $2p^{-1}C < 2(p\rho)^{-1}C$.

11.3 Fixed-Size Confidence Region: Unknown Dispersion Matrix

Suppose that we have available a sequence of independent observations $\boldsymbol{X}_1, \ldots, \boldsymbol{X}_n, \ldots$ from a $N_p(\boldsymbol{\mu}, \boldsymbol{\Sigma})$ distribution. We suppose that $\boldsymbol{\mu}$ is unknown and $\boldsymbol{\Sigma}_{p\times p}$ is p.d., but otherwise it is also completely unknown. A problem of constructing a fixed-size confidence region for $\boldsymbol{\mu}$ is formulated as follows.

We are given two preassigned numbers $0 < d < \infty$ and $0 < \alpha < 1$. Having recorded $n(\geq 2)$ observations $\boldsymbol{X}_1, \ldots, \boldsymbol{X}_n$, we propose a spherical confidence region:

$$\mathcal{R}_n^* = \left\{\boldsymbol{\omega}_{p\times 1} \in I\!R^p : (\overline{\boldsymbol{X}}_n - \boldsymbol{\omega})'(\overline{\boldsymbol{X}}_n - \boldsymbol{\omega}) \leq d^2\right\}, \tag{11.3.1}$$

for $\boldsymbol{\mu}$.

Here, we denote

$$\begin{aligned} \overline{\boldsymbol{X}}_n &= n^{-1}\textstyle\sum_{i=1}^n \boldsymbol{X}_i, \text{ and} \\ S_n &= (n-1)^{-1}\textstyle\sum_{i=1}^n \left(\boldsymbol{X}_i - \overline{\boldsymbol{X}}_n\right)\left(\boldsymbol{X}_i - \overline{\boldsymbol{X}}_n\right)'. \end{aligned} \tag{11.3.2}$$

We note that S_n is a $p \times p$ p.d. matrix that estimates the unknown dispersion matrix $\boldsymbol{\Sigma}$. As usual, we use $\overline{\boldsymbol{X}}_n$ to estimate the unknown mean vector $\boldsymbol{\mu}$. Obviously, $\overline{\boldsymbol{X}}_n$ and S_n are respectively the UMVUEs for $\boldsymbol{\mu}$ and $\boldsymbol{\Sigma}$.

The confidence region $\mathcal{R}_n^*$, centered at $\overline{\boldsymbol{X}}_n$, is said to have a fixed size because it is a p-dimensional sphere having its diameter $2d$ which is preassigned. The volume of $\mathcal{R}_n^*$ can be made as small as one pleases by choosing an appropriately small number d. We also require that the confidence coefficient be at least $1 - \alpha$ which is again preassigned. When $p = 1$, one notes that this construction reduces to fixed-width confidence intervals of a normal mean discussed in Chapter 6.

Before we work out the confidence coefficient, let us first compare two quadratic forms

$$\begin{aligned} &(\overline{\boldsymbol{X}}_n - \boldsymbol{\mu})'\boldsymbol{\Sigma}^{-1}(\overline{\boldsymbol{X}}_n - \boldsymbol{\mu}) \text{ and} \\ &(\overline{\boldsymbol{X}}_n - \boldsymbol{\mu})'(\overline{\boldsymbol{X}}_n - \boldsymbol{\mu}). \end{aligned} \tag{11.3.3}$$

One will find a clue in (11.3.6).

Since $\boldsymbol{\Sigma}$ is p.d., we can express $\boldsymbol{\Sigma} = B'\Delta B$ where $B_{p\times p}$ is an orthogonal matrix and $\Delta_{p\times p} = \text{diagonal}(\lambda_1, \ldots, \lambda_p)$. That is, Δ is filled with $\lambda_1, \ldots, \lambda_p$ along its diagonal. Here, $\lambda_1, \ldots, \lambda_p$ are the eigenvalues of $\boldsymbol{\Sigma}$ and they are all positive. Clearly, $\boldsymbol{\Sigma}^{-1} = B'\Delta^{-1}B$ where $\Delta^{-1} = \text{diagonal}(\lambda_1^{-1}, \ldots, \lambda_p^{-1})$. Let us denote $\lambda_{\max} = \max_{1\le i\le p} \lambda_i$ and $(B\overline{\boldsymbol{X}}_n - B\boldsymbol{\mu})' = (Y_1, \ldots, Y_p)$ so that we have:

$$\begin{aligned} &(\overline{\boldsymbol{X}}_n - \boldsymbol{\mu})'\boldsymbol{\Sigma}^{-1}(\overline{\boldsymbol{X}}_n - \boldsymbol{\mu}) \\ &= (\overline{\boldsymbol{X}}_n - \boldsymbol{\mu})'B'\Delta^{-1}B(\overline{\boldsymbol{X}}_n - \boldsymbol{\mu}) \\ &= (B\overline{\boldsymbol{X}}_n - B\boldsymbol{\mu})'\Delta^{-1}(B\overline{\boldsymbol{X}}_n - B\boldsymbol{\mu}) \\ &= \textstyle\sum_{i=1}^p q_i^{-1} Y_i^2 \\ &\ge q_{\max}^{-1} \textstyle\sum_{i=1}^p Y_i^2. \end{aligned} \tag{11.3.4}$$

Hence, we obtain:

$$\begin{aligned} &\lambda_{\max}(\overline{\boldsymbol{X}}_n - \boldsymbol{\mu})'\boldsymbol{\Sigma}^{-1}(\overline{\boldsymbol{X}}_n - \boldsymbol{\mu}) \\ &\ge \textstyle\sum_{i=1}^p Y_i^2 \\ &= (B\overline{\boldsymbol{X}}_n - B\boldsymbol{\mu})'(B\overline{\boldsymbol{X}}_n - B\boldsymbol{\mu}) \\ &= (\overline{\boldsymbol{X}}_n - \boldsymbol{\mu})'B'B(\overline{\boldsymbol{X}}_n - \boldsymbol{\mu}) \\ &= (\overline{\boldsymbol{X}}_n - \boldsymbol{\mu})'(\overline{\boldsymbol{X}}_n - \boldsymbol{\mu}) \text{ since } B'B = I_{p\times p}. \end{aligned} \tag{11.3.5}$$

Note that this relationship holds whatever be Q as long as Q is p.d., and Q does not have to be the matrix $\boldsymbol{\Sigma}$.

Then, combining (11.3.4) and (11.3.5), we can claim:

$$\begin{aligned}&\mathrm{P}_{\boldsymbol{\mu},\boldsymbol{\Sigma}}\left\{\lambda_{\max}(\overline{\boldsymbol{X}}_n-\boldsymbol{\mu})'\boldsymbol{\Sigma}^{-1}(\overline{\boldsymbol{X}}_n-\boldsymbol{\mu})\le d^2\right\}\\&\quad\le \mathrm{P}_{\boldsymbol{\mu},\boldsymbol{\Sigma}}\left\{(\overline{\boldsymbol{X}}_n-\boldsymbol{\mu})'(\overline{\boldsymbol{X}}_n-\boldsymbol{\mu})\le d^2\right\}.\end{aligned}\tag{11.3.6}$$

In other words, the ellipsoid

$$\lambda_{\max}(\overline{\boldsymbol{X}}_n-\boldsymbol{\mu})'\boldsymbol{\Sigma}^{-1}(\overline{\boldsymbol{X}}_n-\boldsymbol{\mu})\le d^2$$

is completely inside the sphere

$$(\overline{\boldsymbol{X}}_n-\boldsymbol{\mu})'(\overline{\boldsymbol{X}}_n-\boldsymbol{\mu})\le d^2.$$

Hence, the confidence coefficient associated with $\mathcal{R}_n^*$ is expressed as:

$$\begin{aligned}&\mathrm{P}_{\boldsymbol{\mu},\boldsymbol{\Sigma}}\{\boldsymbol{\mu}\in\mathcal{R}_n^*\}\\&\quad= \mathrm{P}_{\boldsymbol{\mu},\boldsymbol{\Sigma}}\left\{(\overline{\boldsymbol{X}}_n-\boldsymbol{\mu})'(\overline{\boldsymbol{X}}_n-\boldsymbol{\mu})\le d^2\right\}\\&\quad\ge \mathrm{P}_{\boldsymbol{\mu},\boldsymbol{\Sigma}}\left\{\lambda_{\max}(\overline{\boldsymbol{X}}_n-\boldsymbol{\mu})'\boldsymbol{\Sigma}^{-1}(\overline{\boldsymbol{X}}_n-\boldsymbol{\mu})\le d^2\right\}\\&\quad= \mathrm{P}_{\boldsymbol{\mu},\boldsymbol{\Sigma}}\left\{(\overline{\boldsymbol{X}}_n-\boldsymbol{\mu})'\boldsymbol{\Sigma}^{-1}(\overline{\boldsymbol{X}}_n-\boldsymbol{\mu})\le \frac{d^2}{\lambda_{\max}}\right\}.\end{aligned}\tag{11.3.7}$$

for all fixed n.

But, it is a simple matter to check that

$$n(\overline{\boldsymbol{X}}_n-\boldsymbol{\mu})'\boldsymbol{\Sigma}^{-1}(\overline{\boldsymbol{X}}_n-\boldsymbol{\mu})\sim\chi_p^2 \text{ under } \boldsymbol{\mu} \text{ and } \boldsymbol{\Sigma}.\tag{11.3.8}$$

By appropriately modifying our derivations from Section 11.7.2, one may directly verify (11.3.8).

As in Section 11.2, let a be such that $F(a)=1-\alpha$, that is a is an upper $100\alpha\%$ point of a χ_p^2 distribution. Since we require that $F\left(nd^2/\lambda_{\max}\right)\ge F(a)=1-\alpha$, we must have $nd^2/\lambda_{\max}\ge a$, so that we determine:

$$n= \text{ the smallest integer } \ge \frac{a\lambda_{\max}}{d^2}=C^*, \text{ say.}\tag{11.3.9}$$

We refer to C^* as an optimal fixed sample size had the dispersion matrix $\boldsymbol{\Sigma}$ been known. But, the magnitude of C^* remains unknown and no fixed-sample-size methodology would help.

We reiterate that Healy (1956) and Chatterjee (1959a,b,1962a) developed analogs of Stein's (1945,1949) two-stage procedure. Srivastava (1967), on the other hand, introduced analogs of Chow and Robbins's (1965) purely sequential procedure.

It is understood that a multistage or sequential strategy would each come up with a final sample size $N\equiv N(d)$. Based on full dataset N, $\boldsymbol{X}_1$, $\boldsymbol{X}_2$, ..., $\boldsymbol{X}_N$, we will propose a fixed-size spherical confidence region

$$\mathcal{R}_N^*=\left\{\boldsymbol{\omega}_{p\times 1}\in \mathbb{R}^p:(\overline{\boldsymbol{X}}_N-\boldsymbol{\omega})'(\overline{\boldsymbol{X}}_N-\boldsymbol{\omega})\le d^2\right\},\tag{11.3.10}$$

for $\boldsymbol{\mu}$.

$\boldsymbol{\Sigma}$ is arbitrary p.d., but unknown: Two-stage strategy (11.3.14) is consistent, but it is asymptotically inefficient.

In order to design suitable stopping times $N \equiv N(d)$, we must plug in an appropriate estimator for $\lambda_{\max}$ in the expression of C^* . So, this is what we intend to do. At any point, having observations $\boldsymbol{X}_1, \ldots, \boldsymbol{X}_n$, we proceed as follows:

$$\begin{array}{l}\text{Estimate } \lambda_{\max}\text{, the largest eigenvalue or}\\ \text{characteristic root of } \boldsymbol{\Sigma}\text{, by } S_{\max(n)}\text{, the}\\ \text{largest eigenvalue or the characteristic root of}\\ S_n\text{, the } p \times p \text{ sample dispersion matrix.}\end{array} \tag{11.3.11}$$

For methodologies included here, one will note that for all fixed $n(\geq m)$, the random variable $I(N = n)$ would be determined by $(S_m, S_{m+1}, \ldots, S_n)$ alone. Thus, $\overline{\boldsymbol{X}}_n$ and $(S_m, S_{m+1}, \ldots, S_n)$ will be independent for all fixed $n(\geq m)$. So, $I(N = n)$ and $\overline{\boldsymbol{X}}_n$ would be independent random variables for all fixed $n(\geq m)$.

Hence, in view of (11.3.7), we can express the confidence coefficient associated with $\mathcal{R}_N^*$ as:

$$\begin{aligned}
&\mathrm{P}_{\boldsymbol{\mu},\boldsymbol{\Sigma}}\{\boldsymbol{\mu} \in \mathcal{R}_N^*\}\\
&= \mathrm{P}_{\boldsymbol{\mu},\boldsymbol{\Sigma}}\{(\overline{\boldsymbol{X}}_N - \boldsymbol{\mu})'(\overline{\boldsymbol{X}}_N - \boldsymbol{\mu}) \leq d^2\}\\
&\geq \mathrm{P}_{\boldsymbol{\mu},\boldsymbol{\Sigma}}\left\{(\overline{\boldsymbol{X}}_N - \boldsymbol{\mu})'\boldsymbol{\Sigma}^{-1}(\overline{\boldsymbol{X}}_N - \boldsymbol{\mu}) \leq \frac{d^2}{\lambda_{\max}}\right\}\\
&= \mathrm{E}_{\boldsymbol{\mu},\boldsymbol{\Sigma}}\left[F\left(\frac{Nd^2}{\lambda_{\max}}\right)\right] \text{ with } F(x) \equiv F(x;p)\\
&\qquad \mathrm{P}\left(U \leq x\right), U \sim \chi_p^2, x > 0.
\end{aligned} \tag{11.3.12}$$

As before, sometimes we drop the index p from $F(x;p)$.

11.3.1 Two-Stage Procedure

The motivation for a two-stage procedure came from Stein (1945,1949). Healy (1956) and Chatterjee (1959a,b) laid the foundations of this multivariate estimation technique.

Let us begin with observations $\boldsymbol{X}_1, \boldsymbol{X}_2, \ldots, \boldsymbol{X}_m$ where $m(\geq p+1)$ is the pilot size. We obtain $\overline{\boldsymbol{X}}_m, S_m$ following (11.3.2) and $S_{\max(m)}$, the largest eigenvalue of $p \times p$ sample dispersion matrix, S_m. Also, let $F_{p,m-p,\alpha}$ be an upper $100\alpha\%$ point of $F_{p,m-p}$ distribution, and denote:

$$b^* \equiv b^*_{p,m,\alpha} = \frac{p(m-1)}{m-p}F_{p,m-p,\alpha}. \tag{11.3.13}$$

Next, we define the following two-stage stopping variable:

$$N \equiv N(d) = \max\left\{m, \left\langle \frac{b^* S_{\max(m)}}{d^2} \right\rangle + 1\right\}, \tag{11.3.14}$$

where the number $b^* \equiv b^*_{p,m,\alpha}$ comes from (11.3.13).

If $N = m$, we do not require any observations at second stage. If $N > m$, we sample the difference at second stage by gathering additional observations $\boldsymbol{X}_{m+1}, \ldots, \boldsymbol{X}_N$.

It is easy to see that N is finite w.p. 1. So, $\mathcal{R}^*_N$ will be a genuine confidence region estimator. Thus, based on *combined* dataset N, $\boldsymbol{X}_1$, $\boldsymbol{X}_2$, ..., $\boldsymbol{X}_N$ from stages 1 and 2, we estimate $\boldsymbol{\mu}$ by $\mathcal{R}^*_N$ from (11.3.10). Analogs of the following result were obtained by Healy (1956) and Chatterjee (1959a,b,1962a).

Theorem 11.3.1: *For the stopping variable N defined by (11.3.14), for all fixed $\boldsymbol{\mu}, \boldsymbol{\Sigma}_{p\times p}, p, m, d$ and α, one has the following properties:*

$$\text{P}_{\boldsymbol{\mu},\boldsymbol{\Sigma}}\{\boldsymbol{\mu} \in \mathcal{R}^*_N\} \ge 1 - \alpha \text{ [\textit{Consistency or Exact Consistency}]}$$

where $b^* \equiv b^*_{p,m,\alpha}$ is taken from (11.3.13) and $\mathcal{R}^*_N$ came from (11.3.10).

Its proof is sketched in Section 11.7.7.

11.3.2 Sequential Procedure

Srivastava (1967) developed a purely sequential methodology in the spirit of Chow and Robbins (1965). Recall that $S_{\max(n)}$ denotes the largest eigenvalue of sample dispersion matrix $S_n, n \ge m$.

We would begin with observations $\boldsymbol{X}_1, \ldots, \boldsymbol{X}_m$ where $m(\ge p+1)$ is pilot size. Then, we will proceed with one additional observation at-a-time, and terminate sampling according to stopping time:

$$\begin{aligned} &N \equiv N(d) \text{ is the smallest integer } n(\ge m) \\ &\text{for which we observe } n \ge \left(\frac{a}{d^2}\right) S_{\max(n)}. \end{aligned} \tag{11.3.15}$$

This sequential strategy is implemented in a usual fashion. Again, we have $\text{P}_{\boldsymbol{\mu},\boldsymbol{\Sigma}}(N < \infty) = 1$. That is, sampling would terminate w.p. 1. Thus, $\mathcal{R}^*_N$ from (11.3.10) will be a genuine confidence region estimator for $\boldsymbol{\mu}$. So, upon termination, we estimate $\boldsymbol{\mu}$ by $\mathcal{R}^*_N$. Note that the associated confidence coefficient is given by (11.3.12).

The following *first-order* results were obtained by Srivastava (1967).

Theorem 11.3.2: *For the stopping variable N defined by (11.3.15), for all fixed $\boldsymbol{\mu}, \boldsymbol{\Sigma}_{p\times p}, p, m$ and α, one has the following properties as $d \to 0$:*

(i) $\lim N/C^* = 1$ *in probability*;

(ii) $\lim \text{E}_{\boldsymbol{\mu},\boldsymbol{\Sigma}}[N/C^*] = 1$ [*Asymptotic First-Order Efficiency*];

(iii) $\lim \text{P}_{\boldsymbol{\mu},\boldsymbol{\Sigma}}\{\boldsymbol{\mu} \in \mathcal{R}^*_N\} = 1 - \alpha$ [*Asymptotic Consistency*];

where $\mathcal{R}_N^*$ *and* $C^* = a\lambda_{\max}/d^2$ *came from (11.3.10) and (11.3.9) respectively.*

We sketch a proof of these results in Section 11.7.8. Our proof of part (ii) is considerably simpler than Srivastava's (1967) original proof. In a technical report, Nagao and Srivastava (1997) looked at some second-order properties of the sequential strategy.

$\boldsymbol{\Sigma}$ is arbitrary p.d., but unknown: Sequential strategy (11.3.15) is asymptotically consistent and asymptotically first-order efficient. But, it does not have consistency property anymore.

11.4 Minimum Risk Point Estimation: Unknown Dispersion Matrix

Suppose that we have available a sequence of independent observations $\boldsymbol{X}_1, \ldots, \boldsymbol{X}_n, \ldots$ from a $N_p(\boldsymbol{\mu}, \boldsymbol{\Sigma})$ distribution. We suppose that $\boldsymbol{\mu}$ is unknown and $\boldsymbol{\Sigma}_{p\times p}$ is p.d., but otherwise it is also completely unknown. A minimum risk point estimation problem for the mean vector $\boldsymbol{\mu}$ is formulated as follows.

We work under a loss function which is composed of estimation error and the cost of sampling. Having recorded $\boldsymbol{X}_1, \ldots, \boldsymbol{X}_n$, we let the loss function for estimating $\boldsymbol{\mu}$ by a sample mean vector $\overline{\boldsymbol{X}}_n$ be:

$$L_n(\boldsymbol{\mu}, \overline{\boldsymbol{X}}_n) = (\overline{\boldsymbol{X}}_n - \boldsymbol{\mu})' A (\overline{\boldsymbol{X}}_n - \boldsymbol{\mu}) + cn. \tag{11.4.1}$$

Here, $c > 0$ is known, and $A_{p\times p}$ is a known p.d. matrix. Also, recall the $p \times p$ sample dispersion matrix from (11.3.2), namely,

$$S_n = (n-1)^{-1} \textstyle\sum_{i=1}^{n} (\boldsymbol{X}_i - \overline{\boldsymbol{X}}_n)(\boldsymbol{X}_i - \overline{\boldsymbol{X}}_n)', n \geq 2. \tag{11.4.2}$$

In (11.4.1), the term cn represents the cost for gathering n observations and $(\overline{\boldsymbol{X}}_n - \boldsymbol{\mu})' A (\overline{\boldsymbol{X}}_n - \boldsymbol{\mu})$ represents the loss due to estimation of $\boldsymbol{\mu}$. If n is small, then the sampling cost cn will be small but $(\overline{\boldsymbol{X}}_n - \boldsymbol{\mu})' A (\overline{\boldsymbol{X}}_n - \boldsymbol{\mu})$ may be large. On the other hand, if n is large, then the sampling cost cn will be large but then $(\overline{\boldsymbol{X}}_n - \boldsymbol{\mu})' A (\overline{\boldsymbol{X}}_n - \boldsymbol{\mu})$ may be small. This estimation problem strikes a balance between the expenses due to sampling and estimation error.

We know that $\overline{\boldsymbol{X}}_n \sim N_p\left(\boldsymbol{\mu}, n^{-1}\boldsymbol{\Sigma}\right)$ and

$$\mathrm{E}_{\boldsymbol{\mu},\boldsymbol{\Sigma}}\left[(\overline{\boldsymbol{X}}_n - \boldsymbol{\mu})(\overline{\boldsymbol{X}}_n - \boldsymbol{\mu})'\right] = n^{-1}\boldsymbol{\Sigma}.$$

Now, fixed-sample-size risk $R_n(c)$ associated with (11.4.1) is expressed as:

$$\begin{aligned} & \mathrm{E}_{\boldsymbol{\mu},\boldsymbol{\Sigma}}\left[L_n(\boldsymbol{\mu},\overline{\boldsymbol{X}}_n)\right] \\ &= \mathrm{E}_{\boldsymbol{\mu},\boldsymbol{\Sigma}}\left[(\overline{\boldsymbol{X}}_n-\boldsymbol{\mu})'A(\overline{\boldsymbol{X}}_n-\boldsymbol{\mu})+cn\right] \\ &= \mathrm{E}_{\boldsymbol{\mu},\boldsymbol{\Sigma}}\left[\mathrm{Trace}(\overline{\boldsymbol{X}}_n-\boldsymbol{\mu})'A(\overline{\boldsymbol{X}}_n-\boldsymbol{\mu})\right]+cn \\ &= \mathrm{E}_{\boldsymbol{\mu},\boldsymbol{\Sigma}}\left[\mathrm{Trace}\left(A(\overline{\boldsymbol{X}}_n-\boldsymbol{\mu})(\overline{\boldsymbol{X}}_n-\boldsymbol{\mu})'\right)\right]+cn \qquad (11.4.3) \\ &= \mathrm{Trace}\left(A\mathrm{E}_{\boldsymbol{\mu},\boldsymbol{\Sigma}}\left[(\overline{\boldsymbol{X}}_n-\boldsymbol{\mu})(\overline{\boldsymbol{X}}_n-\boldsymbol{\mu})'\right]\right)+cn \\ &= \frac{\mathrm{Trace}(A\boldsymbol{\Sigma})}{n}+cn. \end{aligned}$$

Now, our goal is to introduce a purely sequential estimation strategy that would minimize $R_n(c)$ for all p.d. matrices, $\boldsymbol{\Sigma}$. As is customary, let us pretend that n is a continuous variable so that $n \equiv n^*(c)$ would minimize $R_n(c)$ where

$$n^* \equiv n^*(c) = \left(\frac{\mathrm{Trace}(A\boldsymbol{\Sigma})}{c}\right)^{1/2}. \qquad (11.4.4)$$

Even though $n^*(c)$ may not be an integer, we refer to it as an optimal fixed sample size had the $\boldsymbol{\Sigma}$ matrix been known. The minimum fixed-sample-size risk is then given by:

$$R_{n^*}(c) \equiv \frac{\mathrm{Trace}(A\boldsymbol{\Sigma})}{n^*}+cn^* = 2cn^*. \qquad (11.4.5)$$

Surely, the magnitude of optimal fixed sample size remains unknown since the $\boldsymbol{\Sigma}$ matrix is unknown. Indeed, no fixed-sample-size procedure will minimize the risk (11.4.3) under a loss function (11.4.1), uniformly for all p.d. matrix $\boldsymbol{\Sigma}$.

Khan (1968) and Rohatgi and O'Neill (1973) investigated purely sequential estimation strategies in the spirit of Robbins (1959) and Starr (1966b) when $\boldsymbol{\Sigma}_{p\times p}$ was p.d. and diagonal. Sinha and Mukhopadhyay (1976) handled the same problem when $p = 2$, $A_{2\times 2}$ was p.d. and diagonal, but $\boldsymbol{\Sigma}_{2\times 2}$ was p.d. and arbitrary. Wang (1980) considered an analogous problem when $\boldsymbol{\Sigma}_{p\times p} = \sigma^2 H_{p\times p}$ where σ^2 was unknown but H was p.d. and known.

Ghosh et al. (1976) developed the minimum risk point estimation problem for $\boldsymbol{\mu}$ in the most general framework that is under discussion here. They formulated a sequential estimation strategy. In what follows, we discuss their original fundamental methodology and its properties.

11.4.1 Purely Sequential Procedure

The purely sequential stopping rule of Ghosh et al. (1976) is summarized as follows. We begin with $\boldsymbol{X}_1, \ldots, \boldsymbol{X}_m$ where $m(\geq 2)$ is a pilot size. Then, we proceed with one additional $\boldsymbol{X}$ at-a-time according to the stopping time:

$$N \equiv N(c) \text{ is the smallest integer } n(\geq m)$$
$$\text{for which we observe } n \geq \left(\frac{\text{Trace}(AS_n)}{c}\right)^{1/2}. \tag{11.4.6}$$

The sample dispersion matrix S_n was defined in (11.4.2). This sequential strategy is implemented in the same way we did previously for other sequential procedures.

Again, one may check that $\mathrm{P}_{\boldsymbol{\mu},\boldsymbol{\Sigma}}(N < \infty) = 1$, that is sampling would terminate w.p. 1. Upon termination, we estimate $\boldsymbol{\mu}$ by $\overline{\boldsymbol{X}}_N$ obtained from full dataset $N, \boldsymbol{X}_1, \ldots, \boldsymbol{X}_m, \ldots, \boldsymbol{X}_N$.

$\boldsymbol{\Sigma}$ is arbitrary p.d., but unknown: Sequential strategy (11.4.6) is asymptotically second-order efficient and asymptotically risk efficient. Its regret function has an asymptotic second-order expansion.

Next, the risk associated with $\overline{\boldsymbol{X}}_N$ is expressed as:

$$\begin{aligned} R_N(c) &\equiv \mathrm{E}_{\boldsymbol{\mu},\boldsymbol{\Sigma}}\left[L_N(\boldsymbol{\mu}, \overline{\boldsymbol{X}}_N)\right] \\ &= \text{Trace}(A\boldsymbol{\Sigma})\mathrm{E}_{\boldsymbol{\mu},\boldsymbol{\Sigma}}\left[\frac{1}{N}\right] + c\mathrm{E}_{\boldsymbol{\mu},\boldsymbol{\Sigma}}[N], \end{aligned} \tag{11.4.7}$$

since the random variables $I(N = n)$ and $\overline{\boldsymbol{X}}_n$ are independent for all $n \geq m$.

Now, notice from the expression of n^* in (11.4.4) that $\text{Trace}(A\boldsymbol{\Sigma}) = cn^{*2}$ and thus we rewrite (11.4.7) as follows:

$$R_N(c) \equiv cn^{*2}\mathrm{E}_{\boldsymbol{\mu},\boldsymbol{\Sigma}}\left[\frac{1}{N}\right] + c\mathrm{E}_{\boldsymbol{\mu},\boldsymbol{\Sigma}}[N]. \tag{11.4.8}$$

Ghosh and Mukhopadhyay (1976) defined the following measures, along the lines of Robbins (1959), for comparing the sequential estimation strategy $(N, \overline{\boldsymbol{X}}_N)$ from (11.4.6) with the fictitious, and yet optimal fixed-sample-size estimation strategy $(n^*, \overline{\boldsymbol{X}}_{n^*})$. We may pretend for a moment, however, that n^* is a known integer!

$$\begin{aligned} &\textit{RiskEfficiency}: \text{RiskEff}(c) \equiv \frac{R_N(c)}{R_{n^*}(c)} \\ &\qquad = \frac{1}{2}\left\{\mathrm{E}_{\boldsymbol{\mu},\boldsymbol{\Sigma}}\left[\frac{n^*}{N}\right] + \mathrm{E}_{\boldsymbol{\mu},\boldsymbol{\Sigma}}\left[\frac{N}{n^*}\right]\right\}; \\ &\textit{Regret}: \text{Regret}(c) \equiv R_N(c) - R_{n^*}(c) \\ &\qquad = c\mathrm{E}_{\boldsymbol{\mu},\boldsymbol{\Sigma}}\left[\frac{(N-n^*)^2}{N}\right]. \end{aligned} \tag{11.4.9}$$

We first summarize some asymptotic *first-order* results due to Ghosh and Mukhopadhyay (1976) in the following theorem.

Theorem 11.4.1: *For a stopping variable N defined by (11.4.6), for all fixed $\boldsymbol{\mu}, \boldsymbol{\Sigma}_{p\times p}, p(\geq 2), m$ and $A_{p\times p}$, one has following properties:*

(i) $\mathrm{E}_{\boldsymbol{\mu},\boldsymbol{\Sigma}}[N] \leq n^* + m + 1$ *for fixed* c;

(ii) $\lim_{c\to 0} N/n^* = 1$ *in probability*;

(iii) $\mathrm{P}_{\boldsymbol{\mu},\boldsymbol{\Sigma}}(N \leq \varepsilon n^*) = O_e(n^{*-m})$ *for any fixed* ε, $0 < \varepsilon < 1$;

(iv) $\lim_{c\to 0} \mathrm{E}_{\boldsymbol{\mu},\boldsymbol{\Sigma}}[N/n^*] = 1$ [*Asymptotic First-Order Efficiency*];

(v) $\dfrac{N - n^*}{\sqrt{n^*}}$ *converges to* $N\left(0, \frac{1}{2}\tau^2\right)$ *in distribution with*

$\tau^2 \equiv \mathrm{Trace}\left((A\boldsymbol{\Sigma})^2\right)/\mathrm{Trace}^2(A\boldsymbol{\Sigma})$;

(vi) $\lim_{c\to 0} \mathrm{RiskEff}(c) = 1$ [*Asymptotic First-Order Risk Efficiency*];

where $n^* = (\mathrm{Trace}(A\boldsymbol{\Sigma})/c)^{1/2}$ *came from (11.4.4).*

Ghosh and Mukhopadhyay (1976) included proofs for all *first-order* results listed in Theorem 11.4.1. Callahan (1969) independently obtained the risk efficiency result in her unpublished Ph.D. theses assuming that $A = kI_{p\times p}, k > 0$.

Ghosh and Mukhopadhyay (1976) also showed the following *second-order* result in the spirit of Starr and Woodroofe (1969). They gave considerably newer proofs.

Theorem 11.4.2 : *For a stopping variable N defined by (11.4.6), for all fixed $\boldsymbol{\mu}, \boldsymbol{\Sigma}_{p\times p}, p(\geq 2), m$ and $A_{p\times p}$, one has following property as $c \to 0$:*

$$\mathrm{Regret}(c) = O(c).$$

Obviously, the asymptotic first-order risk efficiency property (Theorem 11.4.1, part (vi)) will follow immediately from the stronger assertion in Theorem 11.4.2.

Woodroofe (1977) gave an elegant proof of a result that was considerably sharper than Theorem 11.4.2. He gave asymptotic second-order expansions of both regret function and average sample size. Some crucial second-order results are summarized as follows.

Theorem 11.4.3: *For a stopping variable N defined by (11.4.6), for all fixed $\boldsymbol{\mu}, \boldsymbol{\Sigma}_{p\times p}, p(\geq 2), m$ and $A_{p\times p}$, one has following properties as $c \to 0$:*

(i) $\mathrm{Regret}(c) = \frac{1}{2}\left[\mathrm{Trace}\left((A\boldsymbol{\Sigma})^2\right)/\mathrm{Trace}^2(A\boldsymbol{\Sigma})\right] c + o(c)$ *if* (a) $m \geq 3$ *when* $p = 2$, *and* (b) $m \geq 2$ *when* $p \geq 3$;

(ii) $\mathrm{E}_{\boldsymbol{\mu},\boldsymbol{\Sigma}}[N - n^*] = q^* + o(1)$ *if* $m \geq 2$;

where $n^* = (\mathrm{Trace}(A\boldsymbol{\Sigma})/c)^{1/2}$ *comes from (11.4.4) and* q^* *is an appropriate term involving* $\boldsymbol{\Sigma}$.

Notice that the $O(c)$ term in the expansion of the regret function comes out to

$$\frac{1}{2}\left[\text{Trace}\left((A\boldsymbol{\Sigma})^2\right)/\text{Trace}^2(A\boldsymbol{\Sigma})\right]c \tag{11.4.10}$$

in a p-dimensional case. That is, for arbitrary p, this term would involve both matrices A and $\boldsymbol{\Sigma}$.

But, when $p = 1$, the term from (11.4.10) clearly reduces to $\frac{1}{2}c$ which agrees with a regret expansion that was mentioned in (6.4.14) with $m \geq 4$.

We will sketch a proof of Theorem 11.4.3, part (i) in Section 11.7.9 assuming $m \geq 3$.

11.5 Data Analyses and Conclusions

Simulations of the procedures and related computations descried in Sections 11.2, 11.3 and 11.4 can be carried out with the computer program **Seq11.exe**. The program works with live data input too. Figure 11.5.1 shows the screen input and Figure 11.5.2 presents the output corresponding to the two-stage methodology (11.2.9) implemented by the program.

The main program **Seq11.exe** is divided into three subprograms. The first subprogram evaluates fixed-width confidence region for the mean vector, $\boldsymbol{\mu}$, of a multivariate normal population with covariance matrix of the form $\boldsymbol{\Sigma} = \sigma^2 H$. Here, H is a known positive definite matrix and σ^2 is an unknown positive constant. The program will prompt to select one of the four procedures from Section 11.2, namely: two-stage, modified two-stage, purely sequential and three-stage to construct the confidence region. Figures 11.5.1 and 11.5.2 show respectively the input and output of the program for a simulation study of a 95% confidence region for $\boldsymbol{\mu}$ using two-stage procedure (11.2.9). Data for simulation were generated from a three-dimensional normal distribution with

$$\boldsymbol{\mu} = \begin{pmatrix} 3 \\ 5 \\ 2 \end{pmatrix}, \quad H = \begin{pmatrix} 25 & -2 & 4 \\ -2 & 4 & 1 \\ 4 & 1 & 9 \end{pmatrix}$$

and $\sigma^2 = 4$.

In the example addressed in Figures 11.5.1-11.5.2, we have $p = 3, m = 10$ and $\alpha = 0.05$. So, we have $a \equiv \chi^2_{3,0.05} = 7.8147$ and $b \equiv 3 \times F_{3,27,0.05} = 3 \times 2.9604 = 8.8812$. Now, from Theorem 11.2.1, part (ii), we see that theoretical oversampling percentage ought to be around $100(\frac{b}{a} - 1)\%$ which amounts to 13.647%. From 10000 simulations, however, we have $\overline{n} = 141.92$, but we have $C = \frac{7.8147\times 4}{0.5^2} = 125.04$ since we also fixed $\sigma^2 = 4$ and $d = 0.5$. So, an estimate of oversampling percentage from 10000 simulations should be $100(\frac{\overline{n}}{C} - 1)\%$ which amounts to 13.5%. This estimated oversampling percentage is incredibly close to 13.647%, theoretical oversampling percentage!

```
          Select a program using the number given
          in the left hand side. Details of these
          programs are given in Chapter 11.

*********************************************************
* Number         Program                                *
*  1   Fixed-Size Confidence Region:  Known H Matrix     *
*  2   Fixed-Size Confidence Region:  Unknown Dispersion *
*  3   Minimum Risk Point Estimation: Unknown Dispersion *
*  0   Exit                                              *
*********************************************************
   Enter the program number: 1

   Number  Procedure
     1     Two-Stage
     2     Modified Two-Stage
     3     Purely Sequential
     4     Three-Stage
   Enter procedure number: 1

   Do you want to save the results in a file(Y or N)? Y
   Note that your default outfile name is "SEQ11.OUT".
   Do you want to change the outfile name (Y or N)? N

Enter the following preassigned values for a fixed-size
confidence region for a p-dimensional normal mean vector
   Enter p (integer > 1): 3

Now input matrix H where Cov matrix = sigma^2 H.
Enter the elements of H row by row.
That is, 3 elements in each row

   Enter row  1: 25 -2 4
   Enter row  2: -2  4 1
   Enter row  3:  4  1 9

alpha (for 100(1-alpha)% confidence interval) = .05
d (Fixed-size of the confidence region) = .5
Enter the initial sample size, m: 10
Do you want to change the input parameters?[Y or N] N

Do you want to simulate data or input real data?
Type S (Simulation) or D (Real Data): S

Do you want to store simulated sample sizes (Y or N)? N

Number of simulation replications? 10000

Enter a positive integer(<10) to initialize
          the random number sequence: 1

Enter the following parameters for simulation:
     Mean vector ( 3 values)  = 3 5 2
     Sigma^2 = 4
```

Figure 11.5.1. *Screen input for* ***Seq11.exe****.*

```
Fixed-Size Confidence Region for the Multivariate
    Normal Mean Using Two-Stage Procedure.
    ----------------------------------------

    Data for the simulation were generated from
    3-dimensional normal distribution with
    mean =      3.0000     5.0000     2.0000
    and matrix H:
        25.0000    -2.0000     4.0000
        -2.0000     4.0000     1.0000
         4.0000     1.0000     9.0000

    Input parameters for simulations:
                         alpha =    0.05
                             d =    0.50
                       sigma^2 =    4.00
    Initial sample size, m =       10
    Number of simulations  = 10000

    Optimal sample size (C)              =    125.04
    Average final sample size (n_bar) =    141.92
    Std. dev. of n (s_n)                 =     38.40
    Minimum sample size (n_min)          =     48
    Maximum sample size (n_max)          =    319
    Coverage probability                 =     0.9492
```

Figure 11.5.2. *Results from* ***Seq11.exe***.

```
Fixed-Size Confidence Region for the Multivariate
    Normal Mean Using Three-Stage Procedure
              With rho = 0.50
              ---------------

    Input Parameters:
                         alpha =      0.05
                             d =     0.500
    Initial Sample size, m =        10
    Input Data file Name: S11A.dat

    Input H matrix:
        25.0000    -2.0000     4.0000
        -2.0000     4.0000     1.0000
         4.0000     1.0000     9.0000

    Inverse of H:
         0.0461     0.0290    -0.0237
         0.0290     0.2754    -0.0435
        -0.0237    -0.0435     0.1265

    Second stage sample size (T)=        83
    Final sample size (N)        =       139
    Estimate of sigma^2          =    3.9857
    Sample mean =   4.5079   3.8818   4.0584
```

Figure 11.5.3. ***Seq11.exe*** *program output for a 95% confidence region using a three-stage procedure.*

Additionally, in Figure 11.5.1 we find the estimated coverage probability, $\overline{p} = 0.9492$. Its estimated standard error ought to be $\sqrt{\overline{p}(1-\overline{p})/10000} = 0.0021959$. Let us look at a two-standard deviation interval: $0.9492 \pm 2 \times 0.0021959$ which amounts to $(0.94481, 0.95359)$. This interval includes the target, namely 0.95. In other words, we should feel good about this kind of practical validation of the consistency property (Theorem 11.2.1, part (i)).

The program can also estimate the sample size for a confidence region using live data, for example, via three-stage methodology (11.2.13) - (11.2.14) implemented by the program. Figure 11.5.3 shows the output of this exercise that ran with $\alpha = 0.05, m = 10, d = 0.5$ and $\rho = 0.5$. We find the estimated sample size, sample mean vector and inverse of H. These results were obtained using the data file **S11A.data**.

One can easily construct a three-stage 95% fixed-size confidence region for $\boldsymbol{\mu}$ from all the information compiled by the output in Figure 11.5.3. From (11.2.6) and the results given in Figure 11.5.3 the 95% confidence region for μ can be written as follows:

$$\mathcal{R}_{139} = \left\{\boldsymbol{\mu}_{3\times1} \in I\!R^3 : (\overline{\boldsymbol{x}}_{139} - \boldsymbol{\mu})' H^{-1} (\overline{\boldsymbol{x}}_{139} - \boldsymbol{\mu}) \le 0.5^2 = 0.25\right\},$$

$$\overline{\boldsymbol{x}}_{139} = \begin{pmatrix} 4.5079 \\ 3.8818 \\ 4.0584 \end{pmatrix}, \; H^{-1} = \begin{pmatrix} 0.0461 & 0.0290 & -0.0237 \\ 0.0290 & 0.2754 & -0.0435 \\ -0.0237 & -0.0435 & 0.1265 \end{pmatrix}.$$

'Program number: 2' within **Seq11.exe** would construct a fixed-size confidence region for $\boldsymbol{\mu}$ when the $\boldsymbol{\Sigma}$ is unknown described in Section 11.3. The program will prompt a user to select either the two-stage methodology (11.3.14) or the sequential methodology (11.3.15) to construct the required confidence region.

'Program number: 3' within **Seq11.exe** would obtain minimum risk point estimators for a normal mean vector using the sequential strategy (11.4.6) from Sections 11.4. One may use simulations or live data.

Figure 11.5.4 shows the input screen for our program to simulate a minimum risk point estimation problem for a mean vector. The loss function

$$L_n(\boldsymbol{\mu}, \overline{\boldsymbol{X}}_n) = (\overline{\boldsymbol{X}}_n - \boldsymbol{\mu})' A (\overline{\boldsymbol{X}}_n - \boldsymbol{\mu}) + 0.5n$$

was used for this exercise. The data for simulations were generated from a $N_3(\boldsymbol{\mu}, \boldsymbol{\Sigma})$ distribution where

$$\boldsymbol{\mu} = \begin{pmatrix} 3 \\ 5 \\ 2 \end{pmatrix}, \; \boldsymbol{\Sigma} = \begin{pmatrix} 4 & 3 & 1 \\ 3 & 9 & 2 \\ 1 & 2 & 1 \end{pmatrix} \text{ and } A = \begin{pmatrix} 25 & -2 & 4 \\ -2 & 4 & 1 \\ 4 & 1 & 9 \end{pmatrix}.$$

The results from 10000 simulations with $m = 10$ were averaged and the output is shown in Figure 11.5.5. The second-order term in the expansion of Regret(c) given by (11.4.10) or equivalently in Theorem 11.4.3, part (i) is computed. Also, Regret(c) is estimated from simulations. They are both included in the output shown in Figure 11.5.5. They agree rather well even though n^* is only 17.03 which is understandably quite small!

Possibilities are endless. One should definitely explore many aspects of data analyses with the help of our computer program, ***Seq11.exe***, by changing parameter values as well as the design constants.

```
   Select a program using the number given in
   the left hand side or type 0(zero) to exit.
*********************************************************
* Number          Program                               *
*  1   Fixed-Size Confidence Region:  Known H Matrix    *
*  2   Fixed-Size Confidence Region:  Unknown Dispersion*
*  3   Minimum Risk Point Estimation: Unknown Dispersion*
*  0   Exit                                             *
*********************************************************
   Enter the program number: 3

Enter the following preassigned values for a minimum risk
point estimation for a p-dimensional normal mean vector
   Enter p (integer > 1): 3
   c (positive constant in the loss function) = .5

Now input positive definite matrix A in the loss function
Enter the elements of A row by row.
That is, 3 elements in each row

   Enter row  1: 25 -2 4
   Enter row  2: -2 4 1
   Enter row  3: 4 1 9

Enter the initial sample size, m: 10
Do you want to change the input parameters?[Y or N] N
Do you want to simulate data or input real data?
Type S (Simulation) or D (Real Data): S

Do you want to store simulated sample sizes (Y or N)? N
Number of simulation replications? 10000
Enter a positive integer(<10) to initialize
          the random number sequence: 2

Enter the following parameters for simulation:
     Mean vector ( 3 values)  = 3 5 2
     Covariance matrix: ( 3 elements in each row)

     Enter row  1: 4 3 1
     Enter row  2: 3 9 2
     Enter row  3: 1 2 1
```

Figure 11.5.4. *Screen input for minimum risk point estimation with* ***Seq11.exe***.

```
Minimum Risk Point Estimation of the Multivariate
     Normal Mean Using Sequential Procedure.
     ---------------------------------------

Data for the simulation were generated from
3-dimensional normal distribution with
mean =    3.0000    5.0000    2.0000
and covariance matrix:
    4.0000    3.0000    1.0000
    3.0000    9.0000    2.0000
    1.0000    2.0000    1.0000

Input parameter,       c =   0.50
Initial sample size, m =     10
Number of simulations  = 10000
and Matrix A:
   25.0000   -2.0000    4.0000
   -2.0000    4.0000    1.0000
    4.0000    1.0000    9.0000

Optimal sample size (n*)            =    17.03
Average final sample size (n_bar)   =    17.09
Std. dev. of n (s_n)                =     2.53
Minimum sample size (n_min)         =    10
Maximum sample size (n_max)         =    25
Regret(c) given in Theorem 11.4.3   =   0.1599
Average simulated value of Regret(c) =  0.1643
```

Figure 11.5.5. *Minimum risk sequential point estimation.*

11.6 Other Selected Multistage Procedures

In a context of a fixed-size ellipsoidal confidence region problem for $\boldsymbol{\mu}$ when $\boldsymbol{\Sigma}$ is completely unknown and p.d., one will find appropriate modified two-stage, three-stage, and accelerated sequential methodologies discussed in Ghosh et al. (1997, pp. 193-195). Associated *first-order* properties were included. A three-stage methodology was developed with its *second-order* properties by Lohr (1990). Refer to Datta and Mukhopadhyay (1997) in the non-normal case.

In a minimum risk point estimation problem for $\boldsymbol{\mu}$ when $\boldsymbol{\Sigma}$ is completely unknown and p.d., one will find appropriate three-stage and accelerated sequential methodologies introduced in Ghosh et al. (1997, pp. 195-197). Associated *second-order* properties were included.

In a minimum risk point estimation problem for $\boldsymbol{\mu}$ when $\boldsymbol{\Sigma} = \sigma^2 H$ with unknown σ^2 but known p.d. matrix H, Wang (1980) developed an asymptotically *first-order* risk efficient sequential strategy and related results for finding the distribution of the sample size.

In the spirit of Stein (1955) and James and Stein (1961), Ghosh and Sen (1983) first developed James-Stein estimators of $\boldsymbol{\mu}$ which were strictly better than a sample mean vector under two-stage sampling. Subsequently, Takada (1984) introduced a sequential version of James-Stein estimators.

In these two papers, Ghosh-Sen and Takada considered $p \geq 3$, $A = I_{p\times p}$, and $\mathbf{\Sigma} = \sigma^2 I_{p\times p}$, with σ^2 unknown. Mukhopadhyay (1985b) and Nickerson (1987a) had handled two-sample problems in this same light.

Unfortunately, Ghosh and Sen's (1983) two-stage strategy was not asymptotically *first-order* risk efficient, and Takada's (1984) strategy was not sequential in a strict sense. Ghosh et al. (1987) resolved this anomaly by producing James-Stein estimators of $\boldsymbol{\mu}$ which were strictly better than the sample mean vector under an asymptotically *first-order* risk efficient sequential procedure. Ghosh et al. (1987) considered $p \geq 3$, $A = I_{p\times p}$ and $\mathbf{\Sigma} = \sigma^2 H_{p\times p}$, with σ^2 unknown but H known. In-depth discussions can be found in Ghosh et al. (1997, Section 7.6).

Sequential and multistage *joint estimation* problems for a mean vector $\boldsymbol{\mu}$ and a dispersion matrix $\mathbf{\Sigma}$ were developed by Mukhopadhyay (1979a). Mukhopadhyay (1981) gave a sequential strategy for jointly estimating μ and σ^2 in a $N(\mu, \sigma^2)$ distribution. Aoshima and Mukhopadhyay (1998) compared mean vectors under a number of intraclass correlation normal models.

11.7 Some Selected Derivations

We begin with some geometrical insight into the confidence region $\mathcal{R}_n$ defined in (11.2.1). In the rest of this section, we supply parts of some of the other crucial derivations.

11.7.1 Geometry of $\mathcal{R}_n$ (11.2.1)

First, we provide some geometrical insight into the confidence region $\mathcal{R}_n$ defined in (11.2.1). Clearly, this region will be centered at an observed value of the sample mean, $\overline{\boldsymbol{x}}_n$. Suppose that we write $\overline{\boldsymbol{x}}_n' = (\overline{x}_{1n}, \overline{x}_{2n}, \ldots, \overline{x}_{pn})$.

Now, since H is p.d., we can find an orthogonal matrix $B_{p\times p}$ such that $BHB' = \text{diagonal}(h_1, h_2, \ldots, h_p)$. Here, $h_1, h_2, \ldots, h_p$ are p eigenvalues of H and $\text{diagonal}(h_1, h_2, \ldots, h_p)$ stands for a diagonal matrix with its diagonal elements as shown. Surely, $h_1, h_2, \ldots, h_p$ are all positive. Since H is assumed known, we can explicitly determine B and $h_1, h_2, \ldots, h_p$.

Obviously,

$$B'\text{diagonal}(h_1, h_2, \ldots, h_p)B = B'BHB'B = H$$

$$\Rightarrow H^{-1} = B'\text{diagonal}\left(\frac{1}{h_1}, \frac{1}{h_2}, \ldots, \frac{1}{h_p}\right)B.$$

Now, the region $\mathcal{R}_n$ may be expressed as follows:

$$\mathcal{R}_n = \left\{\boldsymbol{\omega}_{p\times 1} \in I\!R^p : \sum_{i=1}^{p} \frac{y_i^2}{d^2 h_i} \leq 1\right\}, \tag{11.7.1}$$

where $\boldsymbol{y}_{p\times 1} = B(\overline{\boldsymbol{x}}_n - \boldsymbol{\omega})$ and we denote $\boldsymbol{y}' = (y_1, y_2, \ldots, y_p)$.

From this representation of $\mathcal{R}_n$, one ought to realize why the region $\mathcal{R}_n$ is called *ellipsoidal.* When $p = 2$, the region is clearly an ellipse on a plane. When $p = 3$, the region will look like an egg or an American football in three dimensions. Hence, if $p \geq 3$, we call this region $\mathcal{R}_n$ a p-dimensional ellipsoid.

In the original geometry, with usual Cartesian coordinates, $\mathcal{R}_n$ is a p-dimensional ellipsoid is centered at $\overline{\boldsymbol{x}}_n$. But, when we look at (11.7.1), we have the same ellipsoid with a changed coordinate system, and (i) it is centered at the origin, $\mathbf{0}_{p\times 1}$, (ii) the lengths of its p axes are respectively $2d\sqrt{h_i}, i = 1, \ldots, p$.

So, when d is chosen smaller (larger), the lengths $2d\sqrt{h_1}, \ldots, 2d\sqrt{h_p}$ of the p axes of $\mathcal{R}_n$ would all become smaller (larger), thereby making the volume of $\mathcal{R}_n$ correspondingly smaller (larger). This explains how a preassigned value of $d(> 0)$ is directly tied with the notion of a "fixed-size" of the ellipsoidal confidence region under consideration. Fixing some d is equivalent to fixing the volume of the proposed confidence region $\mathcal{R}_n$.

11.7.2 Proof of (11.2.4)

Proof of Part (i):

It is clear that an arbitrary linear function $\boldsymbol{a}'(\overline{\boldsymbol{X}}_n - \boldsymbol{\mu})$ of $(\overline{\boldsymbol{X}}_n - \boldsymbol{\mu})$ is distributed as $N(0, \frac{1}{n}\sigma^2\boldsymbol{a}'H\boldsymbol{a}$ for all non-zero $\boldsymbol{a} \in \mathbb{R}^p$ when $\boldsymbol{\mu}$ and σ^2 are true parameters. That is, $(\overline{\boldsymbol{X}}_n - \boldsymbol{\mu})$ is distributed as $N_p(\mathbf{0}, \frac{1}{n}\sigma^2 H)$.

Next, since H is p.d., one finds a non-singular matrix $B_{p\times p}$ such that $H = BB'$. Let us denote $\boldsymbol{Y} = B^{-1}\left(\overline{\boldsymbol{X}}_n - \boldsymbol{\mu}\right)$ so that $\boldsymbol{Y}$ is distributed as N_p with mean vector $\mathbf{0}$ and a dispersion matrix

$$\frac{1}{n}\sigma^2 B^{-1}HB'^{-1} = \frac{1}{n}\sigma^2 B^{-1}BB'B'^{-1} = \frac{1}{n}\sigma^2 I_{p\times p}.$$

Now, the components of $\boldsymbol{Y}$, namely $Y_1, \ldots, Y_p$, are i.i.d. $N(0, \frac{1}{n}\sigma^2)$. In other words, nY_i^2/σ^2 are i.i.d. χ_1^2 so that

$$n\sum_{i=1}^{p} Y_i^2/\sigma^2 \sim \chi_p^2.$$

But, observe that

$$\begin{aligned} &n(\overline{\boldsymbol{X}}_n - \boldsymbol{\mu})'H^{-1}(\overline{\boldsymbol{X}}_n - \boldsymbol{\mu})/\sigma^2 \\ &= n(\overline{\boldsymbol{X}}_n - \boldsymbol{\mu})'B^{-1\prime}B^{-1}(\overline{\boldsymbol{X}}_n - \boldsymbol{\mu})/\sigma^2 \\ &= n\boldsymbol{Y}'\boldsymbol{Y}/\sigma^2 = n\textstyle\sum_{i=1}^{p} Y_i^2/\sigma^2 \sim \chi_p^2. \end{aligned} \tag{11.7.2}$$

So, we have the desired result. ■

Proof of Part (ii):

Again, we express $H = BB'$ where $B_{p\times p}$ is a non-singular matrix. We denote $\boldsymbol{Y}_i = B^{-1}\boldsymbol{X}_i$, which would be i.i.d. N_p with a mean vector $\boldsymbol{\nu} = B^{-1}\boldsymbol{\mu}$ and dispersion matrix,

$$B^{-1\prime}(\sigma^2 H)B^{-1} = \sigma^2 B^{-1}BB'B^{-1\prime} = \sigma^2 I.$$

Let us write $\boldsymbol{Y}_i' = (Y_{1i}, Y_{2i}, \ldots, Y_{pi})$, $i = 1, \ldots, n$, $\overline{\boldsymbol{Y}}_n' = (\overline{Y}_{1n}, \overline{Y}_{2n}, \ldots, \overline{Y}_{pn})$ and $\boldsymbol{\nu}' = (\nu_1, \nu_2, \ldots, \nu_p)$. Now, we have $\boldsymbol{X}_i = B\boldsymbol{Y}_i$ so that we can write:

$$\begin{aligned}
&p(n-1)S_n^2 \\
&\quad = \textstyle\sum_{i=1}^n (B\boldsymbol{Y}_i - B\overline{\boldsymbol{Y}}_n)'H^{-1}(B\boldsymbol{Y}_i - B\overline{\boldsymbol{Y}}_n) \\
&\quad = \textstyle\sum_{i=1}^n (\boldsymbol{Y}_i - \overline{\boldsymbol{Y}}_n)'B'H^{-1}B(\boldsymbol{Y}_i - \overline{\boldsymbol{Y}}_n) \\
&\quad = \textstyle\sum_{i=1}^n (\boldsymbol{Y}_i - \overline{\boldsymbol{Y}}_n)'(\boldsymbol{Y}_i - \overline{\boldsymbol{Y}}_n) \text{ since } B'H^{-1}B = I_{p\times p} \\
&\quad = \textstyle\sum_{i=1}^n (Y_{1i} - \overline{Y}_{1n})^2 + \sum_{i=1}^n (Y_{2i} - \overline{Y}_{2n})^2 + \ldots \\
&\qquad + \textstyle\sum_{i=1}^n (Y_{pi} - \overline{Y}_{pn})^2.
\end{aligned} \tag{11.7.3}$$

Since $Y_{j1}, \ldots, Y_{jn}$ are i.i.d. $N(\nu_j, \sigma^2)$, we know that $\sum_{i=1}^n (Y_{ji} - \overline{Y}_{jn})^2/\sigma^2 \sim \chi^2_{n-1}$ for each $j = 1, \ldots, p$. But, $\sum_{i=1}^n (Y_{ji} - \overline{Y}_{jn})^2/\sigma^2, j = 1, \ldots, p$, are independent random variables. Thus, we rewrite (11.7.3):

$$p(n-1)S_n^2/\sigma^2 = \sum_{j=1}^p \sum_{i=1}^n (Y_{ji} - \overline{Y}_{jn})^2/\sigma^2 \sim \chi^2_{p(n-1)}.$$

So, we have the desired result. ■

11.7.3 Proof of (11.2.7)

First, we prove that $I(N = n)$ and $\overline{\boldsymbol{X}}_n$ would be independently distributed for all fixed $n(\geq m)$. In what follows, we actually argue that $\overline{\boldsymbol{X}}_n$ and $(S_m^2, S_{m+1}^2, \ldots, S_n^2)$ are independent, for all fixed $n(\geq m)$.

For an arbitrary but fixed value of σ^2, we claim that (i) $\overline{\boldsymbol{X}}_n$ is a complete and sufficient statistic for $\boldsymbol{\mu}$, and (ii) (S_m^2, S_{m+1}^2, ..., S_n^2) is an ancillary statistic for $\boldsymbol{\mu}$.

Now, by Basu's (1955) theorem, $\overline{\boldsymbol{X}}_n$ must be independent of (S_m^2, S_{m+1}^2, ..., S_n^2), for every fixed $0 < \sigma^2 < \infty$ and $n \geq 2$. On the other hand, independence between $\overline{\boldsymbol{X}}_n$ and $(S_m^2, S_{m+1}^2, \ldots, S_n^2)$ does not depend on a particular fixed value of σ^2. Hence, $\overline{\boldsymbol{X}}_n$ and $(S_m^2, S_{m+1}^2, \ldots, S_n^2)$ are independent for all $\boldsymbol{\mu}$, σ^2.

Next, suppose that a multistage methodology is such that $I(N = n)$ is determined by $(S_m^2, S_{m+1}^2, \ldots, S_n^2)$ only, for all fixed $n(\geq m)$. Then, $I(N = n)$ and $\overline{\boldsymbol{X}}_n$ would be independently distributed for all fixed $n(\geq m)$.

Hence, for the methodologies included in Section 11.7.2, we can express the confidence coefficient $\mathrm{P}_{\boldsymbol{\mu},\sigma}\{\boldsymbol{\mu} \in \mathcal{R}_n\}$ associated with $\mathcal{R}_N$ as follows:

$$
\begin{aligned}
&\mathrm{P}_{\boldsymbol{\mu},\sigma}\left\{(\overline{\boldsymbol{X}}_N-\boldsymbol{\mu})'H^{-1}(\overline{\boldsymbol{X}}_N-\boldsymbol{\mu})\le d^2\right\}\\
&\quad=\textstyle\sum_{n=m}^{\infty}\mathrm{P}_{\boldsymbol{\mu},\sigma}\left\{(\overline{\boldsymbol{X}}_N-\boldsymbol{\mu})'H^{-1}(\overline{\boldsymbol{X}}_N-\boldsymbol{\mu})\le d^2\mid N=n\right\}\\
&\qquad\times\mathrm{P}_{\boldsymbol{\mu},\sigma}(N=n)\\
&\quad=\textstyle\sum_{n=m}^{\infty}\mathrm{P}_{\boldsymbol{\mu},\sigma}\left\{(\overline{\boldsymbol{X}}_n-\boldsymbol{\mu})'H^{-1}(\overline{\boldsymbol{X}}_n-\boldsymbol{\mu})\le d^2\mid N=n\right\}\\
&\qquad\times\mathrm{P}_{\boldsymbol{\mu},\sigma}(N=n)\\
&\quad=\textstyle\sum_{n=m}^{\infty}\mathrm{P}_{\boldsymbol{\mu},\sigma}\left\{(\overline{\boldsymbol{X}}_n-\boldsymbol{\mu})'H^{-1}(\overline{\boldsymbol{X}}_n-\boldsymbol{\mu})\le d^2\right\}\mathrm{P}_{\boldsymbol{\mu},\sigma}(N=n)\\
&\qquad\text{since } I(N=n) \text{ and } \overline{\boldsymbol{X}}_n \text{ are independent for all } n\ge m\\
&\quad=\textstyle\sum_{n=m}^{\infty}F\left(\frac{nd^2}{\sigma^2}\right)\mathrm{P}_{\boldsymbol{\mu},\sigma}(N=n) \text{ in view of (11.2.3)}\\
&\quad=\mathrm{E}_{\boldsymbol{\mu},\sigma}\left[F\left(\frac{Nd^2}{\sigma^2}\right)\right].
\end{aligned}
\tag{11.7.4}
$$

Our proof is now complete. ■

11.7.4 Proof of Theorem 11.2.1, Part (i)

Let Y be distributed as χ_p^2 and let it be independent of S_m^2. The confidence coefficient $\mathrm{P}_{\boldsymbol{\mu},\sigma}\{\boldsymbol{\mu}\in\mathcal{R}_N\}$ is:

$$
\begin{aligned}
&\mathrm{E}_{\boldsymbol{\mu},\sigma}\left[F\left(Nd^2/\sigma^2\right)\right]\\
&\quad\ge\mathrm{E}_{\boldsymbol{\mu},\sigma}\left[F\left(bS_m^2/\sigma^2\right)\right] \text{ since } N\ge bS_m^2/d^2 \text{ w.p. 1}\\
&\quad=\mathrm{E}_{\boldsymbol{\mu},\sigma}\left[\mathrm{P}_{\boldsymbol{\mu},\sigma}\left\{Y\le bS_m^2\sigma^{-2}\mid N\right\}\right]\\
&\quad=\mathrm{P}_{\boldsymbol{\mu},\sigma}\left\{p^{-1}Y\div S_m^2\sigma^{-2}\le p^{-1}b\right\}\\
&\quad=\mathrm{P}_{\boldsymbol{\mu},\sigma}\left\{F_{p,p(m-1)}\le p^{-1}b\right\}\\
&\quad=1-\alpha \text{ since } p^{-1}b=F_{p,p(m-1),\alpha}.
\end{aligned}
\tag{11.7.5}
$$

The proof is now complete. ■

11.7.5 Proof of Theorem 11.2.2, Part (ii)

First, note a basic inequality:

$$
\frac{b_{p,m,\alpha}S_m^2}{d^2}\le N\le m+\frac{b_{p,m,\alpha}S_m^2}{d^2} \text{ w.p. 1,} \tag{11.7.6}
$$

and that $m/C\to 0$ as $d\to 0$. Now, we divide throughout (11.7.6) by C in order to write:

$$\frac{b_{p,m,\alpha}S_m^2}{a\sigma^2} \le \frac{N}{C} \le \frac{m}{C} + \frac{b_{p,m,\alpha}S_m^2}{a\sigma^2} \text{ w.p. } 1$$
$$\Rightarrow a^{-1}b_{p,m,\alpha} \le \mathrm{E}_{\boldsymbol{\mu},\sigma}[N/C] \le mC^{-1} + a^{-1}b_{p,m,\alpha} \tag{11.7.7}$$
$$\text{since } \mathrm{E}_{\boldsymbol{\mu},\sigma}\left[S_m^2\right] = \sigma^2.$$

Again, we let Y be distributed as χ_p^2 and let it be independent of S_m^2. From (11.7.5), we obtain:

$$F_{p,p(m-1)} = p^{-1}Y \div S_m^2\sigma^{-2} \text{ and as } d \to 0 \text{(that is, as } m \to \infty),$$
$$F_{p,p(m-1)} \text{ converges in distribution to } p^{-1}Y \Rightarrow \lim_{d\to 0}\frac{b_{p,m,\alpha}}{a} = 1. \tag{11.7.8}$$

Now, the result follows from (11.7.7). ■

11.7.6 Proof of Theorem 11.2.3 Parts (i)-(ii)

Proof of Part (i):

Let $Y_1, \ldots, Y_n, \ldots$ be i.i.d. $\frac{\sigma^2}{p}\chi_p^2$ and $\overline{Y}_n = n^{-1}\sum_{i=1}^n Y_i, n \ge 1$. Observe that we can rewrite the stopping variable from (11.2.12) as follows:

$$N \equiv N(d) \text{ is the smallest integer } n(\ge m) \text{ for which}$$
$$\text{we have } n \ge \left(\frac{a}{d^2}\right)\overline{Y}_{n-1}. \tag{11.7.9}$$

Hence, we may write w.p. 1:

$$N \le m + \left(\frac{a}{d^2}\right)\overline{Y}_{N-2} \Rightarrow (N-m)(N-2) \le \left(\frac{a}{d^2}\right)\sum_{i=1}^{N-2} Y_i$$
$$\Rightarrow N^2 - (m+2)N \le \left(\frac{a}{d^2}\right)\sum_{i=1}^{N} Y_i. \tag{11.7.10}$$

Assume that $\mathrm{E}_{\boldsymbol{\mu},\sigma}[N] < \infty$. We may apply Wald's first equation (Theorem 3.5.4) and claim:

$$\mathrm{E}_{\boldsymbol{\mu},\sigma}\left[\sum_{i=1}^N Y_i\right] = \frac{\sigma^2}{p}p\mathrm{E}_{\boldsymbol{\mu},\sigma}[N] = \sigma^2\mathrm{E}_{\boldsymbol{\mu},\sigma}[N]. \tag{11.7.11}$$

Recall that $C = \dfrac{a\sigma^2}{d^2}$. Next, we take expectations throughout (11.7.10) and combine with (11.7.11) to write:

$$\mathrm{E}^2_{\boldsymbol{\mu},\sigma}[N] - (m+2)\mathrm{E}_{\boldsymbol{\mu},\sigma}[N] \le \frac{a}{d^2}\sigma^2\mathrm{E}_{\boldsymbol{\mu},\sigma}[N] = C\mathrm{E}_{\boldsymbol{\mu},\sigma}[N]$$
$$\Rightarrow \mathrm{E}_{\boldsymbol{\mu},\sigma}[N] - (m+2) \le C. \tag{11.7.12}$$

Thus, we have $\mathrm{E}_{\boldsymbol{\mu},\sigma}[N] \le C + m + 2$.

But, if $\mathrm{E}_{\boldsymbol{\mu},\sigma}[N]$ is not finite, then one may use a truncated stopping variable as follows:

$$N_k = \min\{k, N\}, k = 1, 2, \ldots$$

and claim immediately that $\mathrm{E}_{\boldsymbol{\mu},\sigma}[N_k] \le C+m+2$ for each fixed k. Then, an application of monotone convergence theorem will complete the proof. This also shows that $\mathrm{E}_{\boldsymbol{\mu},\sigma}[N]$ is indeed finite. ∎

Proof of Part (ii):

Suppose that we choose $0 < d_1 < d_2 < \infty$. Observe that $N(d_1)$ cannot be smaller than $N(d_2)$ w.p. 1. That is, $N(d)$ is a non-increasing function of $d > 0$. Actually, $N(d) \uparrow \infty$ as $d \downarrow 0$.

Let us write down a basic inequality from (11.2.12):

$$\frac{a}{d^2}S_N^2 \le N \le m + \frac{a}{d^2}S_{N-1}^2 \text{ w.p. } 1. \tag{11.7.13}$$

We have $\lim_{d\to 0} S_N^2 = \lim_{d\to 0} S_{N-1}^2 = \sigma^2$ since $\lim_{d\to 0} N = \infty$ w.p. 1. Now, we divide throughout (11.7.13) by C and take limits as $d \to 0$ to obtain w.p. 1:

$$1 = \lim_{d\to 0}\frac{S_N^2}{\sigma^2} \le \lim_{d\to 0}\frac{N}{C} \le \lim_{d\to 0}\frac{m}{C} + \lim_{d\to 0}\frac{S_{N-1}^2}{\sigma^2} = 1. \tag{11.7.14}$$

So, $\lim_{d\to 0} N/C = 1$ w.p. 1, which follows immediately from (11.7.14). Hence, Fatou's lemma implies:

$$\liminf{}_{d\to 0}\mathrm{E}_{\boldsymbol{\mu},\sigma}\left[N/C\right] \ge \mathrm{E}_{\boldsymbol{\mu},\sigma}\left[\liminf{}_{d\to 0}N/C\right] = 1. \tag{11.7.15}$$

Next, we rewrite part (i) and obtain:

$$\begin{aligned} &\mathrm{E}_{\boldsymbol{\mu},\sigma}\left[N/C\right] \le 1 + (m+2)C^{-1} \text{ which implies:} \\ &\limsup{}_{d\to 0}\mathrm{E}_{\boldsymbol{\mu},\sigma}\left[N/C\right] \le 1 + \limsup{}_{d\to 0}(m+2)C^{-1} = 1, \end{aligned} \tag{11.7.16}$$

since $\limsup_{d\to 0}(m+2)C^{-1} = 0$. Now, we combining (11.7.15) - (11.7.16):

$$1 \le \liminf{}_{d\to 0}\mathrm{E}_{\boldsymbol{\mu},\sigma}\left[N/C\right] \le \limsup{}_{d\to 0}\mathrm{E}_{\boldsymbol{\mu},\sigma}\left[N/C\right] \le 1.$$

Part (ii) immediately follows. ∎

11.7.7 Proof of Theorem 11.3.1

We follow along (11.3.7) and (11.3.12) to express the confidence coefficient $\mathrm{P}_{\boldsymbol{\mu},\boldsymbol{\Sigma}}\{\boldsymbol{\mu} \in \mathcal{R}_N^*\}$ as follows:

$$\begin{aligned} &\mathrm{P}_{\boldsymbol{\mu},\boldsymbol{\Sigma}}\left\{(\overline{\boldsymbol{X}}_N - \boldsymbol{\mu})'(\overline{\boldsymbol{X}}_N - \boldsymbol{\mu}) \le d^2\right\} \\ &\ge \mathrm{P}_{\boldsymbol{\mu},\boldsymbol{\Sigma}}\left\{(\overline{\boldsymbol{X}}_N - \boldsymbol{\mu})'S_m^{-1}(\overline{\boldsymbol{X}}_N - \boldsymbol{\mu}) \le \frac{d^2}{S_{\max(m)}}\right\} \\ &= \mathrm{P}_{\boldsymbol{\mu},\boldsymbol{\Sigma}}\left\{N(\overline{\boldsymbol{X}}_N - \boldsymbol{\mu})'S_m^{-1}(\overline{\boldsymbol{X}}_N - \boldsymbol{\mu}) \le \frac{Nd^2}{S_{\max(m)}}\right\} \\ &\ge \mathrm{P}_{\boldsymbol{\mu},\boldsymbol{\Sigma}}\left\{N(\overline{\boldsymbol{X}}_N - \boldsymbol{\mu})'S_m^{-1}(\overline{\boldsymbol{X}}_N - \boldsymbol{\mu}) \le b^*\right\}, \end{aligned} \tag{11.7.17}$$

since we have $Nd^2 \ge b^* S_{\max(m)}$ w.p. 1 in view of (11.3.14).

Now, we can claim:

$$N(\overline{\boldsymbol{X}}_N - \boldsymbol{\mu})' S_m^{-1} (\overline{\boldsymbol{X}}_N - \boldsymbol{\mu}) \text{ has Hotelling's } T^2(p, m-p) \text{ distribution under } \boldsymbol{\mu}, \boldsymbol{\Sigma}. \quad (11.7.18)$$

But, we may also rewrite (11.7.18) as follows:

$$\frac{(m-p)N}{p(m-1)} (\overline{\boldsymbol{X}}_N - \boldsymbol{\mu})' S_m^{-1} (\overline{\boldsymbol{X}}_N - \boldsymbol{\mu}) \text{ has } F_{p,m-p} \text{ distribution under } \boldsymbol{\mu}, \boldsymbol{\Sigma}. \quad (11.7.19)$$

Hence, from the last step in (11.7.17), we immediately have:

$$\mathrm{P}_{\boldsymbol{\mu},\boldsymbol{\Sigma}} \{\boldsymbol{\mu} \in \mathcal{R}_N^*\} \ge 1 - \alpha,$$

provided that we choose b^* as follows:

$$\frac{m-p}{p(m-1)} b^* = F_{p,m-p,\alpha} \Leftrightarrow b^* \equiv b^*_{p,m,\alpha} = \frac{p(m-1)}{m-p} F_{p,m-p,\alpha}.$$

This coincides with our choice of b^* given by (11.3.13). ∎

11.7.8 Proof of Theorem 11.3.2

Proof of Part (i):

Suppose that we choose $0 < d_1 < d_2 < \infty$. Observe that $N(d_1)$ cannot be smaller than $N(d_2)$ w.p. 1. In other words, $N(d)$ is a non-increasing function of $d > 0$. Actually, $N(d) \uparrow \infty$ as $d \downarrow 0$.

We write a basic inequality from (11.3.15):

$$\frac{a}{d^2} S_{\max(N)} \le N \le m + \frac{a}{d^2} S_{\max(N-1)} \text{ w.p. } 1. \quad (11.7.20)$$

We have $\lim_{d \to 0} S_{\max(N)} = \lim_{d \to 0} S_{\max(N-1)} = \lambda_{\max}$ since $\lim_{d \to 0} N = \infty$ w.p. 1. Now, we divide throughout (11.7.20) by C^* and take limits as $d \to 0$ to have w.p. 1:

$$1 = \lim_{d \to 0} \frac{S_{\max(N)}}{\lambda_{\max}} \le \lim_{d \to 0} \frac{N}{C^*} \le \lim_{d \to 0} \frac{m}{C^*} + \lim_{d \to 0} \frac{S_{\max(N-1)}}{\lambda_{\max}} = 1. \quad (11.7.21)$$

From (11.7.21) we conclude $\lim_{d \to 0} N/C^* = 1$ in probability. ∎

Proof of Part (ii):

Part (i) together with Fatou's lemma imply:

$$\liminf_{d \to 0} \mathrm{E}_{\boldsymbol{\mu},\boldsymbol{\Sigma}} [N/C^*] \ge \mathrm{E}_{\boldsymbol{\mu},\boldsymbol{\Sigma}} [\liminf_{d \to 0} N/C^*] = \mathrm{E}_{\boldsymbol{\mu},\boldsymbol{\Sigma}} [1] \ge 1. \quad (11.7.22)$$

Now, we reconsider the sample dispersion matrix $S_n \equiv (S_{ijn})_{p \times p}$ from (11.3.2) that is obtained from n observations. Let us denote its p eigenvalues $S_{1(n)} \ge S_{2(n)} \ge \ldots \ge S_{p(n)}$ which are positive w.p. 1. We have estimated $\lambda_{\max}$ with

$$S_{\max(n)} \equiv \max\left\{S_{1(n)}, S_{2(n)}, \ldots, S_{p(n)}\right\}.$$

Thus, for all fixed $n(\geq p+1)$, we have w.p. 1:

$$S_{\max(n)} \leq \textstyle\sum_{i=1}^{p} S_{i(n)} = \text{Trace}\,(S_n) = \sum_{i=1}^{p} S_{iin}, \text{ sum of the diagonal elements of } S_n \equiv \left(S_{ijn}\right)_{p\times p}. \tag{11.7.23}$$

From (11.7.23), for all fixed $n(\geq p+1)$, we immediately claim w.p. 1:

$$\begin{aligned} & S_{\max(n)} \\ & \leq \frac{1}{n-1} \textstyle\sum_{i=1}^{p} \sum_{j=1}^{n} \left(X_{ij} - \overline{X}_{in}\right)^2 \\ & \leq \frac{p+1}{p}\frac{1}{n} \textstyle\sum_{i=1}^{p} \sum_{j=1}^{n} \left(X_{ij} - \mu_i\right)^2 \text{ since } \frac{1}{n-1} \leq \frac{p+1}{pn} \\ & \leq \frac{p+1}{p} W \text{ where } W = \max_{(k\geq 1)} \frac{1}{k} \textstyle\sum_{i=1}^{p} \sum_{j=1}^{k} \left(X_{ij} - \mu_i\right)^2. \end{aligned} \tag{11.7.24}$$

Next, for sufficiently small $0 < d < d_0$, from the right-hand side of basic inequality (11.7.20) and (11.7.24) we obtain w.p. 1:

$$Nd^2 \leq md_0^2 + aS_{\max(N-1)} \leq md_0^2 + a(p+1)p^{-1}W. \tag{11.7.25}$$

Applying Wiener's (1939) ergodic theorem, we note that the random variable W defined in (11.7.24) is integrable. Also, the right-hand side of (11.7.25) does not involve d. Hence, part (i) combined with dominated convergence theorem would imply part (ii). ■

Proof of Part (iii):

From (11.3.12), we recall:

$$\mathrm{P}_{\boldsymbol{\mu},\boldsymbol{\Sigma}}\left\{\boldsymbol{\mu} \in \mathcal{R}_N^*\right\} = \mathrm{E}_{\boldsymbol{\mu},\boldsymbol{\Sigma}}\left[F\left(\frac{Nd^2}{\lambda_{\max}}\right)\right]. \tag{11.7.26}$$

Surely, $F(x)$ is bounded. Next, part (i) implies that $Nd^2/\lambda_{\max}$ converges to a in probability as $d \to 0$. Now, an application of dominated convergence theorem will lead to part (iii). ■

11.7.9 Proof of Theorem 11.4.3, Part (i)

We will sketch a proof assuming $m \geq 3$. One of Woodroofe's (1977) major results is:

$$\frac{(N-n^*)^2}{n^*} \text{ is } \textit{uniformly integrable}. \tag{11.7.27}$$

Theorem 11.4.1, part (v) follows from Ghosh and Mukhopadhyay's (1975) theorem. Next, one may combine that result with (11.7.27) to claim:

$$\mathrm{E}_{\boldsymbol{\mu},\boldsymbol{\Sigma}}\left[\frac{(N-n^*)^2}{n^*}\right] = \frac{1}{2}\tau^2 + o(1) \text{ as } c \to 0, \tag{11.7.28}$$

where $\tau^2 \equiv \text{Trace}\big((A\boldsymbol{\Sigma})^2\big)\,/\text{Trace}^2(A\boldsymbol{\Sigma})$.

Now, let us rewrite the regret function from (11.4.9) as follows:

$$\begin{aligned} & c^{-1}\text{Regret}(c) \\ &= \mathrm{E}_{\boldsymbol{\mu},\boldsymbol{\Sigma}}\left[\frac{(N-n^*)^2}{N} I\left(N > \tfrac{1}{2}n^*\right)\right] \\ &\quad + \mathrm{E}_{\boldsymbol{\mu},\boldsymbol{\Sigma}}\left[\frac{(N-n^*)^2}{N} I\left(N \le \tfrac{1}{2}n^*\right)\right]. \end{aligned} \tag{11.7.29}$$

Next, we observe that

$$\left|\frac{(N-n^*)^2}{N} I\left(N > \frac{1}{2}n^*\right)\right| \le 2\frac{(N-n^*)^2}{n^*} \text{ w.p. 1.}$$

Now, $\dfrac{(N-n^*)^2}{n^*}$ is uniformly integrable (see (11.7.27)), and hence so is $\dfrac{(N-n^*)^2}{N} I\left(N > \tfrac{1}{2}n^*\right)$. But, we also know that

$$\frac{(N-n^{*2})}{N} I\left(N > \frac{1}{2}n^*\right) \text{ converges in distribution to } N(0, \frac{1}{2}\tau^2).$$

Hence, we have:

$$\mathrm{E}_{\boldsymbol{\mu},\boldsymbol{\Sigma}}\left[\frac{(N-n^*)^2}{N} I\left(N > \frac{1}{2}n^*\right)\right] = \frac{1}{2}\tau^2 + o(1). \tag{11.7.30}$$

On the other hand, we may write:

$$\begin{aligned} & \mathrm{E}_{\boldsymbol{\mu},\boldsymbol{\Sigma}}\left[\frac{(N-n^*)^2}{N} I\left(N \le \tfrac{1}{2}n^*\right)\right] \\ &\le \mathrm{E}_{\boldsymbol{\mu},\boldsymbol{\Sigma}}\left[\frac{N^2 + n^{*2}}{N} I\left(N \le \tfrac{1}{2}n^*\right)\right] \\ &\le \mathrm{E}_{\boldsymbol{\mu},\boldsymbol{\Sigma}}\left[\frac{\frac{1}{4}n^{*2} + n^{*2}}{m} I\left(N \le \tfrac{1}{2}n^*\right)\right] \\ &= \frac{5}{4m} n^{*2} \mathrm{P}_{\boldsymbol{\mu},\boldsymbol{\Sigma}}\left(N \le \tfrac{1}{2}n^*\right) \\ &= O\left(n^{*2-m}\right) \text{ in view of Theorem 11.4.1, part(iii)} \\ &= o(1) \text{ if } 2 - m < 0, \text{ that is, if } m \ge 3. \end{aligned} \tag{11.7.31}$$

At this point one combines (11.7.29) - (11.7.31) to claim:

$$c^{-1}\text{Regret}(c) = \mathrm{E}_{\boldsymbol{\mu},\boldsymbol{\Sigma}}\left[\frac{(N-n^*)^2}{N}\right] = \frac{1}{2}\tau^2 + o(1). \tag{11.7.32}$$

The desired result clearly follows from (11.7.32) when $m \geq 3$. ■

11.8 Exercises

Exercise 11.2.1: Show that $\overline{\boldsymbol{X}}_n$ and S_n^2 defined in (11.2.2) are uniformly minimum variance unbiased estimators (UMVUEs) of $\boldsymbol{\mu}, \sigma^2$ respectively.

Exercise 11.2.2: Show that $\overline{\boldsymbol{X}}_n$ and $(n-1)n^{-1}S_n^2$ with $\overline{\boldsymbol{X}}_n, S_n^2$ defined in (11.2.2) are maximum likelihood estimators (MLEs) of $\boldsymbol{\mu}, \sigma^2$ respectively.

Exercise 11.2.3: For the stopping variable N from (11.2.9), show that $\overline{\boldsymbol{X}}_N$ is unbiased for $\boldsymbol{\mu}$. Evaluate $\mathrm{E}_{\boldsymbol{\mu},\sigma}\left[\overline{\boldsymbol{X}}_N^2\right]$ and $\mathrm{V}_{\boldsymbol{\mu},\sigma}\left[\overline{\boldsymbol{X}}_N\right]$. What is the distribution of $\sqrt{N}(\overline{\boldsymbol{X}}_N - \boldsymbol{\mu})$?

Exercise 11.2.4: For all fixed $\boldsymbol{\mu}, \sigma, p, m$ and α, for the stopping variable N from (11.2.9), evaluate $\lim_{d\to 0} \mathrm{E}_{\boldsymbol{\mu},\sigma}\left[N^2/C^2\right]$.

Exercise 11.2.5: For the stopping variable N from (11.2.11), show that $\overline{\boldsymbol{X}}_N$ is unbiased for $\boldsymbol{\mu}$. Evaluate $\mathrm{E}_{\boldsymbol{\mu},\sigma}\left[\overline{\boldsymbol{X}}_N^2\right]$ and $\mathrm{V}_{\boldsymbol{\mu},\sigma}\left[\overline{\boldsymbol{X}}_N\right]$. What is the distribution of $\sqrt{N}(\overline{\boldsymbol{X}}_N - \boldsymbol{\mu})$?

Exercise 11.2.6: For all fixed $\boldsymbol{\mu}, \sigma, p, m$ and α, for the stopping variable N from (11.2.11), show that $\lim_{d\to 0} \mathrm{E}_{\boldsymbol{\mu},\sigma}\left[N^2/C^2\right] = 1$.

Exercise 11.2.7: For all fixed $\boldsymbol{\mu}, \sigma, p, m$ and α, for the stopping variable N from (11.2.14), show that

$$\lim_{d\to 0} \mathrm{E}_{\boldsymbol{\mu},\sigma}\left[N/C\right] = 1 \text{ and } \lim_{d\to 0} \mathrm{E}_{\boldsymbol{\mu},\sigma}\left[N^2/C^2\right] = 1.$$

What is the distribution of $\sqrt{N}(\overline{\boldsymbol{X}}_N - \boldsymbol{\mu})$?

Exercise 11.2.8: For all fixed $\boldsymbol{\mu}, \sigma, p, m$ and α, for the stopping variables N from (11.2.9), (11.2.11), (11.2.12) and (11.2.14), evaluate

$$\mathrm{E}_{\boldsymbol{\mu},\sigma}\left[\left\{\sqrt{N}(\overline{\boldsymbol{X}}_N - \boldsymbol{\mu})' H^{-1}(\overline{\boldsymbol{X}}_N - \boldsymbol{\mu})\right\}^k\right]$$

where $k(> 0)$ is arbitrary but fixed. What will be different if k is arbitrary and fixed but negative?

Exercise 11.2.9: Suppose that we have available i.i.d. k -dimensional observations $\boldsymbol{Y}_i \sim N_k(\boldsymbol{\theta}, \sigma^2 I_{k\times k})$ where $\boldsymbol{Y}_i' = (Y_{1i}, \ldots, Y_{ki}), i = 1, \ldots, n, \ldots$. Here, $\boldsymbol{\theta}_{k\times 1}$ and σ^2 are assumed unknown. We denote $\boldsymbol{\theta}' = (\theta_1, \ldots, \theta_{k-1}, \theta_k)$. This setup may arise as follows: There is a large cohort of subjects with similar age, health status, and other standard covariates. At each time point i, we assign $k-1$ subjects independently to treatments 1 through $k-1$ providing the observations $Y_{1i}, \ldots, Y_{k-1i}$ respectively, and one subject to treatment k (which is a "control") providing an observation Y_{ki}. A set of k new subjects are used at different time points, $i = 1, \ldots, n, \ldots$.

The problem is one of comparing $k-1$ treatment means $\theta_1, \ldots, \theta_{k-1}$ with a control mean θ_k. We require a fixed-size ellipsoidal confidence region for $(k-1)$-dimensional parameters $\theta_k - \theta_1, \ldots, \theta_k - \theta_{k-1}$ with a preassigned confidence coefficient $1-\alpha$.

Formulate this problem in the light of Section 11.2. Write down the methodologies from Section 11.2 explicitly involving Y's.

Exercise 11.2.10: Explain clearly how each methodology described in Exercise 11.2.9 looks and works especially when $k = 2, 3$.

Exercise 11.3.1: Show that $\overline{\boldsymbol{X}}_n$ and S_n defined in (11.3.2) are UMVUEs of $\boldsymbol{\mu}, \boldsymbol{\Sigma}$ respectively.

Exercise 11.3.2: Show that $\overline{\boldsymbol{X}}_n$ and $(n-1)n^{-1}S_n$ with $\overline{\boldsymbol{X}}_n, S_n$ defined in (11.3.2) are MLEs of $\boldsymbol{\mu}, \boldsymbol{\Sigma}$ respectively.

Exercise 11.3.3: For the stopping variable N from (11.3.14), show that $\overline{\boldsymbol{X}}_N$ is unbiased for $\boldsymbol{\mu}$. Evaluate $\mathrm{E}_{\boldsymbol{\mu},\boldsymbol{\Sigma}}\left[\overline{\boldsymbol{X}}_N^2\right]$ and $\mathrm{V}_{\boldsymbol{\mu},\boldsymbol{\Sigma}}\left[\overline{\boldsymbol{X}}_N\right]$. What is the distribution of $\sqrt{N}(\overline{\boldsymbol{X}}_N - \boldsymbol{\mu})$?

Exercise 11.3.4: For all fixed $\boldsymbol{\mu}, \boldsymbol{\Sigma}, p, m, d$ and α, for the stopping variables N from (11.3.14) and (11.3.15), evaluate

$$\mathrm{E}_{\boldsymbol{\mu},\boldsymbol{\Sigma}}\left[\{(\overline{\boldsymbol{X}}_N - \boldsymbol{\mu})'\boldsymbol{\Sigma}^{-1}(\overline{\boldsymbol{X}}_N - \boldsymbol{\mu})\}^k\right]$$

where $k(>0)$ is arbitrary but fixed. What will be different if k is arbitrary and fixed, but negative?

Exercise 11.3.5: For all fixed $\boldsymbol{\mu}, \boldsymbol{\Sigma}, p$ and α, for the stopping variable N from (11.3.14), evaluate $\lim \mathrm{E}_{\boldsymbol{\mu},\boldsymbol{\Sigma}}[N/C^*]$ if we let $m \equiv m(d) \to \infty$ but $m(d)/C^* \to 0$, as $d \to 0$.

Exercise 11.3.6: For all fixed $\boldsymbol{\mu}, \boldsymbol{\Sigma}, p$ and α, for the stopping variable N from (11.3.15), evaluate $\lim_{d\to 0} \mathrm{E}_{\boldsymbol{\mu},\boldsymbol{\Sigma}}\left[N^2/C^{*2}\right]$. What is the distribution of $\sqrt{N}(\overline{\boldsymbol{X}}_N - \boldsymbol{\mu})$?

Exercise 11.3.7: Suppose that we have available i.i.d. k -dimensional observations $\boldsymbol{Y}_i \sim N_k(\boldsymbol{\theta}, \boldsymbol{\Sigma}_{k\times k})$ where $\boldsymbol{Y}_i' = (Y_{1i}, \ldots, Y_{ki}), i = 1, \ldots, n, \ldots$. Here, $\boldsymbol{\theta}_{k\times 1}$ and $\boldsymbol{\Sigma}$ are assumed unknown, but $\boldsymbol{\Sigma}$ is positive definite. We denote $\boldsymbol{\theta}' = (\theta_1, \ldots, \theta_{k-1}, \theta_k)$. This setup may arise as follows: There is a large cohort of subjects with similar age, health status, and other standard covariates. At each time point i, we assign $k-1$ subjects independently to treatments 1 through $k-1$ providing the observations $Y_{1i}, \ldots, Y_{k-1i}$ respectively, and one subject to treatment k (which is a "control") providing an observation Y_{ki}. A set of k new subjects are used at different time points, $i = 1, \ldots, n, \ldots$.

The problem is one of comparing $k-1$ treatment means $\theta_1, \ldots, \theta_{k-1}$ with a control mean θ_k. We require a fixed-size spherical confidence region for $(k-1)$-dimensional parameters $\theta_k - \theta_1, \ldots, \theta_k - \theta_{k-1}$ with a preassigned confidence coefficient $1 - \alpha$.

Formulate this problem in the light of Section 11.3. Write down the methodologies from Section 11.3 explicitly involving Y's.

Exercise 11.3.8: Explain clearly how each methodology described in Exercise 11.3.7 looks and works especially when $k = 2, 3$.

Exercise 11.4.1: For the stopping variable N from (11.4.6), show that $\overline{\boldsymbol{X}}_N$ is unbiased for $\boldsymbol{\mu}$. Evaluate $\mathrm{E}_{\boldsymbol{\mu},\boldsymbol{\Sigma}}\left[\overline{\boldsymbol{X}}_N^2\right]$ and $\mathrm{V}_{\boldsymbol{\mu},\boldsymbol{\Sigma}}\left[\overline{\boldsymbol{X}}_N\right]$. What is the distribution of $\sqrt{N}(\overline{\boldsymbol{X}}_N - \boldsymbol{\mu})$?

Exercise 11.4.2: For all fixed $\boldsymbol{\mu}, \boldsymbol{\Sigma}, p, m, A$ and c, for the stopping variable N from (11.4.6), evaluate

$$\mathrm{E}_{\boldsymbol{\mu},\boldsymbol{\Sigma}}\left[\left\{(\overline{\boldsymbol{X}}_N - \boldsymbol{\mu})'\boldsymbol{\Sigma}^{-1}(\overline{\boldsymbol{X}}_N - \boldsymbol{\mu})\right\}^k\right]$$

where $k(>0)$ is arbitrary but fixed. What will be different if k is arbitrary and fixed, but negative?

Exercise 11.4.3: For all fixed $\boldsymbol{\mu}, \boldsymbol{\Sigma}, p, m$ and A, for the stopping variable N from (11.4.6), evaluate $\lim_{c \to 0} \mathrm{E}_{\boldsymbol{\mu},\boldsymbol{\Sigma}}\left[N^2/n^{*2}\right]$.

Exercise 11.4.4: Suppose that we have available i.i.d. k -dimensional observations $\boldsymbol{Y}_i \sim N_k(\boldsymbol{\theta}, \sigma^2 I_{k\times k})$ where $\boldsymbol{Y}_i' = (Y_{1i}, \ldots, Y_{ki}), i = 1, \ldots, n, \ldots$. Here, $\boldsymbol{\theta}_{k\times 1}$ and σ^2 are assumed unknown. We denote $\boldsymbol{\theta}' = (\theta_1, \ldots, \theta_{k-1}, \theta_k)$. This setup may arise as follows: There is a large cohort of subjects with similar age, health status, and other standard covariates. At each time point i, we assign $k-1$ subjects independently to treatments 1 through $k-1$ providing the observations $Y_{1i}, \ldots, Y_{k-1i}$ respectively, and one subject to treatment k (which is a "control") providing an observation Y_{ki}. A set of k new subjects are used at different time points, $i = 1, \ldots, n, \ldots$.

The problem is one of comparing $k-1$ treatment means $\theta_1, \ldots, \theta_{k-1}$ with a control mean θ_k. We require minimum risk point estimators for $(k-1)$-dimensional parameters $\theta_k - \theta_1, \ldots, \theta_k - \theta_{k-1}$.

Formulate this problem in the light of Section 11.4. In (11.4.1), what matrix should one replace the matrix A with? Write down a purely sequential methodology analogous to (11.4.6) explicitly involving Y's.

Exercise 11.4.5: Explain clearly how the methodology described in Exercise 11.4.4 looks and works especially when $k = 2, 3$.

Exercise 11.5.1: We give 40 four-dimensional observations from a normal population with unknown mean vector $\boldsymbol{\mu}$ in Table 11.8.1. Find the sample size required to construct a 90% confidence region

$$\mathcal{R}_N^* = \left\{\boldsymbol{\mu}_{4\times 1} \in I\!R^4 : (\overline{\boldsymbol{X}}_N - \boldsymbol{\mu})'(\overline{\boldsymbol{X}}_N - \boldsymbol{\mu}) \le 25\right\}.$$

Seq11.exe to explore $\mathcal{R}_N^*$ by employing each methodology in Section 11.3 with data input from Table 11.8.1. Implement

(i) the two-stage procedure with $m = 10$.

(ii) the sequential procedure with $m = 10$.

Obtain the output and critically examine the entries as in Section 11.5.

Table 11.8.1: Multivariate normal data

Obs. no.	x_1	x_2	x_3	x_4	Obs. no.	x_1	x_2	x_3	x_4
1	91	77	59	78	21	91	75	53	72
2	92	75	52	74	22	89	74	51	73
3	83	71	46	70	23	90	78	51	77
4	94	79	55	74	24	98	78	56	75
5	99	81	56	77	25	90	77	50	74
6	88	72	51	73	26	85	68	45	66
7	95	77	50	75	27	87	75	53	76
8	98	74	48	69	28	99	84	56	77
9	88	75	50	73	29	86	73	51	76
10	86	71	46	68	30	90	76	52	74
11	83	66	46	66	31	93	77	51	76
12	91	73	45	67	32	86	71	49	71
13	81	71	48	72	33	90	78	50	73
14	98	79	57	76	34	93	76	46	72
15	87	72	49	71	35	93	75	52	73
16	93	73	48	68	36	90	77	49	76
17	95	78	48	74	37	91	76	49	76
18	92	74	48	67	38	88	77	47	76
19	83	71	48	71	39	90	76	51	75
20	95	80	57	80	40	94	79	53	77

Exercise 11.5.2: Consider a 90% confidence region:

$$\mathcal{R}_N = \left\{\boldsymbol{\mu}_{3\times 1} \in I\!R^3 : (\overline{\boldsymbol{X}}_N - \boldsymbol{\mu})'H^{-1}(\overline{\boldsymbol{X}}_N - \boldsymbol{\mu}) \le 0.4^2 = 0.16\right\}.$$

Here, $\boldsymbol{\mu}$ is the mean of a three-dimensional normal population with its dispersion matrix $\mathbf{\Sigma} = \sigma^2 H$, and

$$H = \begin{pmatrix} 6 & -3 & -1 \\ -3 & 14 & -2 \\ -1 & -2 & 2 \end{pmatrix}.$$

Assume that σ^2 is an unknown parameter. Use the program **Seq11.exe** to simulate $\mathcal{R}_N$ by employing each methodology from Section 11.2. Implement

(i) the two-stage procedure with $m = 10$.
(ii) the modified two-stage procedure with $\gamma = 0.4$.
(iii) the purely sequential procedure with $m = 10$.
(iv) the three-stage procedure with $\rho = 0.6$ and $m = 10$.

Generate data by fixing $\boldsymbol{\mu}' = (3, 4, 2)$, $\sigma^2 = 9$ and obtain output by averaging results from 5000 replications in each case. Critically examine the output entries as in Section 11.5.

Exercise 11.5.3 (Exercise 11.5.1 Continued): Given a loss function

$$L_n(\boldsymbol{\mu}, \overline{\boldsymbol{X}}_n) = (\overline{\boldsymbol{X}}_n - \boldsymbol{\mu})' A (\overline{\boldsymbol{X}}_n - \boldsymbol{\mu}) + 2n$$

where

$$A = \begin{pmatrix} 16 & 9 & 5 & 4 \\ 9 & 10 & 6 & 8 \\ 5 & 6 & 9 & 6 \\ 4 & 8 & 6 & 10 \end{pmatrix}.$$

Find the sequential sample size required to obtain a minimum risk point estimate of $\boldsymbol{\mu}$ using data input from Table 11.8.1. Use **Seq11.exe** by employing the methodology from Section 11.4 with an initial sample size, $m = 10$. Obtain the output and critically examine its entries as in Section 11.5.

CHAPTER 12

Estimation in a Linear Model

12.1 Introduction

In this chapter, we treat exclusively estimation problems for the unknown regression parameters in a linear model. We will assume that the errors are normally distributed. We should make it clear that relevant literature in this field and those cited in Chapter 11 have many commonalities. Thus, we may refrain from citing many sources which were included previously.

Let us begin with a linear model which is also known as the Gauss-Markov setup. Suppose that we observe a sequence of independent observations $Y_1, Y_2, \ldots, Y_n, \ldots$, referred to as a *response* or *dependent* variable. Additionally, we have p covariates $X_1, X_2, \ldots, X_p$, which when fixed at certain levels

$$X_1 = x_{i1}, X_2 = x_{i2}, \ldots, X_p = x_{ip},$$

give rise to the ith response Y_i. The fixed values $x_{i1}, x_{i2}, \ldots, x_{ip}$ constitute the ith design point which supposedly leads to a response $Y_i, i = 1, \ldots, n, \ldots$.

As an example, consider a cohort of individuals. On i^{th} individual, we may record his/her blood pressure (response, Y_i) along with covariates, for example, the individual's age (X_1), height (X_2), weight (X_3), sex ($X_4 = 0$ if male or 1 if female).

A regression surface or model is the following conditional expectation:

$$g(x_1, x_2, \ldots, x_p) \equiv \mathrm{E}\left[Y \mid X_1 = x_1, X_2 = x_2, \ldots, X_p = x_p\right]. \qquad (12.1.1)$$

A *linear regression model* assumes:

$$g(x_1, x_2, \ldots, x_p) = \beta_1 x_1 + \beta_2 x_2 + \ldots + \beta_p x_p, \qquad (12.1.2)$$

where $\beta_1, \beta_2, \ldots, \beta_p$ are called *regression parameters.* In view of (12.1.2), a standard linear regression model is customarily expressed as follows:

$$\begin{aligned} &Y_i = \beta_1 x_{i1} + \beta_2 x_{i2} + \ldots + \beta_p x_{ip} + \varepsilon_i,\ i = 1, 2, \ldots, n, \ldots \\ &\text{with the assumption: } \varepsilon_i\text{'s are i.i.d. } N(0, \sigma^2), 0 < \sigma^2 < \infty. \end{aligned} \qquad (12.1.3)$$

In some situations, one may explicitly want to show an *intercept* term. In that case, one would continue to work with the model (12.1.3) where covariate X_1, for example, is allowed to take the only value 1 so that β_1

will be the intercept. We will not make any distinction whether or not a model (12.1.3) includes an intercept.

Let us introduce more notation. Having recorded n data points or observations

$$(Y_i, x_{i1}, x_{i2}, \ldots, x_{ip}),\ i = 1, 2, \ldots, n, \tag{12.1.4}$$

we present a layout for the linear model (12.1.3) as follows:

$$\boldsymbol{Y}_n = \begin{pmatrix} Y_1 \\ Y_2 \\ \vdots \\ Y_i \\ \vdots \\ Y_n \end{pmatrix}_{n\times 1} \quad \mathrm{X}_n = \begin{pmatrix} x_{11} & x_{12} & \cdots & x_{1p} \\ x_{21} & x_{22} & \cdots & x_{2p} \\ \vdots & \vdots & & \vdots \\ x_{i1} & x_{i2} & \cdots & x_{ip} \\ \vdots & \vdots & & \vdots \\ x_{n1} & x_{n2} & \cdots & x_{np} \end{pmatrix}_{n\times p} \quad \boldsymbol{\varepsilon}_n = \begin{pmatrix} \varepsilon_1 \\ \varepsilon_2 \\ \vdots \\ \varepsilon_i \\ \vdots \\ \varepsilon_n \end{pmatrix}_{n\times 1}$$

$$\text{and define } \boldsymbol{\beta}_{p\times 1} \text{ so that } \boldsymbol{\beta}' = \begin{pmatrix} \beta_1 & \beta_2 & \ldots & \beta_p \end{pmatrix}. \tag{12.1.5}$$

Using these notation, we may can rewrite the linear model (12.1.3) involving matrices and vectors as:

$$\begin{gathered} \boldsymbol{Y}_n = \mathrm{X}_n\boldsymbol{\beta} + \boldsymbol{\varepsilon}_n \text{ with } \boldsymbol{\varepsilon}_n \sim N_n\left(\mathbf{0}_{n\times 1}, \sigma^2 I_{n\times n}\right), \text{ and} \\ \text{rank of X matrix is } \textit{full}, \text{ that is}, R(\mathrm{X}_n) = p, n \geq p+1. \end{gathered} \tag{12.1.6}$$

Here, $\boldsymbol{Y}_n, \mathrm{X}_n, \boldsymbol{\beta}$ and $\boldsymbol{\varepsilon}_n$ are respectively referred to as a response vector, design matrix, regression parameter vector, and error vector. We will discuss relevant material under the linear model (12.1.6).

Healy (1956) and Chatterjee (1962b) developed two-stage fixed-size confidence region problems for $\boldsymbol{\beta}$ under *normal errors* $\boldsymbol{\varepsilon}$ in the spirit of Stein (1945,1949). Gleser (1965), Albert (1966) and Srivastava (1967,1971) developed sequential fixed-size confidence region problems for $\boldsymbol{\beta}$ under *non-normal errors* $\boldsymbol{\varepsilon}$ in the spirit of Chow and Robbins (1965). Mukhopadhyay and Abid (1986b) introduced multistage fixed-size confidence region methodologies along with *second-order* properties under *normal errors* $\boldsymbol{\varepsilon}$. One is also referred to Finster (1983,1985) for details in a general linear model. Section 12.2 is adapted from Mukhopadhyay and Abid (1986b).

A minimum risk point estimation problem for regression parameters $\boldsymbol{\beta}$ was first introduced by Mukhopadhyay (1974b). He developed an asymptotically risk efficient sequential methodology whose regret function was investigated and its rate of convergence found. Finster (1983) included additional *second-order* properties of the Mukhopadhyay procedure. An accelerated sequential version and its asymptotic *second-order* properties were laid out in Ghosh et al. (1997, p. 202). More acceleration techniques with *second-order* properties were developed by Mukhopadhyay and Abid (1999a). The basic material in Section 12.3 is adapted from Mukhopadhyay (1974b) with brief mentions of more recent developments under multistage strategies (Section 12.5). Section 12.4 is devoted to data analyses and concluding remarks. Some selected derivations are provided in Section 12.6.

The papers of Chatterjee (1991), Mukhopadhyay (1991,1993,1995a) and Sinha (1991) as well as Ghosh et al.'s (1997) book may be helpful for reviewing this vast area.

12.2 Fixed-Size Confidence Region

We focus on a linear model described in (12.1.6). Suppose that we have available a sequence of independent observations

$$(Y_i, x_{i1}, x_{i2}, \ldots, x_{ip}),\ i = 1, 2, \ldots, n, \ldots$$

following a linear model. Having recorded $n(\geq p+1)$ observations, we consider the least square estimator of $\boldsymbol{\beta}$ and its distribution:

$$\widehat{\boldsymbol{\beta}}_n = (\mathrm{X}_n'\mathrm{X}_n)^{-1}\mathrm{X}_n'\boldsymbol{Y}_n \sim N_p\left(\boldsymbol{\beta}, \sigma^2(\mathrm{X}_n'\mathrm{X}_n)^{-1}\right). \qquad (12.2.1)$$

The error variance σ^2 is estimated by the *mean squared error* (MSE). The MSE and its distribution are:

$$\begin{gathered} S_n^2 = (n-p)^{-1}(\boldsymbol{Y}_n - \mathrm{X}_n\widehat{\boldsymbol{\beta}}_n)'(\boldsymbol{Y}_n - \mathrm{X}_n\widehat{\boldsymbol{\beta}}_n), \\ \text{and } (n-p)S_n^2 \sim \sigma^2\chi^2_{n-p}. \end{gathered} \qquad (12.2.2)$$

Obviously, $\widehat{\boldsymbol{\beta}}_n$ and S_n^2 respectively estimate $\boldsymbol{\beta}$ and σ^2 unbiasedly and they are the best unbiased estimators (UMVUEs).

We focus on estimating linear regression parameters under usual Gauss-Markov setup with independent normal errors.

The problem of constructing a fixed-size confidence region for $\boldsymbol{\beta}$ is formulated in the spirit of (11.2.1). Given two preassigned numbers $0 < d < \infty$ and $0 < \alpha < 1$, we propose an ellipsoidal confidence region for $\boldsymbol{\beta}$:

$$\mathcal{R}_n = \left\{\boldsymbol{\beta}_{p\times 1} \in I\!R^p : n^{-1}(\widehat{\boldsymbol{\beta}}_n - \boldsymbol{\beta})'(\mathrm{X}_n'\mathrm{X}_n)(\widehat{\boldsymbol{\beta}}_n - \boldsymbol{\beta}) \leq d^2\right\}. \qquad (12.2.3)$$

We have chosen a weight matrix $n^{-1}(\mathrm{X}_n'\mathrm{X}_n)$ since $n^{-1}(\mathrm{X}_n'\mathrm{X}_n)$ is customarily assumed to converge to a positive definite matrix A as $n \to \infty$ in studies of large sample properties of $\widehat{\boldsymbol{\beta}}_n$. However, we avoid making this particular assumption by involving $n^{-1}(\mathrm{X}_n'\mathrm{X}_n)$ as our weight matrix in (12.2.3). Also, the dispersion matrix of $\widehat{\boldsymbol{\beta}}_n$ is $\sigma^2(\mathrm{X}_n'\mathrm{X}_n)^{-1}$ and a weight matrix is often chosen proportional to the inverse of a dispersion matrix.

We also require that the confidence coefficient be at least $1-\alpha$ which is preassigned too. The ellipsoidal confidence region $\mathcal{R}_n$ is said to have a fixed size because its maximum diameter is known and fixed in advance. The volume of $\mathcal{R}_n$ can be made as small as one pleases by choosing an appropriately small number d.

The confidence coefficient associated with $\mathcal{R}_n$ is:

$$\begin{aligned}
&\mathrm{P}_{\boldsymbol{\beta},\sigma}\left\{\boldsymbol{\beta} \in \mathcal{R}_n\right\} \\
&= \mathrm{P}_{\boldsymbol{\beta},\sigma}\left\{(\widehat{\boldsymbol{\beta}}_n - \boldsymbol{\beta})'(\mathrm{X}_n'\mathrm{X}_n)(\widehat{\boldsymbol{\beta}}_n - \boldsymbol{\beta}) \le nd^2\right\} \\
&= F\left(\frac{nd^2}{\sigma^2}; p\right) \text{ with } F(x;p) = \mathrm{P}\left(U \le x\right), \\
&\qquad\qquad\qquad\qquad U \sim \chi^2_p, x > 0.
\end{aligned} \tag{12.2.4}$$

In the sequel, we may write $F(x;p)$ or $F(x)$ interchangeably.

Let "a" satisfy $F(a;p) = 1 - \alpha$, that is a is an upper $100\alpha\%$ point of a χ^2_p distribution. Since we require that $F\left(nd^2/\sigma^2;p\right) \ge F(a;p)$, we must have $nd^2/\sigma^2 \ge a$, so that we determine:

$$n \text{ to be the smallest integer } \ge \frac{a\sigma^2}{d^2} = C, \text{ say.} \tag{12.2.5}$$

We refer to C as an optimal fixed sample size had σ^2 been known. The magnitude of C remains unknown and no fixed-sample-size methodology would help.

The two-stage fixed-size confidence region methodologies of Healy (1956) and Chatterjee (1962b) were consistent.

It is understood that a multistage or sequential methodology would come up with a suitable stopping time $N \equiv N(d)$ and a final dataset

$$N, (Y_i, x_{i1}, x_{i2}, \ldots, x_{ip}),\ i = 1, 2, \ldots, N.$$

Now, analogous to (12.2.3) we propose the fixed-size confidence region for $\boldsymbol{\beta}$ based on full dataset:

$$\mathcal{R}_N = \left\{\boldsymbol{\beta}_{p\times 1} \in I\!R^p : N^{-1}(\widehat{\boldsymbol{\beta}}_N - \boldsymbol{\beta})'(\mathrm{X}_N'\mathrm{X}_N)(\widehat{\boldsymbol{\beta}}_N - \boldsymbol{\beta}) \le d^2\right\}. \tag{12.2.6}$$

We need a crucial result:

$$\begin{aligned}
&\text{For all fixed } n \ge m(\ge p+1), \\
&\widehat{\boldsymbol{\beta}}_n \text{ and } \left(S^2_m, S^2_{m+1}, \ldots, S^2_n\right) \text{ are independent.}
\end{aligned} \tag{12.2.7}$$

In the methodologies summarized here, one will soon notice that for all fixed $n \ge m(\ge p+1)$, the random variables $I(N = n)$ would be determined by $\left(S^2_m, S^2_{m+1}, \ldots, S^2_n\right)$ alone. Thus, $I(N = n)$ and $\widehat{\boldsymbol{\beta}}_n$ would be independent random variables in view of (12.2.7). We sketch a proof of this in Section 12.6.1.

Hence, we express the confidence coefficient associated with $\mathcal{R}_N$ as:

$$
\begin{aligned}
&\mathrm{P}_{\boldsymbol{\beta},\sigma}\{\boldsymbol{\beta}\in\mathcal{R}_n\}\\
&=\mathrm{P}_{\boldsymbol{\beta},\sigma}\left\{N^{-1}(\widehat{\boldsymbol{\beta}}_N-\boldsymbol{\beta})'(\mathrm{X}_N'\mathrm{X}_N)(\widehat{\boldsymbol{\beta}}_N-\boldsymbol{\beta})\le d^2\right\}\\
&=\mathrm{E}_{\boldsymbol{\beta},\sigma}\left[F\left(\frac{Nd^2}{\sigma^2};p\right)\right]\ \text{with}\ F(x;p)=\mathrm{P}\left(U\le x\right),\\
&\qquad\qquad\qquad\qquad U\sim\chi^2_p, x>0.
\end{aligned}
\tag{12.2.8}
$$

We leave its proof out as an exercise.

12.2.1 Two-Stage Procedure

Motivations for this two-stage procedure come from Stein (1945,1949), Healy (1956) and Chatterjee (1962b). Let

$$(Y_i, x_{i1}, x_{i2}, \ldots, x_{ip}),\ i=1,2,\ldots,m$$

be the initial observations where $m(\ge p+1)$ is pilot size. We obtain $\widehat{\boldsymbol{\beta}}_m$ and S_m^2 from (12.2.1) - (12.2.2). Let $F_{p,m-p,\alpha}$ denote an upper $100\alpha\%$ point of a $F_{p,m-p}$ distribution and define:

$$b\equiv b_{p,m,\alpha}=pF_{p,m-p,\alpha}. \tag{12.2.9}$$

Next, we define a two-stage stopping variable:

$$N\equiv N(d)=\max\left\{m,\left\langle\frac{bS_m^2}{d^2}\right\rangle+1\right\}, \tag{12.2.10}$$

where $b\equiv b_{p,m,\alpha}$ is taken from (12.2.9). It is easy to see that N is finite w.p. 1, and thus $\mathcal{R}_N$ is a genuine confidence region.

If $N=m$, we do not require any observation at second stage. But, if $N>m$, then we sample the difference at second stage by gathering additional observations

$$(Y_i, x_{i1}, x_{i2}, \ldots, x_{ip}),\ i=m+1,\ldots,N.$$

Then, based on *combined* set of observations

$$N, (Y_i, x_{i1}, x_{i2}, \ldots, x_{ip}),\ i=1,\ldots,N$$

from stages 1 and 2, we estimate $\boldsymbol{\beta}$ by $\mathcal{R}_N$ from (12.2.6).

The confidence coefficient is given by (12.2.8). The following results were obtained by Mukhopadhyay and Abid (1986b).

Theorem 12.2.1 : *For the stopping variable N defined by (12.2.10), for all fixed $\boldsymbol{\beta},\sigma,p,m$ and α, one has the following properties:*

(i) $\mathrm{P}_{\boldsymbol{\beta},\sigma}\{\boldsymbol{\beta}\in\mathcal{R}_N\}\ge 1-\alpha$ *for all fixed d [Consistency or Exact Consistency]*;

(ii) $\lim_{d\to 0}\mathrm{E}_{\boldsymbol{\beta},\sigma}[N/C]=ba^{-1}$;

(iii) $\lim_{d\to 0} \mathrm{V}_{\boldsymbol{\beta},\sigma}[N]\left\{\frac{1}{2}(m-p)\left(b\sigma^2/d^2\right)^{-2}\right\} = 1$;

(iv) $\lim_{d\to 0} \mathrm{P}_{\boldsymbol{\beta},\sigma}\{\boldsymbol{\beta} \in \mathcal{R}_N\} = 1-\alpha$ [*Asymptotic Consistency*];

where $b \equiv b_{p,m,\alpha}$ *is taken from (12.2.9),* $\mathcal{R}_N$ *came from (12.2.6), and* $C = a\sigma^2/d^2$ *was defined in (12.2.5).*

In part (ii), limiting ratio ba^{-1} practically exceeds one whatever be p, m, σ and α. We referred to this as *asymptotic first-order inefficiency* property in previous chapters. A proof of part (i) is left out as an exercise.

12.2.2 Modified Two-Stage Procedure

Motivated by the results of Mukhopadhyay (1980,1982a), one would first choose and fix a positive and finite number γ. Mukhopadhyay and Abid (1986b) defined a pilot sample size:

$$m \equiv m(d) = \max\left\{p+1, \left\langle\left(\frac{a}{d^2}\right)^{1/(1+\gamma)}\right\rangle + 1\right\}. \tag{12.2.11}$$

One will record initial observations

$$(Y_i, x_{i1}, x_{i2}, \ldots, x_{ip}),\ i = 1, 2, \ldots, m,$$

and determine $\widehat{\boldsymbol{\beta}}_m$ and S_m^2 from (12.2.1) - (12.2.2). Next, define a two-stage stopping variable:

$$N \equiv N(d) = \max\left\{m, \left\langle\frac{bS_m^2}{d^2}\right\rangle + 1\right\}, \tag{12.2.12}$$

where $b \equiv b_{p,m,\alpha}$ is taken from (12.2.9).

At this point, one samples the difference at the second stage by gathering the additional observations

$$(Y_i, x_{i1}, x_{i2}, \ldots, x_{ip}), i = m+1, \ldots, N \text{ if } N > m.$$

Then, based on *combined* set of observations

$$N, (Y_i, x_{i1}, x_{i2}, \ldots, x_{ip}),\ i = 1, \ldots, N$$

from stages 1 and 2, we estimate $\boldsymbol{\beta}$ by $\mathcal{R}_N$ from (12.2.6). Again, N is finite w.p. 1, and thus $\mathcal{R}_N$ is a genuine confidence region. Note that confidence coefficient is given by (12.2.8). The following results were obtained by Mukhopadhyay and Abid (1986b).

Theorem 12.2.2: *For the stopping variable* N *defined by (12.2.11) - (12.2.12), for all fixed* $\boldsymbol{\beta}, \sigma, p, m$ *and* α*, one has the following properties:*

(i) $\mathrm{P}_{\boldsymbol{\beta},\sigma}\{\boldsymbol{\beta} \in \mathcal{R}_N\} \geq 1-\alpha$ *for all fixed* d [*Consistency or Exact Consistency*];

(ii) $\lim_{d\to 0} \mathrm{E}_{\boldsymbol{\beta},\sigma}[N/C] = 1$ [*Asymptotic First-Order Efficiency*];

(iii) $\lim_{d\to 0} \mathrm{V}_{\boldsymbol{\beta},\sigma}[N]\left\{\frac{1}{2}(m-p)\left(b\sigma^2/d^2\right)^{-2}\right\} = 1$;

(iv) $\lim_{d\to 0} \mathrm{P}_{\boldsymbol{\mu},\sigma}\{\boldsymbol{\beta} \in \mathcal{R}_N\} = 1 - \alpha$ [*Asymptotic Consistency*];

where $b \equiv b_{p,m,\alpha}$ *is taken from (12.2.9),* $\mathcal{R}_N$ *came from (12.2.6), and* $C = a\sigma^2/d^2$ *was defined in (12.2.5).*

A proof of part (ii) is left out as an exercise. In connection with part (ii), we may add that although this modified two-stage strategy is asymptotically *first-order* efficient, we fully expect $\lim_{d\to 0} \mathrm{E}_{\boldsymbol{\beta},\sigma}[N - C]$ to be large for all practical purposes. That is, this strategy would fail to be second-order efficient as we had noted a similar feature in previous chapters. More efficient methodologies are summarized next.

The modified two-stage fixed-size confidence region methodology of Mukhopadhyay and Abid (1986b) was both consistent and asymptotically first-order efficient. The Healy (1956) and Chatterjee (1962b) methodologies were consistent but asymptotically inefficient.

12.2.3 Purely Sequential Procedure

Mukhopadhyay and Abid (1986b) included the following purely sequential estimation strategy in the light of Gleser (1965), Albert (1966), and Srivastava (1967,1971). One would start with initial observations

$$(Y_i, x_{i1}, x_{i2}, \ldots, x_{ip}),\ i = 1, 2, \ldots, m,$$

where $m(\geq p+1)$ is pilot size. Then, one will proceed with one at-a-time additional observation as needed and terminate sampling according to the stopping time:

$$N \equiv N(d) \text{ is the smallest integer } n(\geq m) \text{ for which we observe } n \geq \left(\frac{a}{d^2}\right) S_n^2. \qquad (12.2.13)$$

This sequential methodology is implemented in an usual fashion. Again, $\mathrm{P}_{\boldsymbol{\beta},\sigma}(N < \infty) = 1$, that is the sampling process would terminate with probability one. That is, $\mathcal{R}_N$ is a genuine confidence region estimator. Note that the distribution of N involves only σ^2, and not $\boldsymbol{\beta}$. Yet, we may interchangeably write $\mathrm{P}_{\boldsymbol{\beta},\sigma}(\cdot)$, $\mathrm{E}_{\boldsymbol{\beta},\sigma}(\cdot)$ instead of $\mathrm{P}_{\sigma}(\cdot)$, $\mathrm{E}_{\sigma}(\cdot)$.

Upon termination, we estimate $\boldsymbol{\beta}$ by $\mathcal{R}_N$ from (12.2.6) based on full dataset

$$N, (Y_i, x_{i1}, x_{i2}, \ldots, x_{ip}),\ i = 1, \ldots, N.$$

Note that the associated confidence coefficient is given by (12.2.8). The following *first-order* results were obtained by Mukhopadhyay and Abid (1986b).

Theorem 12.2.3 : *For the stopping variable N defined by (12.2.13), for all fixed* $\boldsymbol{\beta}, \sigma, p, m$ *and* α*, one has the following properties:*

(i) $\mathrm{E}_{\boldsymbol{\beta},\sigma}[N] \le C + O(1)$ *for all fixed* d;

(ii) $\lim_{d\to 0} \mathrm{E}_{\boldsymbol{\beta},\sigma}[N/C] = 1$ [*Asymptotic First-Order Efficiency*];

(iii) $\lim_{d\to 0} \mathrm{P}_{\boldsymbol{\beta},\sigma}\{\boldsymbol{\beta} \in \mathcal{R}_N\} = 1 - \alpha$ [*Asymptotic Consistency*];

where $\mathcal{R}_N$ *came from (12.2.6) and* $C = a\sigma^2/d^2$ *was defined in (12.2.5).*

Proofs of parts (i) and (ii) are left out as exercises. The following *second-order* results were also obtained by Mukhopadhyay and Abid (1986b).

One should recall the expressions of $h(1)$ and $f(x;p)$ from (6.2.33). One may recall that $f(x;p)$ stands for the p.d.f. of a χ_p^2 random variable, that is, $f(x;p) = dF(x;p)/dx, x > 0$.

Theorem 12.2.4: *For the stopping variable* N *defined by (12.2.13), for all fixed* $\boldsymbol{\beta}, \sigma, p, m$ *and* α, *one has the following properties as* $d \to 0$:

(i) $\mathrm{E}_{\boldsymbol{\beta},\sigma}[N] = C + \eta_0(p) + o(1)$ *if* $m \ge p + 3$;

(ii) $\mathrm{P}_{\boldsymbol{\beta},\sigma}\{\boldsymbol{\beta} \in \mathcal{R}_N\} = 1 - \alpha + \frac{a}{C}\{\eta_0(p)f(a;p) + af'(a;p)\} + o(C^{-1})$ *if* (a) $m \ge p + 3$ *for* $p = 2$ *or* $p \ge 4$, *and* (b) $m \ge 7$ *for* $p = 3$;

(iii) $\left(\frac{1}{2C}\right)^{1/2}(N - C)$ *converges to* $N(0,1)$ *in distribution;*

where $\mathcal{R}_N$ *came from (12.2.6),* $C = a\sigma^2/d^2$ *was defined in (12.2.5),* $\eta_0(p) = \frac{1}{2} - p - h(1)$, *and* $h(1), f(x;p)$ *came from (6.2.33).*

For its proof, one may refer to Mukhopadhyay and Abid (1986b) and Finster (1983).

Now, it is easy to check:

$$f'(x;p) = \frac{(p-2-x)}{2x} f(x;p) \text{ for any } x > 0.$$

Thus, the second-order term, namely $\frac{a}{C}\{\eta_0(p)f(a;p) + af'(a;p)\}$, can be expressed alternatively as:

$$\frac{a}{C}\left\{\eta_0(p) + \frac{1}{2}(p-2-a)\right\} f(a;p).$$

Sequential and three-stage fixed-size confidence region methodologies of Mukhopadhyay and Abid (1986b) enjoyed asymptotic second-order properties.

12.2.4 Three-Stage Procedure

Mukhopadhyay and Abid (1986b) defined the following three-stage estimation methodology. First, one would choose and fix $0 < \rho < 1$. Next, one would start with initial observations

$$(Y_i, x_{i1}, x_{i2}, \ldots, x_{ip}),\ i = 1, 2, \ldots, m,$$

where $m(\geq p+1)$ is pilot size and determine $\widehat{\boldsymbol{\beta}}_m$ and S_m^2 from (12.2.1)-(12.2.2). We define:

$$T \equiv T(d) = \max\left\{m, \left\langle \rho \frac{aS_m^2}{d^2} \right\rangle + 1\right\}. \tag{12.2.14}$$

We sample the difference at second stage if $T > m$ and denote the combined data from first and second stages,

$$T, (Y_i, x_{i1}, x_{i2}, \ldots, x_{ip}),\ i = 1, 2, \ldots, T.$$

Recall that $C = a\sigma^2/d^2$ where σ^2 is unknown. It is clear that T estimates ρC, a fraction of c.

At this point, one would estimate σ^2 and C by S_T^2 and $aS_T^2 d^{-2}$ respectively. More precisely, we define:

$$N \equiv N(d) = \max\left\{T, \left\langle \frac{aS_T^2}{d^2} + \varepsilon \right\rangle + 1\right\}, \tag{12.2.15}$$

where $\varepsilon = \rho^{-1}\left\{3 - \frac{1}{2}(p-a)\right\} - \frac{1}{2}$. We sample the difference at third stage if $N > T$ and denote the combined data from all three stages,

$$N, (Y_i, x_{i1}, x_{i2}, \ldots, x_{ip}),\ i = 1, 2, \ldots, T, \ldots, N.$$

Now, $\mathcal{R}_N$ is a genuine confidence region estimator. Upon termination, we estimate $\boldsymbol{\beta}$ by $\mathcal{R}_N$ from (12.2.6). The associated confidence coefficient is given by (12.2.8). The following *second-order* results were obtained by Mukhopadhyay and Abid (1986b).

Theorem 12.2.5: *For the stopping variable N defined by (12.2.14) - (12.2.15), for all fixed $\boldsymbol{\beta}, \sigma, p, m$ and α, one has the following properties as $d \to 0$:*

(i) $\mathrm{E}_{\boldsymbol{\beta},\sigma}[N] = C + \rho^{-1}\left\{1 - \frac{1}{2}(p-a)\right\} + o(1)$;

(ii) $\mathrm{P}_{\boldsymbol{\beta},\sigma}\left\{\boldsymbol{\beta} \in \mathcal{R}_N\right\} = 1 - \alpha + o(d^2)$;

(iii) $\left(\dfrac{\rho}{2C}\right)^{1/2}(N - C)$ *converges to* $N(0,1)$ *in distribution;*

where $\mathcal{R}_N$ came from (12.2.6) and $C = a\sigma^2/d^2$ was defined in (12.2.5).

Comparing the results from part (iii) in Theorems 12.2.4 and 12.2.5, we can see that $\mathrm{V}_{\boldsymbol{\beta},\sigma}[N_{\text{sequential}}] \approx 2C$ whereas $\mathrm{V}_{\boldsymbol{\beta},\sigma}[N_{\text{three-stage}}] \approx \frac{2}{\rho}C$ since $(N-C)^2/C$ is uniformly integrable in either case. That is, even though both $N_{\text{sequential}}$ and $N_{\text{three-stage}}$ estimate the same optimal fixed sample size C, the three-stage stopping rule (12.2.14) - (12.2.15) is expected to be more variable.

12.3 Minimum Risk Point Estimation

A minimum risk point estimation problem for $\boldsymbol{\beta}$ in a linear model (12.1.6) was first introduced by Mukhopadhyay (1974b). He developed an asymptotically risk efficient sequential methodology whose regret function was investigated and its rate of convergence found. Finster (1983) included more asymptotic *second-order* properties of the Mukhopadhyay procedure.

Suppose that we have available a sequence of independent observations

$$(Y_i, x_{i1}, x_{i2}, \ldots, x_{ip}),\ i = 1, 2, \ldots, n, \ldots$$

following a linear model in (12.1.6). Having recorded $n(\geq p+1)$ observations, as before, we again consider the least square estimator $\widehat{\boldsymbol{\beta}}_n$ of $\boldsymbol{\beta}$ from (12.2.1) and the estimator S_n^2 of σ^2 from (12.2.2). Now, we introduce a minimum risk point estimation problem for $\boldsymbol{\beta}$ under a loss function which is composed of both estimation error and cost due to sampling:

Having recorded

$$(Y_i, x_{i1}, x_{i2}, \ldots, x_{ip}),\ i = 1, 2, \ldots, n,$$

suppose that the loss in estimating $\boldsymbol{\beta}$ by the least square estimator $\widehat{\boldsymbol{\beta}}_n$ is:

$$L_n\left(\boldsymbol{\beta}, \widehat{\boldsymbol{\beta}}_n\right) = n^{-1}(\widehat{\boldsymbol{\beta}}_n - \boldsymbol{\beta})'(\mathrm{X}_n'\mathrm{X}_n)(\widehat{\boldsymbol{\beta}}_n - \boldsymbol{\beta}) + cn \tag{12.3.1}$$

with $c(>0)$ known. This formulation strikes a balance between cost of sampling and the margin of estimation error. Now, the fixed-sample-size risk function $R_n(c)$ associated with (12.3.1) is expressed as:

$$\begin{aligned}
&\mathrm{E}_{\boldsymbol{\beta},\sigma}\left[L_n\left(\boldsymbol{\beta}, \widehat{\boldsymbol{\beta}}_n\right)\right] \\
&= \mathrm{E}_{\boldsymbol{\beta},\sigma}\left[n^{-1}(\widehat{\boldsymbol{\beta}}_n - \boldsymbol{\beta})'(\mathrm{X}_n'\mathrm{X}_n)(\widehat{\boldsymbol{\beta}}_n - \boldsymbol{\beta})\right] + cn \\
&= n^{-1}\mathrm{Trace}\left\{(\mathrm{X}_n'\mathrm{X}_n)\mathrm{E}_{\boldsymbol{\beta},\sigma}\left[(\widehat{\boldsymbol{\beta}}_n - \boldsymbol{\beta})(\widehat{\boldsymbol{\beta}}_n - \boldsymbol{\beta})'\right]\right\} + cn \\
&= n^{-1}\mathrm{Trace}\left\{(\mathrm{X}_n'\mathrm{X}_n)\sigma^2(\mathrm{X}_n'\mathrm{X}_n)^{-1}\right\} + cn \text{ using (12.2.1)} \\
&= n^{-1}\sigma^2\mathrm{Trace}(I_{p\times p}) + cn \\
&= \frac{p\sigma^2}{n} + cn.
\end{aligned} \tag{12.3.2}$$

Our goal is to design a sequential strategy that would minimize this risk for all $\boldsymbol{\beta}$ and σ^2. Let us pretend that n is a continuous variable so that $n \equiv n^*(c)$ will minimize $R_n(c)$ where

$$n^* \equiv n^*(c) = \left(\frac{p}{c}\right)^{1/2}\sigma. \tag{12.3.3}$$

Even though $n^*(c)$ may not be a (positive) integer, we refer to it as an optimal fixed sample size had σ^2 been known. Clearly, the minimum fixed-sample-size risk is:

$$R_{n^*}(c) \equiv \frac{p\sigma^2}{n^*} + cn^* = 2cn^*. \tag{12.3.4}$$

Surely, the magnitude of n^* remains unknown since σ^2 is unknown. In fact, no fixed-sample-size procedure will minimize the risk (12.3.2) under the loss function (12.3.1), uniformly for all σ^2.

Sequential minmum risk point estimation methodology of Mukhopadhyay (1974b) enjoyed asymptotic second-order properties.

12.3.1 Purely Sequential Procedure

The sequential strategy of Mukhopadhyay (1974b) is now introduced. One would start with initial observations

$$(Y_i, x_{i1}, x_{i2}, \ldots, x_{ip}),\ i = 1, 2, \ldots, m,$$

where $m(\geq p+1)$ is pilot size. Then, one will proceed sequentially with one additional observation at-a-time and terminate sampling according to the stopping time:

$$\begin{aligned} &N \equiv N(c) \text{ is the smallest integer } n(\geq m) \text{ for which} \\ &\qquad \text{we observe } n \geq \left(\frac{p}{c}\right)^{1/2} S_n, \end{aligned} \tag{12.3.5}$$

where S_n^2, the MSE, was defined in (12.2.2).

This sequential methodology is implemented in the same way we did previous sequential procedures. Again, $\mathrm{P}_{\boldsymbol{\beta},\sigma}(N < \infty) = 1$, that is sampling would terminate w.p. 1. Upon termination, we estimate $\boldsymbol{\beta}$ by $\widehat{\boldsymbol{\beta}}_N \equiv (\mathrm{X}_N'\mathrm{X}_N)^{-1}\mathrm{X}_N'\boldsymbol{Y}_N$ obtained from full dataset:

$$N, (Y_i, x_{i1}, x_{i2}, \ldots, x_{ip}),\ i = 1, 2, \ldots, m, \ldots, N.$$

The risk associated with $\widehat{\boldsymbol{\beta}}_N$ is expressed as:

$$R_N(c) \equiv \mathrm{E}_{\boldsymbol{\beta},\sigma}\left[L_N(\boldsymbol{\beta}, \widehat{\boldsymbol{\beta}}_N)\right] = p\sigma^2 \mathrm{E}_{\boldsymbol{\beta},\sigma}\left[\frac{1}{N}\right] + c\mathrm{E}_{\boldsymbol{\beta},\sigma}[N], \tag{12.3.6}$$

since the random variables $I(N = n)$ and $\widehat{\boldsymbol{\beta}}_n$ are independent for all $n \geq m$. Refer to (12.2.7) and its proof given in Section 12.6.1.

Now, notice from (12.3.3) that $p\sigma^2 = cn^{*2}$ and thus we rewrite (12.3.6) as:

$$R_N(c) \equiv cn^{*2}\mathrm{E}_{\boldsymbol{\beta},\sigma}\left[\frac{1}{N}\right] + c\mathrm{E}_{\boldsymbol{\beta},\sigma}[N]. \tag{12.3.7}$$

Mukhopadhyay (1974b) defined the following measures, along the lines of Robbins (1959), for comparing the sequential estimation strategy $(N, \widehat{\boldsymbol{\beta}}_N)$

from (12.3.5) with the fictitious, and yet optimal fixed-sample-size estimation strategy $(n^*, \widehat{\boldsymbol{\beta}}_{n^*})$. We pretend for a moment, however, that n^* is a known integer!

$$\textit{Risk Efficiency}: \text{RiskEff}(c) \equiv \frac{R_N(c)}{R_{n^*}(c)} = \frac{1}{2}\left\{ \text{E}_{\boldsymbol{\beta},\sigma}\left[\frac{n^*}{N}\right] + \text{E}_{\boldsymbol{\beta},\sigma}\left[\frac{N}{n^*}\right]\right\};$$

$$\textit{Regret}: \text{Regret}(c) \equiv R_N(c) - R_{n^*}(c) = c\text{E}_{\boldsymbol{\beta},\sigma}\left[\frac{(N-n^*)^2}{N}\right]. \tag{12.3.8}$$

Now, we summarize some asymptotic *first-order* results due to Mukhopadhyay (1974b) in the following theorem.

Theorem 12.3.1 : *For the stopping variable N defined by (12.3.5), for all fixed $\boldsymbol{\beta}, \sigma, p$ and m, one has the following properties:*

(i) $\lim_{c\to 0} N/n^* = 1$ *in probability;*
(ii) $\lim_{c\to 0} \text{E}_{\boldsymbol{\beta},\sigma}[N/n^*] = 1$ [*Asymptotic First-Order Efficiency*];
(iii) $\lim_{c\to 0} \text{RiskEff}(c) = 1$ [*Asymptotic First-Order Risk Efficiency*];
(iv) $\text{Regret}(c) = O(c)$;

where $n^* = (p/c)^{1/2}\sigma$ was defined in (12.3.3).

Finster (1983,1985) and Mukhopadhyay and Abid (1999b) gave elegant proofs of results that are sharper than those found in Theorem 12.3.1. They provided asymptotic *second-order* expansions of both regret function and average sample size. Some crucial results are summarized next.

Theorem 12.3.2 : *For the stopping variable N defined by (12.3.5), for all fixed $\boldsymbol{\beta}, \sigma, p$ and m, one has the following properties as $c \to 0$:*

(i) $\text{Regret}(c) = \frac{1}{2}c + o(c)$ *if* $m \geq p+3$;
(ii) $\text{E}_{\boldsymbol{\beta},\sigma}[N - n^*] = \eta_3(p) + o(1)$ *if* $m \geq p+2$;

where $n^* = (p/c)^{1/2}\sigma$ was defined in (12.3.3) and

$$\eta_3(p) = -\frac{1}{2}\sum_{n=1}^{\infty}\frac{1}{n}\text{E}\left[\max\left(0, \chi_n^2 - 3n\right)\right]. \tag{12.3.9}$$

We will sketch a proof of this theorem in Section 12.6.2.

12.4 Data Analyses and Conclusions

Now, we examine performances of estimation strategies described in Sections 12.2 and 12.3 using the computer program **Sqe12.exe**. When the program is executed, it will prompt one to select a fixed-size confidence region problem (Section 12.2) or a minimum risk point estimation problem (Section 12.3).

One will be able to analyze the methodologies with either a linear model or a polynomial model (with or without an intercept). One will be prompted to select one or the other (L: linear or P: polynomial). If a model includes an intercept, then β_1 is understood to be that intercept by fixing $X_1 \equiv 1$. Otherwise, β_1 will be the regression parameter that would be the coefficient of the independent variable X_1. In general, β_i is the regression parameter that would be the coefficient of the independent variable X_i, $i = 1, \ldots, p$.

However, in a polynomial model, we interpret:

$$\beta_i = \begin{cases} \text{regression parameter of } x_1^i \text{ if the model has no intercept} \\ \text{regression parameter of } x_1^{i-1} \text{ if the model has an intercept} \end{cases}$$

for $i = 2, \ldots, p$.

Figure 12.4.1 shows screen input for 5000 simulations runs to obtain fixed-size 95% confidence regions for $\boldsymbol{\beta}$ where $\boldsymbol{\beta}' = (\beta_1, \beta_2, \beta_3)$ with $d = 0.5$, under a model:

$$Y_i = \beta_1 + \beta_2 x_{2i} + \beta_3 x_{3i} + \varepsilon_i, \; i = 1, \ldots, N. \qquad (12.4.1)$$

In this exercise, two-stage procedure (Procedure number: 1) described in Section 12.2.1 was used to construct 95% ($\alpha = 0.05$) confidence regions. The data were generated from a linear regression model with following specifications:

- Set $\beta_1 = 3$, $\beta_2 = 1$ and $\beta_3 = 2$;
- The values of ε_i were randomly generated from $N(0, \sigma^2 = 16)$;
- The values for X_2 and X_3 were generated from U$(2,3)$ and U$(1,5)$ distributions respectively.

A summary obtained from averaging 5000 simulations with $m = 10$ is shown in Figure 12.4.2. We note that the optimal sample size, $C = \frac{7.8147 \times 4^2}{0.5^2} \approx 500.15$. But, we see the average simulated value of the sample size $\overline{n} = 837.29$ with a standard deviation $s_n = 448.08$. Since $\frac{\overline{n}}{C} = \frac{837.29}{500.15} \approx 1.6741$, we will say that the two-stage strategy has oversampled by a large margin, 67.41%. Is this oversampling percentage in a ballpark of what should be expected?

This is where "theory" can throw some light. Note that, $F_{3,7,0.05} = 4.3468 \Rightarrow b = 3 \times F_{3,7,0.05} = 13.04$, $a = \chi^2_{3,0.05} = 7.8147$ and $\frac{b}{a} = \frac{13.04}{7.8147} \approx 1.6687$. Now, Theorem 12.2.1, part (ii) suggests that we should expect that $\frac{\overline{n}}{C} \approx 1.6741$ in close proximity of $\frac{b}{a}$. The two numbers are incredibly close to each other.

Also, the estimated coverage probability $\overline{p} = 0.9554$ appears right on target (0.95) or better. Is it really? The estimated standard deviation of $\overline{p}$ should be $\sqrt{\frac{\overline{p}(1-\overline{p})}{5000}} = \sqrt{\frac{0.9554(1-0.9554)}{5000}} = 0.0029193$. Now, let us take a look at $\overline{p} \pm 2 \times 0.0029193 = 0.9554 \pm 0.0058386$ which leads to the interval $(0.94956, 0.96124)$. So, yes, the estimated coverage probability $\overline{p} = 0.9554$ appears right on target (0.95).

```
          Select a program using the number given
          in the left hand side. Details of these
          programs are given in Chapter 12.
        *******************************************
        * Number        Program                   *
        *  1   Fixed-Size Confidence Region for beta *
        *  2   Minimum Risk Point Estimation of beta *
        *  0   Exit                                *
        *******************************************
     Enter the program number: 1

     Type P for a polynomial regression or
          L for a linear regression model: L

     Number  Procedure
       1     Two-Stage
       2     Modified Two-Stage
       3     Purely Sequential
       4     Three-Stage
     Enter procedure number: 1

     Do you want to save the results in a file(Y or N)? Y
     Note that your default outfile name is "SEQ12.OUT".
     Do you want to change the outfile name (Y or N)? N

  Enter the following preassigned values for a fixed-size
  confidence region for a p-dimensional regression
  parameter vector, beta
     Enter p (integer > 1): 3
     Do you want to include an intercept (Y or N)? Y
     alpha (for 100(1-alpha)% confidence interval) = .05
     d (Fixed-size of the confidence region) = .5
     Enter the initial sample size, m: 10
     Do you want to change the input parameters?[Y or N] N

     Do you want to simulate data or input real data?
     Type S (Simulation) or D (Real Data): S
     Do you want to store simulated sample sizes (Y or N)? N
     Number of simulation replications? 5000
     Enter a positive integer(<10) to initialize
               the random number sequence: 3

     Enter the following parameters for simulation:
     Parameter vector, beta ( 3 values)  = 3 1 2
          Sigma^2 = 16

     The values of the covariates of the linear model are
     generated from uniform distributions with user input
     minimum and maximum values.

     Since the model has an intercept, x(1) = 1
     Enter minimum and maximum for x( 2): 2 3
     Enter minimum and maximum for x( 3): 1 5
```

Figure 12.4.1. *Screen input for* ***Seq12.exe***.

```
 Fixed-Size Confidence Region for the Regression
   Parameter Vector Using Two-Stage Procedure.
   -------------------------------------------

Model: Linear with intercept

Data for the simulations were generated from the
following covariates of the model:
   i  beta(i)  min{x(i)}  max{x(i)}
   1    3.00     Intercept
   2    1.00       2.00       3.00
   3    2.00       1.00       5.00

Input parameters for simulations:
                 alpha =   0.05
                     d =   0.50
               sigma^2 =  16.00
Initial sample size, m =     10
Number of simulations  =   5000

Optimal sample size (C)               =     500.15
Average final sample size (n_bar) =     837.29
Std. dev. of n (s_n)                  =     448.08
Minimum sample size (n_min)           =      49
Maximum sample size (n_max)           =    3986
Coverage probability                  =     0.9554
```

Figure 12.4.2. *Output screen from* ***Seq12.exe***.

The program **Seq12.exe** can also be used with live real data. The program reads data only from an ASCII data files and the file should each point on rows. That is, the ith row of the file should have the values

- $y_i, x_{i1}, \ldots, x_{pi}$ if the model is linear and has no intercept;
- $y_i, x_{i2}, \ldots, x_{pi}$ if the model is linear and has an intercept;
- y_i, x_{i1} if the model is polynomial and has no intercept;
- y_i, x_{i1} if the model is polynomial and has an intercept.

Figure 12.4.3 shows results for a 90% fixed-size confidence region estimation using data from Table 12.4.1. In this analysis, the purely sequential methodology was implemented with $d = 1.75$ and initial sample size, $m = 10$. The results show that the analysis had sample size $44(= n)$ at termination. Upon termination, the computed estimates were:

$$\widehat{\sigma}^2 = S_{44}^2 = 6.8337 \text{ and}$$

$$\widehat{\beta}'_{44} = (9.5189, 2.1898, -4.4155, -1.2713, 2.7967, 1.7249).$$

Further the program computed estimated standard errors of the estimated regression parameters and their p-values. A cursory examination of the p-values seen from Figure 12.4.3 may lead one to conclude that the intercept term β_1 is not significant while other regression parameters are clearly significant.

```
 Fixed-Size Confidence Region for the Regression
Parameter Vector Using Purely Sequential Procedure.
---------------------------------------------------

   Model: Linear with intercept

   Input Parameters:
                     alpha =     0.10
                         d =    1.750
   Initial Sample size, m =      10

   Input Data file Name: Reg01.dat

   Final sample size (N)        =       24
   Estimate of sigma^2          =   6.8337

      i  beta_hat(i)  SE(Beta(i))  p-value
      1       9.5189       7.2195   0.2039
      2       2.1898       0.5303   0.0006
      3      -4.4155       0.9190   0.0001
      4      -1.2713       0.5305   0.0276
      5       2.7967       0.6594   0.0005
      6       1.7249       0.4716   0.0018
```

Figure 12.4.3. *Confidence region for regression parameters using data from **Reg01.dat**.*

Table 12.4.1: Dataset **Reg01.dat**

i	Y_i	x_{2i}	x_{3i}	x_{4i}	x_{5i}	x_{6i}
1	38.945	6.48	3.84	1.43	8.46	4.28
2	31.205	3.75	2.60	3.15	6.21	5.69
3	47.030	6.08	2.20	1.86	7.96	6.82
4	32.125	3.11	2.31	3.66	7.72	5.01
5	42.010	5.86	2.35	3.35	8.19	6.11
6	31.695	6.65	3.79	2.24	9.00	2.85
7	28.890	5.99	4.00	2.02	6.70	4.51
8	38.455	5.36	2.18	3.63	7.80	4.55
9	35.005	4.34	2.95	2.34	8.77	2.59
10	38.085	6.42	2.57	4.06	8.56	6.17
11	28.025	3.04	2.10	3.03	8.36	3.00
12	36.655	6.67	3.51	4.68	7.96	5.32
13	27.810	4.41	3.82	3.68	6.97	5.36
14	34.160	3.08	2.04	4.76	8.88	2.73
15	47.525	6.87	2.22	2.31	8.95	5.54
16	21.845	3.76	2.98	4.03	6.59	2.48
17	27.230	5.60	2.42	3.94	6.50	3.72
18	29.070	5.39	3.90	1.24	7.29	4.06
19	31.180	4.68	3.25	1.19	7.99	3.19
20	35.785	5.17	3.06	3.67	8.44	6.83
21	35.685	6.20	2.51	2.76	6.55	5.24
22	35.205	4.19	2.39	1.74	8.92	4.77
23	35.465	3.78	2.38	1.81	8.66	2.76
24	24.480	3.75	3.57	4.53	7.54	4.99
25	36.835	4.71	3.16	2.19	8.37	4.56

The 'program number: 2' within **Seq12.exe** obtains the minimum risk point estimation of regression parameters as described in Section 12.3.1. Again, one may use either a linear model or a polynomial model, with or without an intercept. The data input for the 'program number: 2' stays nearly the same as before, but naturally this program requires one to input a value for c, cost due sampling. One also gets an opportunity to proceed with either simulation or live data input.

The output from minimum risk point estimation of $\boldsymbol{\beta}$ where $\boldsymbol{\beta}' = (\beta_1, \beta_2, \beta_3, \beta_4, \beta_5)$ in a regression model given in (12.4.1) using **Seq12.exe** is shown in Figure 12.4.4. Here, we used data input from Table 12.4.1. Using $c = 0.065, m = 10$, the estimated sample size was $n = 25$. That is, sequential sampling terminated with 25 observations. Upon termination, the program computed estimated values of σ^2 and $\boldsymbol{\beta}$ from the observed data. Interestingly, sentiments expressed from results found in Figure 12.4.4 would be in line with those expressed previously from results in Figure 12.4.3. That ought to make us feel good given that two Figures 12.4.3-12.4.4 report results from two completely different methodologies.

```
 Minimum Risk Point Estimation of the Regression
Parameter Vector Using Purely Sequential Procedure.
---------------------------------------------------

  Model: Linear with intercept

  Input Parameter,      c =   0.0650
  Initial Sample size, m =      10

  Input Data file Name: Reg01.dat

  Final sample size (N)        =      25
  Estimate of sigma^2          =  6.7632

     i  beta_hat(i)  SE(Beta(i))  p-value
     1       9.0032       7.1594   0.2238
     2       2.1165       0.5213   0.0007
     3      -4.3059       0.9061   0.0001
     4      -1.3336       0.5232   0.0196
     5       2.8768       0.6500   0.0003
     6       1.7726       0.4662   0.0012
```

Figure 12.4.4. *Minimum risk point estimation of the regression parameters using data from* ***Reg01.dat***.

The results of a simulation study of a minimum risk sequential point estimation procedure (Section 12.3.1) are shown in Figure 12.4.5. This simulation study with 5000 runs and random number sequence 3, was implemented with data that was generated using a polynomial model:

$$Y_i = \beta_1 + \beta_2 x_{1i} + \beta_3 x_{1i}^2 + \beta_4 x_{1i}^3 + \varepsilon_i,\ i = 1, 2, \ldots.$$

with $\boldsymbol{\beta}' = (3, 1, 5, 2)$, $c = 0.01$, $\sigma^2 = 16$ and $m = 10$. From (12.3.3), we have an optimal sample size:

$$n^* = \sqrt{\frac{p}{c}}\sigma = \sqrt{\frac{4}{0.01}} \times \sqrt{16} = 80.$$

```
  Minimum Risk Point Estimation of the Regression
Parameter Vector Using Purely Sequential Procedure
--------------------------------------------------

  Model: Polynomial with intercept

  Data for the simulations were generated from the
  following covariates of the model:

     i  beta(i)  min{x(i)}  max{x(i)}
     1     3.00     Intercept
     2     1.00       0.00        1.00
     3     5.00     x(2)^2
     4     2.00     x(2)^3

  Input parameter      c  =  0.0100
                 sigma^2  =   16.00
  Initial sample size, m  =    10
  Number of simulations   =  5000

  Optimal sample size (n*)             =      80.00
  Average final sample size (n_bar)    =      79.81
  Std. dev. of n (s_n)                 =       6.72
  Minimum sample size (n_min)          =      10
  Maximum sample size (n_max)          =     103
  Average simulated Regret(c)          =     0.0067
  Standard deviation of the regret     =     0.0698
```

Figure 12.4.5. *A simulation study of the minimum risk point estimation.*

The output shows the n^* value and also the minimum, maximum, average value ($\overline{n}$) of the sample size at termination and the estimated standard deviation (s_n) of sample size. We see that $\overline{n} = 79.81$ which is very close to $n^* = 80$. Let us look at $\overline{n} \pm 2 \times \frac{s_n}{\sqrt{5000}} = 79.81 \pm 0.19007$ which reduces to the interval $(79.61993, 80.10007)$. This interval includes the value $n^* = 80$.

Further, the program gives the average simulated $\text{Regret}(c) = 0.0067$. From Theorem 12.3.2, part (i), as $c \to 0$, we should expect:

$$\text{Regret}(c) \approx \frac{1}{2}c = \frac{0.01}{2} = 0.005.$$

Overall, nearly all simulated results are very close to their theoretical targets, even the ones that were asymptotic to begin with.

12.5 Other Selected Multistage Procedures

In Chaptet 11, we mentioned that Ghosh and Sen (1983) first developed James-Stein estimators of a mean vector which were strictly better than the sample mean vector under a two-stage sampling strategy. We had also cited other important publications. In a linear regression setup, serious efforts went in to embrace similar research initiatives. For examples, one may refer to Nickerson (1987b), Sriram and Bose (1988), Mukhopadhyay and Abid (1999a). Mukhopadhyay (1985b) and Nickerson (1987a) developed analogous two-sample methodologies.

In order to get a glimpse of sequential sampling under generalized linear models, one may refer to Finster (1983,1985), Sriram and Bose (1988), and Mukhopadhyay and Abid (1999b). Finster (1985) included important and interesting design considerations. Martinsek (1990) developed important directions under non-normality of errors.

In a situation when the errors are dependent, Sriram's (1987) paper made a significant contribution. He developed sequential methodologies for estimating the mean of a first-order stationary autoregressive process ($AR(1)$). Subsequently, Sriram (1988) gave analogous results for estimating autoregressive parameters. Mukhopadhyay and Sriram (1992) developed two-sample methodologies.

In the context of sequential estimation problems, Basu and Das (1997) developed methods for a pth-order autoregressive ($AR(p)$) model. Appropriate accelerated versions and their *second-order* properties were developed by Mukhopadhyay and Abid (1999b). Their paper included both generalized linear and $AR(p)$ models.

This line of work has been extended by several authors to a number of problems involving linear processes. Fakhre-Zakeri and Lee (1992) made the first significant headway. In a followup, Fakhre-Zakeri and Lee (1993) developed analogous methodologies for a multivariate linear process. Mukhopadhyay (1995c) introduced multistage methods for the original problem that was formulated by Fakhre-Zakeri and Lee (1992).

12.6 Some Selected Derivations

In this section, we supply parts of some crucial derivations. The derivations may appear sketchy in some places. For fuller understanding, a reader is expected to supply the missing links.

12.6.1 Proof of (12.2.7)

This proof is adapted from Mukhopadhyay (1974b). For any $k, m \le k \le n$, we express *sums of squares of errors* (SSE) as:

$$\begin{aligned} W_k^2 &\equiv \boldsymbol{Y}_k'\boldsymbol{Y}_k - \boldsymbol{Y}_k' X_k \widehat{\boldsymbol{\beta}}_k \\ &= \boldsymbol{Y}_k' \left(I_{k\times k} - X_k (X_k' X_k)^{-1} X_k' \right) \boldsymbol{Y}_k. \end{aligned} \tag{12.6.1}$$

Since $I_{k\times k} - X_k(X_k'X_k)^{-1}X_k'$ is an idempotent matrix of rank $k-p$, it has eigenvalue 0 with multiplicity p and eigenvalue 1 with multiplicity $k-p$.

Suppose that $\boldsymbol{\xi}_1, \ldots, \boldsymbol{\xi}_{k-p}$ are orthonormal eigenvectors corresponding to each eigenvalue 1. Then, we are able to rewrite (12.6.1) as follows:

$$W_k^2 = \boldsymbol{Y}_k' \left(\sum_{i=1}^{k-p} \boldsymbol{\xi}_i \boldsymbol{\xi}_i' \right) \boldsymbol{Y}_k = \sum_{i=1}^{k-p} \left(\boldsymbol{\xi}_i' \boldsymbol{Y}_k \right)^2. \tag{12.6.2}$$

We use a symbol $\mathbf{0}$ for the null vector regardless of its dimension and denote $\boldsymbol{\rho}_i' = (\boldsymbol{\xi}_i' \vdots \mathbf{0}')$, a $1 \times n$ vector so that $W_k^2 = \sum_{i=1}^{k-p} (\boldsymbol{\rho}_i' \boldsymbol{Y}_n)^2$. Also, let $(X_k' \vdots X_{n-k}^{*'})$ be the corresponding partition of X_n.

Next, from (12.2.1), we note that $\widehat{\boldsymbol{\beta}}_n = B\boldsymbol{Y}_n$ where $B = (\mathrm{X}_n'\mathrm{X}_n)^{-1}\mathrm{X}_n'$. A sufficient condition for $B\boldsymbol{Y}_n$ to be independent of all $\boldsymbol{\rho}_i'\boldsymbol{Y}_n, i = 1, \ldots, k-p$ is that $B\boldsymbol{\rho}_i = \mathbf{0}, i = 1, \ldots, k-p$.

In order to verify this sufficient condition, one first notes that $\boldsymbol{\xi}_i$ belongs to the column space of $I_{k \times k} - X_k(X_k'X_k)^{-1}X_k'$. That is, $\boldsymbol{\xi}_i$ belongs to the orthogonal complement of the column space of $X_k(X_k'X_k)^{-1}X_k'$ which coincides with the orthogonal complement of the column space of X_k. This is so because $G_k \equiv (X_k'X_k)^{-1}X_k'$ is a generalized inverse of X_k so that the column space of X_k would coincide with that of X_kG_k.

Hence, we conclude that $X_k'\boldsymbol{\xi}_i = \mathbf{0}$ for all $i = 1, \ldots, k-p$ which implies that $\mathrm{X}_n'\boldsymbol{\rho}_i = \mathbf{0}$ for all $i = 1, \ldots, k-p$. Thus, $B\boldsymbol{\rho}_i = \mathbf{0}, i = 1, \ldots, k-p$. ■

12.6.2 Proof of Theorem 12.3.2

Observe that the stopping time N from (12.3.5) is equivalent to $T + p$ where T is the following stopping time:

$$\begin{aligned} &T \equiv T(c) \text{ is the smallest integer } n (\geq m - p) \text{ for which} \\ &\text{we have } \frac{n(n+p)^2}{n^{*2}} \geq \textstyle\sum_{i=1}^n W_i \text{ with } W\text{'s i.i.d. } \chi_1^2. \end{aligned} \tag{12.6.3}$$

Clearly, the representation of T is exactly the same as that of a stopping variable defined in (A.4.1) in the Appendix with $L_0 = 2p, \delta = 3, r = \frac{1}{2}, \beta^* = \frac{1}{2}, \theta = 1, \tau^2 = 2, n_0^* = n^*, m_0 = m - p, q = \frac{1}{2}$, and

$$\begin{aligned} \nu &= \frac{3}{2} - \textstyle\sum_{n=1}^{\infty} \frac{1}{n} \mathrm{E}\left[\max\left(0, \chi_n^2 - 3n\right)\right] \text{ and} \\ \kappa &= -p - \frac{1}{2} \textstyle\sum_{n=1}^{\infty} \frac{1}{n} \mathrm{E}\left[\max\left(0, \chi_n^2 - 3n\right)\right]. \end{aligned} \tag{12.6.4}$$

See also (A.4.1)-(A.4.5) in the Appendix and (2.4) of Mukhopadhyay (1988a).

Now, from Theorems A.4.1-A.4.2 in the Appendix or Mukhopadhyay (1988a), we quote two very important results:

$$\mathrm{E}_{\boldsymbol{\beta},\sigma}[T] = n^* + \kappa + o(1) \text{ if } m > p + 1, \tag{12.6.5}$$

and

$$\frac{(T - n^*)^2}{n^*} \text{ is uniformly integrable if } m > p + 1. \tag{12.6.6}$$

Proof of Part (i):

From (12.6.6), we conclude the following:

$$\frac{(N-n^*)^2}{n^*} \text{ is uniformly integrable if } m > p+1. \tag{12.6.7}$$

Next, from the theorem of Ghosh and Mukhopadhyay (1975) or Theorem A.4.1, part (ii) in the Appendix, we claim as $c \to 0$:

$$\begin{aligned} & n^{*-1/2}(N-n^*) \text{ converges to } N(0,1/2) \text{ in distribution} \\ & \Rightarrow \frac{(N-n^*)^2}{n^*} \text{ converges to } \frac{1}{2}\chi_1^2 \text{ in distribution.} \end{aligned} \tag{12.6.8}$$

Now, one may combine (12.6.7)-(12.6.8) to write as:

$$\mathrm{E}_{\boldsymbol{\beta},\sigma}\left[\frac{(N-n^*)^2}{n^*}\right] = \frac{1}{2} + o(1) \text{ as } c \to 0, \text{ if } m \geq p+2. \tag{12.6.9}$$

At this point, let us rewrite the regret function:

$$\begin{aligned} c^{-1}\text{Regret}(c) \quad &= \mathrm{E}_{\boldsymbol{\beta},\sigma}\left[\frac{(N-n^*)^2}{N} I\left(N > \tfrac{1}{2}n^*\right)\right] \\ &+ \mathrm{E}_{\boldsymbol{\beta},\sigma}\left[\frac{(N-n^*)^2}{N} I\left(N \leq \tfrac{1}{2}n^*\right)\right]. \end{aligned} \tag{12.6.10}$$

Next, we observe:

$$\left|\frac{(N-n^*)^2}{N} I\left(N > \frac{1}{2}n^*\right)\right| \leq 2\frac{(N-n^*)^2}{n^*}. \tag{12.6.11}$$

But, from (12.6.7), $\dfrac{(N-n^*)^2}{n^*}$ uniformly integrable if $m > p+1$. That is, we have:

$$\frac{(N-n^*)^2}{N} I\left(N > \frac{1}{2}n^*\right) \text{ uniformly integrable if } m > p+1$$

in view of (12.6.11). Also, N/n^* converges to one in probability so that $I(N > \frac{1}{2}n^*)$ converges to one in probability. Hence, in view of (12.6.8), we claim:

$$\frac{(N-n^*)^2}{N} I\left(N > \frac{1}{2}n^*\right) \text{ converges in distribution to } \frac{1}{2}\chi_1^2.$$

Thus, we conclude:

$$\mathrm{E}_{\boldsymbol{\beta},\sigma}\left[\frac{(N-n^*)^2}{N} I\left(N > \frac{1}{2}n^*\right)\right] = \frac{1}{2} + o(1) \text{ if } m > p+1. \tag{12.6.12}$$

On the other hand, we may write:

$$\begin{aligned}
&\mathrm{E}_{\boldsymbol{\beta},\sigma}\left[\frac{(N-n^*)^2}{N}I\left(N\le \tfrac{1}{2}n^*\right)\right]\\
&\quad\le \mathrm{E}_{\boldsymbol{\beta},\sigma}\left[\frac{N^2+n^{*2}}{N}I\left(N\le \tfrac{1}{2}n^*\right)\right]\\
&\quad\le \mathrm{E}_{\boldsymbol{\beta},\sigma}\left[\frac{\frac{1}{4}n^{*2}+n^{*2}}{m}I\left(N\le \tfrac{1}{2}n^*\right)\right]\\
&\quad= \frac{5}{4m}n^{*2}\mathrm{P}_{\boldsymbol{\beta},\sigma}\left(N\le \tfrac{1}{2}n^*\right).
\end{aligned} \tag{12.6.13}$$

However, $\mathrm{P}_{\boldsymbol{\beta},\sigma}\left(N\le \frac{1}{2}n^*\right) = O\left(h^{*(m-p)/2}\right) = O\left(n^{*-m+p}\right)$. See Theorem A.4.1, part (i) in the Appendix. Hence, from (12.6.13), we obtain

$$\begin{aligned}
\mathrm{E}_{\boldsymbol{\beta},\sigma}\left[\frac{(N-n^*)^2}{N}I\left(N\le \tfrac{1}{2}n^*\right)\right] &= O(n^{*2})O\left(n^{*-m+p}\right)\\
&= o(1) \text{ if } m>p+2.
\end{aligned} \tag{12.6.14}$$

At this point one combines (12.6.12) and (12.6.14) to claim:

$$c^{-1}\text{Regret}(c) = \mathrm{E}_{\boldsymbol{\beta},\sigma}\left[\frac{(N-n^*)^2}{N}\right] = \frac{1}{2}+o(1) \text{ if } m\ge p+3.$$

This completes our proof of part (i). ■

Proof of Part (ii):

Recall that $N = T+p$ w.p. 1 and hence if $m>p+1$, we have from (12.6.5):

$$\begin{aligned}
\mathrm{E}_{\boldsymbol{\beta},\sigma}[N] &= \mathrm{E}_{\boldsymbol{\beta},\sigma}[T]+p = n^*+\kappa+p+o(1)\\
&= n^*+\eta_3(p)+o(1),
\end{aligned} \tag{12.6.15}$$

using (12.6.4) with $\eta_3(p)\equiv \kappa+p = -\frac{1}{2}\sum_{n=1}^{\infty}\frac{1}{n}\mathrm{E}\left[\max\left(0,\chi^2_n-3n\right)\right]$. ■

12.7 Exercises

Exercise 12.2.1: Show that $\widehat{\boldsymbol{\beta}}_n$ and S^2_n from (12.2.1)-(12.2.2) are the uniformly minimum variance unbiased estimators (UMVUEs) of $\boldsymbol{\beta},\sigma^2$ respectively.

Exercise 12.2.2: Show that $\widehat{\boldsymbol{\beta}}_n$ defined in (12.2.1) is the maximum likelihood estimator (MLE) of $\boldsymbol{\beta}$.

Exercise 12.2.3: Verify (12.2.8) in the spirit of Section 11.7.3.

Exercise 12.2.4: For the stopping variable N from (12.2.10), show that the final regression estimator $\widehat{\boldsymbol{\beta}}_N$ is unbiased for $\boldsymbol{\beta}$. Evaluate $\mathrm{E}_{\boldsymbol{\beta},\sigma}\left[\widehat{\boldsymbol{\beta}}'_N\widehat{\boldsymbol{\beta}}_N\right]$. What is the distribution of $\sqrt{N}(\widehat{\boldsymbol{\beta}}_N - \boldsymbol{\beta})$?

Exercise 12.2.5: Prove Theorem 12.2.1, part (i) in the spirit of Section 11.7.4.

Exercise 12.2.6: For all fixed $\boldsymbol{\beta}, \sigma, p, m$ and α, for the stopping variable N from (12.2.10), evaluate $\lim_{d\to 0} \mathrm{E}_{\boldsymbol{\beta},\sigma}\left[N^2/C^2\right]$.

Exercise 12.2.7: Prove Theorem 12.2.2, part (ii) in the spirit of Section 11.7.5.

Exercise 12.2.8: For all fixed $\boldsymbol{\beta}, \sigma, p, m$ and α, for the stopping variable N from (12.2.12), show that $\lim_{d\to 0} \mathrm{E}_{\boldsymbol{\beta},\sigma}\left[N^2/C^2\right] = 1$.

Exercise 12.2.9: For the stopping variable N from (12.2.12), show that the final regression estimator $\widehat{\boldsymbol{\beta}}_N$ is unbiased for $\boldsymbol{\beta}$. Evaluate $\mathrm{E}_{\boldsymbol{\beta},\sigma}\left[\widehat{\boldsymbol{\beta}}'_N\widehat{\boldsymbol{\beta}}_N\right]$. What is the distribution of $\sqrt{N}(\widehat{\boldsymbol{\beta}}_N - \boldsymbol{\beta})$?

Exercise 12.2.10: Prove Theorem 12.2.3, parts (i)-(ii) in the spirit of Section 11.7.6.

Exercise 12.2.11: For all fixed $\boldsymbol{\beta}, \sigma, p, m$ and α, for the stopping variable N from (12.2.15), show that $\lim_{d\to 0} \mathrm{E}_{\boldsymbol{\beta},\sigma}\left[N^2/C^2\right] = 1$. What is the distribution of $\sqrt{N}(\widehat{\boldsymbol{\beta}}_N - \boldsymbol{\beta})$?

Exercise 12.2.12: For all fixed $\boldsymbol{\beta}, \sigma, p, m$ and α, for the stopping variables N from (12.2.10), (12.2.12), (12.2.13), and (12.2.15), evaluate

$$\mathrm{E}_{\boldsymbol{\beta},\sigma}\left[\left\{(\widehat{\boldsymbol{\beta}}_N - \boldsymbol{\beta})'\mathrm{X}'_N\mathrm{X}_N(\widehat{\boldsymbol{\beta}}_N - \boldsymbol{\beta})\right\}^k\right]$$

where $k(>0)$ is arbitrary but fixed. What will be different if k is arbitrary, fixed, but negative?

Exercise 12.3.1: For the stopping variable N from (12.3.5), show that the final regression estimator $\widehat{\boldsymbol{\beta}}_N$ is unbiased for $\boldsymbol{\beta}$. Evaluate $\mathrm{E}_{\boldsymbol{\beta},\sigma}\left[\widehat{\boldsymbol{\beta}}'_N\widehat{\boldsymbol{\beta}}_N\right]$. What is the distribution of $\sqrt{N}(\widehat{\boldsymbol{\beta}}_N - \boldsymbol{\beta})$?

Exercise 12.3.2: For all fixed $\boldsymbol{\beta}, \sigma, p$ and m, for the stopping variable N from (12.3.5), evaluate

$$\mathrm{E}_{\boldsymbol{\beta},\sigma}\left[\left\{(\widehat{\boldsymbol{\beta}}_N - \boldsymbol{\beta})'\mathrm{X}'_N\mathrm{X}_N(\widehat{\boldsymbol{\beta}}_N - \boldsymbol{\beta})\right\}^k\right]$$

where $k(>0)$ is arbitrary but fixed. What will be different if k is arbitrary, fixed, but negative?

Exercise 12.3.3: Prove Theorem 12.3.1.

Exercise 12.3.4: For all fixed $\boldsymbol{\beta}, \sigma, p$ and m, for the stopping variable N from (12.3.5), evaluate $\lim_{c\to 0} \mathrm{E}_{\boldsymbol{\beta},\sigma}\left[N^2/n^{*2}\right]$.

Exercise 12.4.1: Let

$$\mathcal{R}_N = \left\{\boldsymbol{\beta}_{6\times 1} \in I\!R^6 : N^{-1}(\widehat{\boldsymbol{\beta}}_N - \boldsymbol{\beta})'(\mathrm{X}_N'\mathrm{X}_N)(\widehat{\boldsymbol{\beta}}_N - \boldsymbol{\beta}) \le 1.75^2\right\}$$

be a 90% fixed-size confidence region for the regression parameter vector $\boldsymbol{\beta}' = (\beta_1, \beta_2, \ \ldots, \beta_6)$ in a regression model:

$$Y = \beta_1 + \beta_2 x_2 + \beta_3 x_3 + \beta_4 x_4 + \beta_5 x_5 + \beta_6 x_6 + \varepsilon.$$

Using the program **Seq12.exe** with input data from Table 12.4.1, show that $\mathcal{R}_N$ requires

(i) $N = 28$ observations if the two-stage procedure with $m = 10$ is used;
(ii) $N = 24$ observations if the modified two-stage procedure with $\gamma = 0.4$ is used;
(iii) $N = 23$ observations if the three-stage procedure with $\rho = 0.5$ and $m = 10$ is used.

Exercise 12.4.2: Consider a 95% confidence region

$$\mathcal{R}_N = \left\{\boldsymbol{\beta}_{3\times 1} \in I\!R^3 : N^{-1}(\widehat{\boldsymbol{\beta}}_N - \boldsymbol{\beta})'(\mathrm{X}_N'\mathrm{X}_N)(\widehat{\boldsymbol{\beta}}_N - \boldsymbol{\beta}) \le 0.5^2\right\}$$

for the regression parameter vector $\boldsymbol{\beta}$ in a model:

$$Y = \beta_1 x_1 + \beta_2 x_2 + \beta_3 x_3 + \varepsilon$$

where $2 \le x_1 \le 5$, $0.5 \le x_2 \le 1.5$, $2.5 \le x_3 \le 5$ and $\varepsilon \sim N(0, 16)$. Compare the sample sizes required to construct $\mathcal{R}_N$ using

(i) the two-stage procedure with $m = 15$;
(ii) the modified two-stage procedure with $\gamma = 0.6$;
(iii) the purely sequential procedure with $m = 15$;
(iv) the three-stage procedure with $\rho = 0.6$ and $m = 15$.

Implement 1000 simulation runs of $\mathcal{R}_N$ assuming $\beta_1 = 5.0$, $\beta_2 = 2.5$ and $\beta_3 = -3.5$ with the help of the computer program **Seq12.exe**.

Exercise 12.4.3 (Exercise 12.4.2 Continued): Repeat Exercise 12.4.2 with a polynomial regression model:

$$Y = \beta_1 x_1 + \beta_2 x_1^2 + \beta_3 x_1^3 + \varepsilon$$

where $2 \leq x_1 \leq 5$ and $\varepsilon \sim N(0, 16)$.

Exercise 12.4.4: Consider a linear regression model:

$$Y = \beta_1 x_1 + \beta_2 x_2 + \beta_3 x_3 + \beta_4 x_4 + \beta_5 x_5 + \varepsilon$$

for the input data from Table 12.4.1. Obtain a minimum risk point estimator of the regression parameter $\boldsymbol{\beta}$ using a loss function:

$$L_n\left(\boldsymbol{\beta}, \widehat{\boldsymbol{\beta}}_n\right) = n^{-1}(\widehat{\boldsymbol{\beta}}_n - \boldsymbol{\beta})'(\mathrm{X}_n'\mathrm{X}_n)(\widehat{\boldsymbol{\beta}}_n - \boldsymbol{\beta}) + 0.1n,$$

and $m = 10$.

Exercise 12.4.5: Consider a minimum risk point estimator of the regression parameter $\boldsymbol{\beta}$ in a linear regression model:

$$Y = \beta_1 + \beta_2 x_2 + \beta_3 x_3 + \beta_4 x_4 + \varepsilon$$

where $\varepsilon \sim N(0, \sigma^2)$ with sampling cost, $c = 0.01$. From (12.3.3), the optimal sample size is:

$$n^* = \left(\frac{4}{0.01}\right)^{1/2} \sigma = 20\sigma.$$

Let N be the sample size required to obtain a minimum point estimator of $\boldsymbol{\beta}$ when the samples are obtained one-at-time sequentially and $\overline{n}$ be the average from 1000 simulated values of N for a given σ.

(i) Use the computer program **Seq12.exe** to obtain $\overline{n}$ for $\sigma = 1, 5$ and 10. You may use $0 \leq x_2 \leq 1$, $0 \leq x_3 \leq 2$, $2 \leq x_4 \leq 5$ and $\boldsymbol{\beta}' = (3.0, 1.0, 4.0, -3.0)$.

(ii) Check whether $\overline{n} \approx 20\sigma = n^*$.

CHAPTER 13

Estimating the Difference of Two Normal Means

13.1 Introduction

In this chapter, we pay attention to two-sample normal problems. Consider two independent sequences $X_{i1}, \ldots, X_{in_i}, \ldots$ of i.i.d. random variables distributed as $N(\mu_i, \sigma_i^2), i = 1, 2$. We assume that all four parameters $\mu_1, \mu_2, \sigma_1, \sigma_2$ are unknown, $-\infty < \mu_1, \mu_2 < \infty$ and $0 < \sigma_1, \sigma_2 < \infty$.

Section 13.2 introduces fixed-width confidence interval problems to compare the two population means, μ_1 and μ_2. We will construct fixed-width confidence intervals for the parameter $\delta \equiv \mu_1 - \mu_2$ in two separate cases. We will include scenario (i) $\sigma_1^2 = \sigma_2^2 = \sigma^2$, but the common variance σ^2 is unknown, and then scenario (ii) $\sigma_1^2 \neq \sigma_2^2$.

In Section 13.3, we include minimum risk point estimation problems for comparing μ_1 and μ_2. We will construct minimum risk point estimators for the parameter $\delta \equiv \mu_1 - \mu_2$ in two separate cases. We will include scenario (i) $\sigma_1^2 = \sigma_2^2 = \sigma^2$, but the common variance σ^2 is unknown, and then scenario (ii) $\sigma_1^2 \neq \sigma_2^2$.

In the literature, the scenario (ii) discussed under either framework is commonly referred to as the *Behrens-Fisher problem* or the *Behrens-Fisher analog*. We would write $\boldsymbol{\theta} = (\mu_1, \mu_2, \sigma^2)$ in case (i) and $\boldsymbol{\theta} = (\mu_1, \mu_2, \sigma_1^2, \sigma_2^2)$ in case (ii). Sometimes, we may continue to write $\mathrm{P}_{\boldsymbol{\theta}}(.)$ or $\mathrm{E}_{\boldsymbol{\theta}}(.)$ under either scenario without explicitly mentioning what $\boldsymbol{\theta}$ is. The parameter vector $\boldsymbol{\theta}$ should be clear from the context.

Section 13.4 briefly mentions analogous investigations including those for (i) two independent negative exponential or exponential distributions, (ii) two multivariate normal distributions, (iii) k-sample comparisons with $k \geq 3$, and (iv) two distribution-free populations. In the end, Section 13.5 provides some selected derivations.

13.2 Fixed-Width Confidence Intervals

Consider independent observations $X_{i1}, \ldots, X_{in_i}$ from a $N(\mu_i, \sigma_i^2)$ distribution, $i = 1, 2$. Also, assume that X_1's and X_2 's are independent. We denote:

$$\begin{aligned}
&\overline{X}_{in_i} = n_i^{-1}\textstyle\sum_{j=1}^{n_i} X_{ij}, S_{in_i}^2 = (n_i-1)^{-1}\sum_{j=1}^{n_i}(X_{ij}-\overline{X}_{in_i})^2, \\
&\quad n_i \geq 2, i=1,2, \delta \equiv \mu_1-\mu_2, \boldsymbol{n}=(n_1,n_2), \text{ and} \\
&T_{\boldsymbol{n}} = \overline{X}_{1n_1}-\overline{X}_{2n_2}, U_{\boldsymbol{n}}^2 = \frac{(n_1-1)S_{1n_1}^2+(n_2-1)S_{2n_2}^2}{n_1+n_2-2}.
\end{aligned} \tag{13.2.1}$$

The sample means $\overline{X}_{1n_1}, \overline{X}_{2n_2}$ respectively estimates the population means μ_1, μ_2 and the sample variances $S_{1n_1}^2, S_{2n_2}^2$ respectively estimates the population variances σ_1^2, σ_2^2. When $\sigma_1^2 = \sigma_2^2 = \sigma^2$, but the common variance σ^2 is unknown, the pooled sample variance $U_{\boldsymbol{n}}^2$ would estimate σ^2.

13.2.1 Common Unknown Variance: Two-Stage Sampling

Since the two population variances are assumed equal, we may decide to take equal number of observations from both distributions. This is theoretically justified later in Section 13.5.1.

For simplicity, let us write $n(\geq 2)$ for the common sample size. Accordingly, we would have $\overline{X}_{1n}, \overline{X}_{2n}, T_n \equiv \overline{X}_{1n} - \overline{X}_{2n}$ and the pooled sample variance $U_n^2 = (S_{1n}^2 + S_{2n}^2)/2$ from (13.2.1).

The distributions of these statistics are now summarized. Under $\boldsymbol{\theta} = (\mu_1, \mu_2, \sigma^2)$, we have:

$$\begin{aligned}
&\sqrt{n}(\overline{X}_{in}-\mu_i)/\sigma \sim N(0,1), i=1,2, \text{ and they are independent;} \\
&(n-1)S_{in}^2/\sigma^2 \sim \chi_{n-1}^2, i=1,2, \text{ and they are independent;} \\
&(\overline{X}_{1n}, \overline{X}_{2n}) \text{ and } (S_{1k}^2, S_{2l}^2 : 2 \leq k, l \leq n) \text{ are independent;} \\
&\sqrt{n}(\overline{X}_{in}-\mu_i)/S_{in} \sim t_{n-1}, i=1,2, \text{ and they are independent;} \\
&2(n-1)U_n^2/\sigma^2 \sim \chi_{2n-2}^2.
\end{aligned} \tag{13.2.2}$$

Next, with $T_n \equiv \overline{X}_{1n} - \overline{X}_{2n}$, we immediately obtain under $\boldsymbol{\theta}$:

$$\begin{aligned}
&\frac{\sqrt{n}(T_n-\delta)}{\sqrt{2}\sigma} \sim N(0,1) \text{ and } \frac{\sqrt{n}(T_n-\delta)}{\sqrt{2}U_n} \sim t_{2n-2}; \\
&\text{and } T_n, (U_k^2 : 2 \leq k \leq n) \text{ are independent.}
\end{aligned} \tag{13.2.3}$$

Now, having a preassigned number $d(>0)$, we wish to construct a fixed-width $(\equiv 2d)$ confidence interval for δ. Consider the confidence interval:

$$J_n = (T_n - d, T_n + d) \tag{13.2.4}$$

for δ.

However, we require that the confidence coefficient, $\mathrm{P}_{\boldsymbol{\theta}}(\delta \in J_n)$, be at least $1-\alpha$ where $0<\alpha<1$ is also preassigned. Thus, we must have:

$$\begin{aligned} &\mathrm{P}_{\boldsymbol{\theta}}\left\{T_n - d < \mu_1 - \mu_2 < T_n + d\right\} \\ &= \mathrm{P}_{\boldsymbol{\theta}}\left\{\frac{\sqrt{n}\mid T_n - \delta\mid}{\sqrt{2}\sigma} < \frac{d\sqrt{n}}{\sqrt{2}\sigma}\right\} \\ &= 2\Phi\left(\frac{d\sqrt{n}}{\sqrt{2}\sigma}\right) - 1. \end{aligned} \tag{13.2.5}$$

Now, the confidence coefficient from (13.2.5) will be at least $1-\alpha$ provided that $d\sqrt{n}(\sqrt{2}\sigma)^{-1} \geq a$ where a is an upper $50\alpha\%$ point of a standard normal distribution. Thus, a common sample size n from each distribution must be

$$\text{the smallest integer } \geq 2a^2\sigma^2/d^2 = C, \text{ say.} \tag{13.2.6}$$

That is, we have an expression for the optimal fixed-sample-size C from each distribution had σ^2 been known. However, the magnitude of C remains unknown.

Along the lines of Stein's (1945,1949) two-stage procedure (6.2.6), we pursue this two-sample problem. We begin with observations $X_{i1}, \ldots, X_{im}$ from the ith distribution, $i=1,2$ where $m(\geq 2)$ is a pilot size. Let $a_{2m-2} \equiv a_{2m-2,\alpha/2}$ stand for an upper $50\alpha\%$ point of a t_{2m-2} distribution.

Next, we define:

$$N \equiv N(d) = \max\left\{m, \left\langle \frac{2a_{2m-2}^2 U_m^2}{d^2} \right\rangle + 1\right\}. \tag{13.2.7}$$

We implement this two-stage procedure in usual fashion by sampling the difference at the second stage from both populations if $N > m$. Then, using the final dataset $N, X_{i1}, \ldots, X_{iN}, i=1,2$ obtained by combining both stages of sampling, the following fixed-width confidence interval

$$J_N = (T_N - d, T_N + d) \text{ with } T_N \equiv \overline{X}_{1N} - \overline{X}_{2N} \tag{13.2.8}$$

is proposed for δ along the lines of (13.2.4). It is easy to see that N is finite w.p. 1, and thus J_N is a genuine confidence interval estimator.

From the last step in (13.2.3), it is clear that $I(N=n)$ and T_n are independent for all fixed $n(\geq m)$. So, in view of (13.2.5), we express the confidence coefficient associated with J_N as:

$$\mathrm{P}_{\boldsymbol{\theta}}\left\{\delta \in (T_N - d, T_N + d)\right\} = \mathrm{E}_{\boldsymbol{\theta}}\left[2\Phi\left(\frac{d\sqrt{N}}{\sqrt{2}\sigma}\right) - 1\right]. \tag{13.2.9}$$

Now, let us summarize some of the crucial properties of this two-stage estimation strategy.

Theorem 13.2.1 : *For the stopping variable N from (13.2.7), for all fixed $\boldsymbol{\theta}, d, m$ and α, one has the following properties:*

(i) $2a_{2m-2}^2\sigma^2 d^{-2} \leq \mathrm{E}_{\boldsymbol{\theta}}(N) \leq m + 2a_{2m-2}^2\sigma^2 d^{-2}$;

(ii) $Q \equiv \sqrt{N}(T_N - \delta)/(\sqrt{2}\sigma)$ *has a standard normal distribution;*
(iii) $\mathrm{P}_{\boldsymbol{\theta}}\{\delta \in J_N\} \ge 1 - \alpha$ [*Consistency or Exact Consistency*];

where a_{2m-2} *is an upper* $50\alpha\%$ *point of a* t_{2m-2} *distribution and* $C = 2a^2\sigma^2/d^2$ *came from (13.2.6).*

We leave its proof out as an exercise. Part (iii) shows that the two-stage methodology (13.2.7) has given us an exact solution to the problem on hand.

> In an unknown but equal variance case, two-stage methodology (13.2.7) is a straightforward Stein-type estimation strategy.

Along the line of (7.2.8), one may show here that

$$\mathrm{E}_{\boldsymbol{\theta}}(N) \ge 2a_{2m-2}^2\sigma^2 d^{-2} > C. \tag{13.2.10}$$

In other words, on an average, the procedure (13.2.7) needs more observations than the optimal fixed sample size C from both populations. This should not be surprising because C is "optimal" when we pretend that σ is known! But, in fact σ is unknown and we end up taking more than C observations. Moreover, there is no implementable fixed-sample-size solution to our problem anyway.

But, then, along the line of (6.2.14), it follows easily that

$$\lim_{d\to 0} \mathrm{E}_{\boldsymbol{\theta}}\left(\frac{N}{C}\right) = \frac{a_{2m-2}^2}{a^2} > 1. \tag{13.2.11}$$

When $d \to 0$, we know that $N \to \infty$ w.p. 1 and $C \to \infty$. So, (13.2.11) points out that $\mathrm{E}_{\boldsymbol{\theta}}(N)$ and C do not agree in the large-sample case, and hence the latter property is more disturbing than (13.2.10). The methodology (13.2.7) is asymptotically *inefficient* in the sense of Chow and Robbins (1965).

The limiting confidence coefficient is described below:

$$\mathrm{P}_{\boldsymbol{\theta}}\{\delta \in J_N\} \to 1 - \alpha \text{ as } d \to 0. \tag{13.2.12}$$

That is, the two-stage strategy (13.2.7) is *asymptotically consistent.* A proof of this result is again left as an exercise.

One may surely think of possibly other multistage estimation strategies including a modified two-stage, three-stage, or accelerated sequential sampling designs. techniques strategy defined appropriately. A sequential strategy was included in Mukhopadhyay and Solanky (1994, pp. 51-52). We have included some of these methodologies in exercises.

13.2.2 Unknown and Unequal Variances: Two-Stage Sampling

Now, we work under the assumption that $\sigma_i^2, i = 1, 2$ are unknown and unequal. This framework is commonly referred to as the *Behrens-Fisher problem* or the *Behrens-Fisher analog.* We denote $\boldsymbol{\theta} = (\mu_1, \mu_2, \sigma_1^2, \sigma_2^2)$.

Suppose we have independent observations $X_{i1}, \ldots, X_{in_i}$ from a $N(\mu_i, \sigma_i^2)$ distribution, $i = 1, 2$. Assume that X_1's and X_2's are independent. We denote:

$$\begin{aligned}
&\overline{X}_{in_i} = n_i^{-1} \textstyle\sum_{j=1}^{n_i} X_{ij},\\
&S_{in_i}^2 = (n_i - 1)^{-1} \textstyle\sum_{j=1}^{n_i} (X_{ij} - \overline{X}_{in_i})^2, \ n_i \geq 2,\\
&\delta \equiv \mu_1 - \mu_2, \boldsymbol{n} = (n_1, n_2),\\
&T_{\boldsymbol{n}} = \overline{X}_{1n_1} - \overline{X}_{2n_2}.
\end{aligned} \tag{13.2.13}$$

The distributions of some of these entities may be summarized easily. Under $\boldsymbol{\theta} = (\mu_1, \mu_2, \sigma_1^2, \sigma_2^2)$, we have:

$$\begin{gathered}
\sqrt{n_i}(\overline{X}_{in_i} - \mu_i)/\sigma_i \sim N(0,1), \ i = 1, 2;\\
(n_i - 1) S_{in_i}^2 / \sigma_i^2 \sim \chi_{n_i - 1}^2, \ i = 1, 2;\\
(\overline{X}_{1n_1}, \overline{X}_{2n_2}), (S_{1k}^2, S_{2l}^2 : 2 \leq k \leq n_1, 2 \leq l \leq n_2) \text{ are independent;}\\
\sqrt{n_i}(\overline{X}_{in_i} - \mu_i)/S_{in_i} \sim t_{n_i - 1}, \ i = 1, 2.
\end{gathered} \tag{13.2.14}$$

Having a preassigned number $d(> 0)$, we wish to obtain a fixed-width $(\equiv 2d)$ confidence interval for δ. Recall that $\boldsymbol{n} = (n_1, n_2)$ and consider the interval:

$$J_{\boldsymbol{n}} = (T_{\boldsymbol{n}} - d, \ T_{\boldsymbol{n}} + d) \tag{13.2.15}$$

for δ.

We also require that the confidence coefficient $\mathrm{P}_{\boldsymbol{\theta}}(\delta \in J_{\boldsymbol{n}})$ be at least $1 - \alpha$ where $0 < \alpha < 1$ is preassigned. Thus, we proceed as follows:

$$\begin{aligned}
&\mathrm{P}_{\boldsymbol{\theta}} \{T_{\boldsymbol{n}} - d < \mu_1 - \mu_2 < T_{\boldsymbol{n}} + d\}\\
&= \mathrm{P}_{\boldsymbol{\theta}} \left\{ | T_{\boldsymbol{n}} - \delta | \left(\frac{\sigma_1^2}{n_1} + \frac{\sigma_2^2}{n_2} \right)^{-1/2} < d \left(\frac{\sigma_1^2}{n_1} + \frac{\sigma_2^2}{n_2} \right)^{-1/2} \right\}\\
&= 2\Phi \left(d \left(\frac{\sigma_1^2}{n_1} + \frac{\sigma_2^2}{n_2} \right)^{-1/2} \right) - 1,
\end{aligned} \tag{13.2.16}$$

since $(T_{\boldsymbol{n}} - \delta) / (n_1^{-1}\sigma_1^2 + n_2^{-1}\sigma_2^2)^{1/2} \sim N(0, 1)$ under $\boldsymbol{\theta}$.

The confidence coefficient from (13.2.16) will be at least $1 - \alpha$ provided that

$$d/(n_1^{-1}\sigma_1^2 + n_2^{-1}\sigma_2^2)^{1/2} \geq a,$$

where a is an upper $50\alpha\%$ point of a standard normal distribution.

It is clear that $(n_1^{-1}\sigma_1^2 + n_2^{-1}\sigma_2^2)^{1/2}$ can be made "small" by choosing n_1, n_2 "large". That is, by increasing n_1 and n_2, the expression $d/(n_1^{-1}\sigma_1^2 + n_2^{-1}\sigma_2^2)^{1/2}$ can be made "large" so that at some point it will exceed a.

However, that may not sound like a very practical thing to do. Since one must work with budgeted resources, one may follow a more practical approach. Minimize a total sample size $n_1 + n_2$ subjected to a restriction that we must have $d/(n_1^{-1}\sigma_1^2 + n_2^{-1}\sigma_2^2)^{1/2} \geq a$ satisfied.

So, we have a *restricted optimization* problem on hand:

$$\text{Minimize } n_1 + n_2 \text{ under the restriction that } \frac{\sigma_1^2}{n_1} + \frac{\sigma_2^2}{n_2} \le \frac{d^2}{a^2}. \quad (13.2.17)$$

In Section 13.5.1, one will find a proof that this problem has the following set of solution:

$$\begin{aligned} n_1 \equiv C_1 &= \frac{a^2}{d^2}\sigma_1(\sigma_1+\sigma_2),\ n_2 \equiv C_2 = \frac{a^2}{d^2}\sigma_2(\sigma_1+\sigma_2) \\ \Rightarrow \frac{C_1}{C_2} &= \frac{\sigma_1}{\sigma_2},\ n \equiv C = C_1 + C_2 = \frac{a^2}{d^2}(\sigma_1+\sigma_2)^2. \end{aligned} \quad (13.2.18)$$

Pretend that C_1, C_2 are integers. Then, these provide expressions of optimal fixed sample sizes required from each distribution. However, their magnitudes remain unknown since σ_1, σ_2 are unknown. In Section 13.2.3, we will make elaborate use of the expressions of C_1, C_2.

Chapman (1950) first came up with his pioneering two-stage estimation strategy so that the associated confidence coefficient could be equal or exceed the preassigned value, $1-\alpha$. By looking at the required restriction $n_1^{-1}\sigma_1^2 + n_2^{-1}\sigma_2^2 \le a^{-2}d^2$ from (13.2.17), we see readily that

$$n_1 \equiv n_1^0 = 2a^2\sigma_1^2/d^2 \text{ and } n_2 \equiv n_2^0 = 2a^2\sigma_2^2/d^2 \quad (13.2.19)$$

clearly fit the bill since

$$(n_1^0)^{-1}\sigma_1^2 + (n_2^0)^{-1}\sigma_2^2 = a^{-2}d^2.$$

It is obvious that n_1^0, n_2^0 do not match the optimal solutions C_1, C_2 from (13.2.18). Yet, Chapman (1950) proceeded to mimic the expressions of n_1^0, n_2^0 from (13.2.19).

Now, we introduce Chapman's (1950) two-stage estimation methodology which took Stein's (1945,1949) two-stage sampling technique (6.2.6) to a different height. One begins with observations $X_{i1}, \ldots, X_{im}$ from the ith population, $i = 1, 2$ where $m(\ge 2)$ is a pilot size. One obtains sample means $\overline{X}_{1m}, \overline{X}_{2m}$ and sample variances S_{1m}^2, S_{2m}^2 from two pilot samples.

We need one other thing before we can define the methodology. Let W_1, W_2 be independent random variables having a common t_{m-1} distribution and let $h \equiv h_{m,\alpha}(>0)$ be a number that satisfies the equation:

$$\text{P}\left\{W_1 - W_2 \le h_{m,\alpha/2}\right\} = 1 - \frac{1}{2}\alpha. \quad (13.2.20)$$

That is, $h_{m,\alpha/2}$ is an upper $50\alpha\%$ point for the probability distribution of $W_1 - W_2$.

Now, we define the stopping variables:

$$N_1 \equiv N_1(d) = \max\left\{m, \left\langle \frac{h^2_{m,\alpha/2} S^2_{1m}}{d^2} \right\rangle + 1\right\},$$
$$N_2 \equiv N_2(d) = \max\left\{m, \left\langle \frac{h^2_{m,\alpha/2} S^2_{2m}}{d^2} \right\rangle + 1\right\}. \quad (13.2.21)$$

We implement this two-stage procedure in an usual fashion by sampling the difference at the second stage from the ith population if $N_i > m, i = 1, 2$. Then, based on final dataset $N_i, X_{i1}, \ldots, X_{iN_i}$ from both stages of sampling from the ith distribution, $i = 1, 2$, we propose the interval:

$$J_{\boldsymbol{N}} = (T_{\boldsymbol{N}} - d, T_{\boldsymbol{N}} + d) \text{ with}$$
$$T_{\boldsymbol{N}} \equiv \overline{X}_{1N_1} - \overline{X}_{2N_2},\ \boldsymbol{N} = (N_1, N_2), \quad (13.2.22)$$

for δ. It is easy to see that N is finite w.p.1. Thus, $J_{\boldsymbol{N}}$ is a genuine confidence interval estimator for δ.

Invoking the third step from (13.2.14), it should be clear that for all fixed $n_1 \geq m, n_2 \geq m, \boldsymbol{n} = (n_1, n_2)$, the random variables $I(N_1 = n_1, N_2 = n_2)$ and $T_{\boldsymbol{n}} \equiv \overline{X}_{1n_1} - \overline{X}_{2n_2}$ are independently distributed. Hence, in view of (13.2.16), we can express the confidence coefficient associated with $J_{\boldsymbol{N}}$ as:

$$\mathrm{P}_{\boldsymbol{\theta}}\left\{T_{\boldsymbol{N}} - d < \mu_1 - \mu_2 < T_{\boldsymbol{N}} + d\right\}$$
$$= \mathrm{E}_{\boldsymbol{\theta}}\left[2\Phi\left(d\left\{\frac{\sigma_1^2}{N_1} + \frac{\sigma_2^2}{N_2}\right\}^{-1/2}\right)\right] - 1. \quad (13.2.23)$$

Now, we summarize the most crucial property of this two-stage estimation strategy. We provide a proof of this result in Section 13.5.2.

Theorem 13.2.2: *For the stopping variable N_1, N_2 defined by (13.2.21), for all fixed $\boldsymbol{\theta}, d, m$ and α, one has:*

$$\mathrm{P}_{\boldsymbol{\theta}}\left\{T_{\boldsymbol{N}} - d < \mu_1 - \mu_2 < T_{\boldsymbol{N}} + d\right\} \geq 1 - \alpha$$

[*Consistency or Exact Consistency*]

where $h_{m,\alpha/2}$ came from (13.2.20), $\boldsymbol{N} = (N_1, N_2)$, and $T_{\boldsymbol{N}} = \overline{X}_{1N_1} - \overline{X}_{2N_2}$.

One should note that for "large" m, the distribution of $W_1 - W_2$ would converge to the distribution of $Y_1 - Y_2$ where Y_1, Y_2 are independent standard normal variables. Clearly, $U = (Y_1 - Y_2)/\sqrt{2}$ would have a standard normal distribution. That is, from (13.2.20), for large m, we may claim:

$$\mathrm{P}\left\{W_1 - W_2 \leq h_{m,\alpha/2}\right\} \approx \mathrm{P}\left\{Y_1 - Y_2 \leq h_{m,\alpha/2}\right\} = \mathrm{P}\left\{U \leq \tfrac{1}{\sqrt{2}} h_{m,\alpha/2}\right\},$$
$$\text{but } \mathrm{P}\{U \leq a\} = 1 - \tfrac{1}{2}\alpha \Rightarrow \tfrac{1}{\sqrt{2}} h_{m,\alpha/2} \approx a,$$

so that $h_{m.\alpha} \approx \sqrt{2}a$.

Chapman (1950) included the required $h_{m,\alpha}$ values for some choices of m and α. This procedure was further investigated by Ghosh (1975a,b) who gave a Cornish-Fisher expansion of $h_{m,\alpha}$. Referring to Chapman's (1950) $h_{m,\alpha}$ values, Ghosh (1975b, p. 463) commented that "many of his values are incorrect."

> The Chapman two-stage methodology (13.2.21) was a breakthrough. It gave an exact solution to Behrens–Fisher problem that remains unsolved in classical non-sequential literature.

Aoshima and Mukhopadhyay (2002) developed a Cornish-Fisher expansion for an analogous upper percentage point under a k-sample scenario when pilot sample sizes were not necessarily all equal. The following approximation is adapted from Theorem 2.1 of Aoshima and Mukhopadhyay (2002) when $k = 2$ and $m_1 = m_2 = m$.

In this approximation, one ought to substitute $k = 2$ and $\nu_1 = \nu_2 = m-1$ and interpret $\sum$ as $\sum_{i=1}^{k}$:

$$
\begin{aligned}
&h_{m,\alpha/2}\\
&\approx a\sqrt{2} + \tfrac{1}{4}k^{-2}a\sqrt{2}(a^2+4k-3)(\textstyle\sum \nu_i^{-1})\\
&+\tfrac{1}{3}k^{-3}a\sqrt{2}\{a^4+2(3k-5)a^2+3(2k^2-6k+5)\}(\textstyle\sum \nu_i^{-2})\\
&+\tfrac{1}{32}k^{-4}a\sqrt{2}\{-9a^4+24(3-k)a^2-16k^2+72k-87\}(\textstyle\sum \nu_i^{-1})^2\\
&+\tfrac{1}{24}k^{-4}a\sqrt{2}\{15a^6+(128k-315)a^4+(264k^2-1280k+1575)a^2\\
&+3(32k^3-264k^2+640k-525)\}(\textstyle\sum \nu_i^{-3})\\
&+\tfrac{1}{12}k^{-5}a\sqrt{2}\{-15a^6+(255-74k)a^4-(90k^2-632k+1035)a^2\\
&-3(8k^3-90k^2+274k-285)\}(\textstyle\sum \nu_i^{-1})(\textstyle\sum \nu_i^{-2})\\
&+\tfrac{1}{128}k^{-6}a\sqrt{2}\{81a^6+9(28k-131)a^4+3(80k^2-672k+1353)a^2\\
&+64k^3-720k^2+2436k-2889\}\left(\textstyle\sum \nu_i^{-1}\right)^3.
\end{aligned} \tag{13.2.24}
$$

Aoshima and Mukhopadhyay (2002) remarked that this approximation of h_m was remarkably accurate.

Again, going back to two sample sizes N_1, N_2 defined in (13.2.21), we can expect the following approximations for average sample sizes:

$$
\mathrm{E}_{\boldsymbol{\theta}}[N_i] \approx h_{m,\alpha/2}^2\sigma_i^2/d^2,\ i=1,2. \tag{13.2.25}
$$

It is abundantly clear from (13.2.25) that the ratio of N_1, N_2 or $\mathrm{E}_{\boldsymbol{\theta}}[N_1]$, $\mathrm{E}_{\boldsymbol{\theta}}[N_2]$ intuitively matches with σ_1^2/σ_2^2. However, the *optimal ratio* C_1/C_2 from (13.2.18) was σ_1/σ_2. So, it is important and relevant to design a sequential strategy where sample sizes and their ratios would closely mimic optimal expressions from (13.2.18). Robbins et al. (1967) did exactly that very elegantly. Srivastava (1970) developed nearly the same sequential methodology.

13.2.3 Unknown and Unequal Variances: Sequential Sampling

Again, we work under the assumption that $\sigma_i^2, i = 1, 2$ are unknown and unequal. We have mentioned earlier that this framework is commonly referred to as the *Behrens-Fisher problem* or the *Behrens-Fisher analog.* We denote $\boldsymbol{\theta} = (\mu_1, \mu_2, \sigma_1^2, \sigma_2^2)$.

Here, the approach is different from anything else that was discussed before in this chapter or in previous chapters. Observe that the two stopping times N_1, N_2 individually should not only mimic the expressions of C_1, C_2, the ratio N_1/N_2 should also resemble the optimal ratio C_1/C_2 very closely. The second point is important especially since both C_1, C_2 involve σ_1, σ_2. That is, an estimator of C_1 would depend on an estimator of C_2 and vice versa. So, at some point, if an additional observation is needed, one should specify which of the two distributions that observation must come from. Thus, a sequential estimation strategy would not only specify when to stop, it must also provide a *sampling* or *allocation* scheme. Robbins et al. (1967) developed the following methodology along a sampling scheme and the stopping rules.

Initial Step:

We begin with observations $X_{i1}, \ldots, X_{im}$ from the ith population, $i = 1, 2$ where $m(\geq 2)$ is a pilot size. Obtain the sample means $\overline{X}_{1m}, \overline{X}_{2m}$ and the sample variances S_{1m}^2, S_{2m}^2.

Sampling (or Allocation) Scheme:

Robbins et al. (1967) proposed this allocation scheme. Suppose that at some point, we have recorded $n_1(\geq m), n_2(\geq m)$ observations from populations 1 and 2 respectively. However, based on accrued data a stopping rule requires that we continue sampling by taking one additional observation. Then, the next observation is chosen

$$\begin{aligned} &\text{from population 1 if } \frac{n_1}{n_2} \leq \frac{S_{1n_1}}{S_{2n_2}}, \\ &\text{from population 2 if } \frac{n_1}{n_2} > \frac{S_{1n_1}}{S_{2n_2}}. \end{aligned} \tag{13.2.26}$$

Let us examine this sampling scheme. At this moment, suppose that we have $\frac{n_1}{n_2} \leq \frac{S_{1n_1}}{S_{2n_2}}$, and a stopping rule asks us not to terminate sampling. That is, we definitely need another observation and proceed as needed. If we may pretend that $\frac{S_{1n_1}}{S_{2n_2}}$ is a "good" estimator of $\frac{\sigma_1}{\sigma_2}$, then we should move the ratio of the updated sample sizes closer to $\frac{\sigma_1}{\sigma_2}$, but how? Clearly, between the two ratios $\frac{n_1+1}{n_2}$ and $\frac{n_1}{n_2+1}$, the ratio $\frac{n_1+1}{n_2}$ will expectedly be closer to $\frac{S_{1n_1}}{S_{2n_2}} \approx \frac{\sigma_1}{\sigma_2}$ since $\frac{n_1}{n_2+1} < \frac{n_1}{n_2} < \frac{n_1+1}{n_2}$. Thus, we would increase n_1 to $n_1 + 1$ by gathering the next observation from population 1.

Instead, suppose that we have $\frac{n_1}{n_2} > \frac{S_{1n_1}}{S_{2n_2}}$, but a stopping rule asks us not to terminate sampling. So, we definitely need another observation and proceed as needed. Again, pretend that $\frac{S_{1n_1}}{S_{2n_2}}$ is a "good" estimator of $\frac{\sigma_1}{\sigma_2}$, and we should move the ratio of the updated sample sizes closer to $\frac{\sigma_1}{\sigma_2}$. There is only one way to accomplish this. Increase n_2 to n_2+1 because this will decrease the present value of $\frac{n_1}{n_2}$ making it move closer to $\frac{S_{1n_1}}{S_{2n_2}} \approx \frac{\sigma_1}{\sigma_2}$. Thus, in this case we will take the next observation from population 2.

We see that the *sampling* or *allocation scheme* introduced in (13.2.26) is indeed very intuitive. More importantly, it appears to accomplish what it should.

Stopping Rules:

Now, we list a series of stopping rules. Again, Robbins et al. (1967) proposed these stopping times.

Rule 1: Stop with the first $n = n_1 + n_2 (\geq 2m)$ such that if n_i observations from the ith population have been taken, $i = 1, 2$, and

$$n \geq \frac{a^2}{d^2}(S_{1n_1} + S_{2n_2})^2 \tag{13.2.27}$$

Rule 2: The same with (13.2.27) replaced by

$$\frac{S_{1n_1}^2}{n_1} + \frac{S_{2n_2}^2}{n_2} \leq \frac{d^2}{a^2} \tag{13.2.28}$$

Rule 3 The same with (13.2.27) replaced by

$$n_1 \geq \frac{a^2}{d^2} S_{1n_1}(S_{1n_1} + S_{2n_2}) \text{ and } n_2 \geq \frac{a^2}{d^2} S_{2n_2}(S_{1n_1} + S_{2n_2}) \tag{13.2.29}$$

One can show that $\text{P}_{\boldsymbol{\theta}} \{N_1 < \infty, N_2 < \infty\} = 1$ under each stopping rule 1-3, for all fixed $\boldsymbol{\theta}, m, d$ and α. We may stress that in practice, one will employ one of these stopping rules. Upon termination under a stopping rule, we will have collected the final dataset consisting of $N_i, X_{i1}, \ldots, X_{iN_i}, i = 1, 2$.

Then, one would propose the fixed-width interval:

$$\begin{aligned} J_{\boldsymbol{N}} &= (T_{\boldsymbol{N}} - d, T_{\boldsymbol{N}} + d) \\ \text{with } T_{\boldsymbol{N}} &\equiv \overline{X}_{1N_1} - \overline{X}_{2N_2}, \boldsymbol{N} = (N_1, N_2), \end{aligned} \tag{13.2.30}$$

for estimating δ. One may check that $J_{\boldsymbol{N}}$ is a genuine confidence interval estimator.

Again, from the third step in (13.2.14), it should be clear that for all fixed $n_1 \geq m, n_2 \geq m$, the random variables $I(N_1 = n_1, N_2 = n_2)$ and $T_{\boldsymbol{n}} \equiv \overline{X}_{1n_1} - \overline{X}_{2n_2}$ are independent. Thus, we can again express the confidence coefficient associated with $J_{\boldsymbol{N}}$ from (13.2.30) as follows:

$$\begin{aligned}\mathrm{P}_{\boldsymbol{\theta}}\left\{T_{\boldsymbol{N}}-d<\mu_1-\mu_2<T_{\boldsymbol{N}}+d\right\}\\=\mathrm{E}_{\boldsymbol{\theta}}\left[2\Phi\left(d\left\{\frac{\sigma_1^2}{N_1}+\frac{\sigma_2^2}{N_2}\right\}^{-1/2}\right)\right]-1.\end{aligned}\tag{13.2.31}$$

Before we go any further, let us briefly explain how one may grasp the boundary conditions found in (13.2.27)-(13.2.29).

From (13.2.18) recall the expression of C, the total of optimal sample sizes C_1, C_2. It is reasonable to stop sampling when total observed sample size $n(=n_1+n_2)$ is equal or larger than the estimated value of C for the first time. We see that the boundary condition in (13.2.27) asks one to compare $n=n_1+n_2$ with $\widehat{C}=\frac{a^2}{d^2}(S_{1n_1}+S_{2n_2})^2$.

From (13.2.17) recall that we are required to satisfy the condition, $n_1^{-1}\sigma_1^2+n_2^{-1}\sigma_2^2\le a^{-2}d^2$. It is reasonable to stop sampling when observed sample sizes n_1, n_2 are such that the estimated value of $n_1^{-1}\sigma_1^2+n_2^{-1}\sigma_2^2$ is equal or smaller than $a^{-2}d^2$ for the first time. Clearly, the boundary condition in (13.2.28) asks one to compare $n_1^{-1}S_{1n_1}^2+n_2^{-1}S_{2n_2}^2$ with $a^{-2}d^2$.

Again from (13.2.18) recall the expression of optimal sample sizes C_1, C_2. It is reasonable to stop sampling when observed sample sizes n_1, n_2 are equal or larger than the estimated values of C_1, C_2 respectively for the first time. We see that the boundary condition in (13.2.29) asks one to compare n_1 with $\widehat{C}_1=\frac{a^2}{d^2}S_{1n_1}(S_{1n_1}+S_{2n_2})$ and n_2 with $\widehat{C}_2=\frac{a^2}{d^2}S_{2n_2}(S_{1n_1}+S_{2n_2})$.

Now, let us summarize some of the crucial properties of this estimation procedure.

Theorem 13.2.3: *For sampling scheme (13.2.26) and any stopping rules from (13.2.27)-(13.2.29) with final sample sizes N_1, N_2 and final total sample size $N\equiv N_1+N_2$, for all fixed $\boldsymbol{\theta}, m$ and α, one has:*

(i) $\mathrm{E}_{\boldsymbol{\theta}}(N)\le C+m+2$ *with d fixed;*

(ii) $N_1/N_2\to\sigma_1/\sigma_2, N_1/C_1\to 1, N_2/C_2\to 1$ *and* $N/C\to 1$ *in probability as* $d\to 0$;

(iii) $\mathrm{E}_{\boldsymbol{\theta}}(N/C)\to 1$ *as* $d\to 0$ *[Asymptotic First-Order Efficiency];*

(iv) $\mathrm{P}_{\boldsymbol{\theta}}\{\delta\in J_{\boldsymbol{N}}\}\to 1-\alpha$ *as* $d\to 0$ *[Asymptotic Consistency];*

where $C_1, C_2, C\equiv C_1+C_2$ came from (13.2.18), $\boldsymbol{N}=(N_1,N_2)$, $T_{\boldsymbol{N}}\equiv\overline{X}_{1N_1}-\overline{X}_{2N_2}$, and $N=N_1+N_2$.

The ingenious ideas from Robbins et al.'s (1967) paper have tremendously influenced subsequent research in addressing Behrens-Fisher analogs via multistage and sequential methodologies.

We may add a remark. Let $N^{(k)}$ be the total sample size upon termination, that is $N^{(k)} = N_1^{(k)} + N_2^{(k)}$, obtained with the stopping rule #k from (13.2.27)-(13.2.29) with associated final sample sizes $N_1^{(k)}, N_2^{(k)}, k = 1, 2, 3$. Then, for all fixed $\boldsymbol{\theta}, m, d$ and α, one may check that

$$N^{(1)} \le N^{(2)} \le N^{(3)} \text{ w.p. } 1. \tag{13.2.32}$$

A complete proof of Theorem 13.2.3 is involved. Thus, we refer our readers to Robbins et al. (1967) and Srivastava (1970) for its proof. Details can also be found in Ghosh et al. (1997, Section 8.3, pp. 256-259). However, assuming part (ii), one can prove Part (iv) easily. We sketch it in Section 13.5.3.

13.2.4 Comparisons and Conclusions

Computations and simulations studies for fixed-width confidence interval methodologies described in Sections 13.2.1 - 13.2.3 can be carried out using the program **Seq13.exe**. This program is divided into two subprograms. : One program ('program number: 1') handles fixed-width confidence interval problems for $\delta = (\mu_1 - \mu_2)$ and another program ('program number: 2') handles minimum risk point estimation problems for δ described in Sections 13.3.1 - 13.3.2. This is explained in Figure 13.2.1.

Consider two datasets, **Pop1.dat** and **Pop2.dat**, provided in Tables 13.2.1 - 13.2.2. We have used these data to illustrate fixed-width confidence interval methodologies with the help of the program, **Seq13.exe**. The data were generated from two normal distributions. Since we were not told anything else, we assumed that means and variances were unknown and that the variances were unequal ($\sigma_1 \neq \sigma_2$).

Table 13.2.1: Dataset **Pop1.dat**

2.5	11.6	15.3	-0.6	-6.7	7.6	21.5	7.1	-6.5	9.4
15.1	4.1	8.3	-5.4	-3.3	-5.5	2.6	-27.4	-0.1	2.7
12.1	0.1	1.0	-2.9	6.5	-16.4	13.4	2.2	2.9	2.6
-5.3	-1.2	-19.7	-2.3	7.8	3.5	7.2	-10.5	6.0	-2.1
9.2	-0.2	15.3	6.5	3.1	9.4	10.3	-2.6	10.2	5.8
-19.6	-2.6	-22.8	10.4	21.6	-3.3	7.8	10.3	21.5	18.2
-11.8	9.1	-7.1	6.5	0.9	22.5	9.9	0.9	22.4	7.9
2.5	4.1	-4.2	1.0	12.8	8.1	1.9	0.1	1.4	1.4
11.5	-15.1	-11.2	-0.1	3.9	-8.0	-4.6	10.8	12.9	14.2
3.6	25.4	-0.9	22.6	-0.9	18.2	0.9	3.8	0.0	25.6
6.4	0.2	-7.1	10.6	2.0	5.0	10.1	4.0	0.7	5.9
16.5	13.0	15.3							

Table 13.2.2: Dataset **Pop2.dat**

6.7	7.7	6.3	1.4	12.0	3.3	1.7	0.2	-4.9	1.9
-0.8	0.4	5.6	6.0	3.6	7.6	1.9	10.5	5.2	3.8
9.0	3.5	10.4	7.9	11.3	4.2				

Now, the problem is one of constructing a 90% confidence interval for the difference of means using (i) two-stage methodology from Section 13.2.2, and (ii) sequential methodology from Section 13.2.3. We set the fixed-width, $2d = 5.0$ (that is, $d = 2.5$) and $m = 20$, a pilot size. Figure 13.2.1 shows the input values used in the program for computation of the confidence interval using two-stage procedure from Section 13.2.2. Figure 13.2.2 shows the output obtained under this procedure. Analogously, Figure 13.2.3 shows the output for a 90% confidence interval of δ using a sequential procedure keeping the width ($2d = 5.0$) and pilot size ($m = 20$) same as before.

```
   Select a program using the number given
   in the left hand side. Details of these
   programs are given in Chapter 13.
 ***************************************
 * Number     Program                  *
 * 1  Fixed-Width Confidence Interval  *
 * 2  Minimum Risk Point Estimation    *
 * 0  Exit                             *
 ***************************************
Enter the program number: 1

Number Procedure
 1  Common Unknown Variance:   Two-Stage Sampling
 2  Unknown Unequal Variances: Two-Stage Sampling
 3  Unknown Unequal Variances: Sequential Sampling

 Enter procedure number: 2

Enter the following preassigned values for the
fixed-width confidence interval computation:
alpha (for 100(1-alpha)% confidence interval) = 0.1
  d (Fixed-width of the confidence interval)  = 2.5
            initial sample size, m = 20

Do you want to change the input parameters (Y or N)? N
Do you want to save the results in a file(Y or N)? Y
Default outfile name is "SEQ13.OUT".
Do you want to change the outfile name (Y or N)? N

Do you want to simulate data or input real data?
Type S (Simulation) or D (Real Data): D

Enter Y, to read data from ASCII files (data from
     two populations must in 2 separate files);
Enter N, to input data from the keyboard.

Type Y or N ? Y

Input data file name (no more than 9 characters)
   Population 1 data: Pop1.dat
   Population 2 data: Pop2.dat
```

Figure 13.2.1. *Screen input for* ***Seq13.exe.***

```
     Fixed-Width Confidence Interval
Unknown and Unequal Variance: Two-Stage Sampling
==================================================

Input parameters:
        alpha =  0.10
          d =  2.50

Population 1 datafile = Pop1.dat
Population 2 datafile = Pop2.dat

Initial sample size    =   20
Final sample size1(N1) =  113
Final sample size2(N2) =   20

Sample mean 1      =   4.081
Sample variance 1 =   99.438
Sample mean 2      =   4.005
Sample variance 2 =   16.036

h(m,alpha/2) from (13.2.24) = 2.4525

Confidence interval for (mu1 - mu2) is:
          ( -2.42,  2.58)
```

Figure 13.2.2. *Output from two-stage sampling using **Seq13.exe**.*

```
     Fixed-Width Confidence Interval
Unknown and Unequal Variance: Sequential Sampling
    Using Stopping Rule 3 Given in (13.2.29)
==================================================

Input parameters:
        alpha =  0.10
          d =  2.50

Population 1 datafile = Pop1.dat
Population 2 datafile = Pop2.dat

Initial sample size    =   20
Final sample size1(N1) =   65
Final sample size2(N2) =   26

Sample mean 1      =   2.666
Sample variance 1 =  106.327
Sample mean 2      =   4.862
Sample variance 2 =   16.806

Confidence interval for (mu1 - mu2) is:
          ( -4.70,  0.30)
```

Figure 13.2.3. *Output from sequential sampling using **Seq13.exe**.*

From Figures 13.2.2 and 13.2.3, it is clear that the two-stage procedure (13.2.19) oversamples compared with the sequential procedure from Section 13.2.3. That should be expected for two-reasons: (i) Two-stage procedures routinely oversample and we have remarked about this a number of times on other occasions; (ii) Sequential procedure closely mimics optimal allocation, but the Chapman two-stage procedure does not.

```
Fixed-Width Confidence Interval Simulation
Common Unknown Variance: Two-Stage Sampling
===========================================

Input parameters:
         alpha =  0.05
           d =  0.50
Initial sample size, m =  10
Number of simulations  = 5000

Simulation parameters:
Population1 : Normal mean =   5.00
                std. dev. =   2.00
Population2 : Normal mean =   2.00
                std. dev. =   2.00

Optimal sample size (C)      =  122.93
Average sample size (n_bar) =  140.46
Std. dev. of n (s_n)        =   46.58
Minimum sample size (n_min) =   27
Maximum sample size (n_max) =  384
Coverage probability (p_bar)=  0.9500
```

Figure 13.2.4. *Simulation results using* ***Seq13.exe*** *for two-stage sampling procedure with common variance.*

Next, we discuss simulation studies of fixed-width confidence interval methodologies for estimating the difference of means, δ, using our program **Seq13.exe**. Figure 13.2.4 shows a summary of output obtained by averaging performances from 5000 simulation runs of the two-stage procedure described in Section 13.2.1. The layout assumes that two population variances are equal and unknown. Data were generated from $N(5, 2^2)$ and $N(2, 2^2)$ distributions and we fixed $\alpha = 0.05$, $d = 0.5$ and $m = 10$. Using (13.2.6), we can evaluate the optimal fixed sample size:

$$C = \frac{2a^2\sigma^2}{d^2} = \frac{2 \times 1.96^2 \times 2^2}{0.5^2} = 122.93.$$

The program obtains C along with the average ($\overline{n} = 140.46$), standard deviation ($s_n = 46.58$), minimum and maximum from 5000 simulated n values determined by the stopping rule in (13.2.7). The coverage probability ($\overline{p} = 0.950$) is the proportion among 5000 fixed-width intervals constructed upon termination that covered the true value of $\delta(= 3)$.

From (13.2.11), we may theoretically expect that oversampling percentage should be close to $100(\frac{a^2_{2m-2}}{a^2}-1)$, that is 14.89%. Empirically, we may estimate the oversampling percentage by $100(\frac{\overline{n}}{C}-1)$, that is 14.26%. The two percentages are very close indeed!

```
     Fixed-Width Confidence Interval Simulation
Unknown and Unequal Variance: Sequential Sampling
    Using Stopping Rule 3 Given in (13.2.29)
=================================================


Input parameters:
         alpha =  0.05
           d =  0.50
Initial sample size, m =  10
Number of simulations = 5000

Simulation parameters:
Population1 : Normal mean =   5.00
                 std. dev. =   2.00
Population2 : Normal mean =   2.00
                 std. dev. =   1.00

h(m,alpha/2) from (13.2.24) =  3.1799

                  Population 1  Population 2
Optimal sample size      92.20         46.10
Average sample size     162.99         41.11
Std. dev. of n           75.16         19.16
Minimum sample size         13            10
Maximum sample size        554           151

Coverage probability  (p_bar) =  0.9530
```

Figure 13.2.5. *Simulation results using* ***Seq13.exe*** *for two-stage sampling procedure.*

Simulation studies for unknown and unequal variances described in Sections 13.2.2-13.2.3 were also carried out by generating observations from $N(5,2^2)$ and $N(2,1^2)$ distributions. Figures 13.2.5 and 13.2.6 show summaries of output obtained from 5000 simulation runs. We fixed $\alpha=0.05$, $d=0.5$ and $m=10$. The results clearly indicate that the two-stage procedure is consistent ($\overline{p}=0.953$) but it has oversampled whereas the sequential procedure undersampled slightly with $\overline{p}=0.946$ which is just a shade under the target 0.95. However, should that make us a little worried about not possibly meeting the target coverage 0.95?

One may explore as follows: Consider the estimated standard error of $\overline{p}$:

$$s_{\overline{p}}=\sqrt{\frac{\overline{p}(1-\overline{p})}{5000}}=\sqrt{\frac{0.946(1-0.946)}{5000}}\approx 0.0031964.$$

Now, obtain the interval:

$$\overline{p}\pm 1.96 s_{\overline{p}}=0.946\pm 1.96\times 0.0031964$$

which gives the following interval $(0.93974, 0.95226)$ for coverage probability. This interval certainly includes the target coverage probability 0.95!

```
Fixed-Width Confidence Interval Simulation
Unknown and Unequal Variance: Sequential Sampling
==================================================

Input parameters:
         alpha =  0.05
             d =  0.50
Initial sample size, m =  10
Number of simulations  = 5000

Simulation parameters:
Population1 : Normal mean =    5.00
                std. dev. =    2.00
Population2 : Normal mean =    2.00
                std. dev. =    1.00

                  Population 1  Population 2
Optimal sample size      92.20         46.10
Average sample size     90.66          45.13
Std. dev. of n          12.33           7.26
Minimum sample size        19             10
Maximum sample size       132             74

Coverage probability  (p_bar) =  0.9460
```

Figure 13.2.6. *Simulation results from* ***Seq13.exe*** *for sequential sampling procedure.*

13.3 Minimum Risk Point Estimation

We begin with two independent normal populations and continue to use the set of notation that became familiar in Section 13.2. Having recorded observations $X_{i1}, \ldots, X_{in_i}$ from i^{th} population, $i = 1, 2$, we suppose that the loss in estimating $\delta(\equiv \mu_1 - \mu_2)$ by a point estimator $T_{\boldsymbol{n}}(\equiv \overline{X}_{1n_1} - \overline{X}_{2n_2})$ is:

$$L_{\boldsymbol{n}}(\delta, T_{\boldsymbol{n}}) = A\,(T_{\boldsymbol{n}} - \delta)^2 + c(n_1 + n_2) \text{ with } A > 0, c > 0, \text{ both } A \text{ and } c \text{ are known.} \tag{13.3.1}$$

Here, $c(n_1 + n_2)$ represents total expenses for gathering $n_1 + n_2$ observations and $(T_{\boldsymbol{n}} - \delta)^2$ represents the loss due to estimation of δ with $T_{\boldsymbol{n}}$. A suitable magnitude of the weight adjuster "A" and its unit of measurement should be appropriately decided on a case by case basis.

The fixed-sample-size risk function associated with (13.3.1) is expressed as:

$$R_{\boldsymbol{n}}(c) \equiv \mathrm{E}_{\boldsymbol{\theta}}\left\{L_{\boldsymbol{n}}(\delta, T_{\boldsymbol{n}})\right\} = A\left(\frac{\sigma_1^2}{n_1} + \frac{\sigma_2^2}{n_2}\right) + c(n_1 + n_2), \tag{13.3.2}$$

since $\mathrm{E}_{\boldsymbol{\theta}}[(T_{\boldsymbol{n}} - \delta)^2]$ is the variance of $\overline{X}_{1n_1} - \overline{X}_{2n_2}$ which is given by $\sigma_1^2 n_1^{-1} + \sigma_2^2 n_2^{-1}$. Now, our goal is to design estimation strategies that would minimize this risk for all $\boldsymbol{\theta}$.

We note that the risk function $R_{\boldsymbol{n}}(c)$ can be "large" when n_1, n_2 are both very small or very large. Pretending that n_1, n_2 are both continuous variables, we may differentiate the risk function $R_{\boldsymbol{n}}(c)$ from (13.3.2) with respect to n_1, n_2 and simultaneously solve the equations:

$$\frac{\partial}{\partial n_i} R_{\boldsymbol{n}}(c) \equiv A\frac{\sigma_i^2}{n_i^2} + c = 0, \ i = 1, 2. \tag{13.3.3}$$

The optimizing solutions are obviously given by:

$$\begin{aligned} & n_i^* = \sqrt{\frac{A}{c}}\sigma_i, \ i = 1, 2 \\ \Rightarrow & \frac{n_1^*}{n_2^*} = \frac{\sigma_1}{\sigma_2} \text{ and } n^* \equiv n_1^* + n_2^* = \sqrt{\frac{A}{c}}(\sigma_1 + \sigma_2) \\ \Rightarrow & R_{\boldsymbol{n}^*}(c) = 2cn_1^* + 2cn_2^* = 2cn^* \text{ where } \boldsymbol{n}^* = (n_1^*, n_2^*). \end{aligned} \tag{13.3.4}$$

We tacitly disregard the fact that the expressions n_1^*, n_2^* and n^* may not be integers. Clearly, the sample sizes $n_i = n_i^*, i = 1, 2$, will minimize the fixed-sample-size risk, $R_{\boldsymbol{n}}(c)$, had σ_1, σ_2 been known. However, the magnitudes of n_1^*, n_2^* and n^* would not be known when σ's remain unknown.

In what follows, we treat the case of a common unknown variance and the case of unknown and unequal variances separately. We should add that we discuss purely sequential estimation strategies only under either scenario.

13.3.1 Common Unknown Variance: Sequential Procedure

Here, we address minimum risk point estimation problem when $\sigma_1^2 = \sigma_2^2 = \sigma^2$, but σ^2 is unknown. Here, $\boldsymbol{\theta}$ would stand for the parameter vector (μ_1, μ_2, σ^2).

Note that the expressions in (13.3.4) would justify gathering the same number of observations from both distributions. Having recorded observations $X_{i1}, \ldots, X_{in}$ from population i, $i = 1, 2$, we work under loss function (13.3.1) for estimating δ by the point estimator, $T_n(\equiv \overline{X}_{1n} - \overline{X}_{2n})$.

The expression from (13.3.2) reduces to:

$$R_n(c) \equiv \frac{2A\sigma^2}{n} + 2cn. \tag{13.3.5}$$

It is minimized when

$$n \equiv n^* = \sqrt{\frac{A}{c}}\sigma, \tag{13.3.6}$$

and the minimum fixed-sample-size risk is:

$$R_{n^*}(c) \equiv 4cn^*. \tag{13.3.7}$$

Since σ is unknown, no fixed-sample-size methodology would minimize the risk function uniformly for all $\boldsymbol{\theta}$. Hence, one may contemplate using one of several available multistage strategies. For brevity, however, we include only a purely sequential technique parallel to what was discussed in Section 6.4.1.

From (13.2.1), we recall the pooled sample variance:

$$U_n^2 = \frac{1}{2}(S_{1n}^2 + S_{2n}^2), n \geq 2.$$

Now, considering the expression of n^* from (13.3.6), we discuss a purely sequential stopping time along the lines of (6.4.5) in the spirit of Robbins (1959).

We begin with $X_{i1}, \ldots, X_{im}$ from the ith population where $m(\geq 2)$ is a pilot size, $i = 1, 2$. Then, we continue by recording one additional observation from both populations at-a-time, but terminate sampling according as:

$$\begin{aligned} &N \equiv N(c) \text{ is the smallest integer } n(\geq m) \text{ for which} \\ &\qquad \text{we observe } n \geq \left(\frac{A}{c}\right)^{1/2} U_n. \end{aligned} \tag{13.3.8}$$

We note that $\mathrm{P}_{\boldsymbol{\theta}}(N < \infty) = 1$, that is sampling would terminate w.p. 1. Upon termination, we estimate δ by $T_N = \overline{X}_{1N} - \overline{X}_{2N}$ obtained from full dataset $N, X_{i1}, \ldots, X_{im}, \ldots, X_{iN}, i = 1, 2$. Observe that T_n and $I(N = n)$ are independently distributed for all fixed $n \geq m$.

We leave it as an exercise to prove that T_N is an unbiased estimator of δ and its variance is $2\sigma^2 \mathrm{E}_{\boldsymbol{\theta}}(N^{-1})$ for all fixed $\boldsymbol{\theta}, m, A, c$.

Next, the risk associated with T_N can be expressed as:

$$R_N(c) \equiv \mathrm{E}_{\boldsymbol{\theta}}\left[L_N(\delta, T_N)\right] = 2A\sigma^2 \mathrm{E}_{\boldsymbol{\theta}}\left[N^{-1}\right] + 2c\mathrm{E}_{\boldsymbol{\theta}}\left[N\right]. \tag{13.3.9}$$

This expression parallels (6.4.6). Again, we address previous measures of comparing a sequential strategy (N, T_N) from (13.3.8) with a fictitious, and yet optimal fixed-sample-size strategy (n^*, T_{n^*}). We write down the notions of risk efficiency and regret in the spirit of Robbins (1959):

$$\begin{aligned} &\textit{Risk Efficiency}: \text{RiskEff}(c) \equiv \frac{R_N(c)}{R_{n^*}(c)} = \tfrac{1}{2}\mathrm{E}_{\boldsymbol{\theta}}\left[N/n^*\right] + \tfrac{1}{2}\mathrm{E}_{\boldsymbol{\theta}}\left[n^*/N\right] \\ &\textit{Regret}: \text{Regret}(c) \equiv R_N(c) - R_{n^*}(c) = 2c\mathrm{E}_{\boldsymbol{\theta}}\left[\frac{(N-n^*)^2}{N}\right]. \end{aligned} \tag{13.3.10}$$

One may embark upon a thorough investigation regarding these measures. First, let us summarize a number of *first-order* results. For all fixed μ_1, μ_2, σ, m and A, one can verify the following results as $c \to 0$:

$$\begin{aligned} &\mathrm{E}_{\boldsymbol{\theta}}\left[N/n^*\right] \to 1; \mathrm{E}_{\boldsymbol{\theta}}\left[n^*/N\right] \to 1; \\ &\text{RiskEff}(c) \to 1 \; [\textit{Asymptotic First-Order Risk Efficiency}]. \end{aligned} \tag{13.3.11}$$

These hold when $m \geq 2$, which we have assumed from the very beginning.

The first-order properties from (13.3.11) may be interpreted as follows: For large n^* values, that is for small c values, we would expect the average sample size $\mathrm{E}_{\boldsymbol{\theta}}[N]$ to hover in a close proximity of n^*. Also, from the last part of (13.3.11) it should be clear that for large n^* values, we would expect the risk efficiency, RiskEff(c), to hover in a close proximity of 1. That is, for large n^* values we would expect the sequential risk $R_N(c)$ and the optimal fixed-sample risk $R_{n^*}(c)$ to be close to one another.

Asymptotic *second-order* results are further involved. We saw that happen before in Section 6.4 and elsewhere. Recall that the tools that are essential in their proofs may be borrowed from Woodroofe (1977,1982), Lai and Siegmund (1977,1979) and Mukhopadhyay (1988a). See also Mukhopadhyay and Solanky (1994, Section 2.4.2) and Ghosh et al. (1997, Section 7.2) for relevant details. We denote:

$$\eta_2 = -\frac{1}{4}\sum_{n=1}^{\infty}\frac{1}{n}\mathrm{E}[\max\{0, \chi^2_{2n} - 6n\}]. \tag{13.3.12}$$

From Woodroofe (1977) and Mukhopadhyay (1988a, Theorem 3), for all fixed μ_1, μ_2, σ, m and A, we summarize the following asymptotic *second-order* results as $c \to 0$:

$$\begin{aligned} &\mathrm{E}_{\boldsymbol{\theta}}(N - n^*) = \eta_2 + o(1);\\ &\frac{N - n^*}{\sqrt{n^*}} \text{ converges to } N(0, \tfrac{1}{4});\\ &\mathrm{V}_{\boldsymbol{\theta}}(N) = \tfrac{1}{4}n^* + o(n^*);\\ \text{Regret}(c) \equiv\ &R_N(c) - R_{n^*}(c) = \tfrac{1}{2}c + o(c) \text{ if } m \geq 3. \end{aligned} \tag{13.3.13}$$

One should note that for the first three results from (13.3.13) to hold, all we need is that $m \geq 2$.

We may contrast these claims with analogous results shown in (6.4.14) in a one-sample case. One may especially note less restrictive demand on pilot sample size m in the two-sample case. We may give a possible reason for this to happen.

In a one-sample problem, the estimate of σ^2 has $n-1$ degrees of freedom when we have n observations on hand. However, in a two-sample problem, the estimate of σ^2 has $2(n-1)$ degrees of freedom when we have n observations from both populations. Thus, in a two-sample problem, σ^2 is more accurately estimated at every step. This added accuracy is eventually "rewarded" and reflected in the economy of our choice for m in (13.3.13).

Sequential procedure (13.3.8) is *asymptotically second-order efficient* and *second-order risk efficient.*

For large n^*, let us interpret the results from (13.3.13). Note that η_2 from (13.3.12) does not involve anything unknown and hence computable. From the first result in (13.3.13), we may expect $\mathrm{E}_{\boldsymbol{\theta}}(N - n^*)$ to lie in a close proximity of η_2. The second part says that the normalized stopping variable, namely $(N - n^*)/\sqrt{n^*}$, should have an approximate normal distribution. The third part says that $\mathrm{V}_{\boldsymbol{\theta}}(N)$ should be close to $\frac{1}{4}n^*$. The last part says that the asymptotic regret amounts to the cost of one-half of one observation. That is, in the absence of any knowledge about σ, we may end up wasting the cost of only one-half of one observation while using the sequential strategy (13.3.8). The purely sequential estimation strategy is that efficient!

These and other interesting features would be explained more with the help of simulations and data analyses in Section 13.3.3.

13.3.2 Unknown and Unequal Variances: Sequential Procedure

Now, we begin with two independent normal distributions and assume that σ_1^2, σ_2^2 are unknown and unequal. So, we go back to the set of notation that we used in the introductory part of Section 13.3. We will denote $\boldsymbol{\theta} = (\mu_1, \mu_2, \sigma_1, \sigma_2)$.

Having recorded observations $X_{i1}, \ldots, X_{in_i}$ from i^{th} population, $i = 1, 2$, we estimate δ by $T_{\boldsymbol{n}} \equiv \overline{X}_{1n_1} - \overline{X}_{2n_2}$ under loss function from (13.3.1).

The fixed-sample-size risk function $R_{\boldsymbol{n}}(c)$ from (13.3.2) was minimized with

$$n_i \equiv n_i^* = \sqrt{A/c}\sigma_i, i = 1, 2.$$

The optimal fixed-sample-size risk was

$$R_{\boldsymbol{n}^*}(c) = 2cn^*$$

with

$$n^* \equiv n_1^* + n_2^* = \sqrt{A/c}(\sigma_1 + \sigma_2).$$

One may refer to (13.3.4).

Since $n_i^* = \sqrt{A/c}\sigma_i, i = 1, 2$, are unknown, no fixed-sample-size methodology will deliver in this case. We have to gather data in at least two-stages. For brevity, however, we discuss only a purely sequential methodology.

Note that we have $n_1^*/n_2^* = \sigma_1/\sigma_2$, and one may like to proceed as in Section 13.2.3. That is, one would propose a suitable sampling or allocation scheme followed by appropriate sequential stopping variables N_1, N_2 along the lines of (13.2.6) and (13.2.27) - (13.2.29) respectively. Mukhopadhyay (1975,1976a,1977) proceeded that way and obtained *first-order* results.

Some crucial results from Woodroofe (1977) were not immediately applicable, especially since the stopping variables N_1, N_2 were necessarily dependent on each other. Thus, Ghosh and Mukhopadhyay (1980) obtained the *second-order* expansion of regret function directly without appealing to Woodroofe's (1977) technique.

However, there is a big difference between the expressions C_1, C_2 from (13.2.4) and the expressions n_1^*, n_2^* from (13.3.4). We observe that n_1^* involves σ_1 only, n_2^* involves σ_2 only, and one has nothing to do with the other. So, one may estimate n_1^*, n_2^* from two independently run sequential stopping times N_1 and N_2. The stopping time N_i could be defined exclusively by means of observations gathered from i^{th} population alone, $i = 1, 2$. It may not be necessary to implement an allocation scheme after all!

Mukhopadhyay and Moreno (1991b) observed this important and interesting point. Such a simple idea made implementation of the Mukhopadhyay methodology much more convenient. Moreover, *second-order* expansion of the associated regret function came out exactly the same as Ghosh and Mukhopadhyay's (1980). In what follows, we describe the sequential methodology of Mukhopadhyay and Moreno (1991b) adapted to our special problem on hand.

We begin with pilot observations $X_{i1}, \ldots, X_{im}$ from i^{th} population where $m(\geq 2)$. We proceed sequentially by taking one additional observation at-a-time from i^{th} population. The stopping time for sampling from i^{th} population is:

$$\begin{aligned}&N_i \equiv N_i(c) \text{ is the smallest integer } n_i(\geq m) \text{ for which}\\&\qquad\text{we observe } n_i \geq \left(\frac{A}{c}\right)^{1/2} S_{in},\end{aligned} \tag{13.3.14}$$

$i = 1, 2$. That is, we implement two separate sequential stopping times, but they run individually and independently of each other.

We terminate sampling when both stopping rules ask us not to record additional observations. Again, $\text{P}_{\boldsymbol{\theta}}(N_i < \infty) = 1, i = 1, 2$. That is, sampling would terminate w.p. 1. Upon termination, we estimate δ by $T_{\boldsymbol{N}} \equiv \overline{X}_{1N_1} - \overline{X}_{2N_2}$ obtained from full dataset $N_i, X_{i1}, \ldots, X_{im}, \ldots, X_{iN_i}$ from population i, $i = 1, 2$.

Observe that for all fixed $n_1 \geq m, n_2 \geq m$, the random variables $T_{\boldsymbol{n}}$ and $I(N_1 = n_1, N_2 = n_2)$ are independently distributed. Also, the stopping variables N_1, N_2 are independently distributed.

We leave it as an exercise to prove that $T_{\boldsymbol{N}}$ is an unbiased estimator of δ and its variance is $\text{E}_{\boldsymbol{\theta}}\left[\sigma_1^2 N_1^{-1} + \sigma_2^2 N_2^{-1}\right]$ for all fixed $\boldsymbol{\theta}, m, A$ and c.

As before, let us denote $n^* = n_1^* + n_2^*$ and $N = N_1 + N_2$. Obviously, we can express the optimal fixed-sample-size risk as:

$$R_{\boldsymbol{n}^*}(c) = 2c(n_1^* + n_2^*) = 2cn^*, \tag{13.3.15}$$

and the sequential risk as:

$$R_{\boldsymbol{N}}(c) = A\sigma_1^2 \text{E}_{\boldsymbol{\theta}}\left[N_1^{-1}\right] + A\sigma_2^2 \text{E}_{\boldsymbol{\theta}}\left[N_2^{-1}\right] + c\text{E}_{\boldsymbol{\theta}}\left[N_1\right] + c\text{E}_{\boldsymbol{\theta}}\left[N_2\right]. \tag{13.3.16}$$

Again, consider the following measures for comparing the sequential strategy $(\boldsymbol{N}, T_{\boldsymbol{N}})$ from (13.3.14) with $(\boldsymbol{n}^*, T_{\boldsymbol{n}^*})$ pretending that $n_i^* =$

$\sqrt{A/c}\sigma_i$, $i = 1, 2$ are known integers, $\boldsymbol{n}^* = (n_1^*, n_2^*)$. Define

$$\begin{aligned}&\textit{Risk Efficiency}\text{: RiskEff}(c) \equiv \frac{R_{\boldsymbol{N}}(c)}{R_{\boldsymbol{n}^*}(c)} \\ &\textit{Regret}\text{: Regret}(c) \equiv R_{\boldsymbol{N}}(c) - R_{\boldsymbol{n}^*}(c).\end{aligned} \tag{13.3.17}$$

Now, we can combine (13.3.15) - (13.3.17) to rewrite the risk efficiency and regret as follows:

$$\text{RiskEff}(c) = \frac{\sigma_1}{2(\sigma_1+\sigma_2)}\text{E}_{\boldsymbol{\theta}}\left[\frac{n_1^*}{N_1}\right] + \frac{\sigma_2}{2(\sigma_1+\sigma_2)}\text{E}_{\boldsymbol{\theta}}\left[\frac{n_2^*}{N_2}\right] + \frac{1}{2}\text{E}_{\boldsymbol{\theta}}\left[\frac{N}{n^*}\right], \tag{13.3.18}$$

and

$$\text{Regret}(c) = c\text{E}_{\boldsymbol{\theta}}\left[\frac{(N_1-n_1^*)^2}{N_1}\right] + c\text{E}_{\boldsymbol{\theta}}\left[\frac{(N_2-n_2^*)^2}{N_2}\right]. \tag{13.3.19}$$

One will notice that in either expression found from (13.3.18) - (13.3.19), there are terms involving N_1, N_2 separately and they resemble those from (6.4.7). Thus, along the lines of (6.4.8), for all fixed $\mu_1, \mu_2, \sigma_1, \sigma_2, m$ and A, we claim the following *first-order* results as $c \to 0$:

$$\begin{aligned}&\text{E}_{\boldsymbol{\theta}}[N_1/n_1^*] \to 1, \text{E}_{\boldsymbol{\theta}}[N_2/n_2^*] \to 1, \text{E}_{\boldsymbol{\theta}}[N/n^*] \to 1, \text{ if } m \geq 2; \\ &\text{E}_{\boldsymbol{\theta}}[n_1^*/N_1] \to 1, \text{E}_{\boldsymbol{\theta}}[n_2^*/N_2] \to 1, \text{E}_{\boldsymbol{\theta}}[n^*/N] \to 1, \text{ if } m \geq 3; \\ &\text{RiskEff}(c) \to 1 \text{ if } m \geq 3 \text{ [\textit{Asymptotic First-Order Risk Efficiency}]}.\end{aligned} \tag{13.3.20}$$

Now, let us refocus on asymptotic *second-order* properties. Recall the expression of $\eta_1(1)$ from (6.4.13). Then, along the lines of (6.4.14), for all fixed $\mu_1, \mu_2, \sigma_1, \sigma_2, m$ and A, we can claim the following *second-order* results as $c \to 0$:

$$\begin{aligned}&\text{E}_{\boldsymbol{\theta}}[N_i - n_i^*] = \eta_1(1) + o(1) \text{ if } m \geq 3, i = 1, 2; \\ &\frac{N_i - n_i^*}{\sqrt{n_i^*}} \text{ converges to } N(0, \tfrac{1}{2}) \text{ in distribution if } m \geq 2, i = 1, 2; \\ &\text{V}_{\boldsymbol{\theta}}(N_i) = \tfrac{1}{2}n_i^* + o(n_i^*) \text{ if } m \geq 3, i = 1, 2; \\ &\text{Regret}(c) \equiv R_{\boldsymbol{N}}(c) - R_{\boldsymbol{n}^*}(c) = c + o(c) \text{ if } m \geq 4.\end{aligned} \tag{13.3.21}$$

Mukhopadhyay (1977) developed *first-order* efficiency and risk-efficiency results shown in (13.3.20) as well as the asymptotic distributions shown in (13.3.21). Ghosh and Mukhopadhyay (1980) developed the exact same *second-order* regret expansion that is given in the last line of (13.3.21). The present regret expansion can be verified immediately in view of our arguments sketched earlier in Section 6.5.3 in a one-sample problem.

> The purely sequential procedure (13.3.14) is *asymptotically second-order efficient* and *second-order risk efficient.*

We observe that the asymptotic regret amounts to the cost of one single observation from either population. So, in the absence of any knowledge about σ_1, σ_2, we may "waste" only one observation in the process of using the purely sequential estimation strategy (13.3.14). However, one ought to remember that there is no suitable fixed-sample-size methodology for this problem!

13.3.3 Comparisons and Conclusions

```
  Select a program using the number given
  in the left hand side. Details of these
  programs are given in Chapter 13.
 ****************************************
 * Number     Program                   *
 * 1  Fixed-Width Confidence Interval   *
 * 2  Minimum Risk Point Estimation     *
 * 0  Exit                              *
 ****************************************
Enter the program number: 2

Number Procedure
 1  Common Unknown Variance: Sequential Sampling
 2  Unknown Unequal Variances: Sequential Sampling
Enter procedure number: 1

Enter the following preassigned values for the
minimum risk point estimation:
      A (Constant in the loss function) = 100
               c (cost due to sampling) = .01
                 initial sample size, m = 10
Do you want to change the input parameters (Y or N)? N

Do you want to save the results in a file(Y or N)? y
Default outfile name is "SEQ13.OUT".
Do you want to change the outfile name (Y or N)? N

Do you want to simulate data or input real data?
Type S (Simulation) or D (Real Data): S
Do you want to store simulated sample sizes (Y or N)? N
Number of simulation replications? 5000
Enter a positive integer(<10) to initialize
     the random number sequence: 3

Enter the following parameters for simulation:
   mean (mu1)    = 5
   mean (mu2)    = 2
   common std dev (sigma) = 2
```

Figure 13.3.1. *Screen input for* ***Seq13.exe.***

Now, we examine sequential procedures from Sections 13.3.1 and 13.3.2 using simulation studies and data analysis. The program **Seq13.exe** can be employed to perform both these tasks. Figure 13.3.1 shows input for 'program number: 2' to run a simulation study of minimum risk point estimation of δ. Here, the variances of the two populations are assumed to equal but unknown. Figure 13.3.1 shows that we fixed $A = 100$, $c = 0.01$, $m = 10$ and number of simulations runs 5000 using random number sequence 3. The data were generated from $N(5, 2^2)$ and $N(2, 2^2)$ distributions. The output is shown in Figure 13.3.2. It provides the average sample size ($\overline{n} = 200.26$) which is remarkably close to the optimal sample size ($n^* = 200.00$).

Summary results from a simulation study for two populations with unknown and unequal variances are given in Figure 13.3.3. This study used the same parameter values as before, but data were generated from $N(5, 2^2)$ and $N(2, 1^2)$ distributions. The output shows estimated average sample sizes ($\overline{n}_1 = 199.85, \overline{n}_2 = 99.89$) and these are very close to the corresponding optimal sample sizes ($n_1^* = 200, n_2^* = 100$).

```
     Minimum Risk Point Estimation
Common Unknown Variance:Sequential Procedure
============================================

     Input parameters:
                         A =  100.00
                         c =   0.010
     Initial sample size, m =     10
     Number of simulations  =   5000

     Simulation parameters:
     Population1 : Normal mean =        5.00
                       std. dev. =      2.00
     Population2 : Normal mean =        2.00
                       std. dev. =      2.00

     Optimal sample size (n*)     =     200.00
     Average sample size (n_bar)  =     200.26
     Std. dev. of n (s_n)         =       7.15
     Minimum sample size (n_min)  =     176
     Maximum sample size (n_max)  =     224
     Average simulated regret(c)  =     0.2565
     Standard error of regret(c)  =     0.0052
```

Figure 13.3.2. *Simulation results from* ***Seq13.exe*** *for common unknown variance.*

13.4 Other Selected Multistage Procedures

We start by mentioning that Stein's (1945,1949) two-stage methodology for a one-sample problem enjoys some robustness properties under mild departures from normality. One may refer to Blumenthal and Govindarajulu (1977) and Ramkaran (1983). One may explore if limited robustness properties would hold for the Chapman procedure from Section 13.2.2.

In the context of the Chapman procedure, one may refer to Ghosh et al. (1997, p. 186) for more details. This procedure was further investigated by

Ghosh (1975a, b) and extended by Aoshima (2001) for the k-sample problem. Mukhopadhyay and de Silva (1997) investigated multistage methodologies beyond two-stage strategies. Aoshima and Mukhopadhyay (2002) generalized this methodology with emphasis on practical implementation. de Silva and Waikar (2006) have proposed ways to make the original methodology more efficient by appropriately incorporating bootstrapping.

```
         Minimum Risk Point Estimation
Unknown and Unequal Variance: Sequential Procedure
==================================================

  Input parameters:
                         A = 100.00
                         c =  0.010
  Initial sample size, m =    10
  Number of simulations  =  5000

  Simulation parameters:
  Population1 : Normal mean =       5.00
                  std. dev. =       2.00
  Population2 : Normal mean =       2.00
                  std. dev. =       1.00

                         Population 1  Population 2
  Optimal sample size          200.00        100.00
  Average sample size          199.85         99.89
  Std. dev. of n                10.06          7.24
  Minimum sample size             165            75
  Maximum sample size             231           125

  Average simulated regret(c) =     1.0508
  Standard error of regret(c) =     0.0000
```

Figure 13.3.3. *Simulation results from* ***Seq13.exe*** *for unknown, unequal variances.*

Two-Stage Optimal Sampling Strategies:

In the case of unknown and unequal variances, it is no secret that the Chapman methodology mimicked expressions of sub-optimal fixed-sample-sizes, n_1^0 and n_2^0, from (13.2.19). We commented upon that in Section 13.2.2. Aoshima et al. (1996) developed a remarkable two-stage fixed-width confidence interval strategy for δ by mimicking expressions of optimal fixed-sample-sizes C_1 and C_2 from (13.2.18). This interval preserved the exact consistency property. For some related developments, one may refer to Dudewicz and Ahmed (1988,1999).

Two-Sample Exponential Problems:

Mukhopadhyay and Hamdy (1984a) developed a two-stage fixed-width confidence interval methodology for estimating the difference of location parameters, $\Delta \equiv \theta_1 - \theta_2$, where the two independent distributions were:

$$\sigma_i^{-1} \exp\{-(x-\theta_i)/\sigma_i\} I(x > \theta_i), i = 1, 2. \tag{13.4.1}$$

They investigated this problem when scale parameters were (i) unknown but equal, and (ii) unknown and unequal. Naturally, the Behrens-Fisher situation, namely when the scale parameters were unknown and unequal, was the real challenging one. The Behrens-Fisher analog was handled in the spirit of Chapman (1950).

Mukhopadhyay and Mauromoustakos (1987) developed a three-stage fixed-width confidence interval strategy for estimating Δ under same distributions where scale parameters were unknown but equal. Associated second-order properties were obtained. Mukhopadhyay and Padmanabhan (1993) discussed extensions.

Mukhopadhyay and Darmanto (1988) dealt with sequential estimation problems for the difference of means, $\mu_1 - \mu_2$, where the two independent distributions were given by (13.4.1). Minimum risk point estimation problems and associated *first-order* properties were introduced when scale parameters were (i) unknown but equal, and (ii) unknown and unequal.

Two-Sample Multivariate Normal Problems:

Mukhopadhyay and Abid (1986a) constructed fixed-size confidence ellipsoidal regions for the difference of mean vectors in p-dimensional normal distributions with dispersion matrices $\sigma_i^2 H_{p\times p}, i = 1, 2$. The matrix H was assumed known and positive definite, but σ_1, σ_2 were unknown and unequal. Mukhopadhyay and Abid (1986a) developed several multistage estimation strategies along with *first-* and/or *second-order* analyses.

Mukhopadhyay (1985b) discussed a two-sample minimum risk point estimation problem for the mean vectors in multivariate normal distributions. He proposed a kind of hybrid James-Stein version of customary estimators based on sample mean vectors and showed the risk-dominance of the James-Stein estimator under a multistage sampling strategy. Nickerson (1987a) also had handled a similar two-sample problem.

k-Sample Univariate Problems:

Mukhopadhyay (1976a) designed an allocation scheme under appropriate sequential rules for estimating a linear function of means of *three* independent normal populations having unknown and unequal variances. He discussed both fixed-width confidence interval and a minimum risk point estimation problems.

Ramkaran et al. (1986) heuristically came up with an allocation scheme in the case of a k-sample problem under a sequential methodology.

Mukhopadhyay and Chattopadhyay (1991) discussed sequential minimum risk point estimation of a linear function of means of $k(\geq 2)$ independent exponential populations. They obtained second-order expansions of the bias and regret associated with the estimator.

Aoshima and Mukhopadhyay (2002) introduced a two-stage strategy for estimating a linear function of the means of $k(\geq 2)$ independent normal populations having unknown and unequal variances. They allowed pilot sample sizes to be unequal, if necessary. Even for $k = 2$, this broadened the practicality of Chapman's (1950) original procedure (13.2.21).

k-Sample Multivariate Problems:

Liberman (1987) and Mukhopadhyay and Liberman (1989), independently of Ramkaran et al. (1986), constructed fixed-size confidence regions for a linear function of $k(\geq 2)$ mean vectors of multivariate normal distributions with special kinds of dispersion matrices. In a sequential case, they developed a sampling scheme with the help of a one-step look-ahead strategy. That sampling scheme was investigated thoroughly.

Two-Sample Distribution-Free Problems:

Mukhopadhyay (1984a) had developed a sequential fixed-width confidence interval procedure for the difference of truncation parameters, namely $\theta_1 - \theta_2$, where the two independent distributions were unknown but θ_1, θ_2 resembled those in (13.4.1).

Mukhopadhyay and Moreno (1991a,b) studied sequential minimum risk point estimation problems for difference of means of two U-statistics or the mean of a generalized U-statistic. They obtained *first-order* properties under appropriate moment conditions when the distributions were unspecified.

Mukhopadhyay and Purkayastha (1994) and Mukhopadhyay (1996) investigated *second-order* properties associated with sequential and accelerated sequential minimum risk point estimation problems respectively for the difference of means of two unspecified distributions.

13.5 Some Selected Derivations

In this section, we supply parts of some crucial derivations.

13.5.1 Proof of (13.2.18)

We temporarily pretend that n_1, n_2 are continuous variables. Now, we wish to minimize $n_1 + n_2$ under the restriction that $\frac{\sigma_1^2}{n_1} + \frac{\sigma_2^2}{n_2} \leq \frac{1}{b}$ where $b = \frac{a^2}{d^2}, n_1 > 0, n_2 > 0$.

With the help of Lagrange's multiplier λ, we define a new function:

$$g(n_1, n_2, \lambda) = n_1 + n_2 + \lambda \left(\frac{\sigma_1^2}{n_1} + \frac{\sigma_2^2}{n_2} - \frac{1}{b} \right). \tag{13.5.1}$$

Now, by successive partial differentiation of $g(n_1, n_2, \lambda)$ with respect to

n_1, n_2 and λ, we have:

$$\begin{aligned}\tfrac{\partial}{\partial n_1} g(n_1, n_2, \lambda) &= 1 - \lambda \frac{\sigma_1^2}{n_1^2} \equiv 0 \Rightarrow n_1 = \sqrt{\lambda}\sigma_1; \\ \tfrac{\partial}{\partial n_2} g(n_1, n_2, \lambda) &= 1 - \lambda \frac{\sigma_2^2}{n_2^2} \equiv 0 \Rightarrow n_2 = \sqrt{\lambda}\sigma_2; \\ \tfrac{\partial}{\partial \lambda} g(n_1, n_2, \lambda) &= \frac{\sigma_1^2}{n_1} + \frac{\sigma_2^2}{n_2} - \frac{1}{b} \equiv 0 \Rightarrow \frac{\sigma_1^2}{n_1} + \frac{\sigma_2^2}{n_2} = \frac{1}{b}.\end{aligned} \tag{13.5.2}$$

At this point, we plug in solutions for n_1, n_2 in the last equation from (13.5.2) to write:

$$\frac{1}{b} = \frac{\sigma_1^2}{\sqrt{\lambda}\sigma_1} + \frac{\sigma_2^2}{\sqrt{\lambda}\sigma_2} = \frac{1}{\sqrt{\lambda}}(\sigma_1 + \sigma_2) \Rightarrow \sqrt{\lambda} = b(\sigma_1 + \sigma_2). \tag{13.5.3}$$

Now, the first two steps in (13.5.2) can be rewritten with the help of (13.5.3) as follows:

$$n_1 = b\sigma_1(\sigma_1 + \sigma_2), n_2 = b\sigma_2(\sigma_1 + \sigma_2)$$

We leave it as an exercise to check that this "solution" minimizes $n_1 + n_2$ under the restriction that $\frac{\sigma_1^2}{n_1} + \frac{\sigma_2^2}{n_2} \le \frac{1}{b}$. The proof is complete. ■

13.5.2 Proof of Theorem 13.2.2

From (13.2.23), recall that $\mathrm{P}_{\boldsymbol{\theta}}\{\delta \in (T_{\boldsymbol{N}} - d, T_{\boldsymbol{N}} + d)\}$ is:

$$\mathrm{E}_{\boldsymbol{\theta}}\left[2\Phi\left(d\left\{\frac{\sigma_1^2}{N_1} + \frac{\sigma_2^2}{N_2}\right\}^{-1/2}\right)\right] - 1. \tag{13.5.4}$$

Now, we utilize the definitions of N_1, N_2 from (13.2.21) to write w.p. 1:

$$\begin{aligned}&\frac{1}{N_i} \le \frac{d^2}{h_{m,\alpha/2}^2 S_{im}^2}, i = 1,2 \Rightarrow \frac{\sigma_1^2}{N_1} + \frac{\sigma_2^2}{N_2} \le \frac{d^2}{h_{m,\alpha/2}^2}\left(\frac{\sigma_1^2}{S_{1m}^2} + \frac{\sigma_2^2}{S_{2m}^2}\right) \\ &\Rightarrow d\left(\frac{\sigma_1^2}{N_1} + \frac{\sigma_2^2}{N_2}\right)^{-1/2} \ge h_{m,\alpha/2}\left(\frac{\sigma_1^2}{S_{1m}^2} + \frac{\sigma_2^2}{S_{2m}^2}\right)^{-1/2}.\end{aligned} \tag{13.5.5}$$

Next, we may combine (13.5.4)-(13.5.5) to obtain:

$$\begin{aligned}&\mathrm{P}_{\boldsymbol{\theta}}\{\delta \in (T_{\boldsymbol{N}} - d, T_{\boldsymbol{N}} + d)\} \\ &\ge \mathrm{E}_{\boldsymbol{\theta}}\left[2\Phi\left(h_{m,\alpha/2}\left\{\frac{\sigma_1^2}{S_{1m}^2} + \frac{\sigma_2^2}{S_{2m}^2}\right\}^{-1/2}\right)\right] - 1.\end{aligned} \tag{13.5.6}$$

At this point, let us think about two i.i.d. t_{m-1} random variables, W_1 and W_2. We can express $W_i = Y_i/\sqrt{V_i/(m-1)}$, $i = 1, 2$ where Y_1, Y_2, V_1, V_2 are independent random variables, Y_1, Y_2 are i.i.d. $N(0, 1)$, and V_1, V_2 are i.i.d. χ^2_{m-1}. Then, we may rewrite (13.2.20) as follows:

$$\begin{aligned}
1-\tfrac{1}{2}\alpha &= \mathrm{P}\left\{W_1 - W_2 \le h_{m,\alpha/2}\right\} \\
&= \mathrm{P}\left\{\frac{Y_1}{\sqrt{V_1/(m-1)}} - \frac{Y_2}{\sqrt{V_2/(m-1)}} \le h_{m,\alpha/2}\right\} \\
&= \mathrm{E}\left[\mathrm{P}\left\{\frac{Y_1}{\sqrt{V_1/(m-1)}} - \frac{Y_2}{\sqrt{V_2/(m-1)}} \le h_{m,\alpha/2} \mid V_1, V_2\right\}\right].
\end{aligned} \quad (13.5.7)$$

Conditionally given V_1, V_2, the random variable $\frac{Y_1}{\sqrt{V_1/(m-1)}} - \frac{Y_2}{\sqrt{V_2/(m-1)}}$ is distributed as $N(0, (m-1)(\frac{1}{V_1} + \frac{1}{V_2}))$. That is, conditionally given V_1, V_2, the random variable

$$\begin{aligned}
&\frac{\frac{Y_1}{\sqrt{V_1/(m-1)}} - \frac{Y_2}{\sqrt{V_2/(m-1)}}}{\sqrt{(m-1)(\frac{1}{V_1} + \frac{1}{V_2})}} \sim N(0,1) \\
&\Rightarrow \mathrm{P}\left\{\frac{Y_1}{\sqrt{V_1/(m-1)}} - \frac{Y_2}{\sqrt{V_2/(m-1)}} \le h_{m,\alpha/2} \mid V_1, V_2\right\} \\
&\quad = \Phi\left(\frac{h_{m,\alpha/2}}{\sqrt{(m-1)(\frac{1}{V_1} + \frac{1}{V_2})}}\right).
\end{aligned} \quad (13.5.8)$$

Now, we may combine (13.5.7) - (13.5.8) to claim that $h_{m,\alpha/2}$ must satisfy:

$$\mathrm{E}\left[\Phi\left(\frac{h_{m,\alpha/2}}{\sqrt{(m-1)(\frac{1}{V_1} + \frac{1}{V_2})}}\right)\right] = 1 - \frac{1}{2}\alpha. \quad (13.5.9)$$

From (13.2.14), one recalls that $(m-1)S^2_{im}/\sigma^2_i \sim \chi^2_{m-1}$, $i = 1, 2$, and they are independent too. Thus, from (13.5.9), we write:

$$\mathrm{E}_{\boldsymbol{\theta}}\left[\Phi\left(h_{m,\alpha/2}\left\{\frac{\sigma_1^2}{S^2_{1m}} + \frac{\sigma_2^2}{S^2_{2m}}\right\}^{-1/2}\right)\right] = 1 - \tfrac{1}{2}\alpha,$$

which implies:

$$\begin{aligned}
&\mathrm{E}_{\boldsymbol{\theta}}\left[2\Phi\left(h_{m,\alpha/2}\left\{\frac{\sigma_1^2}{S^2_{1m}} + \frac{\sigma_2^2}{S^2_{2m}}\right\}^{-1/2}\right)\right] - 1 \\
&\quad = 2\left(1 - \tfrac{1}{2}\alpha\right) - 1 = 1 - \alpha.
\end{aligned} \quad (13.5.10)$$

Now, (13.5.4) combined with (13.5.10) complete the proof. ■

13.5.3 Proof of Theorem 13.2.3 Part (iv)

Recall that:

$$\begin{aligned} &\mathrm{P}_{\boldsymbol{\theta}}\left\{\delta \in (T_{\boldsymbol{N}} - d, T_{\boldsymbol{N}} + d)\right\} \\ &= \mathrm{E}_{\boldsymbol{\theta}}\left[2\Phi\left(d\left\{\frac{\sigma_1^2}{N_1} + \frac{\sigma_2^2}{N_2}\right\}^{-1/2}\right)\right] - 1. \end{aligned} \tag{13.5.11}$$

Now, we have $C_i = a^2\sigma_i(\sigma_1 + \sigma_2)/d^2$, $i = 1, 2$ so that we can rewrite (13.5.11) as:

$$\begin{aligned} &\mathrm{P}_{\boldsymbol{\theta}}\left\{\delta \in (T_{\boldsymbol{N}} - d, T_{\boldsymbol{N}} + d)\right\} \\ &= \mathrm{E}_{\boldsymbol{\theta}}\left[2\Phi\left(a\sqrt{\sigma_1 + \sigma_2}\left\{\sigma_1\frac{C_1}{N_1} + \sigma_2\frac{C_2}{N_2}\right\}^{-1/2}\right)\right] - 1. \end{aligned} \tag{13.5.12}$$

However, $\Phi(.)$ is obviously bounded. Hence, assuming Theorem 13.2.3, part (ii), we can immediately apply dominated convergence theorem to claim that as $d \to 0$:

$$\mathrm{P}_{\boldsymbol{\theta}}\left\{\delta \in (T_{\boldsymbol{N}} - d, T_{\boldsymbol{N}} + d)\right\} \to 2\Phi(a) - 1,$$

which is $1 - \alpha$. ■

13.6 Exercises

Exercise 13.2.1: Prove Theorem 13.2.1 along the lines of proofs of Theorems 6.2.1-6.2.2.

Exercise 13.2.2: For the two-stage strategy (13.2.7), show that

$$\mathrm{E}_{\boldsymbol{\theta}}\left[T_N\right] = \mu_1 - \mu_2 \text{ and } \mathrm{V}_{\boldsymbol{\theta}}\left[T_N\right] = 2\sigma^2\mathrm{E}_{\boldsymbol{\theta}}\left[N^{-1}\right],$$

where $T_N = \overline{X}_{1N} - \overline{X}_{2N}$ that was defined in (13.2.8).

Exercise 13.2.3: For the two-stage strategy (13.2.7), prove that:

$$\lim_{d\to 0} \mathrm{P}_{\boldsymbol{\theta}}\{\mu_1 - \mu_2 \in (T_N - d, T_N + d)\} = 1 - \alpha,$$

for the two-stage strategy (13.2.7). {*Hint:* Improvise on hints that were given in Exercise 7.2.3.}

Exercise 13.2.4: For the fixed-width confidence interval problem from Section 13.2.1, consider implementing a purely sequential stopping time:

$$N \equiv N(d) \text{ is the smallest integer } n(\geq m \geq 2) \text{ for which we observe } n \geq 2z_{\alpha/2}^2 U_n^2/d^2.$$

Finally, propose the confidence interval $J_N = (T_N - d, T_N + d)$ for $\mu_1 - \mu_2$. Recall that $C = 2a^2\sigma^2/d^2$ from (13.2.6). Show that

(i) $\mathrm{E}_{\boldsymbol{\theta}}[N]$ is finite;
(ii) $\mathrm{E}_{\boldsymbol{\theta}}[N] \le C + O(1)$;
(iii) $\mathrm{E}_{\boldsymbol{\theta}}[N/n^*] \to 1$ as $d \to 0$;
(iv) $\mathrm{P}_{\boldsymbol{\theta}}\{\mu_1 - \mu_2 \in (T_N - d, T_N + d)\} \to 1 - \alpha$ as $d \to 0$.

Exercise 13.2.5: For the two-stage strategy (13.2.21), show that

$$\mathrm{E}_{\boldsymbol{\theta}}[T_{\boldsymbol{N}}] = \mu_1 - \mu_2 \text{ and } \mathrm{V}_{\boldsymbol{\theta}}[T_N] = \mathrm{E}_{\boldsymbol{\theta}}\left[\sigma_1^2 N_1^{-1} + \sigma_2^2 N_2^{-1}\right],$$

where $T_{\boldsymbol{N}} = \overline{X}_{1N_1} - \overline{X}_{2N_2}$ that was defined in (13.2.22).

Exercise 13.2.6: Prove that:

$$\lim_{d \to 0} \mathrm{P}_{\boldsymbol{\theta}}\{\mu_1 - \mu_2 \in (T_{\boldsymbol{N}} - d, T_{\boldsymbol{N}} + d)\} = 1 - \alpha,$$

for the two-stage strategy (13.2.21).

Exercise 13.2.7: For the fixed-width confidence interval problem from Section 13.2.2, consider a purely sequential stopping time based on observations from population i:

$$\begin{gathered} N_i \equiv N_i(d) \text{ is the smallest integer } n(\ge m \ge 2) \text{ for which} \\ \text{we observe } n \ge 2z_{\alpha/2}^2 S_{in}^2/d^2, i = 1, 2. \end{gathered}$$

Finally, propose the confidence interval $J_{\boldsymbol{N}} = (T_{\boldsymbol{N}} - d, T_{\boldsymbol{N}} + d)$ for $\mu_1 - \mu_2$. Recall that $n_i^0 = 2a^2\sigma_i^2/d^2$ from (13.2.19). Show that

(i) $\mathrm{E}_{\boldsymbol{\theta}}[N_i]$ is finite, $i = 1, 2$;
(ii) $\mathrm{E}_{\boldsymbol{\theta}}[N_i] \le n_i^0 + O(1)$, $i = 1, 2$;
(iii) $\mathrm{E}_{\boldsymbol{\theta}}[N_i/n_i^0] \to 1$ as $d \to 0$, $i = 1, 2$;
(iv) $\mathrm{P}_{\boldsymbol{\theta}}\{\mu_1 - \mu_2 \in (T_{\boldsymbol{N}} - d, T_{\boldsymbol{N}} + d)\} \to 1 - \alpha$ as $d \to 0$.

Exercise 13.2.8: For any stopping rule from (13.2.27) - (13.2.29), show that

$$\mathrm{E}_{\boldsymbol{\theta}}[T_{\boldsymbol{N}}] = \mu_1 - \mu_2 \text{ and } \mathrm{V}_{\boldsymbol{\theta}}[T_N] = \mathrm{E}_{\boldsymbol{\theta}}\left[\sigma_1^2 N_1^{-1} + \sigma_2^2 N_2^{-1}\right],$$

where $T_{\boldsymbol{N}} = \overline{X}_{1N_1} - \overline{X}_{2N_2}$ was defined in (13.2.30).

Exercise 13.2.9: Let $X_{i1}, \ldots, X_{in_i}, \ldots$ be independent sequences of i.i.d. observations distributed as $N(\mu_i, \sigma_i^2)$, $i = 1, 2, 3$. We want to estimate a contrast (parametric function), $\delta = 2\mu_1 - \mu_2 - \mu_3$, by means of a fixed-width confidence interval of the form $J_{\boldsymbol{n}} = (T_{\boldsymbol{n}} - d, T_{\boldsymbol{n}} + d)$ having preassigned width $2d(> 0)$ and a preassigned confidence level $1 - \alpha, 0 < \alpha < 1$. Here, $T_{\boldsymbol{n}}$ stands for the fixed-sample-size estimator $2\overline{X}_{1n_1} - \overline{X}_{2n_2} - \overline{X}_{3n_3}$ for δ and $\boldsymbol{n} = (n_1, n_2, n_3)$.

(i) Assume that $\sigma_i^2, 1 \le i \le 3$, are unknown but equal. Now, formulate this problem as in Section 13.2.1 and propose J_N, a two-stage fixed-width confidence interval for δ along the line of (13.2.7). State and prove results analogous to those in Theorem 13.2.1.

(ii) Assume that $\sigma_i^2, 1 \le i \le 3$, are unknown and unequal. Now, formulate this problem as in Section 13.2.2 and propose $J_{\boldsymbol{N}}$, a two-stage fixed-width confidence interval for δ along the line of (13.2.21). State and prove a result analogous to that in Theorem 13.2.2.

Exercise 13.2.10: Tables 13.6.1 and 13.6.2 show random samples from independent $N(\mu_1, \sigma_1^2)$ and $N(\mu_2, \sigma_2^2)$ distributions respectively. Here, μ_1, μ_2, σ_1 and σ_2 are unknown parameters. Let $\delta = \mu_1 - \mu_2$.

Table 13.6.1: Dataset **Norm1.dat**

5.38	0.06	8.88	7.52	12.14	7.12	9.66	12.67
5.37	9.05	11.10	10.16	12.77	5.33	8.51	

Table 13.6.2: Dataset **Norm2.dat**

2.06	5.86	8.19	6.82	1.31	7.26	8.95	1.11
9.85	7.26	4.56	4.51	6.26	6.69	2.80	

(i) Consider a 95% fixed-width confidence interval for δ having width, $2d = 1.0$, using the two-stage strategies described in Sections 13.2.1 and 13.2.2. Determine the sample sizes required by these methodologies for constructing confidence intervals assuming that

(a) $\sigma_1 = \sigma_2$ and
(b) $\sigma_1 \neq \sigma_2$

Use program **Seq13.exe** by pretending that given observations in Tables 13.6.1-13.6.2 would serve as pilot samples from respective populations;

(ii) Using program **Seq13.exe**, show that the observations from Tables 13.6.1-13.6.2 are enough to construct a 90% confidence interval for δ under the sequential methodology in Section 13.2.3 with $d = 2.0$ and $m = 10$.

Exercise 13.2.11: Suppose that $N(\mu_1, \sigma_1^2)$ and $N(\mu_2, \sigma_2^2)$ represent two independent populations and let $\delta = \mu_1 - \mu_2$. Consider a simulation study of a 90% confidence interval for δ having fixed-width $2d = 0.5$, $\mu_1 = 10$, $\sigma_1 = 2$, $\mu_2 = 4$, $\sigma_2 = 2$ and $m = 10$. Obtain a summary output from 5000 simulation runs of the confidence interval methodology from **Seq13.exe** using

(i) two-stage strategy with common unknown variance;
(ii) two-stage strategy with unknown and unequal variances;
(iii) sequential procedure with unknown and unequal variances.

Compare three methodologies based on summary output.

Exercise 13.3.1: Verify (13.3.9). Show that

$$\mathrm{E}_{\boldsymbol{\theta}}\left[T_N\right] = \mu_1 - \mu_2 \text{ and } \mathrm{V}_{\boldsymbol{\theta}}\left[T_N\right] = 2\sigma^2 \mathrm{E}_{\boldsymbol{\theta}}\left[N^{-1}\right],$$

where $T_N = \overline{X}_{1N} - \overline{X}_{2N}$ under sequential strategy (13.3.8)

Exercise 13.3.2: Verify (13.3.10).

Exercise 13.3.3: For the stopping rule (13.3.8), show that $\mathrm{E}_{\boldsymbol{\theta}}[N] \le n^* + O(1)$ where $n^* = \sqrt{A/c}\sigma$. Hence, show that $\mathrm{E}_{\boldsymbol{\theta}}[N/n^*] \to 1$ as $c \to 0$.

Exercise 13.3.4: Show that the RiskEff(c) from (13.3.10) under the stopping rule (13.3.8) always exceeds one. {*Hint:* Note that the expression of the risk efficiency can be rewritten as

$$\text{RiskEff}(c) \equiv 1 + \frac{1}{2}\mathrm{E}_{\boldsymbol{\theta}}\left[\left(\sqrt{N/n^*} - \sqrt{n^*/N}\right)^2\right].$$

Does the required result follow from this?}

Exercise 13.3.5: Let $X_{i1}, \ldots, X_{in_i}, \ldots$ be independent sequences of i.i.d. observations distributed as $N(\mu_i, \sigma_i^2)$, $i = 1, 2, 3$. We want to construct a minimum risk point estimator $T_{\boldsymbol{n}}$ for a contrast (parametric function) $\delta = \mu_1 - \frac{1}{2}(\mu_2 + \mu_3)$ under a loss function that is analogous to (13.3.1). Here, $T_{\boldsymbol{n}}$ stands for the fixed-sample-size estimator $\overline{X}_{1n_1} - \frac{1}{2}(\overline{X}_{2n_2} + \overline{X}_{3n_3})$ of δ and $\boldsymbol{n} = (n_1, n_2, n_3)$.

(i) Assume that $\sigma_i^2, 1 \le i \le 3$, are unknown, but equal. Formulate this problem as in Section 13.3.1 and propose T_N, a sequential minimum risk point estimator for δ along the line of (13.3.8). Obtain the expression of $R_N(c)$, the associated sequential risk. State and prove results analogous to those in (13.3.11). Check whether $R_N(c)$ always exceeds one.

(ii) Assume that $\sigma_i^2, 1 \le i \le 3$, are unknown and unequal. Formulate this problem as in Section 13.3.2 and propose $T_{\boldsymbol{N}}$, the sequential minimum risk point estimator for δ along the line of (13.3.14). Obtain the expression of $R_{\boldsymbol{N}}(c)$, the associated sequential risk. State and prove results analogous to those in (13.3.21).

Exercise 13.3.6: Consider the loss function given in (13.3.1) with $A = 100$ and $c = 0.01$. Examine the number of observations that may be required to obtain a minimum risk point estimator of δ by simulating data from normal populations having

(i) $\mu_1 = 25$, $\mu_2 = 10$ and $\sigma_1 = \sigma_2 = 2$.
(ii) $\mu_1 = 25$, $\sigma_1 = 4$, $\mu_2 = 10$ and $\sigma_2 = 2$.

Use program **Seq13.exe** and obtain summary output from 3000 simulation runs with $m = 10$.

Exercise 13.3.7 (Exercise 13.2.10 Continued): Use the loss function given in (13.3.1) with $A = 10$ and $c = 0.6$ to obtain a minimum risk point estimator of δ. Use given data from Tables 13.6.1 and 13.6.2. Consider a pilot size, $m = 10$, from both distributions and implement your estimation strategy under the assumption

(i) $\sigma_1 = \sigma_2$;
(ii) $\sigma_1 \neq \sigma_2$.

CHAPTER 14

Selecting the Best Normal Population

14.1 Introduction

We touched upon selection and ranking briefly in Section 2.4. In this chapter, we discuss more fully a selection problem to identify the "best" one among $k(\geq 2)$ treatments.

The responses from k treatments are assumed independent and normally distributed having unknown means and a common unknown variance. We discuss some methodologies under this setup for three reasons. First, it is easy to appreciate basic principles in this setup. Second, the fundamental formulation and seminal approaches proposed for this problem have led to far reaching developments. Third, it may be a good thing to walk through some of the preliminaries in a familiar context.

This field is vast. Some of the problems tackled under selection and ranking belong to the area of multiple comparisons too. However, the philosophies and approaches of these two areas differ significantly from one another.

Mid 1940s through early 1950s was the golden age of multiple decision theory! Wald-Wolfowitz-Kiefer-Robbins were in the forefront! Young researchers, including Bahadur, Bechhofer, Dunnett, Gupta and Sobel (BBDGS), came aboard.

They challenged the ritualistic practice of *analysis of variance* (ANOVA) techniques. They vehemently argued: When one compared $k(\geq 2)$ treatments, one traditionally ran a designed experiment first, culminating into an ANOVA "F-test". But, the ANOVA F-test rejected a null hypothesis of equal treatment effects nearly every time, especially where data came from real applications of practical importance. Once a null hypothesis was rejected, one customarily turned around and ordered k treatments!

BBDGS forcefully argued that k treatments were known to be different from one another. They challenged the status quo and asked: When multiple comparisons were at stake, what did ANOVA techniques deliver in practice? They discovered that the ANOVA F-test rarely taught anything substantial. That was a revolutionary thought in the 1950s and it remains so today. One may refer to Bahadur (1950) and Bahadur and Robbins (1950).

BBDGS faced the challenge of ranking k treatments head on, bypassing the round-about and awkward approach via ANOVA. Bechhofer (1954) and Gupta (1956) dared and broke new grounds by developing fundamental approaches for selection and ranking problems. These are referred to as Bechhofer's *indifference-zone* approach and Gupta's *subset selection* approach respectively. The rest is history. For brevity, however, we will focus on the first approach in this chapter.

Suppose we have $k(\geq 2)$ independent, normally distributed populations $\Pi_1, \ldots, \Pi_k$ having unknown means $\mu_1, \ldots, \mu_k$ and a common unknown variance $\sigma^2, -\infty < \mu_1, \ldots, \mu_k < \infty, 0 < \sigma^2 < \infty$. Let $\mu_{[1]} \leq \mu_{[2]} \leq \ldots \leq \mu_{[k]}$ be ordered μ-values. We assume no prior knowledge about the pairing between the ordered mean $\mu_{[i]}$ and the unordered mean $\mu_i, i = 1, \ldots, k$.

The observations from a population Π_i, for example, may be the responses (X_i) obtained from using the ith treatment, $i = 1, \ldots, k$. We pretend that a larger response indicates that a treatment is more effective. Now, how should one define which is the "best" treatment?

Note that

$$\begin{array}{c}\max_{1 \leq i \leq k} \mathrm{P}(X_i > c) = \mathrm{P}(X_j > c) \text{ if and only if} \\ \text{the } j^{\text{th}} \text{ treatment mean } \mu_j \text{ coincides with } \mu_{[k]},\end{array} \quad (14.1.1)$$

for any fixed but arbitrary c. The idea is that the best treatment or brand ought to provide a larger response or yield with more probability in the sense of (14.1.1) than any other treatment.Thus, a treatment or brand associated with $\mu_{[k]}$ is referred to as the "best" among its competitors.

In some contexts, a smaller response may be regarded more effective. In those situations, the treatment associated with $\mu_{[1]}$ may be called the "best". One problem serves as a dual of the other. Hence, we will continue to call a treatment associated with $\mu_{[k]}$ as the *best*.

Goal:	To *identify* the best treatment, that is the one associated with the largest mean, $\mu_{[k]}$.	(14.1.2)

The field of selection and ranking may be adequately reviewed from Bechhofer et al. (1968), Bechhofer et al. (1995), Gibbons et al. (1977), Gupta and Panchapakesan (1979), Mukhopadhyay (1993), Mukhopadhyay and Solanky (1994), and other sources. In the area of multiple comparisons, one may refer to following books: Hochberg and Tamhane (1987), Bechhofer et al. (1995) and Hsu (1996). See also Liu (1995).

In Section 14.2, we emphasize Bechhofer's (1954) *indifference-zone* formulation and its underlying approach. In the absence of any fixed-sample-size methodology, Bechhofer et al. (1954) developed their path-breaking selection methodology in the light of Stein's (1945,1949) two-stage strategy from Section 6.2.1. We discuss this selection methodology in Section 14.3. Section 14.4 is devoted to a sequential selection methodology, another pure classic, that was due to Robbins et al. (1968). We provide essential computing tools and data analyses in Section 14.5.

During 1980s and 1990s, many attractive multistage methods evolved.

In Section 14.6, we mention selected references for some closely related multistage selection methodologies. For a comprehensive review, one may refer to Mukhopadhyay and Solanky (1994, Chapter 3) and other original sources. Selected derivations are included in Section 14.7.

14.2 Indifference-Zone Formulation

Having recorded n independent observations $X_{i1}, \ldots, X_{in}$ distributed as $N(\mu_i, \sigma^2)$ from Π_i associated with the ith treatment, let us denote:

$$\begin{aligned}
&\overline{X}_{in} = n^{-1} \textstyle\sum_{j=1}^{n} X_{ij}, \\
&S_{in}^2 = (n-1)^{-1} \textstyle\sum_{j=1}^{n} (X_{ij} - \overline{X}_{in})^2, \ i = 1, \ldots, n, \qquad (14.2.1) \\
&\text{Pooled Sample Variance: } U_n^2 = k^{-1} \left(S_{1n}^2 + \ldots + S_{kn}^2 \right).
\end{aligned}$$

Now, we move forward and adopt the following natural selection rule:

$$\mathcal{R}_n: \begin{array}{l} \text{Select as best the treatment associated with } \Pi_j \\ \text{if } \overline{X}_{jn} \text{ is the largest among all sample means} \\ \overline{X}_{1n}, \ldots, \overline{X}_{kn}. \end{array} \qquad (14.2.2)$$

The phrase, *correct selection* (CS), is regarded equivalent to saying that Π_j has mean $\mu_{[k]}$ under $\mathcal{R}_n$ from (14.2.2). Denote the full parameter space:

$$\Omega = \{\boldsymbol{\mu} = (\mu_1, \ldots, \mu_k) : -\infty < \mu_i < \infty, i = 1, \ldots, k\}.$$

We evaluated $\mathrm{P}_n(CS \mid \mathcal{R}_n)$ over Ω in Chapter 2. From (2.4.4) one will recall that $\mathrm{P}_n(CS \mid \mathcal{R}_n)$ has its minimum value $\frac{1}{k}$. Also, this minimum probability of correct selection is attained when $\mu_1 = \ldots = \mu_k$, that is when all treatment effects are same.

However, we must do better than simply choosing one of the k treatments at random and call it the best. It is reasonable to require:

$$\mathrm{P}_n(CS \mid \mathcal{R}_n) \ge P^*, \text{ a preassigned number, } k^{-1} < P^* < 1.$$

In practical circumstances, an experimenter will specify P^* value close to 1. For example, one would perhaps fix $P^* = 0.90, 0.95$ or 0.99. On the other hand, we cannot hope to equate minimum probability of correct selection over Ω to P^* because we know:

$$\inf_{\boldsymbol{\mu} \in \Omega} \mathrm{P}_n(CS \mid \mathcal{R}_n) = \frac{1}{k}.$$

Bechhofer (1954) ingeniously came up with the idea of an *indifference-zone.* Given $\delta^* > 0$, we define a parameter subspace of Ω as a *preference-zone* defined as follows:

Preference-Zone:

$$\Omega(\delta^*) = \left\{ \boldsymbol{\mu} = (\mu_1, \ldots, \mu_k) \in \Omega : \mu_{[k]} - \mu_{[k-1]} \ge \delta^* \right\}. \qquad (14.2.3)$$

An *indifference-zone* is $\Omega(\delta^*)^c$, that is, the complement of preference-zone $\Omega(\delta^*)$. The preference-zone is a subspace of the full parameter space where the best treatment effect is at least δ^* units ahead of the second best treatment effect. Recall that $\delta^*(>0)$ is preassigned.

Now, a fundamental idea is this: Given P^* with $k^{-1} < P^* < 1$, one wants to claim that $\text{P}_n(CS \mid \mathcal{R}_n)$ is at least P^* whenever $\boldsymbol{\mu}$ is in preference-zone. That is, one will require $\text{P}_n(CS \mid \mathcal{R}_n)$ to be at least P^* whenever best treatment is *sufficiently* better ($\geq \delta^*$) than next best. If the best treatment is not *sufficiently* better than next best, one is not interested in identifying the best treatment. The parameter subspace $\Omega(\delta^*)^c$ is called an *indifference-zone*. This is known as the "P^*-requirement".

At this point, we restate our basic goal from (14.1.2) as:

Goal:	To *identify* the best treatment, the one associated with $\mu_{[k]}$ under the selection rule $\mathcal{R}_n$ so that $\text{P}_n(CS \mid \mathcal{R}_n) \geq P^*$ whenever $\boldsymbol{\mu} \in \Omega(\delta^*)$.	(14.2.4)

This ground breaking formulation is called Bechhofer's (1954) *indifference-zone* approach.

> One may not worry much about selecting the best treatment if it is not sizably better than the next best. Such a simple idea led Bechhofer (1954) to formulate his profoundly influential *indifference-zone* approach.

Let us pretend that σ^2 is known and examine the minimum value of $\text{P}_n(CS \mid \mathcal{R}_n)$ inside $\Omega(\delta^*)$. Recall that $\Phi(.)$ and $\phi(.)$ are respectively the distribution function and p.d.f. of a standard normal random variable.

Denote $\delta_i = \mu_{[k]} - \mu_{[i]}, i = 1, 2, \ldots, k-1$. Note that $\delta_i \geq \delta^*, i = 1, 2, \ldots, k-1$, whenever $\boldsymbol{\mu} \in \Omega(\delta^*)$. From (2.4.3) we recall:

$$\text{P}_n(CS \mid \mathcal{R}_n) = \int_{-\infty}^{\infty} \prod_{i=1}^{k-1} \Phi\left(y + \frac{\sqrt{n}}{\sigma}\delta_i\right) \phi(y)dy, \tag{14.2.5}$$

for all fixed n.

Since Φ is a distribution function, it is non-decreasing. That is, whenever $\boldsymbol{\mu} \in \Omega(\delta^*)$:

$$\begin{gathered} \Phi\left(y + \frac{\sqrt{n}}{\sigma}\delta_i\right) \geq \Phi\left(y + \frac{\sqrt{n}}{\sigma}\delta^*\right) \text{ for all } y, \text{ since } \delta_i \geq \delta^*, \\ \text{for each } i = 1, 2, \ldots, k-1, \text{ which implies} \\ \prod_{i=1}^{k-1} \Phi\left(y + \frac{\sqrt{n}}{\sigma}\delta_i\right) \geq \Phi^{k-1}\left(y + \frac{\sqrt{n}}{\sigma}\delta^*\right) \text{ for all } y. \end{gathered} \tag{14.2.6}$$

Now, combine (14.2.5)-(14.2.6) to conclude:

$$\begin{gathered} \inf_{\boldsymbol{\mu} \in \Omega(\delta^*)} \text{P}_n(CS \mid \mathcal{R}_n) = \int_{-\infty}^{\infty} \Phi^{k-1}\left(y + \frac{\sqrt{n}}{\sigma}\delta^*\right) \phi(y)dy, \\ \text{and the equality holds if and only if} \\ \mu_{[1]} = \ldots = \mu_{[k-1]} = \mu_{[k]} - \delta^*. \end{gathered} \tag{14.2.7}$$

From (14.2.7), note that minimum probability to correctly select the best treatment occurs inside $\Omega(\delta^*)$ when $k-1$ treatment effects pile up at δ^* units below the best. Such a parameter configuration is called the *least favorable configuration* (LFC):

$$\text{LFC: } \mu_{[1]} = \ldots = \mu_{[k-1]} = \mu_{[k]} - \delta^*. \tag{14.2.8}$$

Under Bechhofer's (1954) approach, we will equate the minimum probability of correct selection with P^* and determine the required sample size n accordingly. We define $h \equiv h_{k,P^*} (> 0)$ such that

$$\int_{-\infty}^{\infty} \Phi^{k-1}\left(y + h_{k,P^*}\right) \phi(y)dy = P^*, \tag{14.2.9}$$

and solve for n so that $\frac{\sqrt{n}}{\sigma}\delta^* \geq h_{k,P^*}$. That is, we must determine:

$$n \text{ to be the smallest integer } \geq \frac{h^2_{k,P^*}\sigma^2}{\delta^{*2}} = C, \text{ say.} \tag{14.2.10}$$

This expression, C, is known as the required optimal fixed-sample-size from each treatment under selection rule $\mathcal{R}_n$, had σ^2 been known. If C is not an integer, the number of observations from each treatment ought to be the smallest integer $> C$. Obviously, the magnitude of C remains unknown since σ^2 is unknown!

The number h_{k,P^*} may be alternatively viewed as an upper percentage point of a certain $(k-1)$-dimensional normal distribution having an equi-correlation $\frac{1}{2}$. Some details are shown in Section 14.7.1.

For a given set of values of k and P^*, the solution h_{k,P^*} satisfying (14.2.9) can be found from tables of Tong (1990, p. 237) or from Table A.1 in Gibbons et al. (1977, p. 400). Gupta and Panchapakesan (1979, Chapter 23, Section 2) also includes information in this regard.

In the absence of any fixed-sample-size methodology (Dudewicz, 1971), one must implement a multistage strategy. We will introduce the pioneering two-stage and sequential selection methodologies due to Bechhofer et al. (1954) and Robbins et al. (1968) respectively. There are other multistage selection methodologies that are operationally more convenient than sequential strategy of Robbins et al. (1968) and these preserve comparable *second-order* properties. Mukhopadhyay and Solanky (1994, Chapter 3) discussed these methodologies and *second-order* properties fully.

> No fixed-sample-size selection rule will satisfy P^* criterion inside the preference zone, $\Omega(\delta^*)$.

A multistage selection procedure would have a stopping variable N that would be the number of observations recorded from each treatment. The following theorem (Mukhopadhyay and Solanky, 1994, Theorem 3.2.1, p. 74-75) provides a simple tool for locating the LFC associated with such procedures under mild assumptions. For completeness, a brief sketch of its proof is given in Section 14.7.2.

Theorem 14.2.1: *Suppose that a stopping variable N satisfies the following conditions:*

(i) *The distribution of N does not involve $\boldsymbol{\mu}$;*
(ii) *$I(N = n)$ is independent of $\overline{X}_{1n}, \ldots, \overline{X}_{kn}$ for all fixed $n \geq 2$; and*
(iii) $\mathrm{P}_{\boldsymbol{\mu},\sigma}(N < \infty) = 1$.

Then, under the selection methodology $\mathcal{R}_N$ from (14.2.2), we have

$$\inf_{\boldsymbol{\mu} \in \Omega(\delta^*)} \mathrm{P}_{\boldsymbol{\mu},\sigma}(CS) = \mathrm{E}_\sigma \left[\int_{-\infty}^{\infty} \Phi^{k-1} \left(y + \frac{\sqrt{N}}{\sigma} \delta^* \right) \phi(y) dy \right]$$

where a minimum is attained if and only if $\mu_{[1]} = \ldots = \mu_{[k-1]} = \mu_{[k]} - \delta^*$.

14.3 Two-Stage Procedure

Bechhofer et al. (1954) developed a pioneering selection methodology by advancing Stein's (1945,1949) two-stage sampling strategy (Section 6.2.1). Now, we explain that path-breaking methodology and its implementation.

Bechhofer, Dunnett and Sobel's (1954) two-stage selection methodology (14.3.2)-(14.3.3) is *consistent.*

We begin with pilot observations $X_{ij}, j = 1, \ldots, m\ (\geq 2)$, from i^{th} population Π_i which corresponds to the ith treatment, $i = 1, \ldots, k$. Recall the set of notation from (14.2.1). The initial observations would provide sample mean, $\overline{X}_{im}$, and sample variance, S^2_{im}. In turn, we will have a pooled sample variance, U^2_m, with degrees of freedom $\nu = k(m-1)$.

Now, consider $(W_1, \ldots, W_{k-1})$, a vector of random variables distributed as a $(k-1)$-dimensional t with equi-correlation $\frac{1}{2}$ and degrees of freedom $\nu = k(m-1)$. We define $\tau \equiv \tau_{k,\nu,P^*} (> 0)$ satisfying the equation:

$$\mathrm{P}\left\{ W_i \leq \tau/\sqrt{2}, i = 1, \ldots, k-1 \right\} = P^*. \tag{14.3.1}$$

This is equation also (3.3.4) in Mukhopadhyay and Solanky (1994, p. 77).

For a given set of values of k, ν and P^*, the solution τ_{k,ν,P^*} satisfying (14.3.1) can be found from Gupta and Sobel (1957) and Table A.4 in Gibbons et al. (1977, p. 405). A detailed account of tables available from different sources can be found in Gupta and Panchapakesan (1979, Chapter 23, Section 3) and Gupta et al. (1985).

Next, Bechhofer et al. (1954) would have us determine the required final sample size from each treatment as:

$$N \equiv N(\delta^*) = \max\left\{ m, \langle \tau^2_{k,\nu,P^*} U^2_m / \delta^{*2} \rangle + 1 \right\}. \tag{14.3.2}$$

where $< u >$ continues to stand for the largest integer $< u$.

If $N = m$, one requires no more observations at the second stage from any treatment. However, if $N > m$, one obtains $(N - m)$ additional observations from each treatment at the second stage. Clearly, sampling will terminate w.p.1.

Finally, based on observations $N, X_{i1}, \ldots, X_{iN}$ from both stages, we would obtain i^{th} population's sample mean, $\overline{X}_{iN}, i = 1, \ldots, k$. Now, the selection rule goes like this:

$$\mathcal{R}_N: \begin{array}{l}\text{Select as best the treatment associated with } \Pi_j \\ \text{if } \overline{X}_{jN} \text{ is the largest among all sample means} \\ \overline{X}_{1N}, \ldots, \overline{X}_{kN}.\end{array} \quad (14.3.3)$$

Bechhofer et al. (1954) proved the following extraordinary result under this selection methodology (14.3.2)-(14.3.3). We sketch its proof in Section 14.7.3. Note that the part regarding the form of the LFC follows immediately from Theorem 14.2.1.

Theorem 14.3.1 *: For the stopping variable N defined by (14.3.2), for all fixed σ, δ^*, m and P^*, under $\mathcal{R}_N$ from (14.3.3), one has:*

$$\mathrm{P}_{\boldsymbol{\mu},\sigma}(CS) \geq P^* \text{ for all } \boldsymbol{\mu} \text{ inside } \Omega(\delta^*), \text{ the preference-zone}$$

where τ_{k,ν,P^*} came from (14.3.1) with $\nu = k(m-1)$. The LFC remains the same as in (14.2.8), that is $\mu_{[1]} = \ldots = \mu_{[k-1]} = \mu_{[k]} - \delta^*$.

Note that any influence of unknown σ^2 on the inference procedure has been wiped out! This is the most crucial point that is to be noted. The other crucial point is that the probability of CS is at least P^*, the preassigned target, but we have not invoked any approximation or limiting argument. Observe that the probability distribution of N defined by (14.3.2) does not involve $\boldsymbol{\mu}$.

Under two-stage selection strategy, LFC remains the same as in (14.2.8): $\mu_{[1]} = \ldots = \mu_{[k-1]} = \mu_{[k]} - \delta^*$.

Next, we summarize another set of conclusions. We have left its proof out as an exercise.

Theorem 14.3.2: *For the stopping variable N defined by (14.3.2), for all fixed σ, m and P^*, under $\mathcal{R}_N$ from (14.3.3), one has:*

(i) $\tau^2_{k,\nu,P^*}\sigma^2/\delta^{*2} \leq \mathrm{E}_{\boldsymbol{\mu},\sigma}[N] \leq m + \tau^2_{k,\nu,P^*}\sigma^2/\delta^{*2}$ *for all $\boldsymbol{\mu}$, with δ^* fixed;*
(ii) $\mathrm{E}_{\boldsymbol{\mu},\sigma}[N/C] \to \tau^2_{k,\nu,P^*}/h^2_{k,P^*} (> 1)$ *for all $\boldsymbol{\mu}$, as $\delta^* \to 0$;*
(iii) $\liminf \mathrm{P}_{\boldsymbol{\mu},\sigma}(CS) \geq P^*$ *for all $\boldsymbol{\mu}$ inside $\Omega(\delta^*)$, as $\delta^* \to 0$;*

where τ_{k,ν,P^*} came from (14.3.1) with $\nu = k(m-1)$, h_{k,P^*} came from (14.2.9), and $C = h^2_{k,P^*}\sigma^2/\delta^{*2}$ from (14.2.10).

The parts (i) and (iii) are self explanatory. Part (ii) of this theorem says that compared with the optimal fixed-sample-size C, the two-stage selection methodology $\mathcal{R}_N$ will oversample on an average from all k treatments, even *asymptotically*. We have noticed such an undesirable feature associated with practically every two-stage procedure as long as pilot size m was held fixed. This was certainly the situation with Stein's (1945,1949) two-stage fixed-width confidence interval strategy in Section 6.2.1.

Bechhofer, Dunnett and Sobel's (1954) two-stage selection rule is asymptotically first-order inefficient.

In a number of chapters, we have discussed other sampling techniques to circumvent this unpleasant phenomenon. Robbins et al. (1968) developed a fundamental sequential selection strategy that would make $\mathrm{E}_{\boldsymbol{\mu},\sigma}[N] \approx C$ when δ^* is "small".

14.4 Sequential Procedure

Observe that $\mathrm{E}_{\boldsymbol{\mu},\sigma}[N]$ and C would both be large when δ^* is small. Robbins et al. (1968) developed a pioneering sequential selection methodology in order to have $\mathrm{E}_{\boldsymbol{\mu},\sigma}[N] \approx C$ when δ^* is small. Now, we introduce the Robbins et al. sequential selection strategy. From (14.2.1), we recall the notation for a sample mean, $\overline{X}_{in}$, a sample variance, S_{in}^2, and a pooled sample variance, U_n^2 with degrees of freedom $\nu = k(n-1)$ for $n \geq 2$.

From population Π_i, we begin with pilot observations $X_{ij}, j = 1, \ldots, m$ $(\geq 2), i = 1, \ldots, k$. Subsequently, we record an additional observation vector at-a-time consisting of a response from each treatment. We proceed according to the following purely sequential stopping rule:

$$
\begin{aligned}
&N \equiv N(\delta^*) \text{ is the smallest integer } n(\geq m) \text{ for which} \\
&\qquad\qquad \text{we observe } n \geq \frac{h_{k,P^*}^2 U_n^2}{\delta^{*2}},
\end{aligned}
\tag{14.4.1}
$$

where h_{k,P^*} comes from (14.2.9).

Observe that $N(\delta^*)$, the final sample size from each treatment, estimates $C \equiv h_{k,P^*}^2 \sigma^2/\delta^{*2}$ that comes from (14.2.10). This purely sequential procedure is implemented in an usual fashion. One may verify that $\mathrm{P}_{\boldsymbol{\mu},\sigma}(N < \infty) = 1$, that is, sampling would terminate w.p. 1.

Upon termination, based on all observations $N, X_{i1}, \ldots, X_{iN}$, one would obtain i^{th} population's sample mean, $\overline{X}_{iN}, i = 1, \ldots, k$. Now, the selection rule goes like this:

$$
\begin{aligned}
\mathcal{R}_N: \;& \text{Select as best the treatment associated with } \Pi_j \\
& \text{if } \overline{X}_{jN} \text{ is the largest among all sample means} \\
& \overline{X}_{1N}, \ldots, \overline{X}_{kN}.
\end{aligned}
\tag{14.4.2}
$$

Clearly, this selection rule is exactly the same as in (14.3.3).

Robbins, Sobel and Starr's (1968) sequential selection rule is *asymptotically consistent* and *asymptotically first- order efficient.* It is *not* consistent, that is, it does not satisfy P^* requirement exactly any more inside $\Omega(\delta^*)$.

Under this selection strategy, Robbins et al. (1968) proved the following first-order results. We sketch a proof in Section 14.7.4. Note that the form of the associated LFC follows immediately from Theorem 14.2.1.

Theorem 14.4.1 : *For the stopping variable N defined by (14.4.1), for all fixed σ, m and P^*, under $\mathcal{R}_N$ from (14.4.2), one has:*

(i) $\mathrm{E}_{\boldsymbol{\mu},\sigma}[N] \le C + m + 2$ *for all $\boldsymbol{\mu}$, with δ^* fixed;*
(ii) $\mathrm{E}_{\boldsymbol{\mu},\sigma}[N/C] \to 1$ *for all $\boldsymbol{\mu}$, as $\delta^* \to 0$ [Asymptotic First-Order Efficiency];*
(iii) $\liminf \mathrm{P}_{\boldsymbol{\mu},\sigma}(CS) \ge P^*$ *for all $\boldsymbol{\mu}$ inside $\Omega(\delta^*)$, as $\delta^* \to 0$ [Asymptotic Consistency];*

where h_{k,P^*} came from (14.2.9) and $C = h^2_{k,P^*}\sigma^2/\delta^{*2}$ came from (14.2.10). Part (iii) would have equality under the LFC, that is when $\mu_{[1]} = \ldots = \mu_{[k-1]} = \mu_{[k]} - \delta^*$.

The part (i) is self explanatory. Part (ii) shows that the sequential strategy is asymptotically *first-order efficient.* Comparing this result with Theorem 14.3.2, part (ii), we note that oversampling associated with two-stage selection methodology (14.3.2) is all gone now. However, the consistency or exact consistency property (Theorem 14.3.1) of two-stage strategy is now conspicuously absent. Under sequential sampling, we can claim that $\mathrm{P}_{\boldsymbol{\mu},\sigma}(CS) \approx P^*$ inside $\Omega(\delta^*)$ when δ^* is small. Theorem 14.4.1, part (iii) shows that sequential selection strategy is *asymptotically consistent.*

14.4.1 Second-Order Properties

The second-order results summarized here were originally developed by Mukhopadhyay (1983) when $k = 2$. That special case was not too difficult. The second-order results in the case of arbitrary $k(\ge 2)$ were developed by Mukhopadhyay and Judge (1989). Let us denote:

$$\eta(k) = \tfrac{1}{2} - \tfrac{1}{k} - \tfrac{1}{k}\textstyle\sum_{n=1}^{\infty} n^{-1}\mathrm{E}\left[\max\left(0, \chi^2_{nk} - 2nk\right)\right]. \qquad (14.4.3)$$

Observe that $\eta(k)$ coincides with $\eta(p)$ from (6.2.33) when $p = k$.

We define two functions:

$$\begin{aligned} g(y) &= \textstyle\int_{-\infty}^{\infty} \{\Phi(x+y)\}^{k-1}\phi(x)dx, \\ f(y) &= g(y^{1/2}), y > 0. \end{aligned} \qquad (14.4.4)$$

Now, we are in a position to summarize the *second-order* properties for the Robbins et al. sequential selection methodology. for all $\boldsymbol{\mu}$ inside $\Omega(\delta^*)$

Theorem 14.4.2 : *For the stopping variable N defined by (14.4.1), for all fixed σ, m and P^*, under $\mathcal{R}_N$ from (14.4.2), one has as $\delta^* \to 0$:*

(i) $\mathrm{E}_{\boldsymbol{\mu},\sigma}[N] = C + \eta(k) + o(1)$ *for all fixed* $\boldsymbol{\mu}$ *if* (a) $m \geq 3$ *when* $k = 2$, *and* (b) $m \geq 2$ *when* $k \geq 3$;

(ii) $\mathrm{P}_{\boldsymbol{\mu},\sigma}(CS) \geq P^* + h^2_{k,P^*}C^{-1}\left\{\eta(k)f^{'}(h^2_{k,P^*}) + h^2_{k,P^*}k^{-1}f''(h^2_{k,P^*})\right\} + o(\delta^{*2})$ *for all* $\boldsymbol{\mu}$ *inside* $\Omega(\delta^*)$, *if* (a) $m \geq 4$ *when* $k = 2$, (b) $m \geq 3$ *when* $k = 3, 4, 5$, *and* (c) $m \geq 2$ *when* $k \geq 6$;

(iii) *Part* (ii) *would have equality under LFC, that is when* $\mu_{[1]} = \ldots = \mu_{[k-1]} = \mu_{[k]} - \delta^*$;

where h_{k,P^*} *came from (14.2.9),* $C = h^2_{k,P^*}\sigma^2/\delta^{*2}$ *came from (14.2.10),* $\eta(k)$ *came from (14.4.3), and* $f(y)$ *came from (14.4.4) with*

$$f'(h^2_{k,P^*}) = \frac{d}{dy}f(y)\Big|_{y=h^2_{k,P^*}} \quad \textit{and} \quad f''(h^2_{k,P^*}) = \frac{d^2}{dy^2}f(y)\Big|_{y=h^2_{k,P^*}}.$$

These second-order results are self explanatory and they were fully developed by Mukhopadhyay and Judge (1989). One will find that many details were systematically addressed in Mukhopadhyay and Solanky (1994, chapter 3, pp. 86-91). We give some preliminaries in Section 14.7.5.

Robbins, Sobel and Starr's (1968) sequential selection strategy is *asymptotically second-order efficient.*

14.5 Data Analyses and Conclusions

The two selection methodologies described in this chapter can be simulated and analyzed by using the computer program **Seq14.exe**. Upon execution of this program, a user will be prompted to select either 'program number: 1' or 'program number: 2'. 'Program number: 1' is dedicated to the two-stage selection strategy (Section 14.3) and 'program number: 2' is dedicated to the sequential selection strategy (Section 14.4). **Seq14.exe** program assumes that the maximum number of normal populations is 20 (that is, $2 \leq k \leq 20$) and $0.75 \leq P^* \leq 0.99$. The program can be used to perform simulation studies. It can also work with live data input.

Figure 14.5.1 shows the input screen for 'program number: 1' within **Seq14.exe**. This figure shows the required input to implement two-stage strategy (14.3.2)-(14.3.3) for simulation study involving $5(= k)$ normal populations. We fixed $\delta^* = 0.5, m = 7$ and $P^* = 0.95$. We opted for 5000 simulation runs using random sequence 4. Then, we were asked to input five population mean values and a common standard deviation. We had put in

$$\mu_1 = 10, \mu_2 = \mu_3 = \mu_4 = \mu_5 = 9.5 \text{ and } \sigma = 2.$$

We want to make it clear that we generated five populations assuming that they satisfied the LFC condition knowing fully well that population #1 was the "best" in this group.

```
          Select a program using the number given
          in the left hand side. Details of these
          programs are given in Chapter 14.
           **********************************
           * Number        Program         *
           *  1   Two-Stage Procedure       *
           *  2   Sequential procedure      *
           *  0   Exit                      *
           **********************************
           Enter the program number: 1

     Enter the following preassigned values for
     selecting the normal population:
     [Note: 0.75 =< P* =< 0.99]

     k (Number of normal populations, Max=20) = 5
     P* (Minimum probability of CS)         = 0.95
     delta* (Indifference-zone parameter) = 0.5
     m (Initial sample size)              = 7

     Do you want to change the input parameters (Y or N)? N
     Do you want to save the results in a file(Y or N)? Y

     Default outfile name is "SEQ14.OUT".
     Do you want to change the outfile name (Y or N)? N

     Do you want to simulate data or input real data?
     Type S (Simulation) or D (Real Data): S
     Do you want to store simulated sample sizes (Y or N)? N

     Number of simulation replications? 5000
     Enter a positive integer(<10) to initialize
               the random number sequence: 4

     For simulation of data, enter the mean of
          population number  1: 10
          population number  2: 9.5
          population number  3: 9.5
          population number  4: 9.5
          population number  5: 9.5

          Enter the common std. dev. (sigma): 2
```

Figure 14.5.1. *Input screen for two-stage selection strategy using* ***Seq14.exe***.

We pretended that we did not know the means or the common standard deviation. We let simulations take over by generating data from five specified normal populations. Figure 14.5.2 summarizes the findings from our investigation by averaging performances from all 5000 simulation runs of two-stage selection methodology. It includes the optimal fixed-sample-size $C(= 149.35)$, the estimated average sample size $\overline{n}(= 163.09)$, estimated standard deviation $s_n(= 42.17)$ of n, and the estimated probability $\overline{p}(= 0.9508)$ of correctly selecting the best population (#1).

```
       Selecting the Best Normal Population
 Two-Stage Procedure for Common Unknown Variance
 ===============================================
     Simulation study

     Input parameters:    P*      =  0.950
                       delta*     =  0.500
     Number of populations (k)    =   5
     Initial sample size, m       =   7
     Number of simulations        =  5000

     Simulation parameters:
              common std. dev.    =   2.00

     h(k,P*)       = 3.0552
     tau(k,nu,P*)  =   3.19

     Population mean values:
         10.00      9.50      9.50      9.50
          9.50

     Optimal sample size (C)      =    149.35
     Average sample size (n_bar)  =    163.09
     Std. dev. of n (s_n)         =     42.17
     Minimum sample size (n_min)  =     45
     Maximum sample size (n_max)  =    362
     Estimated probability of CS  =    0.9508
```

Figure 14.5.2. *Output screen for two-stage selection strategy using **Seq14.exe**.*

```
       Selecting the Best Normal Population
 Sequential Procedure for Common Unknown Variance
 ================================================
     Simulation study

     Input parameters:    P*     =  0.950
                       delta*    =  0.500
     Number of populations (k)   =   5
     Initial sample size, m      =   7
     Number of simulations       =  5000

     Simulation parameters:
              common std. dev.   =   2.00

     h(k,P*)       = 3.0552

     Population mean values:
         10.00      9.50      9.50      9.50
          9.50

     Optimal sample size (C)      =    149.35
     Average sample size (n_bar)  =    149.71
     Std. dev. of n (s_n)         =      7.86
     Minimum sample size (n_min)  =    120
     Maximum sample size (n_max)  =    178
     Estimated probability of CS  =    0.9486
```

Figure 14.5.3. *Output screen for sequential selection strategy using **Seq14.exe**.*

The input screen for 'program number: 2' within **Seq14.exe** is very similar to what one finds in Figure 14.5.1. So, we refrain from highlighting it. However, we did opt for 5000 simulation runs using random sequence 4 and implemented sequential selection strategy under identical conditions. Figure 14.5.3 summarizes the findings from our investigation by averaging performances from 5000 simulation runs of sequential selection methodology. It includes the optimal fixed-sample-size $C(= 149.35)$, the estimated average sample size $\overline{n}(= 149.71)$, estimated standard deviation $s_n(= 7.86)$ of n, and the estimated probability $\overline{p}(= 0.9486)$ of correctly selecting the best population (#1).

Table 14.5.1 summarizes findings for two-stage and sequential selection strategies under same parameter configurations for the means of independent normal distributions. We executed the program **Seq14.exe** to generate data using the same "sequence number" under two strategies for each fixed configuration of $\boldsymbol{\mu}$. We fixed $k = 5, \delta^* = 0.5, m = 7, \sigma = 2$ and $P^* = 0.95$. We used 5000 simulation runs to obtain results included in each row of Table 14.5.1.

All simulations that have led to Table 14.5.1 were carried out with five normal populations having means $\mu_1, \mu_2, \mu_3, \mu_4, \mu_5$ belonging to the preference zone, $\Omega(\delta^*)$.

Block 1 shows findings under LFC where the largest mean $\mu_1(= 10)$ stands alone, but $\mu_2, \mu_3, \mu_4, \mu_5$ are all stacked up at $9.5(= \mu_1 - \delta^*)$. One checks easily:

$$C = (3.0552)^2 \times 2^2/(0.5)^2 = 149.35.$$

Two-stage strategy obviously oversamples compared with C. From the column under $\overline{n}/C$, we find that we encountered 9.2% oversampling. Now, we also know that $\tau^2_{5,30,0.95}/h^2_{5,0.95} = 1.0847$, so that we may *asymptotically* expect (Theorem 14.3.2, part (ii)) around 8.47% oversampling. The estimated probability ($\overline{p} = 0.9508$) of CS is larger than $P^*(= 0.95)$ which is a validation of consistency property (Theorem 14.3.1). Overall agreements of empirical oversampling percentage and $\overline{p}$ with their theoretical expectations are remarkable even though $C(= 149.35)$ is not unusually large. We observe that the variation in n under two-stage strategy ($s_n = 42.7$) is much larger than that ($s_n = 7.86$) under sequential strategy.

On the other hand, under LFC, a sequential strategy estimates C very tightly. Theorem 14.4.2, part (ii) generates hope that $\overline{n} - C(= 0.36)$ may hover around $\eta(5)(= 0.24954)$, and it does. If in doubt, consider the interval:

$$\begin{aligned}(\overline{n} - C) \pm 1.96 \times s_n/\sqrt{5000} &= 0.36 \pm 1.96 \times 7.86/\sqrt{5000} \\ &= 0.36 \pm 1.96 \times 7.86/\sqrt{5000} = 0.36 \pm 0.21787,\end{aligned}$$

which includes the number $\eta(5)$.

Table 14.5.1: Comparing two-stage (2stage) and sequential strategies (seq) under normality with 5000 simulation runs each for a fixed set of population means inside $\Omega(\delta^*)$, $k = 5$, $\delta^* = 0.5, P^* = 0.95, m = 7, \sigma = 2, \nu = 30, \eta(5) = 0.24954$ $h_{5,0.95} = 3.0552, \tau_{5,30,0.95} = 3.19, \tau^2_{5,30,0.95}/h^2_{5,0.95} = 1.09$

Block	C	$\overline{n}$	s_n	$\overline{n}/C$	$\overline{n} - C$	$\overline{p}$
1	$\mu_1 = 10, \mu_2 = \mu_3 = \mu_4 = \mu_5 = 9.5$ Random sequence: 4					
2stage	149.35	163.09	42.17	1.0920		0.9508
seq	149.35	149.71	7.86	1.0024	0.36	0.9486
2	$\mu_1 = 10, \mu_2 = \mu_3 = \mu_4 = \mu_5 = 9.0$ Random sequence: 2					
2stage	149.35	162.38	41.85	1.0872		0.9998
seq	149.35	149.53	7.84	1.0012	0.18	1.0000
3	$\mu_1 = 10, \mu_2 = 9.5, \mu_3 = \mu_4 = \mu_5 = 9.0$ Random sequence: 1					
2stage	149.35	162.49	42.00	1.0880		0.9834
seq	149.35	149.69	7.75	1.0023	0.34	0.9818
4	$\mu_1 = 10, \mu_2 = \mu_3 = 9.5, \mu_4 = \mu_5 = 9.0$ Random sequence: 6					
2stage	149.35	163.23	42.49	1.0929		0.9682
seq	149.35	149.72	7.94	1.0025	0.37	0.9692
5	$\mu_1 = 10, \mu_2 = 9.5, \mu_3 = 9.4, \mu_4 = \mu_5 = 9.0$ Random sequence: 7					
2stage	149.35	163.42	41.41	1.0942		0.9796
seq	149.35	149.55	7.71	1.0013	0.20	0.9952
6	$\mu_1 = 10, \mu_2 = 9.5, \mu_3 = 9.4, \mu_4 = \mu_5 = 9.0$ Random sequence: 8					
2stage	149.35	164.06	41.83	1.0985		1.0000
seq	149.35	149.53	7.86	1.0012	0.18	1.0000

Also, under LFC, the sequential strategy came up with $\overline{p} = 0.9486$, a little under $P^*(= 0.95)$. Is this $\overline{p}$ close enough to P^*? Consider the interval:

$$\overline{p} \pm 1.96 \times \sqrt{\overline{p}(1 - \overline{p})/5000}$$

$$= 0.9486 \pm 1.96 \times \sqrt{0.9486(1 - 0.9486)/5000} = 0.9486 \pm 0.0061206,$$

which includes number P^*.

In the same vein, Blocks 2 - 6 summarize performances of other simulation studies executed by **Seq14.exe**. These were carried out under configurations for $\boldsymbol{\mu}$ within the preference zone $\Omega(\delta^*)$ but different from LFC. Block 2 refers to a scenario where the largest mean $\mu_1 = 10$ is fixed and $\mu_2, \mu_3, \mu_4, \mu_5$ stack up at $9.0(< \mu_1 - \delta^*)$. That is, the separation between μ_1 and $\mu_2, \mu_3, \mu_4, \mu_5$ goes up, and as a result we note that $\overline{p}$ has gone up sizably. Block 3 refers to a scenario where the largest mean $\mu_1 = 10$ is again fixed, the next one is $\mu_2 = 9.5$, but μ_3, μ_4, μ_5 stack up at $9.0(< \mu_1 - \delta^*)$. This configuration clearly falls between LFC and one considered in Block 2. One accordingly finds that $\overline{p}$ in Block 3 is bounded from below (above) by $\overline{p}$ in Block 1 (2). This feature remains the same under two-stage and sequential strategies. We leave it as an exercise to critically examine the findings from other blocks.

In Table 14.5.1, we note that we have encountered an oversampling rate between 8.72% to 9.85% under two-stage strategy compared with 8.47% theoretically. Also, under sequential strategy, we see that $\overline{n} - C$ varies between 0.18 and 0.37 whereas we have $\eta(5) = 0.24954$. Overall, we feel good to have this much agreement between the empirical and theoretical entities.

Next, we briefly examine the role of the two selection methodologies when $\boldsymbol{\mu}$ may be outside of $\Omega(\delta^*)$, that is $\boldsymbol{\mu}$ belongs to the indifference zone. Table 14.5.2 summarizes findings from simulation studies with 5000 runs executed by **Seq14.exe**. We considered three parameter configurations of $\boldsymbol{\mu}$ lying outside of $\Omega(\delta^*)$.

Recall the LFC with $\delta^* = 0.5$. In Table 14.5.2, Block 1 presents results when the largest mean $\mu_1 = 10$ is fixed, $\mu_2 = 9.6(> \mu_1 - \delta^*)$ is fixed, and μ_3, μ_4, μ_5 stack up at $8.5(< \mu_1 - \delta^*)$. That is, the second best mean is closer to the largest mean than what LFC had. Under LFC, μ_2 was 9.5 instead. On the other hand, now we have set apart μ_3, μ_4, μ_5 more from μ_1, μ_2 than what LFC had. Yet, in Block 1, the estimated CS probability $\overline{p}$ looks very good under both two-stage and sequential methodologies. In Table 14.5.2, $\overline{p}$ values in Block 2 have gone slightly under P^*. However, $\overline{p}$ values in Block 3 have gone down sizably compared with P^*. Here, in Block 3, the largest and the second largest means are too close for us to call the best population satisfying P^* requirement. One should note, however, that the sampling strategies were designed assuming $\delta^*(= 0.5)$ separation between the best mean and the next best mean, but that is violated "severely" under the configuration of $\boldsymbol{\mu}$ in Block 3.

Whether one considers the entries in Table 14.5.1 or in Table 14.5.2 or both, one feature must have caught everyone's attention. The estimated averages $(\overline{n})$ and standard deviations (s_n) for the sample size n look nearly identical across each selection strategy. That should not be surprising, because the probability distribution of the sample size is not affected by

$\boldsymbol{\mu}$ at all. The probability distribution of the sample size is affected only by σ under either selection strategy.

Table 14.5.2: Comparing two-stage (2stage) and sequential strategies (seq) under normality with 5000 simulation runs each for a fixed set of population means outside $\Omega(\delta^*)$, $k = 5$, $\delta^* = 0.5, P^* = 0.95, m = 7, \sigma = 2, \nu = 30, \eta(5) = 0.24954$ $h_{5,0.95} = 3.0552, \tau_{5,30,0.95} = 3.19, \tau^2_{5,30,0.95}/h^2_{5,0.95} = 1.09$

Block	C	$\overline{n}$	s_n	$\overline{n}/C$	$\overline{n}-C$	$\overline{p}$
1	$\mu_1 = 10, \mu_2 = 9.6, \mu_3 = \mu_4 = \mu_5 = 8.5$ Random sequence: 6					
2stage	149.35	163.23	42.49	1.0929		0.9606
seq	149.35	149.72	7.94	1.0025	0.37	0.9572
2	$\mu_1 = 10, \mu_2 = 9.6, \mu_3 = \mu_4 = \mu_5 = 9.5$ Random sequence: 7					
2stage	149.35	164.42	41.41	1.1009		0.9288
seq	149.35	149.55	7.71	1.0013	0.20	0.9282
3	$\mu_1 = 10, \mu_2 = 9.75, \mu_3 = \mu_4 = \mu_5 = 9.5$ Random sequence: 3					
2stage	149.35	164.48	42.43	1.1013		0.8526
seq	149.35	149.70	7.80	1.0023	0.35	0.8374

```
        Selecting the Best Normal Population
  Two-Stage Procedure for Common Unknown Variance
===============================================
Input parameters:    P*     =    0.95
                 delta*     =    0.25
Number of populations (k) =  6

tau(k,nu,P*)   =    3.26
Datafile name = Norm6.dat

Initial sample size      =      10
Final sample size        =     525
Number of observations in the file = 500

Not enough data in the files to continue
Program will be terminated
```

Figure 14.5.4. *The best normal population from the data in* ***Norm6.dat*** *file using two-stage selection strategy,* $P^* = 0.95$.

Figure 14.5.4 shows the output from **Seq14.exe** when two-stage selection methodology was implemented having six normal populations using live data input from file, **Norm6.dat**. This data file is available from the website. The figure shows observed value of the final sample size ($n = 525$) required for the two-stage procedure with $k = 6, P^* = 0.95, m = 10$ and $\delta^* = 0.25$. Since given data was not sufficiently large, the program terminated after computing the final sample size.

```
      Selecting the Best Normal Population
Two-Stage Procedure for Common Unknown Variance
===============================================
Input parameters:   P*      =     0.90
                 delta*     =     0.25
Number of populations (k) =  6

tau(k,nu,P*)  =   2.78
Datafile name = Norm6.dat

Initial sample size     =      10
Final sample size       =     382

Best population number =   2
           sample mean =       10.048
Pooled sample variance =        3.805

For other populations:
Number  Sample mean
     1        3.123
     3        7.987
     4        3.979
     5        7.047
     6        4.944
```

Figure 14.5.5. *The best normal population from the data in **Norm6.dat** file using two-stage selection strategy,* $P^* = 0.90$.

However, we decided to reduce P^* to 0.90, and then the required final sample size estimate came down from 525 to 382. The output in this case is found in Figure 14.5.5. Finally, the two-stage strategy selected population 2 as the best. By the way, one could hold $P^* = 0.95$ fixed, but increase δ^* to 0.3 instead in order achieve same kind of reduction the sample size.

Figure 14.5.6 shows the output from **Seq14.exe** when sequential selection methodology was implemented with the same six normal populations using live data input from a file, **Norm6.dat**. The figure shows observed value of the final sample size ($n = 453$) required for the strategy under $P^* = 0.90, m = 10$ and $\delta^* = 0.25$. Finally, the sequential strategy also selected population #2 as the best.

```
      Selecting the Best Normal Population
Sequential Procedure for Common Unknown Variance
================================================
Input parameters:   P*     =    0.90
                delta*     =    0.25
Number of populations (k) =  6

h(k,P*)        = 2.7100
Datafile name = Norm6.dat

Initial sample size      =     10
Final sample size        =    453
Best population number =   2
            sample mean =      10.017
Pooled sample variance =        7.707

For other populations:
Number   Sample mean
     1         3.111
     3         7.880
     4         4.037
     5         7.076
     6         4.936
```

Figure 14.5.6. *The best normal population from* ***Norm6.dat*** *file using sequential selection strategy,* $P^* = 0.90$.

14.6 Other Selected Multistage Procedures

For the particular problem discussed in this chapter, a modified two-stage strategy in the spirit of Section 6.2.3 was introduced by Mukhopadhyay and Solanky (1994, chapter 3, pp. 79-83).

An accelerated sequential methodology in the spirit of Section 6.2.4 was developed by Mukhopadhyay and Solanky (1992a). It reduced substantially the sampling operations necessary for implementing the Robbins et al. sequential strategy. The methodology began sequentially and went part of the way which was followed by batch sampling from all k treatments. It had *second-order* properties very similar to those stated in Theorem 14.4.2.

Operationally, the most economical selection methodology would sample in three stages and it would have *second-order* properties very similar to those in Theorem 14.4.2. Such three-stage selection strategies were developed by Mukhopadhyay and Judge (1991). In the context of all pairwise comparisons, one may refer to Liu (1995), Mukhopadhyay and Aoshima (1998) and other sources.

In what follows we include a small number of closely related selection problems and cite some original sources.

14.6.1 Elimination-Type Selection Methodologies

The selection methodology $\mathcal{R}_N$ from (14.4.2) based on Robbins et al.'s (1968) sequential sampling (14.4.1) works well and it enjoys *second-order* properties (Theorem 14.4.2). But, there is room for significant improvement. Consider going through the steps of sampling but suppose that at some point a mean response from one treatment falls way below the largest mean. Then, why bother keeping such "inferior" treatment in the running at all? The point is this: At any point of time, one may look at the most inferior treatment and if it appears practically impossible that it may rise again from the ashes, then discard it from further considerations. Such elimination-type methodologies should be the way to proceed in many practical applications. Such a modification will make Robbins et al.'s (1968) sequential selection rule more ethical and cost effective.

Paulson (1964) first designed elimination-type truncated sequential selection rules and controlled $P(CS)$ at the level of P^* or better whenever $\boldsymbol{\mu}$ came from a preference-zone. This fundamental idea has grown in many directions. One may review from Bechhofer et al. (1995), Gibbons et al. (1977), Gupta and Panchapakesan (1979), Mukhopadhyay and Solanky (1994, chapter 8, pp. 307-310), Solanky (2001,2006) and other sources.

14.6.2 Normal Distributions: Unequal and Unknown Variances

Consider selecting a treatment associated with mean $\mu_{[k]}$ from k independently run treatments where responses from i^{th} treatment are independently and normally distributed having mean μ_i and variance σ_i^2. One assumes that all parameters are unknown, $-\infty < \mu_1, \ldots, \mu_k < \infty$, $0 < \sigma_1, \ldots, \sigma_k < \infty$, and that variances are unequal. Ofusu (1973) and Chiu (1974) first investigated selection problems in a heteroscedastic scenario.

In the spirit of Section 14.2, under an indifference-zone formulation, Dudewicz and Dalal (1975) developed a remarkable two-stage selection methodology. They would control $P(CS)$ at the level of P^* or better whenever $\boldsymbol{\mu}$ came from a preference-zone.

One may refer to Rinott (1978) and Mukhopadhyay (1979b) for important follow-ups. Mukhopadhyay (1979b) included an allocation scheme in the spirit of Section 13.2.3 together with a sequential selection strategy in the light of Robbins et al. (1967) when $k = 2$.

14.6.3 Negative Exponential Locations: A Common but Unknown Scale

A large volume of literature is available for selecting the best (largest) location parameter $\mu_{[k]}$ from k independently run treatments where responses from i^{th} treatment are independently distributed with a common p.d.f.:

$$\sigma^{-1} \exp\left(-(x - \mu_i)/\sigma\right) I(x > \mu_i), i = 1, \ldots, k.$$

In Chapter 7, we denoted such distributions NExpo(μ_i, σ), $i = 1, \ldots, k$.

In the spirit of Section 14.2, Raghavachari and Starr (1970) first gave an optimal fixed-sample-size selection methodology, had σ been known, under an indifference-zone formulation.

However, one often assumes that all parameters $-\infty < \mu_1, \ldots, \mu_k < \infty$ and $0 < \sigma < \infty$ are unknown. In this situation, Desu et al. (1977) developed an appropriate two-stage selection methodology in the light of Bechhofer et al. (1954). Mukhopadhyay (1986) developed modified two-stage and sequential selection methodologies with *first-order* properties under two-stage strategy and *second-order* properties under sequential strategy. An accelerated version of Mukhopadhyay's (1986) sequential methodology was developed by Mukhopadhyay and Solanky (1992b) with *second-order* properties. A three-stage selection strategy was developed by Mukhopadhyay (1987b) with its *second-order* properties.

14.6.4 Negative Exponential Locations: Unknown and Unequal Scales

Consider selecting the largest location parameter $\mu_{[k]}$ from k independently run treatments where responses from the ith treatment are independently distributed with a common p.d.f.:

$$\sigma^{-1} \exp\left(-(x - \mu_i)/\sigma_i\right) I(x > \mu_i), i = 1, \ldots, k.$$

In Chapter 7, we denoted such distributions NExpo(μ_i, σ_i), $i = 1, \ldots, k$.

One would assume that all parameters were unknown, $-\infty < \mu_1, \ldots, \mu_k < \infty$, $0 < \sigma_1, \ldots, \sigma_k < \infty$. The scale parameters $\sigma_1, \ldots, \sigma_k$ are also unequal. In the light of Dudewicz and Dalal (1975), under an indifference-zone formulation, Mukhopadhyay (1984b) developed a two-stage procedure when $k = 2$. Mukhopadhyay and Hamdy (1984b) extended that selection methodology in the case of $k(\geq 3)$ populations. These two-stage selection methodologies would control $P(CS)$ at the level of P^* or better whenever $\boldsymbol{\mu}$ came from a preference-zone. One should refer to Lam and Ng (1990) for interesting follow-ups.

14.6.5 Distribution-Free

The goal of this section is to emphasize that the Robbins et al. sequential methodology from Section 14.4 basically works fine without the normality assumption. Consider $k(\geq 2)$ populations $\Pi_1, \ldots, \Pi_k$ having unknown distributions, discrete or continuous, with unknown means $\mu_1, \ldots, \mu_k$ and a common unknown variance σ^2 with $-\infty < \mu_1, \ldots, \mu_k < \infty$, $0 < \sigma < \infty$. Let us define

$$\begin{aligned} &N \equiv N(\delta^*) \text{ is the smallest integer } n(\geq m) \text{ for which} \\ &\quad \text{we observe } n \geq \frac{h_{k,P^*}^2 (U_n^2 + n^{-1})}{\delta^{*2}}, \end{aligned} \tag{14.6.1}$$

instead of what we had in (14.4.1). As before, h_{k,P^*} comes from (14.2.9).

Observe that $N(\delta^*)$ again estimates $C \equiv h_{k,P^*}^2 \sigma^2/\delta^{*2}$ that comes from (14.2.10). The expression $U_n^2 + n^{-1}$ included in the boundary condition of stopping rule (14.6.1) makes sure that we do not stop too early. Without the term " n^{-1}" in $U_n^2 + n^{-1}$, we may stop too early especially when U_n^2 may be zero with a positive probability. We may encounter such a situation when the distributions are discrete!

One surely notices that the present stopping rule resembles the stopping time of Chow and Robbins (1965) that we introduced in (10.2.4). However, this resemblance should not surprise any one. One should recall that the Chow-Robbins sequential estimation methodology was also developed for handling distribution-free or nonparametric scenarios.

One may verify that the sampling process (14.6.1) will terminate w.p.1. Upon termination, based on all observations $N, X_{i1}, \ldots, X_{iN}$, one would obtain i^{th} population Π_i's sample mean, $\overline{X}_{iN}, i = 1, \ldots, k$. Now, the selection rule would be:

$$\mathcal{R}_N: \quad \begin{array}{l}\text{Select as best the treatment associated with } \Pi_j \\ \text{if } \overline{X}_{jN} \text{ is the largest among all sample means} \\ \overline{X}_{1N}, \ldots, \overline{X}_{kN}.\end{array} \tag{14.6.2}$$

Then, this selection methodology $\mathcal{R}_N$ under sequential sampling via (14.6.1) would have all *first-order* properties stated in Theorem 14.4.1.

The literature on nonparametric sequential selection methodologies is rich. One may get started with Geertsema (1972) and Swanepoel (1977) for quick reviews. Mukhopadhyay and Solanky (1993) developed an accelerated sequential analog of Geertsema's (1972) methodology. The properties that were obtained in these and other closely related contributions included *first-order* asymptotics. Refer to Gupta and Panchapakesan (1979) and Mukhopadhyay and Solanky (1994, chapter 8, pp. 350-359) for additional reviews.

14.7 Some Selected Derivations

In this section, we supply parts of some crucial derivations. Without proper preparation, however, some of the steps may appear a little hard at first pass.

14.7.1 Some Comments on (14.2.9)

Suppose that $Z_1, \ldots, Z_k$ are i.i.d. standard normal variables. Then, for any fixed h, we have:

$$\begin{aligned}&\mathrm{P}\left(Z_k + h > Z_1, Z_k + h > Z_2, \ldots, Z_k + h > Z_{k-1}\right) \\ &\quad = \textstyle\int_{-\infty}^{\infty} \mathrm{P}\left(Z_k + h > Z_1, Z_k + h > Z_2, \ldots, Z_k + h > Z_{k-1} \mid Z_k = z\right) \times \\ &\qquad\qquad \phi(z) dz\end{aligned}$$

$$
\begin{aligned}
&= \int_{-\infty}^{\infty} \mathrm{P}\left(z+h > Z_1, z+h > Z_2, \ldots, z+h > Z_{k-1} \mid Z_k = z\right) \times \\
&\qquad\qquad \phi(z)dz \\
&= \int_{-\infty}^{\infty} \mathrm{P}\left(z+h > Z_1, z+h > Z_2, \ldots, z+h > Z_{k-1}\right) \phi(z)dz \\
&= \int_{-\infty}^{\infty} \textstyle\prod_{i=1}^{k-1} \mathrm{P}\left(Z_i < z+h\right) \phi(z)dz \\
&= \int_{-\infty}^{\infty} \Phi^{k-1}(z+h)\phi(z)dz.
\end{aligned} \tag{14.7.1}
$$

By comparing (14.2.9) with (14.7.1), one can see that $h \equiv h_{k,P^*}$ must equivalently satisfy:

$$
P^* = \mathrm{P}\left(Z_k + h > Z_1, Z_k + h > Z_2, \ldots, Z_k + h > Z_{k-1}\right). \tag{14.7.2}
$$

Next, let us denote $Y_i = (Z_i - Z_k)/\sqrt{2}, i = 1, \ldots, k-1$. Clearly, Y_1, ..., Y_{k-1} jointly have a $(k-1)$-dimensional normal distribution with mean vector **0** and a dispersion matrix:

$$
\mathbf{\Sigma}_{k-1 \times k-1} = \begin{pmatrix} 1 & 1/2 & 1/2 & \ldots & 1/2 & 1/2 \\ 1/2 & 1 & 1/2 & \ldots & 1/2 & 1/2 \\ \vdots & \vdots & \vdots & & \vdots & \vdots \\ 1/2 & 1/2 & 1/2 & \ldots & 1 & 1/2 \\ 1/2 & 1/2 & 1/2 & \ldots & 1/2 & 1 \end{pmatrix}. \tag{14.7.3}
$$

Since its diagonals are all 1, this dispersion matrix is also referred to as a correlation matrix. Its off-diagonal elements are pairwise correlations.

The present structure of $\mathbf{\Sigma}$ is called an equi-correlated case with equi-correlation $\rho = \frac{1}{2}$. We use this to rewrite (14.7.2) as follows:

$$
P^* = \mathrm{P}\left(Y_1 < h/\sqrt{2}, Y_2 < h/\sqrt{2}, \ldots, Y_{k-1} < h\sqrt{2}\right). \tag{14.7.4}
$$

In other words, $h_{k,P^*}/\sqrt{2}$ is an upper equi $100(1 - P^*)\%$ point of a $(k-1)$-dimensional normal distribution with mean vector **0** and dispersion matrix $\mathbf{\Sigma}$ from (14.7.3). By the way, this correlation matrix may also be expressed as:

$$
\begin{aligned}
&\mathbf{\Sigma}_{k-1 \times k-1} \equiv \mathbf{\Sigma}_{k-1 \times k-1}(\tfrac{1}{2}) \text{ where we denote :} \\
&\mathbf{\Sigma}_{k-1 \times k-1}(\rho) = (1-\rho) I_{k-1 \times k-1} + \rho \mathbf{1}\mathbf{1}', \\
&\mathbf{1}' = (1, 1, \ldots, 1)_{1 \times k-1} \text{ and } I_{k-1 \times k-1} \text{ is an identity matrix.}
\end{aligned} \tag{14.7.5}
$$

Many tables are available for multivariate normal distributions. The expression from (14.7.3) specifies clearly how $h_{k,P^*}/\sqrt{2}$, or equivalently h, can be found from a table for an equi-percentage point of a multivariate normal distribution having mean vector **0** and equi-correlation $\frac{1}{2}$. ■

14.7.2 Proof of Theorem 14.2.1

Recall the notation used in (2.4.3) and (14.2.5). Now, let us first rewrite $\mathrm{P}_{\boldsymbol{\mu},\sigma}(CS)$ as:

$$\begin{aligned}
&\mathrm{P}_{\boldsymbol{\mu},\sigma}(CS)\\
&\quad= \textstyle\sum_{n=m}^{\infty} \mathrm{P}_{\boldsymbol{\mu},\sigma}(CS \mid N=n)\mathrm{P}_{\boldsymbol{\mu},\sigma}(N=n)\\
&\quad= \textstyle\sum_{n=m}^{\infty} \mathrm{P}_{\boldsymbol{\mu},\sigma}\left(\overline{X}_{(kn)} > \overline{X}_{(in)}, i=1,\ldots,k-1 \mid N=n\right)\mathrm{P}_{\boldsymbol{\mu},\sigma}(N=n)\\
&\quad= \textstyle\sum_{n=m}^{\infty} \mathrm{P}_{\boldsymbol{\mu},\sigma}\left(\overline{X}_{(kn)} > \overline{X}_{(in)}, i=1,\ldots,k-1\right)\mathrm{P}_{\sigma}(N=n).
\end{aligned} \tag{14.7.6}$$

In the last step, we dropped conditioning event "$N=n$" because of assumption (ii). Also, we dropped $\boldsymbol{\mu}$ from $\mathrm{P}_{\boldsymbol{\mu},\sigma}(N=n)$ and wrote $\mathrm{P}_{\sigma}(N=n)$ in the last step because of assumption (i). The assumption (iii) is needed to make sure that $\mathcal{R}_N$ is a genuine selection rule.

We realize that $\mathrm{P}_{\boldsymbol{\mu},\sigma}(\overline{X}_{(kn)} > \overline{X}_{(in)}, i=1,\ldots,k-1)$ is minimized with respect to $\boldsymbol{\mu}$ within $\Omega(\delta^*)$, for every fixed n, if and only if we work under LFC. The important point is that LFC does not involve n.

Hence, $\mathrm{P}_{\boldsymbol{\mu},\sigma}(CS)$ will be minimized with respect to $\boldsymbol{\mu}$ within $\Omega(\delta^*)$ if and only if we work under a configuration that is same as LFC. Thus, using (14.2.7) and (14.7.6), we write:

$$\begin{aligned}
&\inf_{\boldsymbol{\mu}\in\Omega(\delta^*)} \mathrm{P}_{\boldsymbol{\mu},\sigma}(CS)\\
&\quad= \textstyle\sum_{n=m}^{\infty} \inf_{\boldsymbol{\mu}\in\Omega(\delta^*)} \mathrm{P}_{\boldsymbol{\mu},\sigma}(\overline{X}_{(kn)} > \overline{X}_{(in)}, i=1,\ldots,k-1)\mathrm{P}_{\sigma}(N=n)\\
&\quad= \textstyle\sum_{n=m}^{\infty} \left\{\int_{-\infty}^{\infty} \Phi^{k-1}\left(y+\frac{\sqrt{n}}{\sigma}\delta^*\right)\phi(y)dy\right\}\mathrm{P}_{\sigma}(N=n)\\
&\quad= \mathrm{E}_{\sigma}\left[\int_{-\infty}^{\infty} \Phi^{k-1}\left(y+\frac{\sqrt{N}}{\sigma}\delta^*\right)\phi(y)dy\right].
\end{aligned} \tag{14.7.7}$$

The minimum is obviously attained if and only if $\mu_{[1]} = \ldots = \mu_{[k-1]} = \mu_{[k]} - \delta^*$. ■

14.7.3 Proof of Theorem 14.3.1

If we show that $\inf_{\boldsymbol{\mu}\in\Omega(\delta^*)} \mathrm{P}_{\boldsymbol{\mu},\sigma}(CS) \geq P^*$, the proof will be over. From (14.7.7), we rewrite:

$$\begin{aligned}
\inf_{\boldsymbol{\mu}\in\Omega(\delta^*)} \mathrm{P}_{\boldsymbol{\mu},\sigma}(CS) &= \mathrm{E}_{\sigma}\left[\int_{-\infty}^{\infty} \Phi^{k-1}\left(y+\frac{\sqrt{N}}{\sigma}\delta^*\right)\phi(y)dy\right]\\
&\geq \mathrm{E}_{\sigma}\left[\int_{-\infty}^{\infty} Phi^{k-1}\left(y+\frac{\tau_{k,\nu,P^*}U_m}{\sigma}\right)\phi(y)dy\right],
\end{aligned} \tag{14.7.8}$$

since $N \geq \tau_{k,\nu,P^*}^2 U_m^2/\delta^{*2}$ w.p.1 which follows from (14.3.2). We recall that U_m^2 is a pooled sample variance with degrees of freedom $\nu = k(m-$

1), $\nu U_m^2/\sigma^2 \sim \chi^2_\nu$, and $U_m^2, (\overline{X}_{1n}, \ldots, \overline{X}_{kn})$ are also independently distributed.

Next, we consider $\int_{-\infty}^{\infty} \Phi^{k-1}\left(y + \frac{\tau_{k,\nu,P^*} U_m}{\sigma}\right)\phi(y)dy$, conditionally given U_m. Conditionally given U_m, utilizing (14.7.4), we claim:

$$\begin{aligned} &\int_{-\infty}^{\infty} \Phi^{k-1}\left(y + \frac{\tau_{k,\nu,P^*} U_m}{\sigma}\right)\phi(y)dy \\ &= \mathrm{P}\left(Y_i < \frac{\tau_{k,\nu,P^*} U_m}{\sqrt{2}\sigma}, i = 1, \ldots, k-1\right) \\ &= \mathrm{P}\left(\frac{Y_i}{U_m/\sigma} < \frac{\tau_{k,\nu,P^*}}{\sqrt{2}}, i = 1, \ldots, k-1\right), \end{aligned} \tag{14.7.9}$$

where $Y_1, \ldots, Y_{k-1}$ have a joint $(k-1)$-dimensional normal distribution with mean vector $\mathbf{0}$ and dispersion matrix $\mathbf{\Sigma}_{k-1\times k-1}$ from (14.7.3).

However, U_m is distributed independently of $(\overline{X}_{1n}, \ldots, \overline{X}_{kn})$. So, for every fixed $n \geq m$, from (14.7.9) we claim unconditionally:

$$\begin{aligned} &\mathrm{E}_\sigma\left[\int_{-\infty}^{\infty} \Phi^{k-1}\left(y + \frac{\tau_{k,\nu,P^*} U_m}{\sigma}\right)\phi(y)dy\right] \\ &= \mathrm{P}\left(W_i < \frac{\tau_{k,\nu,P^*}}{\sqrt{2}}, i = 1, \ldots, k-1\right) \end{aligned} \tag{14.7.10}$$

where $W_i \equiv \dfrac{Y_i}{U_m/\sigma}, i = 1, \ldots, k-1$.

It is known that the joint distribution of $W_1, \ldots, W_{k-1}$ is a $(k-1)$-dimensional t with an equi-correlation matrix $\mathbf{\Sigma}(\rho)$ from (14.7.5). Clearly, equi-correlation $\rho = \frac{1}{2}$ and $\tau_{k,\nu,P^*}/\sqrt{2}$ is an upper $100(1-P^*)\%$ equi-coordinate point of this multivariate t distribution.

Now, one would find $\tau_{k,\nu,P^*}/\sqrt{2}$, or equivalently τ_{k,ν,P^*}, in such a way that

$$\mathrm{P}\left(W_i < \frac{\tau_{k,\nu,P^*}}{\sqrt{2}}, i = 1, \ldots, k-1\right) = P^*. \tag{14.7.11}$$

The proof is complete. ■

14.7.4 Proof of Theorem 14.4.1

Proof of Part (i):

From (14.4.1), first observe the basic inequality:

$$\frac{h^2_{k,P^*} U_N^2}{\delta^{*2}} \leq N \leq m + \frac{h^2_{k,P^*} U_{N-1}^2}{\delta^{*2}} \text{ w.p. 1.} \tag{14.7.12}$$

Assume for a moment that $\mathrm{E}_{\boldsymbol{\mu},\sigma}[N] < \infty$. Now, we work from the right-hand side of (14.7.12) to write:

$$N \le m + \frac{h_{k,P^*}^2}{\delta^{*2}} \frac{1}{k(N-2)} \sum_{i=1}^{k} \sum_{j=1}^{N-1} (X_{ij} - \overline{X}_{iN-1})^2 \text{ w.p. 1.}$$

It implies w.p. 1:

$$\begin{aligned} k(N-m)(N-2) &\le \frac{h_{k,P^*}^2}{\delta^{*2}} \textstyle\sum_{i=1}^{k} \sum_{j=1}^{N-1} (X_{ij} - \mu_i)^2 \\ &\le \frac{h_{k,P^*}^2}{\delta^{*2}} \textstyle\sum_{i=1}^{k} \sum_{j=1}^{N} (X_{ij} - \mu_i)^2. \end{aligned}$$

Thus, we have:

$$kN^2 - k(m+2)N \le \frac{h_{k,P^*}^2}{\delta^{*2}} \sum_{i=1}^{k} \sum_{j=1}^{N} (X_{ij} - \mu_i)^2 \text{ w.p. 1.} \quad (14.7.13)$$

Recall that $C = h_{k,P^*}^2 \sigma^2 / \delta^{*2}$. Now, we take expectation throughout the last step in (14.7.13) and use Wald's first equation (Theorem 3.5.4) to obtain:

$$\begin{aligned} k\mathrm{E}_{\boldsymbol{\mu},\sigma}[N^2] - k(m+2)\mathrm{E}_{\boldsymbol{\mu},\sigma}[N] &\le \frac{kh_{k,P^*}^2}{\delta^{*2}} \sigma^2 \mathrm{E}_{\boldsymbol{\mu},\sigma}[N] \\ \Rightarrow \mathrm{E}_{\boldsymbol{\mu},\sigma}^2[N] - (m+2)\mathrm{E}_{\boldsymbol{\mu},\sigma}[N] &\le C\mathrm{E}_{\boldsymbol{\mu},\sigma}[N] \\ \Rightarrow \mathrm{E}_{\boldsymbol{\mu},\sigma}[N] - (m+2) &\le C. \end{aligned} \quad (14.7.14)$$

That is, we have $\mathrm{E}_{\boldsymbol{\mu},\sigma}[N] \le C + m + 2$.

Next, if one cannot assume that $\mathrm{E}_{\boldsymbol{\mu},\sigma}[N]$ is finite, then one may use a truncated stopping variable, namely, $N_r = \min\{r, N\}, r = 1, 2, \ldots$. One may immediately claim that $\mathrm{E}_{\boldsymbol{\mu},\sigma}[N_r] \le C+m+2$ for each fixed r. Then, an application of monotone convergence theorem will complete the proof. This derivation also shows that $\mathrm{E}_{\boldsymbol{\mu},\sigma}[N]$ is finite. ■

Proof of Part (ii):

First, we divide throughout (14.7.12) by C and obtain:

$$\frac{U_N^2}{\sigma^2} \le \frac{N}{C} \le \frac{m}{C} + \frac{U_{N-1}^2}{\sigma^2} \text{ w.p. 1.} \quad (14.7.15)$$

Next, observe that $C \to \infty$, $U_{N-1}^2 \xrightarrow{\text{P}} \sigma^2$, $U_N^2 \xrightarrow{\text{P}} \sigma^2$ as $\delta^* \to 0$. From (14.7.15), it follows that

$$\frac{N}{C} \xrightarrow{\text{P}} 1 \text{ as } \delta^* \to 0. \quad (14.7.16)$$

Hence, Fatou's Lemma will apply and we claim:

$$\liminf\nolimits_{\delta^* \to 0} \mathrm{E}_{\boldsymbol{\mu},\sigma}[N/C] \ge \mathrm{E}_{\boldsymbol{\mu},\sigma}[\liminf\nolimits_{\delta^* \to 0} N/C] = 1.$$

Now, using part (i), we have:

$$\limsup_{\delta^* \to 0} \mathrm{E}_{\boldsymbol{\mu},\sigma}[N/C] \le 1.$$

Hence, part (ii) follows. ■

Proof of Part (iii):

We recall (14.7.7):

$$\inf_{\boldsymbol{\mu} \in \Omega(\delta^*)} \mathrm{P}_{\boldsymbol{\mu},\sigma}(CS) = \mathrm{E}_\sigma \left[\int_{-\infty}^{\infty} \Phi^{k-1} \left(y + \frac{\sqrt{N}}{\sigma}\delta^* \right) \phi y) dy \right].$$

Now, from (14.7.15), we see that $\sqrt{N}\delta^*/\sigma \xrightarrow{\mathrm{P}} h_{k,P^*}$ as $\delta^* \to 0$. Thus, with the help of dominated convergence theorem, we claim as $\delta^* \to 0$:

$$\begin{aligned} \liminf \mathrm{P}_{\boldsymbol{\mu},\sigma}(CS) &= \mathrm{E}_\sigma \left[\int_{-\infty}^{\infty} \Phi^{k-1} \left(y + h_{k,P^*} \right) \phi(y) dy \right] \\ &= \int_{-\infty}^{\infty} \Phi^{k-1} \left(y + h_{k,P^*} \right) \phi(y) dy = P^*, \end{aligned} \tag{14.7.17}$$

by the choice of h_{k,P^*} through (14.2.9). ■

14.7.5 Proof of Theorem 14.4.2

We will not give complete proofs. We will provide the important steps leading to the proofs.

First, one should note that N from (14.4.1) can be alternatively expressed as:

$$\begin{aligned} &N \equiv N(\delta^*) \text{ is the smallest integer } n (\ge m) \text{ for which} \\ &\qquad \text{we observe } \frac{kn(n-1)}{C} \ge \sum_{i=1}^{n-1} W_i, \end{aligned} \tag{14.7.18}$$

where W's are i.i.d. χ^2_k.

Next, observe that N has exactly the same probability distribution as that of $T+1$ where T is expressed as:

$$\begin{aligned} &T \equiv T(\delta^*) \text{ is the smallest integer } n (\ge m-1) \text{ for which} \\ &\qquad \text{we observe } \frac{k(n+1)n}{C} \ge \sum_{i=1}^{n} W_i. \end{aligned} \tag{14.7.19}$$

Now, one can check that T looks exactly like the stopping variable from (A.4.1). The present setup and the setup from (A.4.1)-(A.4.3) are identical if we let:

$$\begin{aligned} \delta = 2, h^* = k/C, L(n) = 1 + n^{-1}, L_0 = 1, \theta = k, \tau^2 = 2k, \\ m_0 = m-1, \beta^* = 1, n_0^* = C, r = k/2 \text{ and } q = 2/k. \end{aligned}$$

Proof of Part (i):

From (A.4.4) we can express:

$$\nu = \frac{1}{2}(k+2) - \sum_{n=1}^{\infty} \frac{1}{n} \mathrm{E}\left\{\max\left(0, \chi^2_{nk} - 2nk\right)\right\}. \tag{14.7.20}$$

It implies that

$$\kappa = \frac{1}{k}\nu - 1 - \frac{2}{k}. \tag{14.7.21}$$

Now, Theorem A.4.2 shows:

$$\mathrm{E}_\sigma(T) = C + \kappa + o(1) \text{ if } m > 1 + 2k^{-1},$$

which implies:

$$\mathrm{E}_\sigma(N) = \mathrm{E}_\sigma(T+1) = C + \eta(k) + o(1) \text{ if } m > 1 + 2k^{-1},$$

where

$$\begin{aligned} \eta(k) &= \kappa + 1 = \tfrac{1}{k}\nu - \tfrac{2}{k} \\ &= \tfrac{1}{2} - \tfrac{1}{k} - \tfrac{1}{k}\textstyle\sum_{n=1}^{\infty} \tfrac{1}{n}\mathrm{E}\left\{\max\left(0, \chi^2_{nk} - 2nk\right)\right\}, \end{aligned} \tag{14.7.22}$$

using (14.7.20)-(14.7.21). ■

Proof of Part (ii):

One needs to expand $\mathrm{P}_{\boldsymbol{\mu},\sigma}(CS)$ only under the LFC. Theorem 14.2.1 gives the following expression of $\mathrm{P}_{\boldsymbol{\mu},\sigma}(CS)$ under the LFC:

$$\mathrm{E}_\sigma\left[\int_{-\infty}^{\infty} \Phi^{k-1}\left(y + \frac{\sqrt{N}}{\sigma}\delta^*\right)\phi(y)dy\right] \equiv \mathrm{E}_\sigma\left[f\left(h^2_{k,P^*}\frac{N}{C}\right)\right] \tag{14.7.23}$$

where $f(.)$ comes from (14.4.4).

Next, let us denote $U^* = (T-C)/\sqrt{C}$. Then, Theorem A.4.2, parts (ii) and (iii) show as $\delta^* \to 0$:

$$\begin{aligned} &U^* \text{ converges in distribution to } N(0,q) \text{ with } q = \tfrac{2}{k}; \text{ and} \\ &U^{*2} \text{ is uniformly integrable if } m > 1 + 2k^{-1}. \end{aligned} \tag{14.7.24}$$

From (14.7.24), it will not be hard to show:

$$\begin{aligned} &(N-C)/\sqrt{C} \text{ converges in distribution to } N(0, \tfrac{2}{k}); \\ &(N-C)^2/C \text{ is uniformly integrable if } m > 1 + 2k^{-1}; \\ &\Rightarrow \mathrm{E}_\sigma\left[\frac{(N-C)^2}{C}\right] = \frac{2}{k} + o(1) \text{ if } m > 1 + 2k^{-1}. \end{aligned} \tag{14.7.25}$$

Now, we expand $f(y)$ around $y = h^2_{k,P^*}$ and then replace y with $h^2_{k,P^*}\frac{N}{C}$

to write:

$$\begin{aligned} &\mathrm{E}_\sigma\left[f\left(h^2_{k,P^*}\frac{N}{C}\right)\right] \\ &= f(h^2_{k,P^*}) + \tfrac{h^2_{k,P^*}}{C}\mathrm{E}_\sigma\left[(N-C)f'(h^2_{k,P^*})\right] \\ &\quad + \tfrac{h^4_{k,P^*}}{2kC}\mathrm{E}_\sigma\left[\frac{(N-C)^2}{C}f''(\xi)\right]. \end{aligned} \tag{14.7.26}$$

where ξ is an appropriate random variable between h^2_{k,P^*} and $h^2_{k,P^*}\frac{N}{C}$.

It should be obvious that $f(h^2_{k,P^*}) = P^*$. From part (i), we can claim:

$$\frac{h^2_{k,P^*}}{C}\mathrm{E}_\sigma\left[(N-C)f'(h^2_{k,P^*})\right] = \frac{h^2_{k,P^*}}{C}\eta(k)f'(h^2_{k,P^*}) + o(1), \tag{14.7.27}$$

where $\eta(k)$ comes from (14.7.22). Next, from the last step in (17.2.25), we can claim:

$$\mathrm{E}_\sigma\left[\frac{(N-C)^2}{C}f''(\xi)\right] = \frac{2}{k}f''(h^2_{k,P^*}) + o(1), \tag{14.7.28}$$

provided that $\frac{(N-C)^2}{C}f''(\xi)$ is uniformly integrable. With some hard work, one can show that $\frac{(N-C)^2}{C}f''(\xi)$ is uniformly integrable with an appropriate condition on m. The details are omitted.

Now, one can combine (14.7.26)-(14.7.28), and obtain:

$$\begin{aligned} &\mathrm{E}_\sigma\left[f\left(h^2_{k,P^*}\frac{N}{C}\right)\right] \\ &= P^* + \tfrac{h^2_{k,P^*}}{C}\eta(k)f'(h^2_{k,P^*}) + \tfrac{h^4_{k,P^*}}{2C}\tfrac{2}{k}f''(h^2_{k,P^*}) + o(C^{-1}) \\ &= P^* + \tfrac{h^2_{k,P^*}}{C}\left\{\eta(k)f'(h^2_{k,P^*}) + k^{-1}h^2_{k,P^*}f''(h^2_{k,P^*})\right\} + o(C^{-1}), \end{aligned} \tag{14.7.29}$$

which is the required expansion under LFC. ■

14.8 Exercises

Exercise 14.1.1: Verify (14.1.1).

Exercise 14.2.1: Show that the pooled sample variance, U^2_n, defined in (14.2.1), is the UMVUE of σ^2.

Exercise 14.2.2: What is the MLE of σ^2? What are the MLEs of σ or $1/\sigma$?

Exercise 14.2.3: Suppose we have two numbers $0 < \delta^*_1 < \delta^*_1 < \infty$. Consider $\Omega(\delta^*_1)$ and $\Omega(\delta^*_2)$ along the line of (14.2.3). Is it true that $\Omega(\delta^*_1) \subset \Omega(\delta^*_2)$ or $\Omega(\delta^*_2) \subset \Omega(\delta^*_1)$?

Exercise 14.2.4: Consider (14.2.9) and h_{2,P^*}. Let z_{1-P^*} be an upper $100(1-P^*)\%$ point of a standard normal distribution. Express h_{2,P^*} in terms of z_{1-P^*}.

Exercise 14.3.1: Consider (14.3.1) and τ_{2,ν,P^*}. Let $t_{\nu,1-P^*}$ be an upper $100(1-P^*)\%$ point of a Student's t distribution with degrees of freedom. Express τ_{2,ν,P^*} in terms of $t_{\nu,1-P^*}$.

Exercise 14.3.2: Show that the distribution of N defined in (14.3.2) would involve the parameter σ^2, but none of those means from $\boldsymbol{\mu}$.

Exercise 14.3.3: Prove Theorem 14.3.2.

Exercise 14.3.4: Suppose that we implement the two-stage selection strategy (14.3.2) where m is determined as follows:

$$m \equiv m(\delta^*) = \max\left\{2, \langle h_{k,P^*}/\delta^*\rangle + 1\right\}.$$

Now, prove the following results.

(i) $\mathrm{P}_{\boldsymbol{\mu},\sigma}(CS) \geq P^*$ for all $\boldsymbol{\mu}$ inside $\Omega(\delta^*)$, the preference-zone with δ^* fixed.

(ii) $\mathrm{E}_{\boldsymbol{\mu},\sigma}[N/C] \to 1$ as $\delta^* \to 0$.

(iii) $\liminf \mathrm{P}_{\boldsymbol{\mu},\sigma}(CS) \geq P^*$ for all $\boldsymbol{\mu}$ inside $\Omega(\delta^*)$ as $\delta^* \to 0$.

Here, τ_{k,ν,P^*} came from (14.3.1) with $\nu = k(m-1)$, h_{k,P^*} came from (14.2.9), and $C = h_{k,P^*}^2\sigma^2/\delta^{*2}$ came from (14.2.10).

Exercise 14.3.5 (Exercise 14.3.4 Continued): Derive the least favorable configuration under this modified two-stage selection methodology.

Exercise 14.4.1: Show that the distribution of N defined in (14.4.1) would involve the parameter σ^2, but none of those means from $\boldsymbol{\mu}$.

Exercise 14.4.2: For the stopping time N defined in (14.4.1), show that

$$\mathrm{P}_\sigma(N = m) = O(C^{-k(m-1)/2}),$$

where $C = h_{k,P^*}^2\sigma^2/\delta^{*2}$ from (14.2.10). {*Hint*: Express $\mathrm{P}_\sigma(N = m)$ in terms of the distribution of a Chi-square random variable and then use a suitable bound.}

Exercise 14.4.3: Suppose that the pooled sample variance, U_n^2, is replaced by the MLE ($\equiv V_n^2$) of σ^2 in the definition of the boundary condition in (14.4.1). Derive LFC under this sequential selection methodology. Show that the results from Theorem 14.4.1 will continue to hold under the corresponding stopping rule N and selection rule $\mathcal{R}_N$.

Exercise 14.4.4: Consider the function $f(y)$ from (14.4.4). Simplify both $f^{'}(h^2_{k,P^*})$ and $f^{''}(h^2_{k,P^*})$ involving integrals. How do these expressions look in their simplest forms when $k = 2$ or 3?

Exercise 14.4.5 (Exercise 14.4.3 Continued): State and prove a version of Theorem 14.4.2 when $k = 2$.

Exercise 14.4.6: Consider the following accelerated sequential selection methodology along the line of Mukhopadhyay and Solanky (1992a). Let $m(\geq 2)$ be a pilot sample size from each treatment. Then, we define:

$$\begin{aligned} &R \equiv R(\delta^*) \text{ is the smallest integer } n(\geq m) \text{ for which} \\ &\qquad \text{we observe } n \geq \rho \frac{h^2_{k,P^*} U^2_n}{\delta^{*2}}, \end{aligned}$$

where $0 < \rho < 1$ is a preassigned number. Here, h_{k,P^*} comes from (14.2.9). Observe that $R(\delta^*)$ estimates $\rho h^2_{k,P^*}\sigma^2/\delta^{*2}$, a fraction of C. At this point, one has R observations from each treatment gathered sequentially. Denote

$$N \equiv N(\delta^*) = \left\langle \frac{R}{\rho} \right\rangle + 1.$$

Note that N estimates C. At this point, one would gather $(N - R)$ observations from each treatment in a single batch. The selection rule goes like this:

$\mathcal{R}_N$: Select as best the treatment associated with Π_j if $\overline{X}_{jN}$ is the largest among all sample means $\overline{X}_{1N}, \ldots, \overline{X}_{kN}$.

Under this accelerated sequential selection methodology, Derive the least favorable configuration under this accelerated sequential selection methodology. State and prove a version of Theorem 14.4.1.

Exercise 14.4.7 (Exercise 14.4.6 Continued): State and prove a version of Theorem 14.4.2 when $k = 2$.

Exercise 14.4.8 (Exercise 14.4.6 Continued): For the stopping time R, show that

$$\mathrm{P}_\sigma(R = m) = O(C^{-k(m-1)/2}),$$

where $C = h^2_{k,P^*}\sigma^2/\delta^{*2}$ from (14.2.10).

Exercise 14.4.9 (Exercise 14.4.6 Continued): Show that the distribution of N would involve the parameter σ^2, but none of those means from $\boldsymbol{\mu}$.

Exercise 14.4.10 (Exercise 14.4.6 Continued): Consider the final sample size N associated with accelerated sequential methodology with $\rho = \frac{1}{2}, \frac{1}{3}$, or $\frac{1}{4}$, In each situation, find an expression of κ so that one may conclude: $E_\sigma[N - C] = \kappa + o(1)$ as $\delta^* \to 0$. What are the conditions on m, if any, that must be satisfied?

Exercise 14.5.1: Revisit the simulation study that gave the output shown in Figure 14.5.4. Now, hold $k = 6, P^* = 0.95, m = 10$ fixed but use a larger $\delta^*(= 0.3)$. Execute **Seq14.exe** with live data input from file, **Norm6.dat** and implement successively

(i) two-stage selection strategy;

(ii) sequential selection strategy.

Prepare a critical summary.

CHAPTER 15

Sequential Bayesian Estimation

15.1 Introduction

Any way one may look, the area of sequential Bayesian estimation is very complicated. Thus, unfortunately, its presentation can quickly get out of hand for an audience at practically any level. It so happens that its in-depth understanding would normally require a high level of mathematical sophistication. However, in a methodological book like ours, it will be a disservice to set a bar for mathematics prerequisite that high. Frankly, we would rather not alienate those who may otherwise embrace the applied flavor of this book. Hence, we only include some selected ideas and concepts of Bayes sequential estimation.

In Section 15.2, we provide a brief review of selected concepts from fixed-sample-size estimation. We include notions of priors, conjugate priors, and marginal and posterior distributions. These are followed by discussions about Bayes risk, Bayes estimator, and highest posterior density (HPD) credible intervals.

For brevity, we do not mention Bayes tests of hypotheses. One may quickly review Bayes tests from Mukhopadhyay (2000, Section 10.6). A fuller treatment is found in Ferguson (1967), Berger (1985) and elsewhere.

Section 15.3 provides an elementary exposition of sequential concepts under a Bayesian framework. First, we discuss risk evaluation and identification of a Bayes estimator under one fixed sequential sampling strategy working in the background. This is followed by the formulation of a Bayes stopping rule. However, identification of a Bayes stopping rule frequently becomes a formidable problem in its own right. Hence, we take liberty to not go into many details at that point other than cite some of the leading references. Section 15.4 includes some data analysis.

Under Bayesian estimation, an unknown parameter ϑ itself is assumed a random entity having a probability distribution.

A common sentiment expressed by many researchers in this area points out that identifying the Bayes stopping rules explicitly is a daunting task in general. There is one exception where considerable simplifications occur which force the Bayes sequential estimation rules to coincide with the fixed-sample-size rules. Without giving too many details, in Section 15.3, we mention a general approach in this direction due to Whittle and Lane (1967). Then, we briefly focus on the binomial and normal examples in Sections 15.4 and 15.5 respectively to highlight sequential Bayesian estimation in the light of Whittle and Lane (1967).

15.2 Selected Fixed-Sample-Size Concepts

The material under review is conceptually different from anything that we have included in previous chapters. We have discussed statistical problems from a frequentist's view point. We have customarily started with a random sample $X_1, \ldots, X_n$ from a distribution having a common probability mass function (p.m.f.) or a probability density function (p.d.f.) $f(x; \vartheta)$ where $x \in \mathcal{X}$ and $\vartheta \in \Theta$, a parameter space. We have treated ϑ as an unknown but fixed parameter. So far, any inference procedure has been largely guided by the *likelihood function*, $\Pi_{i=1}^n f(x_i; \vartheta)$.

In a Bayesian approach, an experimenter additionally assumes that ϑ *is a random variable* having a probability distribution on Θ. Now, $\Pi_{i=1}^n$ $f(x_i; \theta)$ is interpreted as a likelihood function of $X_1, \ldots, X_n$ given $\vartheta = \theta$. Let us denote the p.m.f. or p.d.f. of ϑ by $\xi(\theta)$ which is called a *prior* distribution of ϑ on Θ.

$f(x; \theta)$ denotes a *conditional* p.m.f. or p.d.f. of X given $\vartheta = \theta$ from Θ.

Next, we briefly mention a *posterior* distribution and a *conjugate* prior followed by point estimation, *Bayes* estimator, and *credible interval* estimator of ϑ.

15.2.1 Marginal and Posterior Distribution

For simplicity, we assume that ϑ is a continuous *real valued* random variable on Θ which will customarily be a subinterval of the real line $\mathbb{R}$.

The evidence about ϑ derived from a prior p.d.f. $\xi(\theta)$ is combined with that from a likelihood function $\Pi_{i=1}^n f(x_i; \theta)$ by means of Bayes's theorem. Let us denote $\boldsymbol{X} = (X_1, \ldots, X_n)$.

The distribution of ϑ on a parameter space Θ is called a prior distribution.

Suppose that a statistic $T \equiv T(\boldsymbol{X})$ is (minimal) sufficient for θ given $\vartheta = \theta$. Instead of working with original data $\boldsymbol{X}$, one may equivalently work with an observed value $T = t$ and its p.m.f. or p.d.f. $g(t; \theta)$, given $\vartheta = \theta$. Here, $t \in \mathcal{T}$ where $\mathcal{T}$ is an appropriate subset of $\mathbb{R}$. Again, for simplicity, we will treat T as a continuous variable.

The *joint* p.d.f. of T and ϑ is:

$$g(t; \theta)\xi(\theta) \text{ for all } t \in \mathcal{T} \text{ and } \theta \in \Theta. \tag{15.2.1}$$

The *marginal* p.d.f. of T is:

$$m(t) = \int_{\theta\in\Theta} g(t;\theta)\xi(\theta)d\theta \text{ for all } t \in \mathcal{T}. \tag{15.2.2}$$

The *posterior* distribution of ϑ is the conditional p.d.f. of ϑ given $T = t$:

$$\begin{aligned} k(\theta;t) \equiv k(\theta \mid T = t) = g(t;\theta)\xi(\theta)/m(t) \text{ for all } t \in \mathcal{T} \\ \text{and } \theta \in \Theta \text{ such that } m(t) > 0. \end{aligned} \tag{15.2.3}$$

15.2.2 Conjugate Prior

If a prior $\xi(\theta)$ is such that the integral from (15.2.2) can not be analytically found, it may be nearly impossible to write down a clean expression for a posterior p.d.f. $k(\theta;t)$. However, a certain kind of prior would help us to pin down easily a mathematical expression of a posterior distribution.

A Bayesian combines information about ϑ from both likelihood and prior to arrive at a posterior distribution.

Suppose that $\xi(\theta)$ belongs to a particular family of distributions, Ξ. Then, $\xi(\theta)$ is called a *conjugate* prior for ϑ if and only if the posterior p.d.f. $k(\theta;t)$ also belongs to the same family Ξ. A conjugate prior helps us to identify a mathematically tractable expression of a posterior distribution.

Example 15.2.1 : Consider random variables $X_1, \ldots, X_n$ which are i.i.d. Bernoulli(θ) given that $\vartheta = \theta$, where ϑ is an unknown probability of success, $0 < \vartheta < 1$. Given $\vartheta = \theta$, the statistic $T = \sum_{i=1}^{n} X_i$ is minimal sufficient for θ, and one has $g(t;\theta) = \binom{n}{t}\theta^t(1-\theta)^{n-t}$ for $t \in \mathcal{T} = \{0, 1, \ldots, n\}$. Let us assume that a prior for ϑ on $\Theta = (0,1)$ is described by a Beta(α, β) distribution where $\alpha(> 0)$ and $\beta(> 0)$ are known numbers. From (15.2.2), for $t \in \mathcal{T}$, the marginal p.m.f. of T is:

$$m(t) = \binom{n}{t}\frac{b(t+\alpha, n+\beta-t)}{b(\alpha,\beta)} \text{ where } b(\alpha,\beta) = \frac{\Gamma(\alpha)\Gamma(\beta)}{\Gamma(\alpha+\beta)}. \tag{15.2.4}$$

Thus, posterior p.d.f. of ϑ given the data $T = t$ simplifies to:

$$\begin{aligned} k(\theta;t) = [b(t+\alpha, n-t+\beta)]^{-1}\,\theta^{t+\alpha-1}(1-\theta)^{n+\beta-t-1}, \\ \text{for all } 0 < \theta < 1, \text{ and } t = 0, 1, \ldots, n. \end{aligned} \tag{15.2.5}$$

and fixed values $t \in \mathcal{T}$. A beta prior has led to a beta posterior. In this problem, a conjugate family of priors Ξ consists of beta distributions.

Example 15.2.2 : Let $X_1, \ldots, X_n$ be i.i.d. $N(\theta, \sigma^2)$ given $\vartheta = \theta$, where $-\infty < \vartheta < \infty$ is unknown and $0 < \sigma < \infty$ is assumed known. Given $\vartheta = \theta$, the statistic $T = \sum_{i=1}^{n} X_i$ is minimal sufficient for θ. Let us assume that a prior for ϑ on $\Theta = (-\infty, \infty)$ is described by a $N(\tau, \nu^2)$ where $-\infty < \tau < \infty$ and $0 < \nu < \infty$ are known numbers. In order to find the posterior distribution of ϑ, again there is no real need to determine $m(t)$ first. One can easily check that the posterior distribution of ϑ is $N(\mu, \sigma_0^2)$ where

$$\mu = \left(\frac{t}{\sigma^2} + \frac{\tau}{\nu^2}\right)\left(\frac{n}{\sigma^2} + \frac{1}{\nu^2}\right)^{-1} \text{ and } \sigma_0^2 = \left(\frac{n}{\sigma^2} + \frac{1}{\nu^2}\right)^{-1}. \tag{15.2.6}$$

A normal prior has led to a normal posterior. In this problem, a conjugate family of priors Ξ consists of normal distributions.

We emphasize that a conjugate prior may work well mathematically, but we do not want to create an impression that one must always work with a conjugate prior. In fact a conjugate prior may not be reasonable in every problem. The following example highlights this point.

Let X be distributed as $N(\theta, 1)$ given $\vartheta = \theta$. If we are told that ϑ is positive, clearly there is no point in assuming a normal prior. In some situations, however, it may be reasonable to assume that the prior on $(0, \infty)$ is an exponential distribution, Expo(α), with $\alpha(> 0)$ known. We leave it as an exercise to figure out an analytical expression of posterior in this case. One may refer to Mukhopadhyay (2000, Section 10.7; 2006, Section 10.5) for fuller discussions of such examples.

15.2.3 Point Estimation and Bayes Estimator

How does one approach a point estimation problem? One begins with a loss function. Suppose that a real valued statistic T is (minimal) sufficient for θ given $\vartheta = \theta$. Let $\xi(\theta)$ be a prior distribution of $\vartheta, \theta \in \Theta$. Let $g(t; \theta)$ stand for the p.d.f. of T given $\vartheta = \theta$.

Suppose that the loss in estimating ϑ by a point estimator, denoted by $U \equiv U(T)$, is a non-negative function $L(\vartheta, U)$. In statistics, one often confronts a *squared error loss* function, that is, $L(\vartheta, U) \equiv [U(T) - \vartheta]^2$. However, let us keep $L(\vartheta, U)$ arbitrary and non-negative.

Conditionally given $\vartheta = \theta$, the risk function is given by:

$$R(\theta, U) \equiv \mathrm{E}_{T|\vartheta=\theta}[L(\theta, U)] = \int_{\mathcal{T}} L(\theta, U(t)) g(t; \theta) dt. \tag{15.2.7}$$

This is called the *frequentist* risk. Now, the *Bayesian* risk is given by:

$$r(\vartheta, U) \equiv \mathrm{E}_{\vartheta}[R(\vartheta, U)] = \int_{\Theta} R(\theta, U) \xi(\theta) d\theta. \tag{15.2.8}$$

Next, suppose that $\mathcal{D}$ is the class of all estimators of ϑ whose Bayesian risks are finite. Then, the *best* estimator under a Bayesian paradigm is δ^* from $\mathcal{D}$ such that

$$r(\vartheta, \delta^*) = \inf_{\delta \in \mathcal{D}} r(\vartheta, \delta). \tag{15.2.9}$$

Such an estimator δ^* from $\mathcal{D}$ is called the *Bayes* estimator of ϑ. In many routine problems, a Bayes estimator is essentially unique. The two following results help in identifying the Bayes estimator.

Theorem 15.2.1 : *The Bayes estimate $\delta^* \equiv \delta^*(t)$ is determined in such a way that the posterior risk of $\delta^*(t)$ is the smallest possible, that is*

$$\int_{\Theta} L(\theta, \delta^*(t))k(\theta; t)d\theta = \inf_{\delta \in \mathcal{D}} \int_{\Theta} L(\theta, \delta(t))k(\theta; t)d\theta$$

for all possible observed data, $t \in \mathcal{T}$.

Theorem 15.2.2 : *Under a squared error loss function, the Bayes estimate $\delta^* \equiv \delta^*(t)$ is the mean of the posterior distribution $k(\theta; t)$, that is*

$$\delta^*(t) = \int_{\Theta} \theta k(\theta; t)d\theta \equiv \mathrm{E}_{\theta|T=t}[\vartheta]$$

for all possible observed data, $t \in \mathcal{T}$.

Example 15.2.3 (Example 15.2.1 Continued): With a Beta(α, β) prior for ϑ, recall that the posterior distribution of ϑ given $T = t$ was Beta$(t + \alpha, n - t + \beta)$ for $t = 0, 1, \ldots, n$. Now, in view of Theorem 15.2.2, under a squared error loss function, the Bayes estimate of ϑ is the posterior mean. However, the mean of a Beta$(t + \alpha, n - t + \beta)$ distribution is easily found. We have:

$$\text{The Bayes estimate of } \vartheta \text{ is } \widehat{\vartheta}_B = \frac{(t + \alpha)}{(\alpha + \beta + n)}.$$

Example 15.2.4 (Example 15.2.2 Continued): With a $N(\tau, \nu^2)$ prior for ϑ, recall that the posterior distribution of ϑ given $T = t$ was $N(\mu, \sigma_0^2)$ for $-\infty < t < \infty$, where $\mu = \left(\frac{t}{\sigma^2} + \frac{\tau}{\nu^2}\right)\left(\frac{n}{\sigma^2} + \frac{1}{\nu^2}\right)^{-1}$ and $\sigma_0^2 = \left(\frac{n}{\sigma^2} + \frac{1}{\nu^2}\right)^{-1}$. Now, in view of Theorem 15.2.2, under a squared error loss function, the Bayes estimate of ϑ is the posterior mean. However, the posterior mean is μ. We have:

$$\text{The Bayes estimate of } \vartheta \text{ is } \widehat{\vartheta}_B = \left(\frac{t}{\sigma^2} + \frac{\tau}{\nu^2}\right)\left(\frac{n}{\sigma^2} + \frac{1}{\nu^2}\right)^{-1}.$$

Bayes estimator of ϑ is given by: Posterior mean when $L(\vartheta, \delta) = (\delta - \vartheta)^2$; Posterior median when $L(\vartheta, \delta) = \lvert\delta - \vartheta\rvert$.

15.2.4 Credible Intervals

Having fixed α, $0 < \alpha < 1$, instead of focusing on a $100(1 - \alpha)\%$ confidence interval for ϑ, a Bayesian focuses on a $100(1 - \alpha)\%$ *credible set* for ϑ. A subset Θ^* of Θ is a $100(1-\alpha)\%$ *credible set* for ϑ if and only if posterior probability of Θ^* is at least $(1-\alpha)$, that is:

$$\int_{\Theta^*} k(\theta;t)d\theta \geq 1-\alpha. \tag{15.2.10}$$

In some situations, a $100(1-\alpha)\%$ credible set Θ^* for ϑ may not be a subinterval of Θ. When Θ^* is an interval, we refer to it as a $100(1-\alpha)\%$ *credible interval* for ϑ.

It is conceivable that one will have to choose a $100(1-\alpha)\%$ credible set from a long list of credible sets for ϑ. Naturally, one should form a credible set by including only those values of ϑ which are "most likely" according to a posterior distribution. Also, a credible set should be "small". The concept of the *highest posterior density* (HPD) credible set addresses both issues.

Having fixed α, $0<\alpha<1$, a subset Θ^* of Θ is called a $100(1-\alpha)\%$ HPD credible set for ϑ if and only if Θ^* is given by:

$$\Theta^* = \{\theta \in \Theta : k(\theta;t) \geq a\}, \tag{15.2.11}$$

where $a \equiv a(\alpha,t)$ is the largest number such that the posterior probability of Θ^* is at least $(1-\alpha)$.

If $k(\theta;t)$ is *unimodal*, a $100(1-\alpha)\%$ HPD credible set happens to be an interval. Additionally, if $k(\theta;t)$ is *symmetric*, the HPD $100(1-\alpha)\%$ credible interval Θ^* will be the shortest, equal tailed and symmetric around the posterior mean.

Example 15.2.5 (Example 15.2.4 Continued): The posterior distribution $N(\mu,\sigma_0^2)$ being symmetric about μ, the HPD $100(1-\alpha)\%$ credible interval Θ^* will be:

$$\left\{\left(\frac{n}{\sigma^2}\overline{x}+\frac{1}{\nu^2}\tau\right)\Big/\left(\frac{n}{\sigma^2}+\frac{1}{\nu^2}\right)\right\} \pm z_{\alpha/2}\left(\frac{n}{\sigma^2}+\frac{1}{\nu^2}\right)^{-1/2}. \tag{15.2.12}$$

Centered at the posterior mean ($\widehat{\vartheta}_B$), this interval stretches either way by $z_{\alpha/2}$ times the posterior standard deviation (σ_0).

For brevity, we do not review Bayesian methods for tests of hypotheses. One may quickly review Bayes tests from Mukhopadhyay (2000, Section 10.6). A fuller treatment is in Ferguson (1967) and Berger (1985). An advanced reader may also look at Baron (2004) and Tartakovsky and Veeravalli (2004).

15.3 Elementary Sequential Concepts

From Section 15.2, it is abundantly clear that a fixed-sample-size Bayesian estimation problem would normally include the following important ingredients:

$$\begin{array}{rl} \text{(i)} & \text{a known sample size } n; \\ \text{(ii)} & \text{a p.d.f. or p.m.f. } f(x;\theta) \text{ given } \vartheta = \theta(\in \Theta), \\ & \text{where } f \text{ is assumed known, but } \vartheta \text{ is unknown;} \\ \text{(iii)} & \text{i.i.d. observations } X_1, \ldots, X_n \text{ from } f(x;\theta), \\ & \text{given } \vartheta = \theta(\in \Theta); \\ \text{(iv)} & \text{a loss function } L(\vartheta, \delta), \delta \in \mathcal{D}; \\ \text{(v)} & \text{a prior distribution of } \vartheta,\ \xi(\theta) \in \Xi. \end{array} \tag{15.3.1}$$

One will find numerous references involving a loss function $L(\theta, \delta)$ which is either $|\delta - \theta|$ or $(\delta - \theta)^2$ or zero-one. Incidentally, a "zero-one" loss function may correspond to a problem in interval estimation or test of hypotheses. One may refer to Example 2.3.1.

In a sequential Bayesian estimation problem, one also includes the following additional ingredients:

$$\begin{array}{rl} \text{(vi)} & \text{a known cost } c(>0) \text{ per unit observation;} \\ \text{(vii)} & \text{a stopping variable } N \text{ where the event} \\ & [N > j] \text{ depends only on } X_1, \ldots, X_j, \\ & j = 1, 2, \ldots \text{ that terminates w.p. 1.} \end{array} \tag{15.3.2}$$

Clearly, fewer observations would cost less, but one will encounter larger estimation error. On the other hand, more observations would cost more, but one will encounter smaller estimation error. So, one must strike an appropriate balance.

In a sequential Bayesian estimation problem, one must be mindful about two crucial matters. Appropriate choices must be made about (1) a stopping rule or equivalently a stopping variable N, and (2) a terminal estimator $\delta_N \equiv \delta_N(X_1, \ldots, X_N)$ for ϑ, $\delta_N \in \mathcal{D}$. A Bayesian may focus only on non-randomized terminal estimator δ_N.

Next, let us represent an arbitrary stopping rule or a stopping time as follows: A stopping rule may be viewed more generally as a sequence of functions:

$$\Psi = \{\psi_1(\boldsymbol{X}_1), \psi_2(\boldsymbol{X}_2), \psi_3(\boldsymbol{X}_3), \ldots, \psi_j(\boldsymbol{X}_j), \ldots\} \tag{15.3.3}$$

where $\boldsymbol{X}_j = (X_1, \ldots, X_j)$ and $\psi_j(\boldsymbol{X}_j)$ takes a value 0 (do not stop and continue by taking one more observation beyond $\boldsymbol{X}_j$) or 1 (stop with $\boldsymbol{X}_j$), $j = 1, 2, \ldots$.

In order to clarify this abstract idea, let us revisit the stopping rule from (6.4.5). Based on a sequence $\boldsymbol{X} = \{X_1, X_2, \ldots, X_m, X_{m+1}, X_{m+2}, \ldots\}$ of i.i.d. $N(\mu, \sigma^2)$ observations, we had the following stopping rule:

$$\begin{array}{c} N \text{ is the smallest integer } j(\geq m \geq 2) \text{ for which} \\ \text{we observe } j \geq (A/c)^{1/2} S_j. \end{array} \tag{15.3.4}$$

Here, $\overline{X}_j = j^{-1} \sum_{i=1}^{j} X_i$, $S_j^2 = (j-1)^{-1} \sum_{i=1}^{j} (X_i - \overline{X}_j)^2$ are respectively the sample mean and sample variance obtained from $\boldsymbol{X}_j = (X_1, \ldots, X_j)$, $j = m, m+1, \ldots$.

We begin sampling with pilot observations $X_1, \ldots, X_m$ and thus there is no question about stopping $1, 2, \ldots, m-1$ observations. So, let us define

$$\psi_1(\boldsymbol{X}_1) = \psi_2(\boldsymbol{X}_2) = \ldots = \psi_{m-1}(\boldsymbol{X}_{m-1}) = 0.$$

Next, let us define the following sequence of events:

$$\begin{aligned} A_m &= [m \geq (A/c)^{1/2} S_m], \\ A_{m+j} &= \cap_{l=0}^{j-1} [m + l < (A/c)^{1/2} S_{m+l}] \cap [m + j \geq (A/c)^{1/2} S_{m+j}], \\ &\qquad j = 1, 2, \ldots. \end{aligned} \tag{15.3.5}$$

Thus, we define:

$$\begin{aligned} &\psi_1(\boldsymbol{X}_1) = \psi_2(\boldsymbol{X}_2) = \ldots = \psi_{m-1}(\boldsymbol{X}_{m-1}) = 0, \text{ and} \\ &\psi_{m+j}(\boldsymbol{X}_{m+j}) \equiv I(A_{m+j}) = \begin{cases} 0 & \text{if } A^c_{m+j} \text{ occurs} \\ 1 & \text{if } A_{m+j} \text{ occurs} \end{cases} \\ &\qquad j = 0, 1, 2, \ldots \end{aligned} \tag{15.3.6}$$

One notes that (15.3.6) specifies a full Ψ-sequence mentioned in (15.3.3). The particular Ψ-sequence from (15.3.6) describes the exact same stopping rule or N defined in (15.3.4). By the way, another role of the Ψ-sequence from (15.3.6) may be understood as follows:

$$\begin{aligned} \mathrm{P}_\theta(N = j) &= \mathrm{E}_\theta[\psi_j(\boldsymbol{X}_j)] \equiv 0, j = 1, \ldots, m-1, \text{ but} \\ \mathrm{P}_\theta(N = m + l) &= \mathrm{E}_\theta[\psi_{m+l}(\boldsymbol{X}_{m+l})], l = 0, 1, 2, \ldots. \end{aligned} \tag{15.3.7}$$

In general, the role of a Ψ-sequence mentioned in (15.3.3) is no different from what is understood via (15.3.7). We may consider a class of arbitrary stopping times where each stopping time N will be associated with a unique Ψ-sequence such that

$$P_\theta(N = j) = \mathrm{E}_\theta[\psi_j(\boldsymbol{X}_j)], j = 1, 2, \ldots.$$

15.3.1 *Risk Evaluation and Bayes Estimator Under One Fixed Sequential Sampling Strategy*

We suppose that we are all set with the ingredients listed in (15.3.1)-(15.3.2). We will work interchangeably with an arbitrary stopping time N or its associated Ψ-sequence as needed.

First, let us focus on a fixed stopping time N (or Ψ) and an estimator $\delta_N \equiv \delta_N(X_1, \ldots, X_N)$ for ϑ. We suppose that $\mathcal{D}$, the collection of all possible estimators for ϑ, is such that each estimator has a finite Bayes risk.

Given $\vartheta = \theta$, along the line of (15.2.7), the "frequentist" risk will be:

$$\begin{aligned} & R(\theta, (\Psi, \delta_N)) \\ & = \textstyle\sum_{j=1}^{\infty} \mathrm{E}_{\boldsymbol{X}|\theta,\Psi} \left[\{L(\theta, \delta_j(\boldsymbol{X}_j)) + cj\} I[N = j]\right] \\ & = \textstyle\sum_{j=1}^{\infty} \mathrm{E}_{\boldsymbol{X}|\theta} \left[\{L(\theta, \delta_j(\boldsymbol{X}_j)) + cj\} \psi_j(\boldsymbol{X}_j))\right]. \end{aligned} \tag{15.3.8}$$

Now, along the line of (15.2.8), an expression of the Bayesian risk follows from (15.3.8):

$$\begin{aligned} & r(\vartheta, (\Psi, \delta_N)) \\ & = \textstyle\sum_{j=1}^{\infty} \int_{\Theta} \mathrm{E}_{\boldsymbol{X}|\theta} \left[\{L(\theta, \delta_j(\boldsymbol{X}_j)) + cj\} \psi_j(\boldsymbol{X}_j)\right] \xi(\theta) d\theta \\ & = \textstyle\sum_{j=1}^{\infty} \mathrm{E}_{\boldsymbol{X},\theta} \left[\{L(\theta, \delta_j(\boldsymbol{X}_j)) + cj\} \psi_j(\boldsymbol{X}_j)\right]. \end{aligned} \tag{15.3.9}$$

Suppose that $\delta_j^0 \equiv \delta_j^0(\boldsymbol{X}_j)$ is the Bayes estimator of ϑ based on $\boldsymbol{X}_j$ for each fixed $j = 1, 2, \ldots$. According to the notion of a Bayes estimator found in (15.2.9), what that means is this:

$$\begin{aligned} & \mathrm{E}_{\boldsymbol{X},\theta} \left[\{L(\theta, \delta_j(\boldsymbol{X}_j)) + cj\} \psi_j(\boldsymbol{X}_j)\right] \\ & \quad \geq \mathrm{E}_{\boldsymbol{X},\theta} \left[\left\{L(\theta, \delta_j^0(\boldsymbol{X}_j)) + cj\right\} \psi_j(\boldsymbol{X}_j)\right], \end{aligned} \tag{15.3.10}$$

for all $\delta_j(\boldsymbol{X}_j) \in \mathcal{D}$, for each fixed $j = 1, 2, \ldots$.

Next, combining (15.3.9) and (15.3.10), for fixed Ψ and $\delta_N \in \mathcal{D}$, we have:

$$r(\vartheta, (\Psi, \delta_N)) \geq \sum_{j=1}^{\infty} \mathrm{E}_{\boldsymbol{X},\theta} \left[\left\{L(\theta, \delta_j^0(\boldsymbol{X}_j)) + cj\right\} \psi_j(\boldsymbol{X}_j))\right]. \tag{15.3.11}$$

This implies that the sequential estimator

$$\delta^0 \equiv \left\{\delta_1^0(\boldsymbol{X}_1), \delta_2^0(\boldsymbol{X}_2), \ldots, \delta_j^0(\boldsymbol{X}_j), \ldots\right\}$$

is the Bayes estimator of ϑ whatever may be Ψ, a fixed stopping rule.

We have actually proved a very important result in Bayesian sequential estimation. The result is this: Whatever may be the specifics of a fixed sequential sampling strategy, suppose that it terminates with j observations recorded as $\boldsymbol{X}_j$. Then, we need to estimate ϑ by the fixed-sample-size Bayes estimator $\delta_j^0(\boldsymbol{X}_j)$. At this point, how exactly data $\boldsymbol{X}_j$ was gathered becomes irrelevant whatever be $j = 1, 2, \ldots$.

Example 15.3.1 (Example 15.2.4 Continued): Let $X_1, \ldots, X_j$ be i.i.d. $N(\theta, \sigma^2)$ given $\vartheta = \theta$, where $-\infty < \vartheta < \infty$ is unknown and $0 < \sigma < \infty$ is assumed known. With a $N(\tau, \nu^2)$ prior for ϑ, recall from (15.2.6) that the posterior distribution of ϑ was $N(\mu, \sigma_0^2)$, where

$$\mu = \left(j\sigma^{-2}\overline{x}_j + \nu^{-2}\tau\right)\left(j\sigma^{-2} + \nu^{-2}\right)^{-1} \text{ and } \sigma_0^2 = \left(j\delta^{-2} + \nu^{-2}\right)^{-1}.$$

Now, having observed data $\boldsymbol{x}_j = (x_1, \ldots, x_j)$, under a squared error loss function $L(\vartheta, \delta) = (\delta - \vartheta)^2$, we know that the Bayes estimator of ϑ is the posterior mean:

$$\delta_j^0 = \left(\frac{j\overline{x}_j}{\sigma^2} + \frac{\tau}{\nu^2}\right)\left(\frac{j}{\sigma^2} + \frac{1}{\nu^2}\right)^{-1}. \qquad (15.3.12)$$

Now, whatever may be the sequential strategy leading to sample size j and data $\overline{x}_j$ at termination, that information becomes redundant at data analysis stage. For a sequential problem, the Bayes estimator of ϑ will be

$$\delta^0 \equiv \left\{\delta_1^0(\boldsymbol{X}_1), \delta_2^0(\boldsymbol{X}_2), \ldots, \delta_j^0(\boldsymbol{X}_j), \ldots\right\},$$

where $\delta_j^0, j = 1, 2, \ldots$, would come from (15.3.12).

In other words, a sequentially generated data $(j, \overline{x}_j)$ will be analyzed by a Bayesian pretending that this was a fixed-sample-size data. Some authors take this phenomenon to heart and refer to it as the *stopping rule principle* (Berger 1985, pp. 502-503). There is no doubt that trusting this "principle" does make parts of some sequential Bayesian calculations, for example, finding the Bayes estimator, rather easy. However, this principle has its critics too including P. Armitage and staunch Bayesians such as D. Basu and others. Armitage (1961) and Basu (1975) discussed a very damaging example that seriously questioned the logic behind the "*principle*".

15.3.2 Bayes Stopping Rule

In Section 15.3.1, we have shown that

$$\inf_{\delta \in \mathcal{D}} r(\vartheta, (\Psi, \delta_N)) = r(\vartheta, (\Psi, \delta^0)) \qquad (15.3.13)$$

where

$$\delta^0 \equiv \left\{\delta_1^0(\boldsymbol{X}_1), \delta_2^0(\boldsymbol{X}_2), \ldots, \delta_j^0(\boldsymbol{X}_j), \ldots\right\}$$

is the Bayes estimator of ϑ whatever may be Ψ, that is the fixed stopping rule. Such optimal decision rule or estimator does not involve Ψ.

Now, one would like to find a stopping rule (or a stopping time), or equivalently, a Ψ-sequence denoted by Ψ^0, such that it minimizes the Bayes risk from (15.3.13). That is, we want to identify Ψ^0 such that

$$\inf_{\Psi} r(\vartheta, (\Psi, \delta^0)) = r(\vartheta, (\Psi^0, \delta^0)). \qquad (15.3.14)$$

Once Ψ^0 is identified, the corresponding stopping rule together with δ^0 will constitute the *optimal Bayesian solution* to the original sequential estimation problem.

Intuitively speaking, a Bayes stopping rule will be guided by a very practical idea: At every stage j, we would compare the posterior Bayes risk (A) of terminating and estimating ϑ on the basis of $\boldsymbol{X}_j$ with expected posterior Bayes risk (B) that will be accrued if more data are collected. We will go one way or the other depending on which route is cheaper. Take the first (second) route if $A < (>) B$.

While identifying δ^0 looked simple enough, the problem of finding Ψ^0 satisfying (15.3.14) is far away from anything that we may call "simple". Berger (1985, Section 7.4.1, p. 442) writes, "While Bayesian analysis in fixed sample size problems is straightforward (robustness considerations aside), Bayesian sequential analysis is very difficult." Ghosh et al. (1997, Section 5.4, p. 125) write, "From the discussion of the previous sections, it is clear that finding Bayes stopping rules explicitly is usually a formidable task except in those special cases where the Bayes rules reduce to fixed-sample-size rules." It is a topic that is quite frankly out of scope for this methodologically oriented book. For a thorough treatment, one may refer to Ghosh et al. (1997, Chapter 5).

Whittle and Lane (1967) developed general theory to cover a number of scenarios where the Bayes sequential estimation and stopping rules would reduce to fixed-sample-size rules. They gave sufficient conditions for this to happen with illustrations.

There is a large amount of literature out there for approximate and asymptotic Bayesian sequential solutions. This is especially so since exact solutions are very rare and they are normally extremely hard to come by. The series of papers by Bickel and Yahav (1967,1968,1969a,b) and their general approach to find *asymptotically pointwise optimal* (APO) stopping rules provided opportunities for researchers in this area for a very long time. One will find many other references from Ghosh et al. (1997, Chapter 5).

15.4 Data Analysis

The Bayes estimation of the parameter ϑ described in Examples 15.2.3 and 15.2.4 can be simulated and analyzed by using the computer program **Seq15.exe**. Upon execution of this program, a user will be prompted to select either 'program number: 1' or 'program number: 2'. 'Program number: 1' is dedicated to the estimation problem for Bernoulli success probability ϑ described in Example 15.2.3. 'Program number: 2' is dedicated to the estimation problem for the normal mean ϑ described in Example 15.2.4. These programs can be used to perform simulation studies or with live data input.

```
    Select a program using the number given
    in the left hand side. Details of these
    programs are given in Chapter 15.

*******************************************************
* Number         Program                              *
*                                                     *
*  1   Estimation of Bernoulli parameter, theta       *
*      using beta conjugate prior                     *
*                                                     *
*  2   Estimation of normal mean, theta using         *
*      normal conjugate prior                         *
*                                                     *
*  0   Exit                                           *
*******************************************************
     Enter the program number: 1

Do you want to save the results in a file(Y or N)? Y

Note that your default outfile name is "SEQ15.OUT".
Do you want to change the outfile name?
Y(yes) or N(no)]: N

This program computes the Bayesian estimator of the
Bernoulli parameter, theta using beta conjugate prior.

Enter the following parameters of beta prior:
          alpha = 3
          beta  = 2

Enter sample size, n = 40
Do you want to change the input parameters?[Y or N]: N

Do you want to simulate data or input real data?
 Type S (Simulation) or D (Real Data): S

Number of replications for simulation? 5000

Enter a positive integer(<10) to initialize
          the random number sequence: 3

Do you want to store simulated theta values?
Type Y (Yes) or N (No): N

For the data simulation, enter theta = 0.35
```

Figure 15.4.1. *Input screen for a simulation study of estimation of Bernoulli* ϑ *using* ***Seq15.exe***.

Figure 15.4.1 shows the input screen for 'program number: 1' within **Seq15.exe**. This figure highlights the required input to implement a simulation study for the estimation of ϑ in a Bernoulli(ϑ) distribution under a Beta(α, β) prior. We fixed $\alpha = 3, \beta = 2$ and the sample size, $n = 40$. We opted for 5000 simulation runs using the random sequence 3. Then, the value of ϑ was set at 0.35. Thus, simulations were performed by generating data from Bernoulli($\vartheta = 0.35$) population. Figure 15.4.2 summarizes our findings by averaging performances from all 5000 simulation runs of the Bayes estimation problem. It includes the average simulated value of $\widehat{\vartheta}(= 0.3770)$ and its simulated standard error $(= 0.0010)$ along with the minimum $(= 0.1556)$ and maximum $(= 0.6222)$ values.

```
Bayes Estimator of Bernoulli Parameter Theta
  Using Beta(alpha,beta) Conjugate Prior
============================================

Simulation study

Input parameters:  alpha =    3.000
                   beta  =    2.000
         sample size, n =    40

Simulation parameters:
                  theta = 0.3500
   number of simulations =    5000

Average simulated theta   =  0.3770
Simulated standard error =  0.0010
Minimum simulated theta   =  0.1556
Maximum simulated theta   =  0.6222
```

Figure 15.4.2. *Output results for the simulation given in Figure 15.4.1.*

Table 15.4.1: **Ber6.dat**, a dataset from a Bernoulli population

0	1	0	1	0	1	0	1	1	0
1	1	1	0	1	0	1	1	1	0
1	1	1	1	1	1	1	0	1	1
1	1	1	0	0	1	1	1	0	1

```
Bayes Estimator of Bernoulli Parameter Theta
  Using Beta(alpha,beta) Conjugate Prior
============================================

Estimation using input data file: ber6.dat

Input parameters:  alpha =   2.500
                   beta  =   1.500

Sample size, n =   40

Estimated value:
    theta_hat  =  0.6477
```

Figure 15.4.3. *Output results for data given in Table 15.4.1 using* ***Seq15.exe****.*

Consider a Bernoulli dataset, **Ber6.dat** provided in Table 15.4.1. The data were obtained from a Bernoulli population with unknown θ. We used the program **Seq15.exe** to obtain $\widehat{\vartheta}$, the Bayes estimate of θ with Beta($\alpha = 2.5, \beta = 1.5$) prior. Figure 15.4.3 gives the output of the program on these data. It shows that $\widehat{\vartheta} = 0.6477$ with prior Beta(2.5, 1.5).

The input screen for 'program number: 2' within **Seq15.exe** is similar to what one finds in Figure 15.4.1. So, we refrain from highlighting it. We opted for 5000 simulation runs using the random sequence 4. We implemented the Bayesian estimation method for a normal mean ϑ with a normal conjugate prior $N(\tau, \nu^2)$ given in Example 15.2.4 by fixing $\tau = 3$, $\nu = 2$, $n = 50$ and $\sigma = 4$.

```
Bayes Estimator of Normal Mean, Theta
Using Normal(tau,delta^2) Conjugate Prior
=========================================
Simulation study

Input parameters:      tau    = 3.000
                     delta    = 2.000

Population std dev (sigma) = 4.000
          sample size, n   =    50

Posterior std dev (sigma0) = 0.544

Simulation parameters:
                     theta  = 5.000
     number of simulations =  5000

Average simulated theta   = 4.848
Simulated standard error = 0.007
Minimum simulated theta   = 2.952
Maximum simulated theta   = 6.866
```

Figure 15.4.4. *Simulation summary results for the Bayesian estimation of a normal mean.*

Table 15.4.2: **Inc.dat**, a dataset from a normal population with $\sigma = 15$

30.5188	38.4435	72.1393	37.4989	42.1953	24.5941	72.3075
57.5129	61.6863	50.5715	38.7393	69.5922	72.0587	46.3513
48.5052	32.2928	54.4011	66.6335	43.3433	54.5920	62.1059
49.5175	55.9080	68.6105	45.4381	31.0215	68.3956	56.0156
51.3662	45.7886	37.7619	51.3655	80.1450	35.9669	57.6301
68.4749	64.7568	58.1053	55.9939	69.6683	56.1571	34.4603
52.7069	23.5020	49.5279	55.0199	60.8312	39.7401	52.4627
46.6824	46.6881	74.2213	70.1652	80.6352	51.7996	43.9706
64.4379	69.8026	42.8364	49.0709	44.9183	68.4548	80.2041
60.3784	47.0002	51.8936	21.3821	49.0744	52.5003	52.0141
47.2946	63.3401	47.6967	66.9386	51.4478	44.8396	41.3007
34.4352	23.3663	43.9954				

Data for simulations were randomly generated from $N(\vartheta = 5, \sigma^2 = 4^2)$. Figure 15.4.4 summarizes our findings by averaging performances from 5000 simulation runs of the Bayes estimate of ϑ. It includes the average simulated $\widehat{\vartheta}(= 4.848)$ and its simulated standard error $(= 0.007)$ along with the minimum $(= 2.952)$ and maximum $(= 6.866)$ values. It also computed the posterior standard deviation $\sigma_0(= 0.544)$.

Next, Table 15.4.2 consists of $80(= n)$ observations obtained from a normal population with $\sigma = 15$. The mean, ϑ, for this population is unknown and we used the program **Seq15.exe** under a prior $N(\tau = 5, \nu^2 = 3^2)$ to obtain the Bayes estimator for ϑ. The output is highlighted in Figure 15.4.5. It includes the Bayes estimator $\widehat{\vartheta} = 41.050$ and the posterior standard deviation $\sigma_0 = 1.464$.

```
Bayes Estimator of Normal Mean, Theta
Using Normal(tau,delta^2) Conjugate Prior
================================================

Estimation using input data file: inc.dat

Input parameters:     tau   =    5.000
                      delta =    3.000

Population std dev (sigma) =   15.000

             Sample size, n =   80

Posterior std dev (sigma0) =    1.464

Estimated value, theta_hat =  41.050
```

Figure 15.4.5. *Output results for data* ***Inc.dat*** *given in Table 15.4.2 using* ***Seq15.exe.***

15.5 Exercises

Exercise 15.2.1: Let $X_1, \ldots, X_n$ be i.i.d. Poisson(θ) given $\vartheta = \theta$ where $\vartheta(> 0)$. Given $\vartheta = \theta$, show that the statistic $T = \sum_{i=1}^{n} X_i$ is minimal sufficient for θ. Under a Gamma(α, β) prior where $\alpha(> 0)$ and $\beta(> 0)$ are known numbers, show that

(i) the marginal p.m.f. is given by $\dfrac{n^t \beta^t \Gamma(\alpha + t)}{t!(n\beta + 1)^{\alpha + t}\Gamma(\alpha)}, t = 0, 1, \ldots.$

(ii) the posterior distribution is Gamma$(t + \alpha, \beta(n\beta + 1)^{-1})$.

(iii) $\widehat{\vartheta}_B = \dfrac{(t + \alpha)}{(\alpha + \beta + n)}$ under a squared error loss.

Exercise 15.2.2: Prove Theorem 15.2.1. {*Hint*: Express Bayesian risk in the following form:

$$r(\vartheta,\delta) = \int_\Theta \int_\mathcal{T} L(\theta,\delta(t))g(t;\theta)\xi(\theta)dtd\theta$$
$$= \int_\mathcal{T} \left[\int_\Theta L^*(\theta,\delta(t))k(t;\theta)d\theta\right] m(t)dt.$$

Argue that the integral $\int_\Theta \int_\mathcal{T}$ can be changed to $\int_\mathcal{T} \int_\Theta \cdot$}

Exercise 15.2.3: Prove Theorem 15.2.2. {*Hint*: One needs to minimize

$$\int_\Theta L^*(\theta,\delta(t))k(\theta;t)d\theta$$

with respect to δ, for every fixed $t \in \mathcal{T}$. Rewrite:

$$\int_\Theta L^*(\theta,\delta(t))k(\theta;t)d\theta = \int_\Theta [\theta^2 - 2\theta\delta + \delta^2]k(\theta;t)d\theta$$
$$= a(t) - 2\delta \mathrm{E}_{\theta|T=t}[\vartheta] + \delta^2,$$

where $a(t) = \int_\Theta \theta^2 k(\theta;t)d\theta$. Look at $a(t) - 2\delta \mathrm{E}_{\theta|T=t}[\vartheta] + \delta^2$ as a function of $\delta \equiv \delta(t)$ and minimize it with respect to δ.}

Exercise 15.2.4 (Example 15.2.3 Continued): Show that Bayes estimator of ϑ, $\widehat{\vartheta}_B = (t+\alpha)(\alpha+\beta+n)^{-1}$, is a weighted average of sample mean and prior mean. Which one between these two means receives more weight when

(i) n becomes large, but α, β are held fixed.
(ii) α or β become large, but n remains fixed.

Explain your thoughts in layman's terms.

Exercise 15.2.5 (Example 15.2.4 Continued): Show that Bayes estimator of ϑ, $\widehat{\vartheta}_B = \left(n\sigma^{-2}\overline{x} + \nu^{-2}\tau\right) / \left(n\sigma^{-2} + \nu^{-2}\right)$, is a weighted average of sample mean and prior mean. Which one between these two means receives more weight when

(i) n becomes large, but τ, ν are held fixed.
(ii) τ, ν become large, but n remains fixed.

Explain your thoughts in layman's terms.

Exercise 15.2.6: Suppose that $X_1, \ldots, X_n$ are i.i.d. with a common p.d.f. $f(x;\theta) = \theta\exp(-\theta x)I(x>0)$ given $\vartheta = \theta$ where $\vartheta(>0)$ is an unknown parameter. Assume prior density $\xi(\theta) = \alpha\exp(-\alpha\theta)I(\theta>0)$ where $\alpha(>0)$ is known. Derive posterior p.d.f. of ϑ given $T=t$ where $T=\sum_{i=1}^n X_i, t>0$. Under a squared error loss function, obtain the expression of Bayes estimator of ϑ.

Exercise 15.2.7: Suppose that $X_1, \ldots, X_{10}$ are i.i.d. Poisson(θ) given $\vartheta = \theta$ where $\vartheta(>0)$ is an unknown parameter. Assume prior density $\xi(\theta) = \frac{1}{4}\exp(-\theta/4)I(\theta > 0)$. Draw and compare the posterior p.d.f.s of ϑ given that $\sum_{i=1}^{10} X_i = t$ where $t = 30, 40, 50$. Under a squared error loss function, obtain the expression of Bayes estimator of ϑ.

Exercise 15.2.8: Suppose that $X_1, \ldots, X_n$ are i.i.d. Uniform$(0, \theta)$ given $\vartheta = \theta$ where $\vartheta(>0)$ is an unknown parameter. We say that ϑ has a *Pareto* prior, denoted by Pareto(α, β), when prior p.d.f. is given by

$$\xi(\theta) = \beta\alpha^{\beta}\theta^{-(\beta+1)}I(\alpha < \theta < \infty).$$

Here, α, β are known positive numbers. Denote the minimal sufficient statistic $T = X_{n:n}$, the largest order statistic, given $\vartheta = \theta$. Show that posterior distribution of ϑ given $T = t$ turns out to be Pareto$(\max(t, \alpha), n + \beta)$. Under a squared error loss function, obtain the expression of Bayes estimator of ϑ.

Exercise 15.2.9: Suppose that $X_1, \ldots, X_n$ are i.i.d. with a common p.d.f. $f(x; \theta) = \frac{1}{\theta}\exp(-x/\theta)I(x > 0)$ given $\vartheta = \theta$ where $\vartheta(>0)$ is an unknown parameter. We referred to this distribution as Expo(θ) in Chapter 8. Assume prior density

$$\xi(\theta) = \frac{1}{\Gamma(\alpha)\beta^{\alpha}}\theta^{-\alpha-1}\exp\left(-1/(\beta\theta)\right)I(\theta > 0),$$

where $\alpha(>0)$ and $\beta(>0)$ are known. This prior is called inverse gamma and it is referred to as IG(α, β) distribution.

(i) Show that the prior mean of ϑ is $1/\left(\beta(\alpha-1)\right)$ if $\alpha > 1$ and the prior variance of ϑ is $1/\left(\beta^2(\alpha-1)^2(\alpha-2)\right)$ if $\alpha > 2$.

(ii) Show that the posterior distribution ϑ given $T \equiv \sum_{i=1}^{n} X_i = t(>0)$ is IG$(n + \alpha, \beta/(t\beta + 1))$.

(iii) Under a squared error loss function, obtain the expression of Bayes estimator of ϑ.

Exercise 15.3.1: Verify (15.3.7) in the context of N from (15.3.4) where the specific sequence of Ψ-functions was defined in (15.3.6).

Exercise 15.3.2: Define an explicit sequence of Ψ-functions in the context of N from (6.2.6). Then, verify an equation that is similar in spirit with (15.3.7).

Exercise 15.3.3: Define an explicit sequence of Ψ-functions in the context of N from (6.2.17). Then, verify an equation that is similar in spirit with (15.3.7).

Exercise 15.3.4: Define an explicit sequence of Ψ-functions in the context of N from (7.3.5). Then, verify an equation that is similar in spirit with (15.3.7).

Exercise 15.3.5: Define an explicit sequence of Ψ-functions in the context of N from (8.2.5). Then, verify an equation that is similar in spirit with (15.3.7).

Exercise 15.4.1 (Exercise 15.2.9 Continued): Consider the program **Seq08.exe** from Chapter 8. First, execute the sequential minimum risk point estimation rule by fixing $A = 100, c = 0.04, \vartheta = 2, m = 10$ and number of simulations $= 1$. The program will implement the purely sequential stopping rule from (8.2.5) one time only and upon termination, the screen output will show the values of n and $\widehat{\vartheta}$ among other entries. Recall that this $\widehat{\vartheta}$ is the sample mean so that $t \equiv \sum_{i=1}^{n} x_i = n\widehat{\vartheta}$. Having observed n and t, now revisit Exercise 15.2.9 under a prior $\text{IG}(3, \frac{1}{4})$ and evaluate the Bayes estimate of ϑ under a squared error loss function.

Exercise 15.4.2 (Exercise 15.4.1 Continued): For the observed data from Exercise 15.4.1, repeat evaluation of a Bayes estimate of ϑ under a squared error loss function successively using priors $\text{IG}(3, \frac{1}{4.2})$, $\text{IG}(3, \frac{1}{5})$ and $\text{IG}(3, \frac{1}{6})$. What do you discover? Did you expect to see what you see? Why or why not?

Exercise 15.4.3 (Exercise 15.4.2 Continued): Repeat Exercise 15.4.1 with other choices of A, c, ϑ, m, but fix the number of simulations $= 1$. Try different choices of inverse gamma priors successively, evaluate Bayes estimates and critically examine the findings.

Exercise 15.4.4 (Exercises 15.4.1 - 15.4.3 Continued): Repeat these exercises with a stopping rule dedicated to the bounded risk point estimation problem implemented under **Seq08.exe** from Chapter 8.

CHAPTER 16

Selected Applications

16.1 Introduction

Mahalanobis (1940) envisioned the importance of sampling in multiple steps and developed sampling designs for estimating the acreage of jute crop in Bengal. This pathbreaking applied work in large-scale survey sampling of national importance in India is regarded as the forerunner of sequential analysis.

Wald and his collaborators built the foundation of sequential analysis and brought it to the forefront of statistical science. This was especially crucial during the World War II when sequential analysis began its march in response to demands for more efficient testing of anti-aircraft gunnery. Applications involving inventory, queuing, reliability and quality control were immensely important. The introduction in Ghosh and Sen (1991) included many historical details.

In the 1960s through 1970s, many researchers in clinical trials saw the importance of emerging sequential designs and optimal stopping. The area of clinical trials has continued to be one of the major beneficiaries of basic research in sequential methodologies.

In the past decade, we have noticed added vigor in applying sequential methodologies in contemporary areas of statistical science. More recently, we find substantial applications in agriculture, clinical trials, data mining, finance, gene mapping, micro-arrays, multiple comparisons, surveillance, tracking and other areas. The edited volume of Mukhopadhyay et al. (2004) would support this sentiment.

We emphasize that sequential methodologies continue to remain relevant and important. They provide valuable tools to run statistical experiments in important contemporary areas. In this chapter, we include a selection from recent practical applications.

Section 16.2 briefly touches upon clinical trials. In Section 16.3, we quickly review the literature on integrated pest management and summarize key findings from Mukhopadhyay and de Silva (2005) with regard to insect infestation. Attention is given to Mexican bean beetle data analysis. Section 16.4 introduces an interesting problem from experimental psychology dealing with human perception of 'distance'. A summary and some selected data analyses are provided from Mukhopadhyay (2005b). Section 16.5 includes a problem that originally came from a horticulturist.

A summary, some data analyses, and real data are provided that come from Mukhopadhyay et al. (2004). We end with Section 16.6 by including snapshots of a number of other contemporary areas where sequential and multistage sampling designs have made a real difference.

16.2 Clinical Trials

Sequential designs where one adopts an allocation scheme using the accumulated data came into existence at an opportune time. Robbins (1956) suggested a play-the-winner (PW) rule. Milton Sobel started working with G. H. Weiss on adaptive sampling followed by selection of the better treatment. Their 1970 paper is regarded as a breakthrough for achieving ethical allocation of subjects to one of two treatments in a clinical trial. Sobel and Weiss (1970) also included crucial properties of their procedure. This fundamental work led to a series of extremely important collaborations, for example, Sobel and Weiss (1971a,b,1972) and Hoel and Sobel (1971). Armitage (1975) has been especially influential. These have made significant impacts on how adaptive designs for modern clinical trials are constructed. The monographs of Whitehead (1997), Jennison and Turnbull (1999), and Rosenberger and Lachin (2002) would provide a broad review in this area.

For specific practical applications, one may refer to Biswas and Dewanji (2004) and Lai (2004b). Biswas and Dewanji (2004) propose a binary longitudinal response design in a clinical trial with *Pulsed Electro-Magnetic Field* (PEMF) therapy for rheumatoid arthritic patients. Lai (2004b) includes recent commentaries on interim and terminal analysis of *Beta-Blocker Heart Attack Trial* (BHAT).

A recent article of Shuster and Chang (2008) has drawn attention to the issue of second-guessing clinical trial designs. They propose a methodology that may help biostatistical reviewers of nonsequential randomized clinical trials. Armed only with a single summary statistic from a trial, the Shuster-Chang article boldly addresses the 'what if' question 'could a group sequential design have reached a definitive conclusion earlier?' Thought provoking views expressed in this article are critically examined by a number of leading experts in discussion pieces which are then followed by responses from Shuster and Chang.

16.3 Integrated Pest Management

The practical uses of a negative binomial distribution in modeling data from biological and agricultural studies are quite numerous to list them all. Early on, working with insect counts, Anscombe (1949) emphasized the role of negative binomial modeling. Bliss and Owen (1958) considered some related issues as well. Kuno (1969,1972) developed sequential sampling procedures for estimating the mean of a population whose variance is a quadratic function of the population mean. Binns (1975), on the other hand, considered sequential confidence interval procedures for the mean of a negative binomial model.

The works of Willson (1981) and Willson and Folks (1983), however, constitute a comprehensive study of modern sequential sampling approaches for estimating the mean of a negative binomial model. Since purely sequential sampling can sometimes be operationally unattractive, Mukhopadhyay and Diaz (1985) developed a two-stage estimation methodology in the case of the point estimation problem.

References to a negative binomial model in agricultural experiments, including monitoring weeds and pests, are also quite frequent. We may simply mention articles of Marshall (1988), Mulekar and Young (1991,2004), Berti et al. (1992), Wiles et al. (1992), Mulekar et al. (1993), Young (1994), and Johnson et al. (1995). Among these articles, several sequential sampling strategies were proposed and investigated in Berti et al. (1992), Mulekar et al. (1993), and Young (1994,2004).

16.3.1 Two-Stage Estimation Methodologies

Let us suppose that we can observe a sequence of responses $\{X_1, X_2, \ldots\}$ of independent random variables, each having the p.m.f.

$$\mathrm{P}(X = x; \mu, \kappa) = \binom{\kappa + x - 1}{\kappa - 1} \left(\frac{\mu}{\mu + \kappa}\right)^x \left(\frac{\kappa}{\mu + \kappa}\right)^\kappa, x = 0, 1, 2, \ldots. \tag{16.3.1}$$

The parameters μ, κ are both assumed finite and positive. Briefly, we write that X has a $\mathrm{NB}(\mu, \kappa)$ distribution. A negative binomial distribution in all computer programs uses this parametrization.

Here, X may, for example, stand for the count of insects on a sampling unit of plants or it may be the count of some kind of weed on a sampling unit of an agricultural plot. In these examples, for the distribution (16.3.1), the parameter μ stands for the average infestation or average number of weeds per sampling unit, whereas the parameter κ points toward the degree of clumping of infestation or weeds per sampling unit. A small (large) κ indicates heavy (light) clumping. The parameterization laid down in (16.3.1) was introduced by Anscombe (1949,1950). For the distribution (16.3.1), it turns out that the mean and variance are given by

$$\mathrm{E}_\mu[X] = \mu \text{ and } \sigma^2 = \mathrm{V}_\mu[X] = \mu + \kappa^{-1}\mu^2. \tag{16.3.2}$$

Known Clumping Parameter κ:

First, we assume that the mean $\mu(> 0)$ is unknown but the clumping parameter $\kappa(> 0)$ is known. Having recorded n observations $X_1, \ldots, X_n$, along the line of Willson (1981), we suppose that the loss incurred in estimating μ by the sample mean $\overline{X}_n (= n^{-1} \sum_{i=1}^n X_i)$ is given by

$$L_n = \mu^{-2}(\overline{X}_n - \mu)^2. \tag{16.3.3}$$

The risk function associated with (16.3.3) is then given by $\mathrm{E}_\mu[L_n] = \sigma^2(n\mu^2)^{-1}$. Observe that this risk function may be interpreted as the square of the CV and hence we may refer to the associated methodology as *the CV approach.*

Willson (1981) and Willson and Folks (1983) investigated purely sequential point and interval estimation of μ when κ was known. Mukhopadhyay and Diaz (1985) came up with a two-stage sampling strategy for the point estimation problem considered earlier by Willson (1981) and Willson and Folks (1983) because a two-stage sampling design is operationally simpler. Mukhopadhyay and Diaz (1985) continued to assume that κ was known. With regard to the loss function, the main theme in these papers revolved around controlling the *coefficient of variation* (CV) associated with the mean estimator.

In a recent paper, Mukhopadhyay and de Silva (2005) discussed a very striking property of the two-stage sampling strategy of Mukhopadhyay and Diaz (1985). Mukhopadhyay and de Silva (2005) proved that for large sample sizes, the difference between the average sample size and the optimal fixed-sample-size remains bounded! This showed that the two-stage procedure of Mukhopadhyay and Diaz (1985) was *asymptotically second-order efficient* in the sense of Ghosh and Mukhopadhyay (1981).

Unknown Clumping Parameter κ:

One may be tempted to plug-in an appropriate estimate of κ in the two-stage procedure of Mukhopadhyay and Diaz (1985) and hope to proceed accordingly. But, that approach does not work very well because estimation of κ when μ is also unknown is far from trivial. The method of maximum likelihood or the method of moment estimation of κ may cross one's mind, and yet either estimator fails to estimate κ well in a large segment of its parameter space! Willson (1981) investigated and reported some of the ramifications when κ has to be estimated.

When κ was unknown, Mukhopadhyay and de Silva (2005) pursued a completely different approach. Even if κ remains unknown, a practitioner may be able to reasonably combine the associated uncertainty in the form of a suitable weight function $g(k)$ where $k(>0)$ is a typical value of κ. Then, while estimating the mean μ, instead of trying to control the coefficient of variation, Mukhopadhyay and de Silva (2005) make it their goal to control a new criterion, namely the *integrated coefficient of variation* (ICV). Here, averaging CV is carried out with the help of a weight function, $g(.)$.

Computer Simulations:

With the help of extensive sets of computer simulations, Mukhopadhyay and de Silva (2005) evaluated the methodologies both when κ was known or unknown. The role of the weight function $g(.)$ was critically examined.

They also suggested guidelines for a practitioner to choose an appropriate weight function $g(.)$ that is to be used in implementing the methodology.

16.3.2 Mexican Bean Beetle Data Analysis

Several sets of the Mexican bean beetle data that were collected by Jose Alexandre Freitas Barrigossi for his Ph.D. research (Barrigossi, 1997). One bit of information about the data came along these lines: The parameter κ for these data was anticipated to be quite small, near 0.3 or 0.4, but it could be smaller. Mukhopadhyay and de Silva (2005) have applied the original two-stage methodology of Mukhopadhyay and Diaz (1985) for this data assuming first that κ was known to be 0.3 or 0.4. Then, they proceeded to apply their new two-stage methodology with the *mean integrated coefficient of variation* (MICV) approach assuming that κ was unknown. In this latter situation, the weight function $g(.)$ was chosen so that the weighted average of κ came close to 0.3 or 0.4 but then κ was also "allowed" to be smaller. The sense of uncertainty about κ was thus built within the weight function $g(.)$. For much more details, one should refer to Mukhopadhyay and de Silva (2005).

16.4 Experimental Psychology: Cognition of Distance

An individual's perception of "distance" can be an important piece of information in many situations. For example, when a car rear-ends another, how far apart they were before the accident would be an important input in determining their traveling speed. Here is another example. The credibility of an eyewitness's description of a fleeing bank robber depends on the distance of a witness from a crime scene when the crime took place. So, how "distance" is perceived by individuals is an interesting and important subject area to consider.

Mukhopadhyay (2005b) designed and implemented a number of simple sequential experiments and what he reported appears to be an eye-opener. He had opted for sequential experiments over fixed-sample-size experiments because he required rather narrow confidence intervals having both predetermined width (or proportional closeness) and confidence level.

16.4.1 A Practical Example

This is one experiment that Mukhopadhyay (2005b) had designed, implemented, and reported. He experimented with a particular 15″ wooden ruler by asking his subjects to guess the length of the ruler when he showed the back side of the ruler to each subject standing 5 feet away!

A thought that may come to mind is that many subjects would probably get an impression of "seeing" a 12″ wooden ruler simply because a 12″ wooden ruler is such a common thing to have around. Generally, the feeling is that practically everyone can "surely tell" a 12″ wooden ruler from the other side of a room! At the least, that is the perception one may have. Mukhopadhyay (2005b) wanted to examine whether a 15″ wooden ruler could fool some brains!

It is not hard to imagine that the margin of error in estimating the length of a straight line or a ruler would probably go up if its length increased. For example, if a ruler was 6″ long, the subjects would likely be able to estimate the length more accurately than they would if the ruler was 20″ long. In other words, construction of a fixed-width confidence interval for μ may not quite serve us well. As usual, let $\overline{X}_n$ stand for the average of n successive sample values and we may try to achieve the following:

$$\mathrm{P}\left\{\mu : \left|\frac{\overline{X}_n}{\mu} - 1\right| \le d\right\} \equiv P\left\{\mu : \left|\overline{X}_n - \mu\right| \le d\left|\mu\right|\right\} \approx 1-\alpha, \qquad (16.4.1)$$

the prescribed confidence level. Here $d(>0)$ is a small positive number. For example, if $d = 0.125$, it means that one is attempting to estimate μ within 12.5% of $|\mu|$. On the other hand, if $d = 0.02$, it means that one is attempting to estimate μ within 2% of $|\mu|$. If μ is believed to be large, then d may be chosen little "large", but if μ is small, then d would be chosen "small" too.

If the sample size n is not too small, the central limit theorem will kick in, and then the optimal fixed-sample-size will be approximately given by $n^* = z^2_{\alpha/2}\sigma^2/(d^2\mu^2)$. Nadas (1969) investigated a purely sequential sampling technique mimicking this expression of n^*. For operational convenience, however, Mukhopadhyay (2005b) ran a two-stage sampling design that was specifically constructed for handling this practical problem.

Since the subjects were supposedly extremely familiar with 12″ rulers, one may feel strongly that there is no way that μ would exceed 15″, that is one may assume that $\mu \le 15''(\equiv \mu_U)$. Mukhopadhyay (2005b) also assumed a'priori that $\sigma > 0.5''(\equiv \sigma_L)$ for all practical purposes.

One may fix $\alpha = 0.05$ (that is, $z_{\alpha/2} = 1.96$) and $d = 0.01$, that is one may wish to estimate μ within 1% of what it is. Hence, Mukhopadhyay (2005b) had determined the pilot sample size as:

$$m \equiv m(d) = \max\left\{30, \left\langle z^2_{\alpha/2}\sigma^2_L/(d^2\mu^2_U)\right\rangle + 1\right\}. \qquad (16.4.2)$$

By the way, the pilot sample size m was picked at least 30 for approximate validity of the central limit theorem in (16.4.1).

For the given set of design specifications, we find

$$\begin{aligned} m &= \max\left\{30, \left\langle (1.96 \times 0.5/(0.01 \times 15.0))^2 \right\rangle + 1\right\} \\ &= \max\left\{30, \langle 42.684\rangle + 1\right\} = 43. \end{aligned}$$

So, Mukhopadhyay (2005b) collected initial data from 43 subjects and that pilot data gave $\overline{X}_m = 12.4186$ and $S_m = 1.4012$.

Then, by mimicking the expression of $n^* \equiv z^2_{\alpha/2}\sigma^2/(d^2\mu^2)$, the sample size N was calculated as follows:

$$N \equiv N(d) = \max\left\{m, \left\langle z_{\alpha/2}^2 S_m^2/(d^2\overline{X}_m^2)\right\rangle + 1\right\}. \tag{16.4.3}$$

Finally, based on all N observations $X_1, \ldots, X_N$, one would propose the $100(1-\alpha)\%$ *fixed-size confidence set,*

$$\{\mu : |\overline{X}_N - \mu| \le d\,|\mu|\} \tag{16.4.4}$$

for μ where $\overline{X}_N = N^{-1}\sum_{k=1}^N X_k$.

In this example, we find

$$\begin{aligned} N &= \max\left\{43, \left\langle (1.96 \times 1.40123/(0.01 \times 12.4186))^2 \right\rangle + 1\right\} \\ &= \max\{43, \langle 489.09\rangle + 1\} = 490. \end{aligned}$$

Thus, Mukhopadhyay (2005b) needed additional observations from 447 new subjects. The full data from both stages of sampling are shown in Table 16.4.1.

Table 16.4.1: Frequency data for estimating the ruler-length (inches)

	Pilot Data			Number of Subjects
Estimated length	12″	15″	18″	Asked
Frequency	39	2	2	43
	Second-Stage Data			Number of Subjects
Estimated length	12″	15″	18″	Asked
Frequency	417	9	21	447

From the combined data, we found $\overline{X}_N = 12.3490$ so that the 95% fixed-size confidence set from (16.4.4) became

$$\{\mu : |12.3490 - \mu| \le d\,|\mu|\}$$

$$\Leftrightarrow \frac{12.3490}{1.01} \le \mu \le \frac{12.3490}{0.99} \Leftrightarrow 12.227 \le \mu \le 12.474 \text{ inches.}$$

In other words, these 490 subjects estimated the ruler-length approximately around 12″ on an average! We are tempted to conclude that these subjects were not really guessing the ruler's length, rather a overwhelming majority of this group was giving answers that were "consistent" with what they thought they were "seeing".

Mukhopadhyay (2005b) ran a number of other experiments and he summarized his concluding thoughts as follows: "These experiments and their associated data analyses, however, have a clear message to deliver: When a subject was asked to guess the length of an object (for example, a *straight line* or a *ruler* or a *measuring tape*), we found time and again that the answers were commonly affected by (i) landmarks or reference points, if any, and (ii) what one *thought* of seeing rather than the actual item that was shown. We feel a serious need for more investigations, perhaps to be run in a computer-lab under close supervision, where subjects may be randomly drawn from an appropriate population."

16.5 A Problem in Horticulture

New Englanders make flower-beds that are bordered with marigolds because these are colorful and hardy. It is also fairly common to make flower-beds with marigolds alone. To remain competitive, a nursery would usually maintain a steady supply of these plants.

It is well-known that customers prefer selecting plants having some buds or blossoms. Thus, if a variety takes too long for first bud(s) to appear, then a nursery would eventually lose customers and money.

In early spring, seeds from different marigold varieties are customarily planted in large number (in thousands!) in greenhouses under a controlled environment. The number of days (X) a variety takes from planting seeds to the point when the first bud appears on $4-6$ inches tall plants is of considerable importance because plants with bud(s) are quickly put on display for buyers!

This interesting practical situation guided Mukhopadhyay et al. (2004) to formulate a new sampling design to gather observations and analyze the data. They became involved in comparing three varieties (#1, #2, and #3) of marigold with respect to the response variable X. They were advised that X could be recorded within one-half day, that is one would watch for the growth of bud(s) on each plant in the morning and evening. A normal distribution for the response variable was suggested by the horticulturist.

The idea was to estimate the maximum average waiting time between "seeding" and "budding". It was decided that a 99% confidence interval for the largest variety mean, $\mu_{[3]}$, having width one day would suffice. These specifications took into account a steady demand for marigold plants, the length of a normal season of spring and early summer to start flower-beds and borders, as well as the time it should usually take to move supplies from greenhouse to store. The variety associated with $\mu_{[3]}$ was interpreted as the *worst* by the horticulturist, but he was more keen on estimating $\mu_{[3]}$.

One may refer to Chapter 21 of Gupta and Panchapakesan (1979) and Chapters 5 and 7 of Mukhopadhyay and Solanky (1994), among other sources, for reviewing the area of ordered parameter estimation. The fixed-width confidence interval estimation problem for the largest mean was studied by Chen and Dudewicz (1976), Saxena and Tong (1969), Saxena

(1976), and Tong (1970,1973). Mukhopadhyay et al. (1993) gave second-order approximations for various multi-stage estimation strategies.

In general, let us suppose that $X_{i1}, X_{i2}, \ldots, X_{in_i}, \ldots$ denote independent observations, corresponding to the ith variety, having a common probability density function

$$\phi(x; \mu_i, \sigma_i) = \{\sqrt{2\pi}\sigma_i\}^{-1} \exp\{-\frac{1}{2}(x-\mu_i)^2/\sigma_i^2\} \tag{16.5.1}$$

where all parameters are unknown, $\sigma_i^2 \neq \sigma_j^2, i \neq j = 1, \ldots, k(\geq 2)$. We also suppose that X_i is independent of X_j for all $i \neq j = 1, \ldots, k$. Let us denote $\mu_{[1]} \leq \mu_{[2]} \leq \ldots \leq \mu_{[k]}$ to be the ordered variety averages. The primary concern is to develop an appropriate two-stage sampling design to estimate the parameter $\mu_{[k]} = \max\{\mu_1, \ldots, \mu_k\}$ with prescribed accuracy.

When variances are unknown and unequal, the existing two-stage estimation methodology is due to Chen and Dudewicz (1976). This method exclusively demands equal pilot sample sizes. But, this setup is not immediately applicable for the type of problem on hand since the horticulturist provided positive lower bounds for the variances which made the pilot sample sizes unequal. We are not aware of a methodology with unequal pilot sample sizes that would readily apply here. Accordingly, a new two-stage sampling design was developed and implemented by Mukhopadhyay et al. (2004). Their paper gave important exact as well as large-sample properties of the new methodology. For large sample sizes, the methodology was proven to be theoretically superior to the existing methodology provided that the pilot sizes could be "chosen" equal. For the real marigold data, superiority of the proposed methodology was also indicated by Mukhopadhyay et al. (2004).

16.5.1 Implementation and Data Analyses

We briefly mention how Mukhopadhyay et al. (2004) implemented their methodology and summarize some partial data analyses borrowed from their presentation. As mentioned earlier, a local horticulturist had three marigold varieties to consider. The response times (namely, X_1, X_2 and X_3) from seeding to first-time budding for each variety was believed to be normal with average times μ_1, μ_2, and μ_3 . The varieties #1 and #3 were on-site hybrids whereas variety #2 came from a supplier. For varieties #1 and #3, the horticulturist felt confident that buds would show within $\mu_i \pm 4$ days, $i = 1, 3$, and we were assured that σ_1 would be at least $1.5(= \sigma_{1L})$ days and σ_3 would be at least $1.0(= \sigma_{3L})$ day. For variety #2, the supplier's asking price was considerably less than the cost of other two local varieties! The horticulturist did not sound too confident but he felt that σ_2 would be at least 4 days considering the price differential across varieties. In view of such uncertainty, we opted for a conservative lower bound $2.0(= \sigma_{2L})$

days for σ_2. Since plants with buds will be in demand, we were told that the initial sample size from each variety could be at least 25.

Mukhopadhyay et al. (2004) had fixed $d = 0.5$, a "small" number, so that they would be in a position to clearly partition the observed average response times across varieties. They opted for a 99% fixed-width (= $2d$) confidence level $J_{\boldsymbol{N}}$ covering $\mu_{[3]}$, that is they also fixed $\alpha = 0.01$.

At this point, adaptation of Mukhopadhyay et al.'s (2004) two-stage estimation methodology became straightforward, but not trivial. They determined:

$$m_1 = 67, m_2 = 118, \text{ and } m_3 = 30. \tag{16.5.2}$$

Three $10' \times 1'$ shelves were setup inside a greenhouse under controlled environment. A total of 215 identical $3''$ pots were prepared of which 67 were seeded with variety #1, 118 were seeded with variety #2, 30 were seeded with variety #3, and these were placed on shelves #1, #2, #3 respectively around 6 AM on March 15th, 2001. Each pot was appropriately labelled. With regard to time-line, 6 AM on March 16th was recorded as day 1, 6 AM on March 17th was recorded as day 2, and so on. The seeded pots were cared for under exact same conditions (e.g., exposure to sunlight, temperature, humidity, watering, feeding). The setup was such that the time a variety took for the first bud to appear could be recorded within -12 hours, because daily rounds of inspection were scheduled at 6 AM and 6 PM. During inspection, any plant found with bud(s) was immediately transferred to the store.

From pilot data, reproduced in Table 16.5.1, Mukhopadhyay et al. (2004) reported:

1 $x_{1\,\min} = 25.5$ days, $x_{1\,\max} = 34$ days, $\overline{x}_{1m_1} = 30.336$ days, $s_{1m_1} = 2.036$ days for variety #1, that is final responses from this variety became available around April 18-19.

2 $x_{2\,\min} = 27.5$ days, $x_{2\,\max} = 45$ days, $\overline{x}_{2m_2} = 35.169$ days, $s_{2m_2} = 3.79$ days for variety #2, that is final responses from this variety became available around April 29-30.

3 $x_{3\,\min} = 23.0$ days, $x_{3\,\max} = 31$ days, $\overline{x}_{3m_3} = 27.917$ days, $s_{3m_3} = 2.248$ days for variety #3, that is final responses from this variety became available around April 15-16.

Let $G_{\nu_i}(t)$ denote the cumulative distribution function of a Student's t variate with $\nu_i = m_i - 1$ degrees of freedom, $i = 1, \ldots, k$, and one would determine the positive number $h_{\boldsymbol{m}} \equiv h(k, \alpha, \boldsymbol{m})$ such that

$$\Pi_{i=1}^{k} G_{\nu_i}(h_{\boldsymbol{m}}) - \Pi_{i=1}^{k} G_{\nu_i}(-h_{\boldsymbol{m}}) = 1 - \alpha. \tag{16.5.3}$$

From the initial responses $X_{i1}, \ldots, X_{im_i}$ on variety $\#i$, one would obtain the sample variance $S_{im_i}^2$ and determine

$$N_i \equiv N_i(d) = \max\left\{m_i, \langle h_{\boldsymbol{m}}^2 S_{im_i}^2 / d^2 \rangle + 1\right\} \tag{16.5.4}$$

to be the final sample size obtained from the ith variety, $i = 1, \ldots, k$.

Mukhopadhyay and Aoshima (2004) gave a general expression for $h_{\boldsymbol{m}}$ that was very accurate.

Now, with $\nu_1 = 66, \nu_2 = 117$, and $\nu_3 = 29$, Mukhopadhyay et al. (2004) had found $h_{\boldsymbol{m}} = 2.8289$. The observed values of N_1, N_2 and N_3 were respectively given by

$$\begin{aligned} n_1 &= \max\left\{67, \left\langle (2.8289)^2(2.036)^2/(0.5)^2 \right\rangle + 1\right\} = 133, \\ n_2 &= \max\left\{118, \left\langle (2.8289)^2(3.79)^2/(0.5)^2 \right\rangle + 1\right\} = 460, \\ n_3 &= \max\left\{30, \left\langle (2.8289)^2(2.248)^2/(0.5)^2 \right\rangle + 1\right\} = 162. \end{aligned} \tag{16.5.5}$$

It is clear from (16.5.5) that in the second stage of the experiment, the horticulturist had to plant variety #1 seeds in 66(= 133 − 67) new pots, variety #2 seeds in 342(= 460 − 118) new pots, and variety #3 seeds in 132(= 162 − 30) new pots. Six shelves were setup under exact same conditions as before. For logistical reasons, it was decided to start the second stage of sampling for all three varieties at the same time.

A total of 540 pots were prepared of which 66 pots were seeded with variety #1 and placed on shelf #1, 342 pots were seeded with variety #2 and placed on shelves #2 - #4, and 132 pots were seeded with variety #3 and placed on shelves #5 - #6 around 6 AM on May 1st, 2001. Each pot was appropriately labeled. Now, 6 AM on May 2nd was recorded as day 1, 6 AM on May 3rd was recorded as day 2, and so on. The data was then gathered using the exact same protocol that was followed during the pilot phase and the associated responses are laid out in Table 16.5.2 that is reproduced from Mukhopadhyay et al. (2004).

Here, Mukhopadhyay et al. (2004) showed by example how an appropriate sequential sampling strategy *must* be designed and implemented to solve one practical problem. They had gathered data according to proposed design, and their paper included the dataset. That way, researchers will get the opportunity to revisit, reuse, and reanalyze this live data at any time in the future. With that in mind, we reproduce Mukhopadhyay et al.'s (2004) dataset.

16.5.2 The Fixed-Width Confidence Interval

First, Mukhopadhyay et al. (2004) painstakingly validated all the assumptions made during the design and implementation phases of their exercise.

They observed $\boldsymbol{n} = (133, 460, 162)$ and $\overline{x}^*_{\boldsymbol{n}} = \max\{\overline{x}_{1n_1}, \overline{x}_{2n_2}, \overline{x}_{3n_3}\} = 35.135$. The 99% fixed-width confidence interval for $\mu_{[3]}$ with $d = 0.5$ turned out to be

$$J_{\boldsymbol{n}} = [35.135 \pm 0.5].$$

With 99% confidence, that is, perhaps for all practical purposes, the maximum average budding time $\mu_{[3]}$ was reported to lie between 34.635 days and 35.635 days. It did appear that the maximum average budding time far exceeded the average budding times associated with the other two varieties (#1, #3) by at least 4.5 days! The horticulturist did not reconsider variety #2 any further even though this variety was considerably cheaper than the two other varieties.

Table 16.5.1: Time to first bud(s) in the first-stage of the experiment

				Variety #1					
31.5	33.5	32.0	31.5	26.0	27.5	27.0	27.0	28.5	28.5
34.0	31.0	31.5	28.5	31.0	29.0	28.5	30.5	32.5	28.5
33.5	30.0	32.5	29.5	33.0	27.5	32.5	30.0	31.5	34.0
31.0	29.0	31.5	28.0	31.0	32.5	31.5	27.5	32.5	30.0
30.0	33.5	28.5	29.0	32.5	31.5	30.0	30.0	29.5	31.0
30.5	32.5	27.0	30.5	31.5	29.0	28.5	32.0	28.0	30.0
29.0	32.5	29.5	33.0	30.5	31.0	25.5			
				Variety #2					
37.5	38.0	32.0	29.5	40.5	42.0	34.5	34.5	36.5	31.0
36.5	30.0	36.5	41.0	45.0	35.0	31.5	32.5	39.0	34.5
31.5	40.0	34.0	36.5	40.5	36.5	39.0	32.0	36.0	38.0
35.0	32.0	35.0	41.0	40.0	34.5	37.5	37.5	39.5	36.0
28.5	36.0	36.5	31.5	35.0	29.5	36.5	31.0	41.5	34.5
33.0	33.0	42.0	27.5	31.5	34.5	33.0	31.0	40.5	30.5
37.5	37.0	39.0	35.0	36.0	33.5	36.0	32.5	35.0	38.5
30.0	28.5	31.5	32.5	36.0	34.5	31.0	34.0	35.5	31.0
39.0	37.5	39.0	31.5	36.0	41.5	28.0	37.5	29.5	35.0
39.5	32.5	38.5	35.0	33.0	34.0	30.5	30.5	38.0	30.5
42.5	35.5	31.5	35.0	37.5	28.5	34.0	42.5	29.5	37.5
36.5	37.0	41.0	34.5	35.5	37.0	38.5	28.5		
				Variety #3					
31.0	28.5	26.0	23.0	31.0	29.5	27.0	28.5	31.0	29.5
30.5	28.0	27.0	29.5	28.5	27.5	27.0	24.5	31.0	26.5
27.0	26.5	29.5	27.5	29.0	23.5	26.5	30.0	29.0	24.0

Table 16.5.2: Time to first bud(s) in the second-stage of the experiment

				Variety #1					
32.0	30.5	28.0	33.0	32.5	31.5	25.0	35.5	29.0	29.5
30.5	34.5	28.5	28.0	32.0	27.5	31.5	29.0	30.0	29.0
27.5	26.5	26.5	29.5	30.5	30.0	29.5	32.0	29.0	32.0
32.0	29.5	30.0	28.0	32.0	28.5	29.0	31.5	31.0	33.5
26.5	31.5	30.0	29.0	32.0	29.0	30.0	26.5	26.0	27.5
28.0	29.0	28.0	27.0	30.0	31.0	30.0	30.0	31.0	30.0
29.0	26.0	30.0	31.5	33.5	29.5				

Table 16.5.2 (Continued): Time to first bud(s) in the second-stage of the experiment

Variety #2

31.5	37.5	38.5	41.0	36.5	35.5	30.5	29.5	33.5	34.0
38.0	32.5	33.0	36.0	42.0	31.5	35.5	34.0	32.5	34.5
32.0	22.0	35.0	27.5	29.0	36.5	37.5	34.5	29.0	32.5
33.5	29.5	30.5	32.0	34.5	33.5	28.5	38.0	37.0	43.5
33.5	32.5	35.5	43.0	29.0	29.0	34.0	32.5	30.0	30.5
40.5	34.5	39.0	35.0	35.5	32.0	35.0	34.5	41.0	34.0
31.0	34.5	37.0	36.5	33.0	40.5	37.0	28.5	33.0	31.0
36.0	36.5	30.5	40.5	29.0	33.0	30.5	25.5	34.0	40.0
31.5	36.0	35.0	32.0	30.0	25.0	34.5	31.0	30.5	37.0
34.0	31.0	32.0	33.5	43.5	38.5	31.5	38.5	41.5	37.5
35.0	37.0	31.5	38.5	34.5	43.5	38.5	37.5	34.5	33.0
32.5	36.5	39.0	34.5	28.5	36.5	39.5	34.5	31.5	35.0
32.5	39.5	33.0	40.0	40.0	37.0	35.0	39.0	39.5	40.0
36.0	34.0	34.0	35.5	36.5	37.0	36.5	30.0	35.5	30.5
38.5	34.0	33.5	33.5	38.5	37.0	33.5	32.5	33.0	31.0
42.0	46.0	30.5	39.0	30.0	32.5	31.0	36.5	39.5	33.0
37.5	35.5	34.0	35.5	36.5	39.0	27.0	36.5	29.0	39.0
38.0	33.0	40.0	40.0	31.0	28.5	37.0	37.5	29.5	33.0
42.0	30.0	36.0	34.5	39.5	28.5	41.5	32.0	29.0	38.0
35.0	35.0	40.5	31.0	29.0	24.5	33.0	39.0	39.5	36.0
34.0	35.0	40.0	34.0	44.0	39.5	40.0	32.0	30.5	33.5
39.5	33.0	38.0	39.5	35.5	34.0	34.0	33.5	29.5	29.0
39.0	39.0	42.0	34.0	31.5	35.0	38.0	34.5	30.0	30.5
44.0	39.5	34.5	36.5	37.0	32.5	43.0	30.5	30.0	36.0
37.5	40.0	35.5	32.5	31.0	33.0	32.5	31.0	35.5	35.5
37.5	37.5	34.5	40.5	32.0	37.5	44.5	27.5	46.5	29.0
34.0	40.5	34.0	34.5	36.0	30.0	36.0	33.0	32.5	39.0
38.0	35.0	29.5	38.5	39.0	32.0	41.5	40.5	43.0	32.5
37.0	36.5	38.0	31.5	31.5	40.5	35.5	30.5	35.5	34.0
31.5	44.0	38.5	36.0	38.5	30.5	30.5	46.5	34.0	27.0
39.0	34.0	34.5	36.5	35.0	26.5	37.0	30.0	32.0	36.0
30.5	40.5	40.0	37.5	38.0	30.5	34.5	33.0	40.0	38.5
29.5	38.0	38.0	40.0	39.0	34.0	36.5	40.5	29.5	36.5
37.5	39.0	37.0	39.0	32.5	37.5	37.5	35.5	41.5	41.5
33.0	39.5								

Variety #3

26.5	28.5	27.0	29.5	29.0	25.0	32.5	28.5	29.0	30.0
28.0	29.0	31.5	25.5	28.0	28.0	27.5	28.5	27.5	29.0
29.5	29.0	29.0	29.0	26.5	28.0	27.0	26.0	32.5	30.0
29.0	26.5	26.5	28.5	26.5	30.0	26.0	30.0	34.5	25.0
32.5	27.5	24.5	30.5	28.5	27.5	32.5	27.5	29.5	31.0
28.5	29.5	27.5	30.5	22.5	30.5	31.0	27.5	30.0	31.5
29.0	28.5	28.0	30.5	26.5	28.5	24.5	31.0	29.5	30.0
28.5	27.0	23.5	25.0	28.5	27.0	28.5	25.5	28.5	29.0
27.5	28.0	26.0	27.0	27.0	23.5	26.5	28.0	28.5	28.5
26.0	31.5	27.0	29.5	27.5	25.5	28.0	31.5	28.5	29.0
27.0	32.0	27.0	26.0	26.5	29.5	28.5	28.5	26.0	27.0
32.0	30.0	26.0	25.0	29.5	27.0	29.5	28.5	26.5	29.0
23.5	25.0	33.5	30.0	31.0	29.5	26.0	30.0	29.5	29.0
27.5	26.5								

16.6 Other Contemporary Areas of Applications

In this section we cite very briefly a few applications in some of the other contemporary areas of statistical science that could not be included earlier.

Change-Point Problems:

A very recent article of Tartakovsky et al. (2006) on sequential detection of intrusions or information-breach, with eleven discussions, is especially noteworthy because of its timeliness, relevance, and immediate applications for the line of defense in the area of homeland security. Naturally, such detection algorithms should go into effect as soon as possible after an intrusion, but at the same time, one must keep the number of false alarms at a minimum possible level.

Computer Simulation Designs:

In the paper of Mukhopadhyay and Cicconetti (2004b), the authors demonstrated how sequential and multistage sampling techniques could help enormously in designing many aspects of computer simulations. They particularly emphasized how to determine the number of replications sequentially and very accurately in order to achieve some predetermined level of accuracy efficiently. The authors introduced many delicate points through formulation of a problem, implementation of both sequential and multistage simulation methods, followed by in-depth analyses of real data.

Data Mining:

Chang and Martinsek (2004) used the ideas of clustering (tree) analysis and then developed and employed sequential methods for variable selection. They used data from in-patient records of 1172 hospitals during a calendar year. There were more than 2 million records and 75 variables under consideration. Chang and Martinsek (2004) demonstrated that sequential sampling were more efficient than complete sample analysis with regard to CPU time.

Environmental Sampling:

Mukhopadhyay et al. (1992) developed multistage sampling techniques for ascertain the acidity in lake water. Mukhopadhyay's (2002) article in the *Encyclopedia of Environmetrics* will be a good place to find some of the references for applications in environmental sequential sampling and analyses of real data. Mulekar and Young (2004) and Young (2004) respectively presented full syntheses as well as appraisals of many applications of sequential methodologies in both estimation and tests of hypotheses.

Financial Applications:

The paper of Efromovich (2004) gave sequential nonparametric curve estimation techniques. Paraphrasing the author, we would say that this paper emphasizes an entirely data-driven sequential methodology (software) that allows an investor to analyze existing relationship between the excess rates of returns on an asset and the returns on the market using the shortest period of historical data. A number of examples based on real datasets were included.

Nonparametric Density Estimation:

de Silva and Mukhopadhyay (2004) developed two-stage nonparametric kernel density estimation methods and their smooth bootstrapped versions having some predetermined level of accuracy. Delicate analyses were presented with regard to the initial sample size selection as well as the band-width selection. Such intricate details were developed especially to apply the proposed methodologies for estimating the density of the fineness of the diameter (in microns) of wool fiber samples obtained by the *Australian Wool Testing Authority* (AWTA).

Appendix: Selected Reviews, Tables, and Other Items

A.1 Introduction

In Section A.2, we review the notions of $O(.)$ and $o(.)$. Section A.3 briefly reviews some key elementary concepts and results from probability theory. Subsection A.3.1 summarizes some selected inequalities.

In order to review fundamental ideas of large-sample methods and approximations systematically, one may refer to Sen and Singer (1993) and Serfling (1984).

Section A.4 summarizes the basic formulation on which the non-linear renewal theory of Woodroofe (1977,1982) and Lai and Siegmund (1977,1979) stand. However, we adopt the layout found in Mukhopadhyay (1988a) and then summarize some important results in the light of Mukhopadhyay and Solanky (1994, Section 2.4.2). Second-order asymptotic properties of purely sequential stopping times and associated estimators in numerous circumstances follow from these basic results and other related tools. Pertinent details can be found in Ghosh et al. (1997, Sections 2.8-2.9) and cited references including Siegmund (1985).

Section A.5 summarizes some of the abbreviations and notations that have been used throughout this textbook.

Section A.6 includes some standard statistical tables. Tables for the standard normal, Chi-square, t, and F distributions have been provided. We acknowledge that these tables have been borrowed from two textbooks, Mukhopadhyay (2000,2006).

A.2 Big O(.) and Little o(.)

Consider two terms a_n and b_n, both real valued and obviously both depending on n, $n = 1, 2, \ldots$. Then, the term a_n is called $O(b_n)$, expressed by writing $a_n = O(b_n)$, provided that a_n and b_n have the same order. That is, $a_n/b_n \to k$, a constant not involving n, as $n \to \infty$.

On the other hand, we write $a_n = o(b_n)$ provided that $a_n/b_n \to 0$ as $n \to \infty$.

Example A.2.1 : Let $a_n = n - 2, b_n = 3n + \sqrt{n}$ and $c_n = -5n^{8/7} + n, n = 1, 2, \ldots$. Observe that both a_n, b_n tend to ∞ as $n \to \infty$ at the same rate.

One can check that $a_n/b_n \to \frac{1}{3}$ as $n \to \infty$. So, we can write $a_n = O(b_n)$. When we write $a_n = O(b_n)$, it just means that their rates of convergence (to whatever it may be) are similar.

One can also check that $a_n/c_n \to 0$ as $n \to \infty$ so that we can write $a_n = o(c_n)$. Observe that $a_n \to \infty, c_n \to -\infty$ as $n \to \infty$, but $c_n \to -\infty$ at a faster rate since we clearly have $a_n = O(n)$ and $c_n = O(n^{8/7})$.

Again, since $b_n/c_n \to 0$ as $n \to \infty$, we can write $b_n = o(c_n)$. Also, one can write, for example, $b_n = O(n)$, $c_n = o(n^2)$ or $c_n = o(n^{9/7})$.

A.3 Some Probabilistic Notions and Results

In this section, we review some notions and results from probability theory with the hope that these may help in one's basic understanding of large-sample approximations.

Definition A.3.1: *Consider a sequence of real valued random variables $\{U_n; n \geq 1\}$. Then, U_n is said to converge to a real number u in probability as $n \to \infty$, denoted by $U_n \xrightarrow{P} u$, if and only if the following condition holds:*

$$\mathrm{P}\{| U_n - u | > \varepsilon\} \to 0 \text{ as } n \to \infty, \text{ for every fixed } \varepsilon (> 0).$$

Intuitively speaking, $U_n \xrightarrow{P} u$, means that U_n may be expected to lie in a close proximity of u with probability nearly one when n is large.

Theorem A.3.1: *Let $\{T_n; n \geq 1\}$ be a sequence of real valued random variables such that with some real numbers $r(> 0)$ and a, one can claim that $\xi_{r,n} \equiv \mathrm{E}\{| T_n - a |^r\} \to 0$ as $n \to \infty$. Then, $T_n \xrightarrow{P} a$ as $n \to \infty$.*

One frequently abbreviates a *weak law of large numbers* by WLLN. In what follows, we first state a weaker version of WLLN.

Theorem A.3.2 [Weak WLLN]: *Let $X_1, \ldots, X_n, \ldots$ be i.i.d. real valued random variables with $\mathrm{E}(X_1) = \mu$ and $\mathrm{V}(X_1) = \sigma^2$, $-\infty < \mu < \infty$, $0 < \sigma < \infty$. Write $\overline{X}_n (= n^{-1} \sum_{i=1}^n X_i)$ for the sample mean, $n \geq 1$. Then, $\overline{X}_n \xrightarrow{P} \mu$ as $n \to \infty$.*

The following result gives a version of WLLN that is significantly stronger than the weak WLLN since it leads to the same conclusion without requiring a finite variance of the X's.

Theorem A.3.3 [Khintchine's WLLN]: *Let $X_1, \ldots, X_n$ be i.i.d. real valued random variables with $\mathrm{E}(X_1) = \mu$, $-\infty < \mu < \infty$. Then, $\overline{X}_n \xrightarrow{P} \mu$ as $n \to \infty$.*

Theorem A.3.4: *Suppose that we have a sequence of real valued random variables $\{U_n; n \geq 1\}$ and that $U_n \xrightarrow{P} u$ as $n \to \infty$. Let $g(.)$ be a real valued continuous function. Then, $g(U_n) \xrightarrow{P} g(u)$ as $n \to \infty$.*

Definition A.3.2: *Consider a sequence of real valued random variables* $\{U_n; n \geq 1\}$ *and another real valued random variable* U *with respective distribution functions* $F_n(u) = \text{P}(U_n \leq u)$, $F(u) = \text{P}(U \leq u)$, $u \in I\!R$. *Then,* U_n *is said to converge in distribution to* U *as* $n \to \infty$, *denoted by* $U_n \xrightarrow{\pounds} U$, *if and only if* $F_n(u) \to F(u)$ *pointwise at all continuity points* u *of* $F(.)$. *The distribution of* U *is referred to as the limiting or asymptotic (as* $n \to \infty$*) distribution of* U_n.

Let us suppose that the moment generating functions of the real valued random variables U_n and U are respectively denoted by

$$M_n(t) \equiv \text{E}\left[e^{tU_n}\right] \text{ and } M(t) \equiv \text{E}\left[e^{tU}\right].$$

We assume that both $M_n(t)$ and $M(t)$ are finite for all t such that $| t |< h$ where $h(> 0)$ is a fixed number.

Theorem A.3.5: *Suppose that* $M_n(t) \to M(t)$ *for* $| t |< h$ *as* $n \to \infty$. *Then,*

$$U_n \xrightarrow{\pounds} U \text{ as } n \to \infty.$$

The following result provides some basic tools for combining the two types of convergence, namely the notions of convergence in probability and in distribution.

Theorem A.3.6 [Slutsky's Theorem]: *Let us consider two sequences of real valued random variables* $\{U_n, V_n; n \geq 1\}$, *another real valued random variable* U, *and a fixed real number* v. *Suppose that* $U_n \xrightarrow{\pounds} U$, $V_n \xrightarrow{\text{P}} v$ *as* $n \to \infty$. *Then, we have as* $n \to \infty$:

(i) $U_n \pm V_n \xrightarrow{\pounds} U \pm v$;

(ii) $U_n V_n \xrightarrow{\pounds} vU$;

(iii) $U_n V_n^{-1} \xrightarrow{\pounds} Uv^{-1}$ *provided that* $\text{P}\{V_n = 0\} = 0$ *for all* n *and* $v \neq 0$.

The following result is referred to as the *central limit theorem* (CLT). It is one of the most celebrated results in probability theory.

Theorem A.3.7 [Central Limit Theorem]: *Let* $X_1, \ldots, X_n$ *be i.i.d. real valued random variables having the common mean* μ *and variance* σ^2, $-\infty < \mu < \infty$ *and* $0 < \sigma < \infty$. *Denote* $\overline{X}_n = n^{-1}\sum_{i=1}^{n} X_i$ *for* $n \geq 1$. *Then, we have:*

$$\frac{\sqrt{n}(\overline{X}_n - \mu)}{\sigma} \xrightarrow{\pounds} N(0,1) \text{ as } n \to \infty.$$

Theorem A.3.8 [CLT for the Sample Variance]: *Let* $X_1, \ldots, X_n$ *be i.i.d. real valued random variables with the common mean* μ, *variance* $\sigma^2(> 0)$, $\mu_4 = \text{E}\{(X_1 - \mu)^4\}$, *and we assume that* $0 < \mu_4 < \infty$ *as well as* $\mu_4 > \sigma^4$. *Denote* $\overline{X}_n = n^{-1}\sum_{i=1}^{n} X_i$ *and* $S_n^2 = (n-1)^{-1}\sum_{i=1}^{n}(X_i - \overline{X}_n)^2$ *for* $n \geq 2$. *Then, we have:*

$$\sqrt{n}\left(S_n^2 - \sigma^2\right) \xrightarrow{\pounds} N(0, \mu_4 - \sigma^4) \text{ as } n \to \infty.$$

For an elementary proof of the CLT for a sample variance, one may refer to Mukhopadhyay (2000, pp. 262-263).

Now, suppose that we have already established the CLT for $\{T_n; n \geq 1\}$. Can we then claim a CLT for a function of T_n? The following result assures us that the answer is in the affirmative in many circumstances.

Theorem A.3.9 [Mann-Wald Theorem]: *Suppose that $\{T_n; n \geq 1\}$ is a sequence of real valued random variables such that $\sqrt{n}(T_n - \theta) \xrightarrow{\pounds} N(0, \sigma^2)$ as $n \to \infty$ where $0 < \sigma^2 < \infty$ and σ^2 may involve θ. Let $g(.)$ be a continuous real valued function such that $\frac{d}{d\theta}g(\theta)$, denoted by $g'(\theta)$, is finite and non-zero. Then, we have:*

$$\sqrt{n}\{g(T_n) - g(\theta)\} \xrightarrow{\pounds} N\left(0, \{\sigma g'(\theta)\}^2\right) \text{ as } n \to \infty.$$

Theorem A.3.10: *Suppose that we have a sequence of real valued random variables $\{U_n; n \geq 1\}$ and another real valued random variable U. Suppose also that $U_n \xrightarrow{\pounds} U$ as $n \to \infty$. Let $g(.)$ be a continuous real valued function. Then, $g(U_n) \xrightarrow{\pounds} g(U)$ as $n \to \infty$.*

In general, if $U_n \xrightarrow{\text{P}} u$, we may not be able to conclude that

$$\lim_{n \to \infty} \text{E}[g(U_n)] = g(u).$$

The following results specify sufficient conditions under which we can claim that $\lim_{n\to\infty} \text{E}[g(U_n)]$ must coincide with $\text{E}[\lim_{n\to\infty} g(U_n)]$, that is, the limiting operation and expectation may be interchanged.

Theorem A.3.11 [Monotone Convergence Theorem]: *Let $\{U_n; n \geq 1\}$ be a sequence of real valued random variables. Suppose that $U_n \xrightarrow{\text{P}} u$ as $n \to \infty$. Let $g(x), x \in \mathbb{R}$, be an increasing or decreasing real valued function. Then, $\text{E}[g(U_n)] \to g(u)$ as $n \to \infty$.*

Theorem A.3.12 [Dominated Convergence Theorem]: *Let $\{U_n; n \geq 1\}$ be a sequence of real valued random variables and another random variable U. Suppose that $U_n \xrightarrow{\text{P}} U$ as $n \to \infty$, that is $U_n - U \xrightarrow{\text{P}} 0$. Let $g(x), x \in \mathbb{R}$, be a continuous real valued function such that $|g(U_n)| \leq W$ such that $\text{E}[W]$ is finite. Then, $\text{E}[g(U_n)] \to \text{E}[g(U)]$ as $n \to \infty$.*

Theorem A.3.13 [Fatou's Lemma]: *Consider $\{U_n; n \geq 1\}$, a sequence of real valued random variables. Suppose that $U_n \geq W$ for all $n \geq 1$ and that $\text{E}[W]$ is finite. Then,*

$$\liminf_{n\to\infty} \text{E}[U_n] \geq \text{E}\left[\liminf_{n\to\infty} U_n\right].$$

In general, if $U_n \xrightarrow{\pounds} U$, we may not be able to conclude that $\lim_{n\to\infty} \text{E}[U_n] = \text{E}[U]$. The following result plays a very crucial role in this regard.

Theorem A.3.14: *Consider $\{U_n; n \geq 1\}$, a sequence of real valued random variables. Suppose that $U_n \xrightarrow{\pounds} U$ as $n \to \infty$ where U is a real valued* *random variable. Then, we have*

$$\lim_{n\to\infty} \mathrm{E}[U_n] = \mathrm{E}[U] \text{ if and only if } |U_n| \text{ is uniformly integrable.}$$

In order to verify the uniform integrability condition for a sequence of random variables, the following result helps in some situations.

Theorem A.3.15: *Consider $\{U_n; n \ge 1\}$, a sequence of real valued random variables. Then, $|U_n|$ is uniformly integrable if $\mathrm{E}[|U_n|^{1+p}]$ is finite for some fixed $p > 0$.*

Definition A.3.3: *Consider a sequence of real valued random variables $\{U_n;\ n \ge 1\}$ defined on a probability space $(\Omega, \mathcal{A}, \mathrm{P})$. Then, U_n is said to converge to a real number u with probability 1 (w.p. 1) as $n \to \infty$ if and only if the following condition holds:*

$$\mathrm{P}\left\{\omega \in \Omega\colon \lim_{n\to\infty} U_n = u\right\} = 1.$$

Intuitively speaking, $U_n \to u$ w.p. 1 means that U_n would lie in a close proximity of u with probability one when n is large.

A.3.1 Some Standard Probability Inequalities

Theorem A.3.16 [Markov Inequality]: *Suppose that W is a real valued random variable such that $\mathrm{P}(W \ge 0) = 1$ and $E(W)$ is finite. Then, for any fixed ε (> 0), one has:*

$$\mathrm{P}(W \ge \varepsilon) \le \varepsilon^{-1} E(W).$$

Theorem A.3.17 [Tchebysheff's Inequality]: *Suppose that X is a real valued random variable with the finite second moment. Let us denote its mean μ and variance $\sigma^2 (> 0)$. Then, for any fixed real number $\varepsilon (> 0)$, one has*

$$\mathrm{P}\{|X - \mu| \ge \varepsilon\} \le \sigma^2/\varepsilon^2.$$

Theorem A.3.18 [Cauchy-Schwarz Inequality]: *Suppose that we have two real valued random variables X_1 and X_2, such that $E[X_1^2], \mathrm{E}[X_2^2]$ and $\mathrm{E}[X_1X_2]$ are all finite. Then, we have*

$$\mathrm{E}^2[X_1X_2] \le \mathrm{E}[X_1^2]\mathrm{E}[X_2^2]. \tag{A.3.1}$$

In (A.3.1), equality holds if and only if $X_1 = kX_2$ w.p. 1 for some constant k.

Here is a restatement of Cauchy-Schwarz inequality that sometimes goes by a different name.

Theorem A.3.19 [Covariance Inequality]: *Suppose that we have two real valued random variables X_1, X_2, such that $\mathrm{E}[X_1^2], \mathrm{E}[X_2^2]$ and $\mathrm{E}[X_1X_2]$ are all finite. Then, we have*

$$\mathrm{Cov}^2(X_1, X_2) \le \mathrm{V}[X_1]\mathrm{V}[X_2]. \tag{A.3.2}$$

In (A.3.2), equality holds if and only if $X_1 = a + bX_2$ *w.p. 1 for some constants a and b.*

Definition A.3.4: *Consider a function* $f : I\!R \to I\!R$. *The function f is called convex if and only if*

$$f(\alpha u + (1-\alpha)v) \le \alpha f(u) + (1-\alpha) f(v)$$

for all $u, v \in I\!R$ *and* $0 \le \alpha \le 1$. *A function f is called concave if and only if* $-f$ *is convex.*

> Suppose that $f(x)$ is twice differentiable at all points $x \in I\!R$. Then, $f(x)$ is *convex* (*concave*) if $d^2 f(x)/dx^2$ is *positive* (*negative*) for all $x \in I\!R$. This is a *sufficient condition.*

Theorem A.3.20 [Jensen's Inequality]: *Suppose that X is a real valued random variable and let* $f(x), x \in I\!R$ *be a convex function. Assume that* $\mathrm{E}[X]$ *is finite. Then, one has*

$$\mathrm{E}[f(X)] \ge f(\mathrm{E}[X]).$$

> In the statement of Jensen's inequality, equality holds only when $f(x)$ is linear in $x \in I\!R$. Also, inequality gets reversed when $f(x)$ is concave.

Theorem A.3.21 [Hölder's Inequality]: *Let X and Y be two real valued random variables. Then, with* $r > 1, s > 1$ *such that* $r^{-1} + s^{-1} = 1$, *one has*

$$|\mathrm{E}[XY]| \le \mathrm{E}^{1/r}[|X|^r]\mathrm{E}^{1/s}[|Y|^s],$$

provided that $\mathrm{E}[|X|^r]$ *and* $\mathrm{E}[|Y|^s]$ *are finite.*

Theorem A.3.22 [Central Absolute Moment Inequality]: *Suppose that we have i.i.d. real valued random variables* $X_1, X_2, \ldots$ *having a common mean* μ. *Let us also assume that* $\mathrm{E}[|\, X_1 \,|^{2\xi}] < \infty$ *for some* $\xi \ge \frac{1}{2}$. *Denote* $\overline{X}_n = n^{-1}\sum_{i=1}^n X_i, n \ge 1$. *Then, we have*

$$E\left[|\, \overline{X}_n - \mu \,|^{2\xi}\right] \le kn^{-\tau},$$

where k does not depend on n, and $\tau = 2\xi - 1$ *or* ξ *according as* $\frac{1}{2} \le \xi < 1$ *or* $\xi \ge 1$ *respectively.*

A.4 A Glimpse at Nonlinear Renewal Theory

Here, we summarize the basic formulation on which the non-linear renewal theory of Woodroofe (1977,1982) and Lai and Siegmund (1977,1979) stand. However, we adopt the layout found in Mukhopadhyay (1988a) and summarize some important results in the light of Mukhopadhyay and Solanky

(1994, Section 2.4.2). Pertinent details can be found in Ghosh et al. (1997, Sections 2.8-2.9) and cited references including Siegmund (1985).

Let us begin with a sequence of i.i.d. positive and continuous random variables $W_1, W_2, \ldots$. For simplicity, let us *assume* that these random variables have all positive moments finite, that is, $\mathrm{E}[W^k] < \infty$ for every fixed $k > 0$. Now, let us define the following stopping time:

$$Q \equiv Q(h^*) = \text{ smallest integer } n(\geq m_0) \text{ such that} \sum_{i=1}^{n} W_i \leq h^* n^\delta L(n) \tag{A.4.1}$$

with

$$\begin{gathered} \delta > 1, h^* > 0, L(n) = 1 + L_0 n^{-1} + o(n^{-1}) \text{ as } n \to \infty \\ \text{with } -\infty < L_0 < \infty, \text{ and the initial sample size } m_0(\geq 1). \end{gathered} \tag{A.4.2}$$

The final sample size "N" required by a large majority of the purely sequential methodologies described in this book can be equivalently expressed as a simple linear function $Q(h^*) + a$ where a is a fixed integer and $Q(h^*)$ has the same representation shown by (A.4.1).

We denote

$$\begin{gathered} \theta = \mathrm{E}[W_1],\ \tau^2 = \mathrm{E}[W_1^2] - \theta^2,\ \beta^* = (\delta - 1)^{-1}, \\ n_0^* = (\theta h^{*-1})^{\beta^*}, \text{ and } q = \beta^{*2}\tau^2\theta^{-2}, \end{gathered} \tag{A.4.3}$$

and

$$\begin{gathered} \nu = \frac{\beta^*}{2\theta}\left\{(\delta-1)^2\theta^2 + \tau^2\right\} - \textstyle\sum_{n=1}^{\infty} \dfrac{1}{n}\mathrm{E}\left[\max\left\{0, \textstyle\sum_{i=1}^{n} W_i - n\delta\theta\right\}\right], \\ \text{and } \kappa = \beta^*\theta^{-1}\nu - \beta^* L_0 - \frac{1}{2}\delta\beta^{*2}\tau^2\theta^{-2}. \end{gathered} \tag{A.4.4}$$

We also assume that the distribution function of W_1 satisfies the following condition:

$$\begin{gathered} \mathrm{P}(W_1 \leq u) \leq Bu^r \text{ for all } u > 0, \text{ with some } r > 0 \text{ and } B > 0, \\ \text{but } r, B \text{ cannot involve } u. \end{gathered} \tag{A.4.5}$$

Now, we can summarize the following sets of crucial results.

Theorem A.4.1: *For the stopping variable $Q(h^*)$ from (A.4.1), we have as $h^* \to 0$:*

(i) $\mathrm{P}\{Q(h^*) \leq \varepsilon n_0^*\} = O(h^{*m_0 r})$ *if* $0 < \varepsilon < 1$ *fixed.*

(ii) $U^* \xrightarrow{\pounds} N(0, q)$ *where* $U^* \equiv (Q(h^*) - n_0^*) / \sqrt{n_0^*}$.

(iii) $|U^*|^\omega$ *is uniformly integrable if* $m_0 r > \frac{1}{2}\beta^*\omega$ *where* ω *is an arbitrary, but fixed positive number.*

with m_0 *from (A.4.1),* n_0^*, β^*, q *from (A.4.3), and* r *from (A.4.5).*

One may apply the fundamental results from Chow and Robbins (1965) or use the basic inequality approach directly to claim that $Q(h^*)/n_0^* \xrightarrow{P} 1$ as $h^* \to 0$. In other words, for small values of h^*, we would expect $Q(h^*)$ to be in a close neighborhood of n_0^* with high probability. So, surely $\mathrm{P}\{Q(h^*) \le \varepsilon n_0^*\}$ should converge to zero for any fixed $0 < \varepsilon < 1$ since εn_0^* falls strictly under n_0^*. But, Theorem A.4.1, part (i) gives the rate of convergence of $\mathrm{P}\{Q(h^*) \le \varepsilon n_0^*\}$ to zero for small values of h^* and any fixed $0 < \varepsilon < 1$. Theorem A.4.1, part (ii) specifies a limiting normal distribution for the normalized stopping variable $U^* \equiv (Q(h^*) - n_0^*)\sqrt{n_0^*}$. This part follows immediately from Ghosh and Mukhopadhyay's (1975) theorem which was stated as Theorem 2.4.3 in Mukhopadhyay and Solanky (1994, p. 47) and in Ghosh et al. (1997, exercise 2.7.4, p. 66).

Theorem A.4.1, part (iii) specifies that $|U^*|^\omega$ is uniformly integrable when m_0 is sufficiently large. If we substitute $\omega = 2$, we can claim that U^{*2} is uniformly integrable when $m_0 > \beta^*$. From part (ii), we already know that the asymptotic distribution of U^* is $N(0, q)$. So, we may apply Theorem A.3.10 to claim that $U^{*2} \xrightarrow{\pounds} q\chi_1^2$. At this point, one may apply Theorem A.3.14 to conclude that $\lim_{h^* \to 0} \mathrm{E}\left[U^{*2}\right] = q$, that is $\mathrm{E}\left[U^{*2}\right] = q + o(1)$ if $m_0 > \beta^*$.

The following result gives the *second-order* expansion of various moments of the stopping variable $Q(h^*)$. We state the result along the lines of Mukhopadhyay (1988a) and Mukhopadhyay and Solanky (1994, section 2.4.2). The fundamental results from Lai and Siegmund (1977,1979) and Woodroofe (1977,1982) led to Mukhopadhyay's (1988a) proof of this theorem.

Theorem A.4.2: *For the stopping variable $Q(h^*)$ from (A.4.1), we have as $h^* \to 0$:*

$$\mathrm{E}\left[\{Q(h^*)/n_0^*\}^\omega\right] = 1 + \left\{\omega\kappa + \frac{1}{2}\omega(\omega - 1)q\right\} n_0^{*-1} + o(n_0^{*-1})$$

provided that

(i) $m_0 > (3-\omega)\beta^*/r$ *for* $\omega \in (-\infty, 2) - \{-1, 1\}$,
(ii) $m_0 > \beta^*/r$ *for* $\omega = 1$ *and* $\omega \ge 2$,
(iii) $m_0 > 2\beta^*/r$ *for* $\omega = -1$,

with m_0 from (A.4.1), n_0^*, β^*, q from (A.4.3), and r from (A.4.5).

If we substitute $\omega = 1$, then for $m_0 > \beta^*/r$, we can claim from Theorem A.4.2, part (ii):

$$\begin{aligned} \mathrm{E}\left[Q(h^*)/n_0^*\right] &= 1 + \kappa n_0^{*-1} + o(n_0^{*-1}) \\ \Rightarrow \mathrm{E}\left[Q(h^*)\right] &= n_0^* + \kappa + o(1) \end{aligned} \tag{A.4.6}$$

which was originally proved by Woodroofe (1977,1979) and Lai and Siegmund (1977).

Next, if we substitute $\omega = 2$, then for $m_0 > \beta^*/r$, we can again claim from Theorem A.4.2, part (ii):

$$\begin{aligned} \mathrm{E}\left[\{Q(h^*)/n_0^*\}^2\right] &= 1 + (2\kappa + q)n_0^{*-1} + o(n_0^{*-1}) \\ \Rightarrow \mathrm{E}\left[\{Q(h^*)\}^2\right] &= n_0^{*2} + (2\kappa + q)n_0^* + o(n_0^*). \end{aligned} \tag{A.4.7}$$

This was originally proved by Woodroofe (1977) and Lai and Siegmund (1977,1979).

A.5 Abbreviations and Notations

Some abbreviations and notations are summarized here.

beta(α, β), $\alpha > 0, \beta > 0$	beta function; $\Gamma(\alpha)\Gamma(\beta)/\Gamma(\alpha + \beta)$
CLT	central limit theorem
d.f. or c.d.f.	distribution function
i.i.d.	independent and identically distributed
lhs	left-hand side
$\log(a), \ln(a), a > 0$	natural (base e) logarithm of a
m.g.f.	moment generating function
MLE	maximum likelihood estimator
p.d.f.	probability density function
p.m.f.	probability mass function
rhs	right-hand side
UMVUE	uniformly minimum variance unbiased estimator
WLLN	weak law of large numbers
w.p. 1	with probability one
$a \approx b$	a, b are approximately same
$\langle u \rangle$	largest integer $< u$
$I(A), I_A$	indicator function of the set A
$\Re, I\!R$	real line; $(-\infty, \infty)$
$\Re^+, I\!R^+$	positive half of $\Re$ or $I\!R$; $(0, \infty)$
$\Re^p, I\!R^p$	p-dimensional Euclidean space
$\in; x \in A$	belongs to; x belongs to the set A
■	indicates the end of a proof
$\boldsymbol{x}, \boldsymbol{X}, \boldsymbol{\theta}$	vectors of dimension more than one
$a \equiv b$	a, b are equivalent
$X_n \xrightarrow{\mathrm{P}} a$	X_n converges to a in probability as $n \to \infty$
$X_n \xrightarrow{\pounds} X$	X_n converges to X in distribution as $n \to \infty$
$\phi(.), \Phi(.)$	standard normal p.d.f. and c.d.f. respectively
z_α	upper $100\alpha\%$ point of $N(0,1)$
$\chi^2_{\nu,\alpha}$	upper $100\alpha\%$ point of χ^2_ν
$t_{\nu,\alpha}$	upper $100\alpha\%$ point of t_ν
$F_{\nu_1,\nu_2,\alpha}$	upper $100\alpha\%$ point of F_{ν_1,ν_2}

A.6 Statistical Tables

This section provides some of the standard statistical tables. These are borrowed from Mukhopadhyay (2000,2006).

Tables A.6.1a-A.6.1b correspond to the distribution function of a standard normal distribution. Tables A.6.2, A.6.3 and A.6.4 respectively correspond to the percentage points of the Chi-square, Student's t, and F distribution.

A.6.1 The Standard Normal Distribution Function

In Tables A.6.1a-A.6.1b, the first column and first row respectively designate the "first" and "second" decimal point of z. The tables provide the associated values of $\Phi(z)$.

Table A.6.1a: Values of the $N(0,1)$ distribution function $\Phi(z) = \int_{-\infty}^{z} (\sqrt{2\pi})^{-1} e^{-x^2/2} dx$

z	**.00**	**.01**	**.02**	.03	**.04**
0.0	.50000	.50399	.50798	.51197	.51595
0.1	.53983	.54380	.54776	.55172	.55567
0.2	.57926	.58317	.58706	.59095	.59483
0.3	.61791	.62172	.62552	.62930	.63307
0.4	.65542	.65910	.66276	.66640	.67003
0.5	.69146	.69497	.69847	.70194	.70540
0.6	.72575	.72907	.73237	.73565	.73891
0.7	.75804	.76115	.76424	.76730	.77035
0.8	.78814	.79103	.79389	.79673	.79955
0.9	.81594	.81859	.82121	.82381	.82639
1.0	.84134	.84375	.84614	.84849	.85083
1.1	.86433	.86650	.86864	.87076	.87286
1.2	.88493	.88686	.88877	.89065	.89251
1.3	.90320	.90490	.90658	.90824	.90988
1.4	.91924	.92073	.92220	.92364	.92507
1.5	.93319	.93448	.93574	.93699	.93822
1.6	.94520	.94630	.94738	.94845	.94950
1.7	.95543	.95637	.95728	.95818	.95907
1.8	.96407	.96485	.96562	.96638	.96712
1.9	.97128	.97193	.97257	.97320	.97381
2.0	.97725	.97778	.97831	.97882	.97932
2.1	.98214	.98257	.98300	.98341	.98382
2.2	.98610	.98645	.98679	.98713	.98745
2.3	.98928	.98956	.98983	.99010	.99036
2.4	.99180	.99202	.99224	.99245	.99266
2.5	.99379	.99396	.99413	.99430	.99446
2.6	.99534	.99547	.99560	.99573	.99585
2.7	.99653	.99664	.99674	.99683	.99693
2.8	.99744	.99752	.99760	.99767	.99774
2.9	.99813	.99819	.99825	.99831	.99836
3.0	.99865	.99869	.99874	.99878	.99882
3.1	.99903	.99906	.99910	.99913	.99916
3.2	.99931	.99934	.99936	.99938	.99940
3.3	.99952	.99953	.99955	.99957	.99958
3.4	.99966	.99968	.99969	.99970	.99971
3.5	.99977	.99978	.99978	.99979	.99980

Table A.6.1b: Values of the N(0,1) distribution function $\Phi(z) = \int_{-\infty}^{z} (\sqrt{2\pi})^{-1} e^{-x^2/2} dx$

z	.05	.06	.07	.08	.09
0.0	.51994	.52392	.52790	.53188	.53586
0.1	.55962	.56356	.56749	.57142	.57535
0.2	.59871	.60257	.60642	.61026	.61409
0.3	.63683	.64058	.64431	.64803	.65173
0.4	.67364	.67724	.68082	.68439	.68793
0.5	.70884	.71226	.71566	.71904	.72240
0.6	.74215	.74537	.74857	.75175	.75490
0.7	.77337	.77637	.77935	.78230	.78524
0.8	.80234	.80511	.80785	.81057	.81327
0.9	.82894	.83147	.83398	.83646	.83891
1.0	.85314	.85543	.85769	.85993	.86214
1.1	.87493	.87698	.87900	.88100	.88298
1.2	.89435	.89617	.89796	.89973	.90147
1.3	.91149	.91309	.91466	.91621	.91774
1.4	.92647	.92785	.92922	.93056	.93189
1.5	.93943	.94062	.94179	.94295	.94408
1.6	.95053	.95154	.95254	.95352	.95449
1.7	.95994	.96080	.96164	.96246	.96327
1.8	.96784	.96856	.96926	.96995	.97062
1.9	.97441	.97500	.97558	.97615	.97670
2.0	.97982	.98030	.98077	.98124	.98169
2.1	.98422	.98461	.98500	.98537	.98574
2.2	.98778	.98809	.98840	.98870	.98899
2.3	.99061	.99086	.99111	.99134	.99158
2.4	.99286	.99305	.99324	.99343	.99361
2.5	.99461	.99477	.99492	.99506	.99520
2.6	.99598	.99609	.99621	.99632	.99643
2.7	.99702	.99711	.99720	.99728	.99736
2.8	.99781	.99788	.99795	.99801	.99807
2.9	.99841	.99846	.99851	.99856	.99861
3.0	.99886	.99889	.99893	.99896	.99900
3.1	.99918	.99921	.99924	.99926	.99929
3.2	.99942	.99944	.99946	.99948	.99950
3.3	.99960	.99961	.99962	.99964	.99965
3.4	.99972	.99973	.99974	.99975	.99976
3.5	.99981	.99981	.99982	.99983	.99983

A.6.2 Percentage Points of Chi-Square Distribution

Table A.6.2 provides the lower $100\gamma\%$ point $\chi^2_{\nu,1-\gamma}$ for the χ^2_ν distribution for different values of ν and γ.

Table A.6.2: Lower $100\gamma\%$ point $\chi^2_{\nu,1-\gamma}$ for the χ^2_ν distribution $\mathrm{P}(\chi^2_\nu \leq \chi^2_{\nu,1-\gamma}) = \gamma$

ν	γ=.01	γ=.025	γ=.05	γ=.95	γ=.975	γ=.99
1	.00016	.00098	.00393	3.8415	5.0239	6.6349
2	.02010	.05064	.10259	5.9915	7.3778	9.2103
3	.11490	.21570	.35185	7.8147	9.3484	11.345
4	.29711	.48442	.71072	9.4877	11.143	13.277
5	.55430	.83121	1.1455	11.070	12.833	15.086
6	.87209	1.2373	1.6354	12.592	14.449	16.812
7	1.2390	1.5643	2.1673	14.067	16.013	18.475
8	1.6465	2.1797	2.7326	15.507	17.535	20.090
9	2.0879	2.7004	3.3251	16.919	19.023	21.666
10	2.5582	3.2470	3.9403	18.307	20.483	23.209
11	3.0535	3.8157	4.5748	19.675	21.920	24.725
12	3.5706	4.4038	5.2260	21.026	23.337	26.217
13	4.1069	5.0088	5.8919	22.362	24.736	27.688
14	4.6604	5.6287	6.5706	23.685	26.119	29.141
15	5.2293	6.2621	7.2609	24.996	27.488	30.578
16	5.8122	6.9077	7.9616	26.296	28.845	32.000
17	6.4078	7.5642	8.6718	27.587	30.191	33.409
18	7.0149	8.2307	9.3905	28.869	31.526	34.805
19	7.6327	8.9065	10.117	30.144	32.852	36.191
20	8.2604	9.5908	10.851	31.410	34.170	37.566
21	8.8972	10.283	11.591	32.671	35.479	38.932
22	9.5425	10.982	12.338	33.924	36.781	40.289
23	10.196	11.689	13.091	35.172	38.076	41.638
24	10.856	12.401	13.848	36.415	39.364	42.980
25	11.524	13.120	14.611	37.652	40.646	44.314
26	12.198	13.844	15.379	35.563	41.923	45.642
27	12.879	14.573	16.151	40.113	43.195	46.963
28	13.565	15.308	16.928	41.337	44.461	48.278
29	14.256	16.047	17.708	42.550	45.731	49.590
30	14.953	16.791	18.493	43.780	46.980	50.890

A.6.3 Percentage Points of Student's t Distribution

Table A.6.3 provides the lower $100\gamma\%$ point $t_{\nu,1-\gamma}$ for Student's t_ν distribution for different values of ν and γ.

Table A.6.3: Lower $100\gamma\%$ point $t_{\nu,1-\gamma}$ for Student's t_ν distribution $\mathrm{P}(t_\nu \leq t_{\nu,1-\gamma}) = \gamma$

ν	γ=.90	γ=.95	γ=.975	γ=.99	γ=.995
1	3.0777	6.3140	12.706	31.821	63.657
2	1.8856	2.9200	4.3027	6.9646	9.9248
3	1.6377	2.3534	3.1824	4.5407	5.8409
4	1.5332	2.1318	2.7764	3.7469	4.6041
5	1.4759	2.0150	2.5706	3.3649	4.0321
6	1.4398	1.9432	2.4469	3.1427	3.7074
7	1.4149	1.8946	2.3646	2.9980	3.4995
8	1.3968	1.8595	2.3060	2.8965	3.3554
9	1.3830	1.8331	2.2622	2.8214	3.2498
10	1.3722	1.8125	2.2281	2.7638	3.1693
11	1.3634	1.7959	2.2010	2.7181	3.1058
12	1.3562	1.7823	2.1788	2.6810	3.0545
13	1.3502	1.7709	2.1604	2.6503	3.0123
14	1.3450	1.7613	2.1448	2.6245	2.9768
15	1.3406	1.7531	2.1314	2.6025	2.9467
16	1.3368	1.7459	2.1199	2.5835	2.9208
17	1.3334	1.7396	2.1098	2.5669	2.8982
18	1.3304	1.7341	2.1009	2.5524	2.8784
19	1.3277	1.7291	2.0930	2.5395	2.8609
20	1.3253	1.7247	2.0860	2.5280	2.8453
21	1.3232	1.7207	2.0796	2.5176	2.8314
22	1.3212	1.7171	2.0739	2.5083	2.8188
23	1.3195	1.7139	2.0687	2.4999	2.8073
24	1.3178	1.7109	2.0639	2.4922	2.7969
25	1.3163	1.7081	2.0595	2.4851	2.7874
26	1.3150	1.7056	2.0555	2.4786	2.7787
27	1.3137	1.7033	2.0518	2.4727	2.7707
28	1.3125	1.7011	2.0484	2.4671	2.7633
29	1.3114	1.6991	2.0452	2.4620	2.7564
30	1.3104	1.6973	2.0423	2.4573	2.7500
35	1.3062	1.6896	2.0301	2.4377	2.7238
40	1.3031	1.6839	2.0211	2.4233	2.7045
45	1.3006	1.6794	2.0141	2.4121	2.6896
50	1.2987	1.6759	2.0086	2.4033	2.6778
100	1.2901	1.6602	1.9840	2.3642	2.6259

A.6.4 Percentage Points of F Distribution

Table A.6.4 provides the lower $100\gamma\%$ point $F_{\nu_1,\nu_2,1-\gamma}$ for the F_{ν_1,ν_2} distribution for different values of ν_1, ν_2 and γ.

Table A.6.4: Lower $100\gamma\%$ point $F_{\nu_1,\nu_2,1-\gamma}$ for the F_{ν_1,ν_2} distribution $\mathrm{P}(F_{\nu_1,\nu_2} \leq F_{\nu_1,\nu_2,1-\gamma}) = \gamma$ where ν_1, ν_2 are respectively the numerator and denominator degrees of freedom

					ν_2			
ν_1	γ	**1**	**2**	**5**	**8**	**10**	**15**	**20**
	.90	39.863	8.5263	4.0604	3.4579	3.2850	3.0732	2.9747
	.95	161.45	18.513	6.6079	5.3177	4.9646	4.5431	4.3512
1	**.975**	647.79	38.506	10.007	7.5709	6.9367	6.1995	6.1995
	.99	4052.2	98.503	16.258	11.259	10.044	8.6831	8.0960
	.995	16211	198.50	22.785	14.688	12.826	10.798	9.9439
	.90	49.500	9.0000	3.7797	3.1131	2.9245	2.6952	2.5893
	.95	199.50	19.000	5.7861	4.4590	4.1028	3.6823	3.4928
2	**.975**	799.50	39.000	8.4336	6.0595	5.4564	4.765	4.4613
	.99	4999.5	99.000	13.274	8.6491	7.5594	6.3589	5.8489
	.995	20000	199.00	18.314	11.042	9.4270	7.7008	6.9865
	.90	57.240	9.2926	3.453	2.7264	2.5216	2.273	2.1582
	.95	230.16	19.296	5.0503	3.6875	3.3258	2.9013	2.7109
5	**.975**	921.85	39.298	7.1464	4.8173	4.2361	3.5764	3.2891
	.99	5763.6	99.299	10.967	6.6318	5.6363	4.5556	4.1027
	.995	23056	199.30	14.940	8.3018	6.8724	5.3721	4.7616
	.90	59.439	9.3668	3.3393	2.5893	2.3772	2.1185	1.9985
	.95	238.88	19.371	4.8183	3.4381	3.0717	2.6408	2.4471
8	**.975**	956.66	39.373	6.7572	4.4333	3.8549	3.1987	2.9128
	.99	5981.1	99.374	10.289	6.0289	5.0567	4.0045	3.5644
	.995	23925	199.37	13.961	7.4959	6.1159	4.6744	4.0900
	.90	60.195	9.3916	3.2974	2.538	2.3226	2.0593	1.9367
	.95	241.88	19.396	4.7351	3.3472	2.9782	2.5437	2.3479
10	**.975**	968.63	39.398	6.6192	4.2951	3.7168	3.0602	2.7737
	.99	6055.8	99.399	10.051	5.8143	4.8491	3.8049	3.3682
	.995	24224	199.40	13.618	7.2106	5.8467	4.4235	3.8470
	.90	61.220	9.4247	3.2380	2.4642	2.2435	1.9722	1.8449
	.95	245.95	19.429	4.6188	3.2184	2.8450	2.4034	2.2033
15	**.975**	984.87	39.431	6.4277	4.1012	3.5217	2.8621	2.5731
	.99	6157.3	99.433	9.7222	5.5151	4.5581	3.5222	3.0880
	.995	24630	199.43	13.146	6.8143	5.4707	4.0698	3.5020
	.90	61.740	9.4413	3.2067	2.4246	2.2007	1.9243	1.7938
	.95	248.01	19.446	4.5581	3.1503	2.7740	2.3275	2.1242
20	**.975**	993.10	39.448	6.3286	3.9995	3.4185	2.7559	2.4645
	.99	6208.7	99.449	9.5526	5.3591	4.4054	3.3719	2.9377
	.995	24836	199.45	12.903	6.6082	5.274	3.8826	3.3178

References

Albert, A. (1966). Fixed size confidence ellipsoids for linear regression parameters. *Ann. Math. Statist.,* **37**, 1602-1630.

Anscombe, F. J. (1949). The statistical analysis of insect counts based on the negative binomial distribution. *Biometrics*, **5**, 165-173.

Anscombe, F. J. (1950). Sampling theory of the negative binomial and logarithmic series distribution. *Biometrika*, **37**, 358-382.

Anscombe, F. J. (1952). Large sample theory of sequential estimation. *Proc. Camb. Phil. Soc.*, **48**, 600-607.

Ansell, J. and Phillips, M. (1989). Practical problems in the statistical analysis of reliability data. *Appl. Statist.*, **38**, 205-247.

Aoshima, M. (2001). Sample size determination for multiple comparisons with components of a linear function of mean vectors. *Commun. Statist. Theory & Method, Series A*, **30**: 1773-1788.

Aoshima, M., Hyakutake, H., and Dudewicz, E. J. (1996). An asymptotically optimal fixed-width confidence interval for the difference of two normal means. *Sequential Analysis,* **15**, 61-70.

Aoshima, M. and Mukhopadhyay, N. (1998). Fixed-width simultaneous confidence intervals for multinormal means in several intraclass correlation models. *J. Multvar. Anal.*, **66**, 46-63.

Aoshima, M. and Mukhopadhyay, N. (2002). Two-stage estimation of a linear function of normal means with second-order approximations. *Sequential Analysis*, **21**, 109-144.

Aras, G. (1987). Sequential estimation of the mean exponential survival time under random censoring. *J. Statist. Plan. Inf.*, **16**, 147-158.

Aras, G. (1989). Second order sequential estimation of the mean exponential survival time under random censoring. *J. Statist. Plan. Inf.*, **21**, 3-17.

Aras, G. and Woodroofe, M. (1993). Asymptotic expansions for the moment of a randomly stopped average. *Ann. Statist.,* **21**, 503-519.

Armitage, P. (1961). Contribution to the discussion of C. A. B. Smith, "Consistency in statistical inference and decision." *J. Roy. Statist. Soc., Series B*, **23**, 1-37.

Armitage, P. (1975). *Sequential Medical Trials*, 2nd edition. Blackwell Scientific Publications: Oxford.

Arnold, K. J. (1951). Tables to facilitate sequential t-tests. *Nat. Bur. Standards Appl. Math., Series 7*, 82 pages.

Bahadur, R. R. (1950). On a problem in the theory of k populations. *Ann. Math. Statist.*, **21**, 362-375.

Bahadur, R. R. (1958). A note on the fundamental identity of sequential analysis. *Ann. Math. Statist.*, **29**, 534-543.

Bahadur, R. R. (1966). A note on quantiles in large samples. *Ann. Math. Statist.*, **37**, 577-580.

Bahadur, R. R. and Robbins, H. (1950). The problem of the greater mean. *Ann. Math. Statist.*, **21**, 469-487. Corrections (1951), *Ann. Math. Statist.*, **22**, 310.

Bain, L. J. (1978). *Statistical Analysis of Reliability and Life Testing Models.* Marcel Dekker: New York.

Balakrishnan, N. and Basu, A. P. (1995). *The Exponential Distribution*, edited volume. Gordon and Breach: Amsterdam.

Bar-Lev, S. K. and Enis, P. (1986). Reproducibility and natural exponential families with power variance function. *Ann. Statist.*, **14**, 1507-1522.

Bar-Lev, S. K. and Reiser, B. (1982). An exponential subfamily which admits UMPU test based on a single test statistic. *Ann. Statist.*, **10**, 979-989.

Barlow, R. and Proschan, F. (1975). *Statistical Theory of Reliability and Life Testing.* Holt, Rinehart and Winston: New York.

Barndorff-Nielsen, O. (1978). *Information and Exponential Families in Statistical Theory.* Wiley: New York.

Baron, M. (2004). Sequential methods for multistate processes. In *Applied Sequential Methodologies*, N. Mukhopadhyay, S. Datta, and S. Chattopadhyay, eds., pp. 53-67. Marcel Dekker: New York.

Barrigossi, J. A. F. (1997). *Development of an IPM System for the Mexican Bean Beetle (Epilachna Varivestis Mulsant) As a Pest of Dry Bean (Phaseolus Vulgaris L.). Ph.D. thesis*, Department of Biometry, University of Nebraska, Lincoln.

Basu, A. K. and Das, J. K. (1997). Sequential estimation of the autoregressive parameters in $AR(p)$ model. *Sequential Analysis*, **16**, 1-24.

Basu, A. P. (1971). On a sequential rule for estimating the location of an exponential distribution. *Naval Res. Logist. Qr.*, **18**, 329-337.

Basu, A. P. (1991). Sequential methods in reliability and life testing. In *Handbook of Sequential Analysis*, B. K. Ghosh and P. K. Sen, eds., pp. 581-592. Marcel Dekker: New York.

Basu, D. (1955). On statistics independent of a complete sufficient statistic. *Sankhya, Series A*, **15**, 377-380.

Basu, D. (1975). Statistical information and likelihood (with discussions). *Sankhya, Series A*, **37**, 1-71.

Bechhofer, R. E. (1954). A single-stage multiple decision procedure for ranking means of normal populations with variances. *Ann. Math. Statist.*, **25**, 16-39.

Bechhofer, R. E., Dunnett, C. W., and Sobel, M. (1954). A two-sample multiple decision procedure for ranking means of normal populations with a common unknown variance. *Biometrika*, **41**, 170-176.

Bechhofer, R. E., Kiefer, J., and Sobel, M. (1968). *Sequential Identification and Ranking Procedures.* University of Chicago Press: Chicago.

Bechhofer, R. E., Santner, T. J., and Goldsman, D. M. (1995). *Design and Analysis of Experiments for Statistical Selection, Screening, and Multiple Comparisons.* Wiley: New York.

Beg, M. (1980). Estimation of $P(Y < X)$ for exponential families. *IEEE Trans. Reliab.*, **29**, 158-160.

Beg, M and Singh, N. (1979). Estimation of $P(Y < X)$ for the Pareto distribution. *IEEE Trans. Reliab.*, **28**, 411-414.

Berger, J. O. (1985). *Statistical Decision Theory and Bayesian Analysis.* Springer-Verlag: New York.

Berti, A., Zanin, G., Baldoni, G., Grignani, C., Mazzoncini, Montemurro, Tei, F., Vazzana, C., and Viggiani, P. (1992). Frequency distribution of weed counts and applicability of a sequential sampling method to integrated weed management. *Weed Research*, **32**, 39-44.

Bickel, P. and Yahav, J. A. (1967). Asymptotically pointwise optimal procedures in sequential analysis. *Proc. Fifth Berkeley Symp. Math. Statist. Probab.*, **1**, 401-413. University of California Press: Berkeley.

Bickel, P. and Yahav, J. A. (1968). Asymptotically optimal Bayes and minimax procedures in sequential estimation. *Ann. Math. Statist.*, **39**, 442-456.

Bickel, P. and Yahav, J. A. (1969a). On an A.P.O. rule in sequential estimation with quadratic loss. *Ann. Math. Statist.*, **40**, 417-426.

Bickel, P. and Yahav, J. A. (1969b). Some contributions to the asymptotic theory of Bayes solutions. *Zeit. Wahhrsch. verw. Geb.*, **11**, 257-276.

Binns, D. (1975). Sequential estimation of the mean of a negative binomial distribution. *Biometrika*, **62**, 433-440.

Birnbaum, A. and Healy, W. C., Jr. (1960). Estimates with prescribed variance based on two-stage sampling. *Ann. Math. Statist.*, **31**, 662-676.

Biswas, A. and Dewanji, A. (2004). Sequential adaptive designs for clinical trials with longitudinal responses. In *Applied Sequential Methodologies*, N. Mukhopadhyay, S. Datta, and S. Chattopadhyay, eds., pp. 69-84. Marcel Dekker: New York.

Bliss, C. L. and Owen, A. R. C. (1958). Negative binomial distributions with a common k. *Biometrika*, **45**, 37-38.

Blumenthal, S. and Govindarajulu, Z. (1977). Robustness of Stein's two-stage procedure for mixtures of normal populations. *J. Amer. Statist. Assoc.*, **72**, 192-196.

Bose, A. and Boukai, B. (1993a). Sequential estimation results for a two-parameter exponential family of distributions. *Ann. Statist.*, **21**, 484-582.

Bose, A. and Boukai, B. (1993b). Sequential estimation of the mean of NEF-PVF distributions. Personal communication.

Bose, A. and Mukhopadhyay, N. (1994a). Sequential estimation by accelerated stopping times in a two-parameter exponential family of distributions. *Statistics & Decisions*, **12**, 281-291.

Bose, A. and Mukhopadhyay, N. (1994b). Sequential estimation via replicated piecewise stopping number in a two-parameter exponential family of distributions. *Sequential Analysis*, **13**, 1-10.

Bose, A. and Mukhopadhyay, N. (1995a). Sequential estimation of the mean of an exponential distribution via replicated piecewise stopping number. *Statistics & Decisions*, **13**, 351-361.

Bose, A. and Mukhopadhyay, N. (1995b). Sequential interval estimation via replicated piecewise stopping times in a class of two-parameter exponential family of distributions. *Sequential Analysis*, **14**, 287-305.

Bose, A. and Mukhopadhyay, N. (1995c). A note on accelerated sequential estimation of the mean of NEF-PVF distributions. *Ann. Inst. Statist. Math.*, **47**, 99-104.

Brown, R. (1977). Evaluation of $P(X > Y)$ when both X and Y are from 3-parameter Weibull Distributions. *IEEE Trans. Reliab.*, **22**, 100-105.

Callahan, J. (1969). *On Some Topics in Sequential Multiparameter Estimation. Ph.D. thesis*, Department of Statistics, Johns Hopkins University, Baltimore.

Chang, Y.-c., I. and Martinsek, A. T. (2004). Sequential approaches to data mining. In *Applied Sequential Methodologies*, N. Mukhopadhyay, S. Datta, and S. Chattopadhyay, eds., pp. 85-103. Marcel Dekker: New York.

Chapman, D. G. (1950). Some two-sample tests. *Ann. Math. Statist.*, **21**, 601-606.

Chatterjee, S. K. (1959a). On an extension of Stein's two-sample procedure to the multinormal problem. *Calcutta Statist. Assoc. Bul.*, **8**, 121-148.

Chatterjee, S. K. (1959b). Some further results on the multinormal extension of Stein's two-sample procedure. *Calcutta Statist. Assoc. Bul.*, **9**, 20-28.

Chatterjee, S. K. (1962a). Simultaneous confidence intervals of predetermined length based on sequential samples. *Calcutta Statist. Assoc. Bul.*, **11**, 144-159.

Chatterjee, S. K. (1962b). Sequential inference procedures of Stein's type for a class of multivariate regression problems. *Ann. Math. Statist.*, **33**, 1039-1064.

Chatterjee, S. K. (1990). Multi-step sequential procedures for a replicable linear model with correlated variables. In *Probability, Statistics and Design of Experiments* (Proceedings of R. C. Bose Symposium 1988), R. R. Bahadur, ed., pp. 217- 226. Wiley Eastern: New Delhi.

Chatterjee, S. K. (1991). Two-stage and multistage procedures. In *Handbook of Sequential Analysis*, B. K. Ghosh and P. K. Sen, eds., pp. 21-45. Marcel Dekker: New York.

Chaturvedi, A. and Shukla, P. S. (1990). Sequential point estimation of location parameter of a negative exponential distribution. *J. Indian Statist. Assoc.*, **28**, 41-50.

Chen, H. J. and Dudewicz, E. J. (1976). Procedures for fixed-width interval estimation of the largest normal mean. *J. Amer. Statist. Assoc.*, **71**, 752-756.

Chernoff, H. (1972). *Sequential Analysis and Optimal Design.* CBMS #8. SIAM: Philadelphia.

Chiu, W. K. (1974). Selecting the m populations with largest means from k normal populations with unequal variances. *Australian J. Statist.*, **16**, 144-147.

Choi, S. C. (1971). Sequential test for correlation coefficient. *J. Amer. Statist. Assoc.*, **66**, 575-576.

Chow, Y. S. and Martinsek, A. T. (1982). Bounded regret of a sequential procedure for estimation of the mean. *Ann. Statist.*, **10**, 909-914.

Chow, Y. S. and Robbins, H. (1965). On the asymptotic theory of fixed width sequential confidence intervals for the mean. *Ann. Math. Statist.*, **36**, 457-462.

Chow, Y. S., Robbins, H., and Siegmund, D. (1971). *Great Expectations: The Theory of Optimal Stopping.* Houghton Mifflin: Boston.

Chow, Y. S., Robbins, H., and Teicher, H. (1965). Moments of randomly stopped sums. *Ann. Math. Statist.*, **36**, 789-799.

Chow, Y. S. and Yu, K. F. (1981). The performance of a sequential procedure for the estimation of the mean. *Ann. Statist.*, **9**, 184-188.

Cicconetti, G. (2002). *Contributions to Sequential Analysis. Ph.D. thesis*, Department of Statistics, University of Connecticut, Storrs.

Costanza, M. C., Hamdy, H. I., and Son, M. S. (1986). Two stage fixed width confidence intervals for the common location parameter of several exponential distributions. *Commun. Statist. Theory & Method, Series A*, **15**, 2305-2322.

Dantzig, G. B. (1940). On the non-existence of tests of Student's hypothesis having power functions independent of σ. *Ann. Math. Statist.*, **11,** 186-192.

Darling, D. A. and Robbins, H. (1968). Some nonparametric sequential tests with power one. *Proc. Nat. Acad. Sci. USA*, **61**, 804-809.

Datta, S. and Mukhopadhyay, N. (1997). On sequential fixed-size confidence regions for the mean vector. *J. Multvar. Anal.*, **60**, 233-251.

de Silva, B. M. and Mukhopadhyay, N. (2004). Kernel density of wool fiber diameter. In *Applied Sequential Methodologies*, N. Mukhopadhyay, S. Datta, and S. Chattopadhyay, eds., 141-170. Marcel Dekker: New York.

de Silva, B. M. and Waikar, V. B. (2006). A sequential approach to Behrens-Fisher problem. *Sequential Analysis*, Milton Sobel Memorial Issue, **25**, 311-326.

Desu, M. M., Narula, S. C., and Villarreal, B. (1977). A two-stage procedure for selecting the best of k exponential distributions. *Commun. Statist. Theory & Method, Series A*, **6**, 1231-1243.

Dudewicz, E. J. (1971). Non-existence of a single-sample selection procedure whose P(CS) is independent of the variance. *S. Afr. Statist. J.*, **5**, 37-39.

Dudewicz, E. J. and Ahmed, S. U. (1998). New exact and asymptotically optimal solution to the Behrens-Fisher problem with tables. *Amer. J. Math. Management Sci.*, **18**, 359-426.

Dudewicz, E. J. and Ahmed, S. U. (1999). New exact and asymptotically heteroscedastic statistical procedures and tables II. *Amer. J. Math. Management Sci.*, **19**, 157-180.

Dudewicz, E. J. and Dalal, S. R. (1975). Allocation of observations in ranking and selection with unequal variances. *Sankhyā, Series B*, **37**, 28-78.

Efromovich, S. (2004). Financial applications of sequential nonparametric curve estimation. In *Applied Sequential Methodologies*, N. Mukhopadhyay, S. Datta, and S. Chattopadhyay, eds., pp. 171-192. Marcel Dekker: New York.

Epstein, B. and Sobel, M. (1953). Life testing. *J. Amer. Statist. Assoc.*, **48**, 486-502.

Epstein, B. and Sobel, M. (1954). Some theorems relevant to life testing from an exponential distribution. *Ann. Math. Statist.*, **25**, 373-381.

Epstein, B. and Sobel, M. (1955). Sequential procedures in life testing from an exponential distribution. *Ann. Math. Statist.*, **26**, 82-93.

Fakhre-Zakeri, I. and Lee, S. (1992). Sequential estimation of the mean of a linear process. *Sequential Analysis*, **11**, 181-197.

Fakhre-Zakeri, I. and Lee, S. (1993). Sequential estimation of the mean vector of a multivariate linear process. *J. Multvar. Anal.*, **47**, 196-209.

Farrell, R. H. (1966). Bounded length confidence intervals for the p-point of a distribution function III. *Ann. Math. Statist.*, **37**, 586-592.

Ferguson, T. S. (1967). *Mathematical Statistics.* Academic Press: New York.

Finster, M. (1983). A frequentist approach to sequential estimation in the general linear model. *J. Amer. Statist. Assoc.*, **78**, 403-407.

Finster, M. (1985). Estimation in the general linear model when the accuracy is specified before data collection. *Ann. Statist.*, **13**, 663-675.

Gardiner, J. C. and Susarla, V. (1983). Sequential estimation of the mean survival time under random censorship. *Sequential Analysis*, **2**, 201-223.

Gardiner, J. C. and Susarla, V. (1984). Risk efficient estimation of the mean exponential survival time under random censoring. *Proc. Nat. Acad. Sci. USA*, **81**, 5906-5909.

Gardiner, J. C. and Susarla, V. (1991). Time-sequential estimation. In *Handbook of Sequential Analysis*, B. K. Ghosh and P. K. Sen, eds., pp. 613-631. Marcel Dekker: New York.

Gardiner, J. C., Susarla, V., and van Ryzin. J. (1986). Time sequential estimation of the exponential mean under random withdrawals. *Ann. Statist.*, **14**, 607-618.

Geertsema, J. C. (1970). Sequential confidence intervals based on rank tests. *Ann. Math. Statist.*, **41**, 1016-1026.

Geertsema, J. C. (1972). Nonparametric sequential procedure for selecting the best of k populations. *J. Amer. Statist. Assoc.*, **67**, 614-616.

Ghosh, B. K. (1970). *Sequential Tests of Statistical Hypotheses.* Addison-Welsley: Reading.

Ghosh, B. K. (1975a). A two-stage procedure for the Behrens-Fisher problem. *J. Amer. Statist. Assoc.*, **70**, 457-462.

Ghosh, B. K. (1975b). On the distribution of the difference of two t-variables. *J. Amer. Statist. Assoc.*, **70**, 463-467.

Ghosh, B. K. and Sen, P. K. (1991). *Handbook of Sequential Analysis*, edited volume. Marcel Dekker: New York.

Ghosh, J. K. (1960a). On some properties of sequential t-test. *Calcutta Statist. Assoc. Bul.*, **9**, 77-86.

Ghosh, J. K. (1960b). On the monotonicity of the OC of a class of sequential probability ratio tests. *Calcutta Statist. Assoc. Bul.*, **9**, 139-144.

Ghosh, M. (1980). Rate of convergence to normality for random means: Applications to sequential estimation. *Sankhya, Series A*, **42**, 231-240.

Ghosh, M. and Dasgupta, R. (1980). Berry-Esseen theorems for U-statistics in the non iid case. In *Proc. Nonparam. Statist. Inf.*, pp. 219-313. Budapest, Hungary.

Ghosh, M. and Mukhopadhyay, N. (1975). Asymptotic normality of stopping times in sequential analysis. Unpublished report.

Ghosh, M. and Mukhopadhyay, N. (1976). On two fundamental problems of sequential estimation. *Sankhya, Series B*, **38**, 203-218.

Ghosh, M. and Mukhopadhyay, N. (1979). Sequential point estimation of the mean when the distribution is unspecified. *Commun. Statist. Theory & Method, Series A*, **8**, 637-652.

Ghosh, M. and Mukhopadhyay, N. (1980). Sequential point estimation of the difference of two normal means. *Ann. Statist.*, **8**, 221-225.

Ghosh, M. and Mukhopadhyay, N. (1981). Consistency and asymptotic efficiency of two-stage and sequential procedures. *Sankhya, Series A*, **43**, 220-227.

Ghosh, M. and Mukhopadhyay, N. (1989). Sequential estimation of the percentiles of exponential and normal distributions. *S. Afr. Statist. J.*, **23**, 251-268.

Ghosh, M. and Mukhopadhyay, N. (1990). Sequential estimation of the location parameter of an exponential distribution. *Sankhya, Series A*, **52**, 303-313.

Ghosh, M., Mukhopadhyay, N., and Sen, P. K. (1997). *Sequential Estimation.* Wiley: New York.

Ghosh, M., Nickerson, D. M., and Sen, P. K. (1987). Sequential shrinkage estimation. *Ann. Statist.*, **15**, 817-829.

Ghosh, M. and Sen, P. K. (1971). Sequential confidence intervals for the regression coefficients based on Kendall's tau. *Calcutta Statist. Assoc. Bul.*, **20**, 23-36.

Ghosh, M. and Sen, P. K. (1972). On bounded length confidence intervals for the regression coefficient based on a class of rank statistics. *Sankhya, Series A*, **34**, 33-52.

Ghosh, M. and Sen, P. K. (1973). On some sequential simultaneous confidence interval procedures. *Ann. Inst. Statist. Math.*, **25**, 123-135.

Ghosh, M. and Sen, P. K. (1983). On two-stage James-Stein estimators. *Sequential Analysis*, **2**, 359-367.

Ghosh, M. and Sen, P. K. (1984). On asymptotically risk-efficient versions of generalized U-statistics. *Sequential Analysis*, **3**, 233-252.

Ghosh, M., Sinha, B. K., and Mukhopadhyay, N. (1976). Multivariate sequential point estimation. *J. Multvar. Anal.*, **6**, 281-294.

Ghurye, S. G. (1958). Note on sufficient statistics and two-stage procedures. *Ann. Math. Statist.*, **29**, 155-166.

Gibbons, J. D., Olkin, I., and Sobel, M. (1977). *Selecting and Ordering Populations: A New Statistical Methodology.* Wiley: New York. Reprinted 1999, SIAM: Philadelphia.

Gleser, L. J. (1965). On the asymptotic theory of fixed-size sequential confidence bounds for linear regression parameters. *Ann. Math. Statist.*, **36**, 463-467. Corrections: 1966, *Ann. Math. Statist.*, **37**, 1053-1055.

Govindarajulu, Z. (1981). *The Sequential Statistical Analysis.* American Sciences Press: Columbus.

Govindarajulu, Z. (1985). Exact expressions for the stopping time and confidence coefficient in point and interval estimation of scale parameter of exponential distribution with unknown location. *Tech. Report 254*, Department of Statistics, University of Kentucky.

Govindarajulu, Z. (2004). *Sequential Statistics.* World Scientific Publishing: Singapore.

Govindarajulu, Z. and Sarkar, S. C. (1991). Sequential estimation of the scale parameter in exponential distribution with unknown location. *Utilitas Math.*, **40**, 161-178.

Graybill, F. A. and Connell, T. L. (1964). Sample size for estimating the variance within d units of the true value. *Ann. Math. Statist.*, **35**, 438-440.

Grubbs, F. E. (1971). Approximate fiducial bounds on reliability for the two parameter negative exponential distribution. *Technometrics*, **13**, 873-876.

Gupta, S. S. (1956). *On a Decision Rule for a Problem in Ranking Means. Ph.D. thesis*, Department of Statistics, University of North Carolina, Chapel Hill.

Gupta, S. S. and Panchapakesan, S. (1979). *Multiple Decision Procedures: Theory and Methodology of Selecting and Ranking Populations.* Wiley: New York. Reprinted 2002, SIAM: Philadelphia.

Gupta, S. S., Panchapakesan, S., and Sohn, J. K. (1985). On the distribution of the Studentized maximum of equally correlated normal random variables. *Commun. Statist. Simul. & Comp., Series B*, **14**, 103-135.

Gupta, S. S. and Sobel, M. (1957). On a statistic which arises in selection and ranking problems. *Ann. Math. Statist.*, **28**, 957-967.

Gut, A. (1988). *Stopped Random Walks: Limit Theorems and Applications.* Springer-Verlag: New York.

Hall, P. (1981). Asymptotic theory of triple sampling for estimation of a mean. *Ann. Statist.*, **9**, 1229-1238.

Hall, P. (1983). Sequential estimation saving sampling operations. *J. Roy. Statist. Soc., Series B*, **45**, 219-223.

Hall, W. J. (1965). Methods of sequentially testing composite hypotheses with special reference to the two-sample problem. *Inst. of Statist. Mimeo Ser. 441*, University of North Carolina, Chapel Hill.

Hall, W. J., Wijsman, R. A., and Ghosh, J. K. (1965). The relationship between sufficiency and invariance with applications in sequential analysis. *Ann. Math. Statist.*, **36**, 575-614.

Hamdy, H. I. (1988). Remarks on the asymptotic theory of triple stage estimation of the normal mean. *Scand. J. Statist.*, **15**, 303–310.

Hamdy, H. I., Mukhopadhyay, N., Costanza, M. C., and Son, M. S. (1988). Triple stage point estimation for the exponential location parameter. *Ann. Inst. Statist. Math.*, **40**, 785-797.

Healy, W. C., Jr. (1956). Two-sample procedures in simultaneous estimation. *Ann. Math. Statist.*, **27**, 687-702.

Hilton, G. F. (1984). *Sequential and Two-Stage Point Estimation Problems for Negative Exponential Distributions. Ph.D. thesis*, Department of Statistics, Oklahoma State University, Stillwater.

Hochberg, Y. and Tamhane, A. C. (1987). *Multiple Comparison Procedures.* Wiley: New York.

Hoeffding, W. (1953). A lower bound for the average sample number of a sequential test. *Ann. Math. Statist.*, **24**, 127-130.

Hoel, D. G. and Sobel, M. (1971). Comparison of sequential procedures for selecting the best binomial population. *Proc. Sixth Berkeley Symp. Math. Statist. Probab.*, **4**, 53-69. University of California Press: Berkeley.

Hoel, P. G. (1954). On a property of the sequential t-test. *Skandinavisk Aktuarietidskrift*, **37**, 19-22.

Hsu, J. C. (1996). *Multiple Comparisons: Theory and Methods.* Chapman & Hall/CRC: Boca Raton.

Isogai, E. and Uno, C. (1994). Sequential estimation of a parameter of an exponential distribution. *Ann. Inst. Statist. Math.*, **46**, 77-82.

Isogai, E. and Uno, C. (1995). On the sequential point estimation of the mean of a gamma distribution. *Statist. Probab. Lett.*, **22**, 287-293.

James, W. and Stein, C. (1961). Estimation with quadratic loss. *Proc. Fourth Berkeley Symp. Math. Statist. Probab.*, **1**, 361-379. University of California Press: Berkeley.

Jennison, C. and Turnbull, B. W. (1999). *Group Sequential Methods with Applications in Clinical Trials.* Chapman & Hall: London.

Johnson, G., Mortensen, D. A., Young, L. J., and Martin, A. R. (1995). The stability of weed seedling population models and parameters in eastern Nebraska corn (Zea mays) and soybean (Glycine max) fields. *Weed Science*, **43**, 604-611.

Johnson, N. L. and Kotz, S. (1970). *Continuous Univariate Distributions-2*. Wiley: New York.

Khan, R. A. (1968). Sequential estimation of a mean vector of a multivariate normal distribution. *Sankhya, Series A*, **30**, 331-334.

Khan, R. A. (1969). A general method of determining fixed-width confidence intervals. *Ann. Math. Statist.*, **40**, 704-709.

Kowalski, C. J. (1971). The OC and ASN functions of some SPRT's for the correlation coefficient. *Technometrics*, **13**, 833-840.

Kubokawa, T. (1989). Improving on two-stage estimators for scale families. *Metrika*, **36**, 7-13.

Kuno, E. (1969). A new method of sequential sampling to obtain population estimates with a fixed level of accuracy. *Researches in Population Ecology*, **11**, 127-136.

Kuno, E. (1972). Some notes on population estimation by sequential sampling. *Researches in Population Ecology*, **14**, 58-73.

Lai, T. L. (2004a). Likelihood ratio identities and their applications to sequential analysis (with discussions). *Sequential Analysis*, **23**, 467-497.

Lai, T. L. (2004b). Interim and terminal analyses of clinical trials with failure-time endpoints and related group sequential designs. In *Applied Sequential Methodologies*, N. Mukhopadhyay, S. Datta, and S. Chattopadhyay, eds., pp. 193-218. Marcel Dekker: New York.

Lai, T. L. and Siegmund, D. (1977). A nonlinear renewal theory with applications to sequential analysis I. *Ann. Statist.*, **5**, 946-954.

Lai, T. L. and Siegmund, D. (1979). A nonlinear renewal theory with applications to sequential analysis II. *Ann. Statist.*, **7**, 60-76.

Lam, K. and Ng, C. K. (1990). Two-stage procedures for comparing several exponential populations with a control when the scale parameters are unknown and unequal. *Sequential Analysis*, **9**, 151-164.

Lawless, J. F. and Singhal, K. (1980). Analysis of data from lifetest experiments under an exponential model. *Naval Res. Logist. Qr.*, **27**, 323-334.

Lehmann, E. L. (1951). *Notes on the Theory of Estimation*. University of California Press: Berkeley.

Liberman, S. (1987). *Some Sequential Aspects for the Multivariate Behrens-Fisher Problem. Ph.D. thesis*, Department of Statistics, University of Connecticut, Storrs.

Liu, W. (1995). Fixed-width simultaneous confidence intervals for all pairwise comparisons. *Computational Statist. & Data Anal.*, **20**, 35-44.

Lohr, S. (1990). Accurate multivariate estimation using triple sampling. *Ann. Statist.,* **18**, 1615-1633.

Lombard, F. and Swanepoel, J. W. H. (1978). On finite and infinite confidence sequences. *S. Afr. Statist. J.*, **12**, 1-24.

Lomnicki, Z. (1973). Some aspects of the statistical approach to reliability. *J. Roy. Statist. Soc., Series A*, **136**, 395-420.

Mahalanobis, P. C. (1940). A sample survey of acreage under jute in Bengal, with discussion on planning of experiments. *Proc. Second Indian Statist. Conference*. Statistical Publishing Society: Calcutta.

Marshall, E. J. P. (1988). Field-scale estimates of grass weed population in arable land. *Weed Research*, **28**, 191-198.

Martinsek, A. T. (1983). Second order approximation to the risk of a sequential procedure. *Ann. Statist.*, **11**, 827-836.

Martinsek, A. T. (1988). Negative regret, optional stopping and the elimination of outliers. *J. Amer. Statist. Assoc.*, **83**, 160-163.

Martinsek, A. T. (1990). Sequential point estimation in regression models with nonnormal errors. *Sequential Analysis*, **9**, 243-268.

Morris, C. N. (1982). Natural exponential families with quadratic variance functions. *Ann. Statist.*, **10**, 65-80.

Morris, C. N. (1983). Natural exponential families with quadratic variance functions: Statistical theory. *Ann. Statist.*, **11**, 515-529.

Moshman, J. (1958). A method for selecting the size of the initial sample in Stein's two-sample procedure. *Ann. Math. Statist.,* **29**, 667-671.

Mukhopadhyay, N. (1974a). Sequential estimation of location parameter in exponential distributions. *Calcutta Statist. Assoc. Bul.*, **23**, 85-95.

Mukhopadhyay, N. (1974b). Sequential estimation of regression parameters in Gauss-Markoff setup. *J. Indian Statist. Assoc.*, **12**, 39-43.

Mukhopadhyay, N. (1975). *Sequential Methods in Estimation and Prediction. Ph.D. thesis*, Indian Statistical Institute, Calcutta.

Mukhopadhyay, N. (1976a). Sequential estimation of a linear function of means of three normal populations. *J. Amer. Statist. Assoc.*, **71**, 149-153.

Mukhopadhyay, N. (1976b). Fixed-width confidence intervals for the mean using a three-stage procedure. Unpublished report.

Mukhopadhyay, N. (1977). Remarks on sequential estimation of a linear function of two means: The normal case. *Metrika*, **24**, 197-201.

Mukhopadhyay, N. (1978). Sequential point estimation of the mean when the distribution is unspecified. *Statist. Tech. Report No. 312*, University of Minnesota, Minneapolis.

Mukhopadhyay, N. (1979a). Fixed size simultaneous confidence region for mean vector and dispersion of a multinormal distribution. *Calcutta Statist. Assoc. Bul.*, **28**, 147-150.

Mukhopadhyay, N. (1979b). Some comments on two-stage selection procedures. *Commun. Statist. Theory & Method, Series A*, **8**, 671-683.

Mukhopadhyay, N. (1980). A consistent and asymptotically efficient two-stage procedure to construct fixed-width confidence intervals for the mean. *Metrika*, **27**, 281-284.

Mukhopadhyay, N. (1981). Sequential approach to simultaneous estimation of the mean and variance. *Metrika*, **28**, 133-136.

Mukhopadhyay, N. (1982a). Stein's two-stage procedure and exact consistency. *Skandinavisk Aktuarietdskrift*, 110-122.

Mukhopadhyay, N. (1982b). A study of the asymptotic regret while estimating the location of an exponential distribution. *Calcutta Statist. Assoc. Bul.*, **31**, 207-213.

Mukhopadhyay, N. (1983). Theoretical investigations of some sequential and two-stage procedures to select the larger mean. *Sankhya, Series A*, **45**, 346-356.

Mukhopadhyay, N. (1984a). Nonparametric two-sample sequential problems for truncation parameters of unknown distributions. *Calcutta Statist. Assoc. Bul.*, **33**, 27-34.

Mukhopadhyay, N. (1984b). Sequential and two-stage procedures for selecting the better exponential population covering the case of unknown and unequal scale parameters. *J. Statist. Plan. Inf.*, **9**, 33-43.

Mukhopadhyay, N. (1985a). A note on three-stage and sequential point estimation procedures for a normal mean. *Sequential Analysis*, **4**, 311-320.

Mukhopadhyay, N. (1985b). A note on two-stage James-Stein estimators for two-sample problems. *J. Indian Statist. Assoc.*, **23**, 19-26.

Mukhopadhyay, N. (1986). On selecting the best exponential population. *J. Indian Statist. Assoc.*, **24**, 31-41.

Mukhopadhyay, N. (1987a). Minimum risk point estimation of the mean of a negative exponential distribution. *Sankhya, Series A*, **49**, 105-112.

Mukhopadhyay, N. (1987b). Three-stage procedures for selecting the best exponential population. *J. Statist. Plan. Inf.*, **16**, 345-352.

Mukhopadhyay, N. (1988a). Sequential estimation problems for negative exponential populations. *Commun. Statist. Theory & Method, Series A*, **17**, 2471-2506.

Mukhopadhyay, N. (1988b). Fixed precision estimation of a positive location parameter of a negative exponential population. *Calcutta Statist. Assoc. Bul.*, **37**, 101-104.

Mukhopadhyay, N. (1990). Some properties of a three-stage procedure with applications in sequential analysis. *Sankhya, Series A*, **52**, 218-231.

Mukhopadhyay, N. (1991). Parametric sequential point estimation. In *Handbook of Sequential Analysis*, B. K. Ghosh and P. K. Sen, eds., pp. 245-267. Marcel Dekker: New York.

Mukhopadhyay, N. (1993). Nonlinear renewal theory and beyond: Reviewing the roles in selection and ranking. In *Multiple Comparisons, Selection, and Applications in Biometry*, F. M. Hoppe, ed., pp. 293-313. Marcel Dekker: New York.

Mukhopadhyay, N. (1994). Improved sequential estimation of means of exponential distributions. *Ann. Inst. Statist. Math.*, **46**, 509-519.

Mukhopadhyay, N. (1995a). Two-stage and multi-stage estimation. In *The Exponential Distribution*, N. Balakrishnan and A. P. Basu, eds., pp. 429-452. Gordon and Breach: Amsterdam.

Mukhopadhyay, N. (1995b). Second-order approximations in the time-sequential point estimation methodologies for the mean of an exponential distribution. *Sequential Analysis*, **14**, 133-142.

Mukhopadhyay, N. (1995c). Sequential estimation of means of linear processes. *Metrika*, **42**, 279-290.

Mukhopadhyay, N. (1996). An alternative formulation of accelerated sequential procedures with applications to parametric and nonparametric estimation. *Sequential Analysis*, **15**, 253-269.

Mukhopadhyay, N. (1999a). Higher than second-order approximations via two-stage sampling. *Sankhya, Series A*, **61**, 254-269.

Mukhopadhyay, N. (1999b). Second-order properties of a two-stage fixed-size confidence region for the mean vector of a multivariate normal distribution. *J. Multivar. Anal.*, **68**, 250-263.

Mukhopadhyay, N. (2000). *Probability and Statistical Inference.* Marcel Dekker: New York.

Mukhopadhyay, N. (2002). Sequential sampling. In *The Encyclopedia of Environmetrics*, Volume 4, A. H. Shaarawi and W. W. Piegorsch, eds., pp. 1983-1988. Wiley: Chichester.

Mukhopadhyay, N. (2005a). A new approach to determine the pilot sample size in two-stage sampling. *Commun. Statist., Theory & Method*, Special Sobel Memorial Issue, **34**, 1275-1295.

Mukhopadhyay, N. (2005b). Estimating perception of distance: Applied sequential methodologies in experimental psychology. In *Proceedings of 5th International Triennial Calcutta Symposium*, U. Bandyopadhyay, K. Das, and R. Mukherjee, eds., *Calcutta Statist. Assoc. Bul.*, **56**, 35-55.

Mukhopadhyay, N. (2006). *Introductory Statistical Inference.* Chapman & Hall/CRC: New York.

Mukhopadhyay, N. and Abid, A. (1986a). Fixed-size confidence regions for the difference of means of two multinormal populations. *Sequential Analysis*, **5**, 169-191.

Mukhopadhyay, N. and Abid, A. (1986b). On fixed-size confidence regions for the regression parameters. *Metron*, **44**, 297-306.

Mukhopadhyay, N. and Abid, A. (1999a). Accelerated sequential shrinkage estimation. *Metron*, **57**,159-171.

Mukhopadhyay, N. and Abid, A. (1999b). Accelerated sequential sampling for generalized linear and AR(p) models. *Stoch. Modelling & Applications*, **2**, 1-15.

Mukhopadhyay, N. and Al-Mousawi, J. S. (1986). Fixed-size confidence regions for the mean vector of a multinormal distribution. *Sequential Analysis*, **5**, 139-168.

Mukhopadhyay, N. and Aoshima, M. (1998). Multivariate multistage methodologies for simultaneous all pairwise comparisons. *Metrika*, **47**, 185-201.

Mukhopadhyay, N. and Aoshima, M. (2004). Percentage points of the largest among Student's t random variable. *Methodology & Computing in Appl. Probab.*, **6**, 161-179.

Mukhopadhyay, N., Bendel, B., Nikolaidis, N., and Chattopadhyay, S. (1992). Efficient sequential sampling strategies for environmental monitoring. *Water Resources Research*, **28**, 2245-2256.

Mukhopadhyay, N. and Chattopadhyay, S. (1991). Sequential methodologies for comparing exponential mean survival times. *Sequential Analysis*, **10**, 139-148.

Mukhopadhyay, N., Chattopadhyay, S., and Sahu, S. K. (1993). Further developments in estimation of the largest mean of k normal populations. *Metrika*, **40**, 173-183.

Mukhopadhyay, N. and Cicconetti, G. (2000). Estimation of the reliability function after sequential experimentation. *Second International Conference on Mathematical Methods in Reliability, MMR'2000, Abstract Book*, **2**, M. Nikulin and N. Limnios, eds., pp. 788-791. Universite Victor Segalen: Bordeaux.

Mukhopadhyay, N. and Cicconetti, G. (2004a). Applications of sequentially estimating the mean in a normal distribution having equal mean and variance. *Sequential Analysis*, **23**, 625-665.

Mukhopadhyay, N. and Cicconetti, G. (2004b). How many simulations should one run? In *Applied Sequential Methodologies*, N. Mukhopadhyay, S. Datta, and S. Chattopadhyay, eds., pp. 261-292. Marcel Dekker: New York.

Mukhopadhyay, N. and Cicconetti, G. (2005). Estimating reliabilities following purely sequential sampling from exponential populations. In *Advances in Selection and Ranking, Multiple Comparisons and Reliability*, S. Panchapakesan 70th Birthday Festschrift, N. Balakrishnan, N. Kannan, and H. N. Nagaraja, eds., pp. 303-332. Birkhauser: Boston.

Mukhopadhyay, N. and Darmanto, S. (1988). Sequential estimation of the difference of means of two negative exponential populations. *Sequential Analysis*, **7**, 165-190.

Mukhopadhyay, N. and Datta, S. (1994). Replicated piecewise multistage sampling with applications. *Sequential Analysis*, **13**, 253-276.

Mukhopadhyay, N. and Datta, S. (1995). On fine-tuning a purely sequential procedure and associated second-order properties. *Sankhya, Series A*, **57**, 100-117.

Mukhopadhyay, N. and Datta, S. (1996). On sequential fixed-width confidence intervals for the mean and second-order expansions of the associated coverage probabilities. *Ann. Inst. Statist. Math.*, **48**, 497-507.

Mukhopadhyay, N., Datta, S., and Chattopadhyay, S. (2004). *Applied Sequential Methodologies*, edited volume. Marcel Dekker: New York.

Mukhopadhyay, N. and de Silva, B.M. (1997). Multistage fixed-width confidence intervals in the two-sample problem: The normal case. *J. Statist. Research*, **31**, 1-20.

Mukhopadhyay, N. and de Silva, B. M. (1998a). Multistage partial piecewise sampling and its applications. *Sequential Analysis*, **17**, 63-90.

Mukhopadhyay, N. and de Silva, B. M. (1998b). Sequential estimation of the mean of an exponential distribution via partial piecewise sampling. In *Applied Statistical Science, III: Nonparametric Statistics and Related Topics*, A. K. Md. E. Saleh 65th Birthday Festschrift, S. Ejaz Ahmed, M. Ahsanullah, and B. K. Sinha, eds., pp. 215-223. Nova Science Publishers.

Mukhopadhyay, N. and de Silva, B. M. (2005). Two-stage estimation of mean in a negative binomial distribution with applications to Mexican bean beetle data. *Sequential Analysis*, **24**, 99-137.

Mukhopadhyay, N. and Diaz, J. (1985). Two-stage sampling for estimating the mean of a negative binomial distribution. *Sequential Analysis*, **4**, 1-18.

Mukhopadhyay, N. and Duggan, W. T. (1997). Can a two-stage procedure enjoy second-order properties? *Sankhya, Series A*, **59**, 435-448.

Mukhopadhyay, N. and Duggan, W. T. (2000a). On a two-stage procedure having second-order properties with applications. *Ann. Inst. Statist. Math.*, **51**, 621-636.

Mukhopadhyay, N. and Duggan, W. T. (2000b). New results on two-stage estimation of the mean of an exponential distribution. In *Perspectives in Statistical Science*, A. K. Basu, J. K. Ghosh, P. K. Sen, and B. K. Sinha, eds., pp. 219-231. Oxford University Press: New Delhi.

Mukhopadhyay, N. and Duggan, W. T. (2001). A two-stage point estimation procedure for the mean of an exponential distribution and second-order results. *Statistics & Decisions*, **19**, 155-171.

Mukhopadhyay, N. and Ekwo, M. E. (1987). A note on minimum risk point estimation of the shape parameter of a Pareto distribution. *Calcutta Statist. Assoc. Bul.*, **36**, 69-78.

Mukhopadhyay, N. and Hamdy, H. I. (1984a). On estimating the difference of location parameters of two negative exponential distributions. *Canadian J. Statist.*, **12**, 67-76.

Mukhopadhyay, N. and Hamdy, H. I. (1984b). Two-stage procedures for selecting the best exponential population when the scale parameters are unknown and unequal. *Sequential Analysis*, **3**, 51-74.

Mukhopadhyay, N., Hamdy, H. I., Al-Mahmeed, M., and Costanza, M. C. (1987). Three-stage point estimation procedures for a normal mean. *Sequential Analysis*, **6**, 21-36.

Mukhopadhyay, N. and Hilton, G. F. (1986). Two-stage and sequential procedures for estimating the location parameter of a negative exponential distribution. *S. Afr. Statist. J.*, **20**, 117-136.

Mukhopadhyay, N. and Judge, J. (1989). Second-order expansions for a sequential selection procedure. *Sankhya, Series A*, **51**, 318-327.

Mukhopadhyay, N. and Judge, J. (1991). Three-stage procedures for selecting the largest normal mean. In *Modern Sequential Statistical Analysis*, H. Robbins 70th Birthday Festschrift, Z. Govindarajulu, ed., pp. 241-261. American Sciences Press: Syracuse.

Mukhopadhyay, N. and Liberman, S. (1989). Sequential estimation of a linear function of mean vectors. *Sequential Analysis*, **8**, 381-395.

Mukhopadhyay, N. and Mauromoustakos, A. (1987). Three-stage estimation for the negative exponential distributions. *Metrika*, **34**, 83-93.

Mukhopadhyay, N. and Moreno, M. (1991a). Multistage point estimation procedures for the mean of a U-statistic and some associated moderate sample size comparisons. *Calcutta Statist. Assoc. Bul.*, H. K. Nandi Memorial, **40**, 283-298.

Mukhopadhyay, N. and Moreno, M. (1991b). Multistage and sequential minimum risk point estimation procedures for the means of generalized U-statistics. *Sankhya, Series A*, **53**, 220-254.

Mukhopadhyay, N. and Padmanabhan, A. R. (1993). A note on three-stage confidence intervals for the difference of locations: The exponential case. *Metrika*, **40**, 121-128.

Mukhopadhyay, N., Padmanabhan, A. R., and Solanky, T. K. S. (1997). On estimation of the reliability after sequentially estimating the mean: The exponential case. *Metrika*, **45**, 235-252.

Mukhopadhyay, N. and Pepe, W. (2006). Exact bounded risk estimation when the terminal sample size and estimator are dependent: The exponential case. *Sequential Analysis*, **25**, 85-110.

Mukhopadhyay, N. and Purkayastha, S. (1994). On sequential estimation of the difference of means. *Statistics & Decisions*, **12**, 41-52.

Mukhopadhyay, N. and Sen, P. K. (1993). Replicated piecewise stopping numbers and sequential analysis. *Sequential Analysis*, **12**, 179-197.

Mukhopadhyay, N. and Solanky, T. K. S. (1991). Second order properties of accelerated stopping times with applications in sequential estimation. *Sequential Analysis*, **10**, 99-123.

Mukhopadhyay, N. and Solanky, T. K. S. (1992a). Accelerated sequential procedure for selecting the largest mean. *Sequential Analysis*, **11**, 137-148.

Mukhopadhyay, N. and Solanky, T. K. S. (1992b). Accelerated sequential procedure for selecting the best exponential population. *J. Statist. Plan. Inf.*, **32**, 347-361.

Mukhopadhyay, N. and Solanky, T. K. S. (1993). A nonparametric accelerated sequential procedure for selecting the largest center of symmetry. *J. Nonparam. Statist.*, **3**, 155-166.

Mukhopadhyay, N. and Solanky, T. K. S. (1994). *Multistage Selection and Ranking Procedures: Second-Order Asymptotics.* Marcel Dekker: New York.

Mukhopadhyay, N., Son, M. S., and Ko, Y. C. (2004). A new two-stage sampling design for estimating the maximum average time to flower. *J. Agric. Biol. Environ. Statist.*, **9**, 479-499.

Mukhopadhyay, N. and Sriram, T. N. (1992). On sequential comparisons of means of first-order autoregressive models. *Metrika*, **39**, 155-164.

Mukhopadhyay, N. and Vik, G. (1985). Asymptotic results for stopping times based on U-statistics. *Sequential Analysis*, **4**, 83-110.

Mukhopadhyay, N. and Vik, G. (1988a). Convergence rates for two-stage confidence intervals based on U-statistics. *Ann. Inst. Statist. Math.*, **40**, 111-117.

Mukhopadhyay, N. and Vik, G. (1988b). Triple sampling to construct fixed-width confidence intervals for estimable parameters based on U-statistics. *Metron*, **46**, 165-174.

Mulekar, M. S. and Young, L. J. (1991). Approximations for a fixed sample size selection procedure for negative binomial populations. *Commun. Statist. Theory & Method, Series A*, **20**, 1767-1776.

Mulekar, M. S. and Young, L. J. (2004). Sequential estimation in agricultural Sciences. Approximations for a fixed sample size selection procedure for negative binomial populations. In *Applied Sequential Methodologies*, N. Mukhopadhyay, S. Datta, and S. Chattopadhyay, eds., pp. 293-318. Marcel Dekker: New York.

Mulekar, M. S., Young, L. J., and Young, J. H. (1993). Introduction to 2-SPRT for testing insect population densities. *Environmental Entomology*, **22**, 346-351.

Nadas, A. (1969). An extension of a theorem of Chow and Robbins on sequential confidence intervals for the mean. *Ann. Math. Statist.*, **40**, 667-671.

Nagao, H. and Srivastava, M. S. (1997). Fixed width confidence region for the mean of a multivariate normal distribution. *Tech. Report No. 9713*, Department of Statistics, University of Toronto, Toronto.

Neyman, J. and Pearson, E. S. (1933). On the problem of the most efficient tests of statistical hypotheses. *Phil. Trans. Roy. Soc., Series A*, **231**, 289-337. Reprinted In *Breakthroughs in Statistics Volume I*, S. Kotz and N. L. Johnson, eds., 1992. Springer-Verlag: New York.

Nickerson, D. M. (1987a). Sequential shrinkage estimation of the difference between two multivariate normal means. *Sequential Analysis*, **6**, 325-350.

Nickerson, D. M. (1987b). Sequential shrinkage estimation of linear regression parameters. *Sequential Analysis*, **6**, 93-118.

Ofusu, J. B. (1973). A two-stage procedure for selecting mean from several normal populations with unknown variances. *Biometrika*, **60**, 117-124.

Paulson, E. A. (1964). Sequential procedure for selecting the population with the largest mean from k normal populations. *Ann. Math. Statist.*, **35**, 174-180.

Phatarfod, R. M. (1971). A sequential test for gamma distributions. *J. Amer. Statist. Assoc.*, **66**, 876-878.

Pradhan, M. and Sathe, Y. S. (1975). An unbiased estimator and a sequential test for the correlation coefficient. *J. Amer. Statist. Assoc.*, **70**, 160-161.

Raghavachari, M. and Starr, N. (1970). Selection problems for some terminal distributions. *Metron*, **28**, 185-197.

Ramkaran (1983). The robustness of Stein's two-stage procedure. *Ann. Statist.*, **11**, 1251-1256.

Ramkaran, Chaturvedi, A. C. and Akbar, S. A. (1986). Sequential estimation of a linear function of k-normal means. *J. Indian Soc. Agric. Statist.*, **38**, 395-402.

Rao, C. R. (1973). *Linear Statistical Inference and Its Applications*, 2nd edition. Wiley: New York.

Ray, W. D. (1957). Sequential confidence intervals for the mean of a normal population with unknown variance. *J. Roy. Statist. Soc., Series B*, **19**, 133-143.

Rinott, Y. (1978). On two-stage selection procedures and related probability inequalities. *Commun. Statist. Theory & Method, Series A*, **7**, 799-811.

Robbins, H. (1956). A sequential decision problem with a finite memory. *Proc. Nat. Acad. Sci. USA*, **42**, 920-923.

Robbins, H. (1959). Sequential estimation of the mean of a normal Population. In *Probability and Statistics*, H. Cramer Volume, U. Grenander, ed., pp. 235-245. Almquist and Wiksell: Uppsala.

Robbins, H. (1970). Statistical methods related to the law of the iterated logarithm. *Ann. Math. Statist.*, **41**, 1397-1409.

Robbins, H., Simons, G., and Starr, N. (1967). A sequential analogue of the Behrens-Fisher problem. *Ann. Math. Statist.*, **38**, 1384-1391.

Robbins, H., Sobel, M., and Starr, N. (1968). A sequential procedure for selecting the best of k populations. *Ann. Math. Statist.*, **39**, 88-92.

Robbins, H. and Starr, N. (1965). Remarks on sequential hypothesis testing. *Statist. Tech. Rep. No. 68*, University of Minnesota, Minneapolis.

Rohatgi, V. K. and O'Neill, R. T. (1973). On sequential estimation of the mean of a multinormal population. *Ann. Inst. Statist. Math.*, **25**, 321-325.

Rosenberger, W. F. and Lachin, J. L. (2002). *Randomization in Clinical Trials: Theory and Practice.* Wiley: New York.

Rushton, S. (1950). On a sequential t-test. *Biometrika*, **37**, 326-333.

Sacks, J. (1965). A note on the sequential t-test. *Ann. Math. Statist.*, **36**, 1867-1869.

Samuel, E. (1966). Estimators with prescribed bound on the variance for the parameter in the binomial and Poisson distributions based on two-stage sampling. *J. Amer. Statist. Assoc.*, **61**, 220-227.

Saxena, K. M. L. (1976). A single-sample procedure for the estimation of the largest mean. *J. Amer. Statist. Assoc.*, **71**, 147-148.

Saxena, K. M. L. and Tong, Y. L. (1969). Interval estimation of the largest mean of k normal populations with known variances. *J. Amer. Statist. Assoc.*, **64**, 296-299.

Scheffé, H. (1943). On solutions of Behrens-Fisher problem based on the t-distribution. *Ann. Math. Statist.*, **14**, 35-44.

Scheffé, H. (1970). Practical solutions of the Behrens-Fisher problem. *J. Amer. Statist. Assoc.*, **65**, 1501-1508.

Schmitz, N. (1972). *Optimal Sequentially Planned Decision Procedures.* Lecture Notes in Statistics, Springer-Verlag: New York.

Seelbinder, D. M. (1953). On Stein's two-stage procedure. *Ann. Math. Statist.*, **24**, 640-649.

Sen, P. K. (1980). On time-sequential estimation of the mean of an exponential distribution. *Commun. Statist. Theory & Method, Series A*, **9**, 27-38.

Sen, P. K. (1981). *Sequential Nonparametrics.* Wiley: New York.

Sen, P. K. (1985). *Theory and Applications of Sequential Nonparametrics.* CBMS # 49. SIAM Publications: Philadelphia.

Sen, P. K. and Ghosh, M. (1981). Sequential point estimation of estimable parameters based on U-statistics. *Sankhya, Series A*, **43**, 331-344.

Sen, P. K. and Singer, J. M. (1993). *Large Sample Methods in Statistics: An Introduction with Applications.* Chapman & Hall: New York.

Serfling, R. J. (1984). *Approximation Theorems of Mathematical Statistics.* Wiley: New York.

Shiryaev, A. N. (1978). *Optimal Stopping Rules.* Springer-Verlag: New York.

Shuster, J. and Chang, M. (2008). Second guessing clinical trial designs (with discussions). *Sequential Analysis*, **27**, 2-57.

Siegmund, D. (1985). *Sequential Analysis: Tests and Confidence Intervals.* Springer-Verlag: New York.

Simons, G. (1968). On the cost of not knowing the variance when making a fixed-width confidence interval for the mean. *Ann. Math. Statist.*, **39**, 1946-1952.

Sinha, B. K. (1991). Multivariate problems. In *Handbook of Sequential Analysis,* B. K. Ghosh and P. K. Sen, eds., pp. 199-227. Marcel Dekker: New York.

Sinha, B. K. and Mukhopadhyay, N. (1976). Sequential estimation of a bivariate normal mean vector. *Sankhya, Series B*, **38**, 219-230.

Sobel, M. (1956). Statistical techniques for reducing experiment time in reliability. *Bell. System Tech. J.*, **36**, 179-202.

Sobel, M. and Weiss, G. H. (1970). Play-the-winner sampling for selecting the better of two binomial populations. *Biometrika*, **57**, 357-365.

Sobel, M. and Weiss, G. H. (1971a). Play-the-winner rule and inverse sampling in selecting the better of two binomial populations. *J. Amer. Statist. Assoc.*, **66**, 545-551.

Sobel, M. and Weiss, G. H. (1971b). A comparison of play-the-winner and vector-at-a-time sampling for selecting the better of two binomial populations with restricted parameter values. *Trabajos de Estadistica Y de Investigation Operativa*, **22**, 195-206.

Sobel, M. and Weiss, G. H. (1972). Recent results on using play-the-winner sampling rule with binomial selection problems. *Proc. Sixth Berkeley Symp. Math. Statist. Probab.*, **1**, 717-736. University of California Press: Berkeley.

Solanky, T. K. S. (2001). A sequential procedure with elimination for partitioning a set of normal populations having a common unknown variance. *Sequential Analysis*, **20**, 279-292.

Solanky, T. K. S. (2006). A two-stage procedure with elimination for partitioning a set of normal populations with respect to a control. *Sequential Analysis*, Special Milton Sobel Memorial Issue-II, **25**, 297-310.

Sproule, R. N. (1969). *A Sequential Fixed-Width Confidence Interval for the Mean of a U-statistic. Ph.D. thesis*, Department of Statistics, University of North Carolina, Chapel Hill.

Sproule, R. N. (1974). Asymptotic properties of U-statistics. *Trans. Amer. Math. Soc.*, **199**, 55-64.

Sriram, T. N. (1987). Sequential estimation of the mean of a first-order stationary autoregressive process. *Ann. Statist.*, **15**, 1079-1090.

Sriram, T. N. (1988). Sequential estimation of the autoregressive parameter in a first-order autoregressive process. *Sequential Analysis*, **7**, 53-74.

Sriram, T. N. and Bose, A. (1988). Sequential shrinkage estimation in the general linear model. *Sequential Analysis*, **7**, 149-163.

Srivastava, M. S. (1967). On fixed width confidence bounds for regression parameters and mean vector. *J. Roy. Statist. Soc., Series B*, **29**, 132-140.

Srivastava, M. S. (1970). On a sequential analogue of the Behrens-Fisher problem. *J. Roy. Statist. Soc., Series B*, **32**, 144-148.

Srivastava, M. S. (1971). On fixed-width confidence bounds for regression parameters. *Ann. Math. Statist.*, **42**, 1403-1411.

Starr, N. (1966a). The performance of a sequential procedure for fixed-width interval estimate. *Ann. Math. Statist.*, **36**, 36-50.

Starr, N. (1966b). On the asymptotic efficiency of a sequential procedure for estimating the mean. *Ann. Math. Statist.*, **36**, 1173-1185.

Starr, N. and Woodroofe, M. (1969). Remarks on sequential point estimation. *Proc. Nat. Acad. Sci. USA*, **63**, 285-288.

Starr, N. and Woodroofe, M. (1972). Further remarks on sequential estimation: The exponential case. *Ann. Math. Statist.*, **43**, 1147-1154.

Stein, C. (1945). A two sample test for a linear hypothesis whose power is independent of the variance. *Ann. Math. Statist.*, **16**, 243-258.

Stein, C. (1946). A note on cumulative sums. *Ann. Math. Statist.*, **17**, 498-499.

Stein, C. (1949). Some problems in sequential estimation (abstract). *Econometrica*, **17**, 77-78.

Stein, C. (1955). Inadmissibility of the usual estimator for the mean of a multivariate normal distribution. *Proc. Third Berkeley Symp. Math. Statist. Probab.*, University of California Press: Berkeley, **1**, 197-206.

Swanepoel, J. W. H. (1977). Nonparametric elimination selection procedures based on robust estimators. *S. Afr. Statist. J.*, **11**, 27-41.

Swanepoel, J. W. H. and van Wyk, J. W. J. (1982). Fixed-width confidence intervals for the location parameter of an exponential distribution. *Commun. Statist. Theory & Method, Series A,* **11**, 1279-1289.

Takada, Y. (1984). Inadmissibility of a sequential estimation rule of the mean of a multivariate normal distribution. *Sequential Analysis*, **3**, 267-271.

Takada, Y. (1986). Non-existence of fixed sample size procedures for scale families. *Sequential Analysis*, **5**, 93-101.

Tartakovsky, A. G., Rozovskii, B. L., Blažek, R. B., and Kim, H. (2006). Detection of intrusions in information systems by sequential change-point methods (with discussions). *Statistical Methodology*, **3**, 252-340.

Tartakovsky, A. G. and Veeravalli, V. V. (2004). Change-point detection in multichannel and distributed systems. In *Applied Sequential Methodologies*, N. Mukhopadhyay, S. Datta, and S. Chattopadhyay, eds., pp. 339-370. Marcel Dekker: New York.

Tong, H. (1977). On estimation of $\mathrm{P}(Y < X)$ for exponential families. *IEEE Trans. Reliab.*, **26**, 54-57.

Tong, Y. L. (1970). Multi-stage interval estimations of the largest mean of k normal populations. *J. Roy. Statist. Soc., Series B*, **32**, 272-277.

Tong, Y. L. (1973). An asymptotically optimal sequential procedure for the estimation of the largest mean. *Ann. Statist.*, **1**, 175-179.

Tong, Y. L.. (1990). *The Multivariate Normal Distribution.* Wiley: New York.

Wald, A. (1947). *Sequential Analysis.* Wiley: New York.

Wald, A. and Wolfowitz, J. (1948). Optimum character of the sequential probability ratio test. *Ann. Math. Statist.*, **19**, 326-339.

Wang, Y. H. (1980). Sequential estimation of the mean vector of a multivariate population. *J. Amer. Statist. Assoc.,* **75**, 977-983.

Wetherill, G. B. (1975). *Sequential Methods in Statistics*, 2nd edition. Chapman & Hall: London.

Whitehead, J. (1997). *The Design and Analysis of Sequential Clinical Trials*, 2nd edition. Wiley: Chichester.

Whittle, P. and Lane, R. O. D. (1967). A class of procedures in which a sequential procedure is nonsequential. *Biometrika*, **54**, 229-234.

Wiener, N. (1939). The ergodic theorem. *Duke Math. J.*, **5**, 1-18.

Wijsman, R. A. (1971). Exponentially bounded stopping time of sequential probability ratio test for composite hypotheses. *Ann. Math. Statist.*, **42**, 1859-1869.

Wiles, L. J., Oliver, G. W., York, A. C., Gold, H. J., and Wilkerson, G. G. (1992). Spatial distribution of broadleaf weeds in north carolina soybean (Glycine max). *Weed Science*, **40**, 554-557.

Willson, L. J. (1981). *Estimation and Testing Procedures for the Parameters of the Negative Binomial Distribution. Ph.D. thesis*, Department of Statistics, Oklahoma State University, Stillwater.

Willson, L. J. and Folks, J. L. (1983). Sequential estimation of the mean of the negative binomial distribution. *Sequential Analysis*, **2**, 55-70.

Woodroofe, M. (1977). Second order approximations for sequential point and interval estimation. *Ann. Statist.*, **5**, 984-995.

Woodroofe, M. (1982). *Nonlinear Renewal Theory in Sequential Analysis.* CBMS #39. SIAM: Philadelphia.

Woodroofe, M. (1985). Asymptotic local minimaxity in sequential point estimation. *Ann. Statist.*, **13**, 676-688.

Woodroofe, M. (1987). Asymptotically optimal sequential point estimation in three stages. In *New Perspectives in Theoretical and Applied Statistics*, M. L. Puri, ed., pp. 397-411. Wiley: New York.

Young, L. J. (1994). Computations of some exact properties of Wald's SPRT when sampling from a class of discrete distributions. *Biometrical J.*, **36**, 627-637.

Young, L. J. (2004). Sequential testing in the agriculture sciences. In *Applied Sequential Methodologies*, N. Mukhopadhyay, S. Datta, and S. Chattopadhyay, eds., pp. 381-410. Marcel Dekker: New York.

Zacks, S. and Mukhopadhyay, N. (2006a). Bounded risk estimation of the exponential parameter in a two-stage sampling. *Sequential Analysis*, **25**, 437-452.

Zacks, S. and Mukhopadhyay, N. (2006b). Exact risks of sequential point estimators of the exponential parameter. *Sequential Analysis*, **25**, 203-226.

Zelen, M. (1966). Application of exponential models to problems in cancer research. *J. Roy. Statist. Soc.*, **129**, 368-398.

Zou, G. (1998). *Weed Population Sequential Sampling Plan and Weed Seedling Emergence Pattern Prediction. Ph.D. thesis*, Department of Plant Science, University of Connecticut, Storrs.

Author Index

Subject Index

C

D

E

F

N

O

P

Q

R

S

T

Milton Keynes UK
Ingram Content Group UK Ltd.
UKHW021914071024
449327UK00022B/1664

9 780367 386535